Global Ch

Springer

Berlin
Heidelberg
New York
Hong Kong
London
Milan
Paris
Tokyo

Guy P. Brasseur · Ronald G. Prinn · Alexander A. P. Pszenny (Eds.)

Atmospheric Chemistry in a Changing World

An Integration and Synthesis of a Decade of Tropospheric Chemistry Research

The International Global Atmospheric Chemistry Project of the International Geosphere-Biosphere Programme

With 109 Figures and 22 Tables

Springer

Editors

Guy P. Brasseur

Max Planck Institute for Meteorology
Hamburg, Germany

Ronald G. Prinn

Massachusetts Institute of Technology
Cambridge, MA, USA

Alexander A. P. Pszenny

Mount Washington Observatory
North Conway, NH, USA

ISSN 1619-2435
ISBN 3-540-43050-4 Springer-Verlag Berlin Heidelberg New York

Cataloging-in-Publication Data applied for

A catalog record for this book is available from the Library of Congress.

Bibliographic information published by Die Deutsche Bibliothek
Die Deutsche Bibliothek lists this publication in the Deutsche Nationalbibliografie;
detailed bibliographic data is available in the Internet at http://dnb.ddb.de

Springer-Verlag Berlin Heidelberg New York
is a part of Springer Science+Business Media
springeronline.com

Printed in Germany

Cover Design: Erich Kirchner, Heidelberg
Dataconversion: Büro Stasch, Bayreuth

SPIN: 10973554 3111 – 5 4 3 2 1 – Printed on acid-free paper

Preface

Since the pre-industrial era, Earth's population has grown from 1.6 billion to more than 6 billion, and our energy consumption has increased by approximately two orders of magnitude. Agricultural practices and industrial activities have profoundly perturbed the chemical composition of the atmosphere, especially in Europe, North America, and Japan. Rapid population growth and economic development in other parts of the world, especially in the Third World, have contributed significantly to these atmospheric changes. Today, about half of the world's population is concentrated in urban areas and it is estimated that more than a billion people are exposed to air pollution (Schwela 1995).

Several global or regional environmental issues directly related to changes in the chemical composition of the atmosphere have been identified during the 20th century. These include:

- the acidification of precipitation with damaging impacts on aquatic life and plants;
- increasing levels of atmospheric oxidants (including tropospheric ozone) and aerosols with adverse effects on health and crop productivity and with potential modifications in the atmosphere's self-cleansing efficiency;
- stratospheric ozone depletion with enhanced intensities of solar ultraviolet radiation and associated health effects; and
- increasing concentrations of radiatively active compounds with expected effects on Earth's climate.

The problem of air quality, initially regarded as a local or a regional problem, has become a global issue that must be addressed in an international context. International agreements have been adopted in the last 15 years to minimize the magnitude and the effects of global environmental perturbations. This is the case of the *Montreal Protocol on Substances That Deplete the Ozone Layer* (1987) and subsequent amendments, as well as the *Kyoto Protocol to the United Nations Framework Convention on Climate Change* (1997). No global convention to limit long-range transport of air pollution has yet been adopted, although some regional agreements (such as the *Convention on Long-Range Transport* and *Atmospheric Pollution in Europe*) have been ratified. However, there is growing recognition that climate change and long-range transport of air pollutants have the potential to significantly affect global air quality in the coming decades (NRC 2001).

The International Global Atmospheric Chemistry (IGAC) Project of the International Geosphere-Biosphere Programme (IGBP) was created in the late 1980s at the suggestion of the International Association of Meteorology and Atmospheric Sciences (IAMAS) Commission on Atmospheric Chemistry and Global Pollution (CACGP). IGAC was designed to address growing international concerns over rapid changes observed in Earth's atmosphere. The initial science plan for IGAC was developed at a CACGP workshop held at Dookie College, Victoria, Australia in November 1988 (Galbally 1989). The overall objectives of IGAC are:

- to develop a fundamental understanding of the processes that determine the composition of the atmosphere;

- to understand the interactions between atmospheric chemical composition and physical, biospheric, and climatic processes; and
- to predict the impact of natural and anthropogenic forcings on the chemical composition of the atmosphere.

IGAC has helped define goals, objectives, and priorities, and has acted as initiator, catalyst, and coordinator for much research. It has provided an international framework for many national research initiatives. Through a large number of projects and activities, IGAC has created a worldwide community of scientists, thereby enhancing international cooperation towards understanding the role of atmospheric processes in the entire Earth system. Although the primary foci of IGAC are the composition (gases and aerosols) and chemistry of the atmosphere and biosphere-atmosphere interactions, it promotes research that recognizes that the atmosphere, the hydrosphere, the biosphere, and the pedosphere form an interacting system whose components collectively determine the future evolution of our planet.

IGAC has facilitated the organisation of several large international field campaigns (see Appendix A.3), has encouraged the development of new instrumentation and its evaluation through intercomparisons, and has developed jointly with SPARC (Stratospheric Processes and their Role in Climate), a laboratory kinetics activity. Simultaneously, the Project has considered several cross-cutting or synthesis issues, such as the development of a global database for surface emissions of chemical compounds, and the evaluation and use of chemical transport models to assist in data analyses. IGAC has also promoted educational activities in different parts of the world, and has organized several international scientific conferences and workshops. One major achievement of IGAC has been the creation of stronger links among atmospheric chemistry research communities on different continents.

During the early years of its existence, IGAC was organized as five regional foci (tropical, polar, boreal, midlatitude, marine) plus a fundamental and a global focus. Since 1998, it has refocused around three major themes: biosphere-atmosphere interactions, photochemical oxidants, and atmospheric aerosols. Much of the work organised under these foci has been published in the scientific literature. Books and special issues of professional journals (see Appendix A.4) as well as individual articles report on the specific findings or summarize numerous presentations made at IGAC scientific conferences. After more than ten years of intense activity, however, it is timely to synthesize the work performed by the scientific community and to integrate it in the broad context of global change. The purpose of this book is therefore to address a simple question: What has the scientific community in atmospheric chemistry learned over the last 10 to 15 years, and what are the major remaining research challenges? This integration and synthesis effort has involved approximately 150 lead authors and co-authors as well as more than 40 reviewers from many countries.

The present report is a synthesis of the major accomplishments of a broad atmospheric chemistry research community, although it highlights specifically the research conducted within the IGAC framework. It can be regarded as the first international assessment of global tropospheric chemistry. In contrast to most recent assessments by intergovernmental panels, however, this report does not attempt to be comprehensive, nor does it provide a consensus view. The present document should rather be regarded as a summary of ten years of intense scientific research by the international community contributing to IGAC's objectives.

The preparation of the present report was initiated at a meeting in Toulouse, France in November 1998. A draft was reviewed, modified, and augmented at a workshop that took place at the Aspen Institute in Aspen, Colorado in April 2000 that involved nearly a hundred participants. It was further reviewed anonymously in late 2000 to early 2001 and completed in April 2001 at a meeting of the lead authors at the Joint Research Centre (JRC) in Ispra, Italy.

We would like to thank all organizations and agencies that supported the preparation of the present report, in particular: the Stockholm Office of IGBP; the European

Commission in Brussels; the European IGAC Project Office in Ispra, Italy; the Bundesministerium fuer Bildung and Forschung (BMBF) in Bonn, Germany; the Centre National de la Recherche Scientifique (CNRS) in France; NASDA in Tokyo, Japan; the National Center for Atmospheric Research (NCAR) in Boulder, Colorado; the Max Planck Institute for Meteorology in Hamburg, Germany; and Medias-France in Toulouse, France. We express our gratitude in particular to Harriet Barker and Edmund Carlevale, who coordinated the technical preparation of the book, and to Stanislaw Cieslik, Jean-Pierre Lacaux, and Hanna Stadelhofer, who helped organise the Toulouse, Aspen, and Ispra meetings. The IGAC Core Project Office is supported in the US by the National Science Foundation (NSF), the National Aeronautics and Space Administration (NASA), the National Oceanic and Atmospheric Administration (NOAA), and the US Department of Energy (DOE).

January 2003

Guy P. Brasseur
Chair IGAC-SSC (1996–2001)

Ronald G. Prinn
Chair IGAC-SSC (1990–1995)

Alexander A. P. Pszenny
IGAC Core Project Officer (1992–2001)

Permissions and Acknowledgements

The authors and publisher would like to thank those who provided illustrative material and are grateful to the following for permission to reproduce copyright material (for which acknowledgement and citation are made in figure captions and lists of references): Academic Press for 3.26; American Association for the Advancement of Science for 5.2; American Geophysical Union for 1.6, 2.6, 2.7, 2.10, 2.11, 2.12, 2.13, 2.15, 2.19, 2.20, 3.5, 3.10, 3.13, 3.14, 3.15, 3.17, 3.18, 3.19, 3.20, 3.21, 3.27, 3.28, 3.29, 3.30, 3.31a, 3.32, 3.33a–c, 3.39, 3.41, 3.42, 3.43, 4.9, 5.1, 6.1, 7.1, 7.2, and 7.11; Annual Reviews for 2.17 and 2.18; Blackwell Science for 2.8; Cambridge University Press for 1.7; Elsevier Science/ Pergamon for 1.2, 1.3, 2.1, 2.9, 3.25, 3.26, and 6.2; Kluwer Academic for 3.24; National Center for Atmospheric Research for 3.4a–c; National Oceanic and Atmospheric Administration for 3.7, 3.8, 3.9, and 3.11; Nature for 1.5, 2.21, 2.22, 2.23, and 3.16; The Netherlands National Institute of Public Health and Environment for 2.4, and 2.5; Springer-Verlag for 3.38; The Stockholm Environment Institute for 2.16; John Wiley & Sons for 2.22, and the World Meteorological Organization for 3.1.

Every effort has been made to trace and acknowledge copyright holders. Should any infringements have occurred, apologies are tendered and omissions will be rectified in the event of a reprint of the book.

Contents

Chapter 1

Changes in the Chemical Composition of the Atmosphere and Potential Impacts

Lead authors: Donald J. Wuebbles · Guy P. Brasseur · Henning Rodhe
Co-authors: Leonard A. Barrie · Paul J. Crutzen · Robert J. Delmas · Daniel J. Jacob · Charles Kolb · Alex Pszenny
Will Steffen · Ray F. Weiss

1.1 Introduction

1.1.1 Atmospheric Chemistry and Life on Earth

The story of the importance of atmospheric chemistry begins with the origin and evolution of life on Earth. The accumulation of greenhouse gases in Earth's atmosphere allowed surface temperatures to be maintained above the freezing point of water. Reactions involving carbon, hydrogen, and nitrogen compounds in the primeval soup led to the formation of self-replicating molecules, and, about 500 million years ago, the rise of atmospheric oxygen led to the formation of the stratospheric ozone layer, which protects life on Earth from extremely harmful levels of solar ultraviolet radiation. Running in the shadow of these major events has been a full suite of atmospheric chemistry processes affecting and being affected by the evolving nature of both terrestrial and oceanic life. Some of these processes include biogeochemical cycling of elements, long-range transport of nutrients, regulation of temperature, and exposure to air pollution.

It is a source of wonder that the chemistry of Earth's atmosphere has been sufficiently stable to sustain some form of life on our planet for at least the past 3.5 billion years despite large changes in the physical and chemical climate. Our sister planets, Venus and Mars, show what happens when such stability does not exist. Venus's proximity to the sun has prevented the formation of an ocean and led to a runaway greenhouse effect, resulting in a very high surface temperature and a thick atmosphere consisting mainly of carbon dioxide. The Martian atmosphere is also dominated by carbon dioxide but is thin and has a very weak greenhouse effect. However, one has only to contemplate Earth's historical record to see that changes in atmospheric chemistry have played a crucial role in defining and altering the forms of sustainable life. Indirect sources of information – from ice cores and lake sediments, for example – have shown that large changes in atmospheric composition have occurred mainly during the agro-industrial era (the 'anthropocene'), i.e. the past 200 years.

Over the past century, humanity has been altering the chemical composition of the atmosphere in an unprecedented way, over an astonishingly short time. Resulting changes in concentrations of airborne toxic gases and fine particles pose a potential threat to human health, reduce agricultural productivity, and challenge the viability of sensitive plant and animal species. The net effects of the buildup of radiatively active trace gases and the changing burden of atmospheric particles appear to be responsible for much of the climate trend observed during the 20th century, particularly the warming over the last few decades (IPCC 2001).

World-wide emissions from growing industrial and transportation activity and more intensive agriculture have caused widespread increases in concentrations of photochemical oxidants, acid gases, fine particles, and other toxic chemical species. Many of these air pollutants are known to have detrimental impacts on human health and/or natural and managed ecosystem viability.

Furthermore, higher fossil fuel consumption coupled with agriculturally driven increases in biomass burning, fertiliser usage, crop by-product decomposition, and production of animal based food and fibre have led to increasing emissions of key greenhouse gases, such as carbon dioxide, methane, and nitrous oxide. The resulting increases in absorption, and re-radiation back to Earth's surface, of infrared radiation by these gases (referred to as radiative forcing), coupled with changes in the atmospheric burden of sulphate and other components of small particles, seem to be causing significant modification of Earth's climate. Predicted impacts of climate change include disruptions of agricultural productivity, fresh water supplies, ecosystem stability, and disease patterns. Significant increases in sea level and changes in the frequency of severe weather events are also forecast. The resulting effects of all these stresses on biogeochemical cycles could exacerbate changing atmospheric composition and result in further effects on climate.

If current trends are unchecked, much more significant warming is predicted, potentially driving a wide range of perturbations in other components of the climate system. What are and will be the consequences for life on Earth? We need to understand the full range of possible effects from small perturbations lending them-

Box 1.1. A brief look at the historical development of atmospheric chemistry[1]

Although Aristotle (384–322 B.C.) recognised water vapour as a distinct component of air, it was not until the 18th century that knowledge of the composition of the atmosphere began to be known, and our understanding has evolved considerably since then. In 1750, Joseph Black discovered carbon dioxide, and a few years later Daniel Rutherford discovered molecular nitrogen. Molecular oxygen was first isolated in 1773 by Carl Wilhelm Scheele and a year later by Joseph Priestly. This gas was named oxygen by Antoine-Laurent Lavoisier in 1789.

As new observational techniques became available, less abundant chemical constituents were discovered. For example, in 1862, J. B. Boussingault discovered that methane was a component of the air. About 30 years later, Armand Gautier discovered several natural and human related sources of methane. Scientists continue to find new minor atmospheric constituents to this day, as instrument sensitivity continues to increase.

Urban air pollution connected with large cities has been known from at least the time of the Romans. Harold Des Voeux coined the term "smog" from the words smoke and fog based on his observations in London. M. Ducros introduced the term acid rain in 1845.

In 1840, Christian F. Schönbein identified ozone by its peculiar odour following electrical discharge. André Houzeau made the first measurements of ground level ozone in 1858. The measurement of the ozone absorption spectrum by Walter Noel Hartley in 1881 led him to suggest that the observed cut-off of UV radiation ($\lambda < 300$ nm) was due to ozone. J. Chappuis, and William Huggins later, discovered additional ozone absorption bands. The first ozone column measurements, made by Charles Fabry and Henri Buisson in 1920, noted that the column was about 3 mm thick (at atmospheric pressure) with large variations. Also in the 1920s, G. M. B. Dobson developed the spectrophotometer for measuring total ozone that is still used today. Dobson made many of the first measurements showing the global variability of ozone spatially and seasonally.

In 1929, at a conference in Paris, Sidney Chapman proposed the first chemical mechanism to explain the origin of ozone, suggesting that stratospheric ozone was formed by the photolysis of molecular oxygen. It was not until the 1950s that additional reactions were added to the mechanism described by Chapman. In 1950, David Bates and Marcel Nicolet proposed the potential importance of hydrogen oxides (HO_x) chemistry in the mesosphere, while in 1965 J. Hampson highlighted the role of HO_x for the chemistry of the stratosphere. In 1970, Paul Crutzen stressed the importance of nitrogen compounds (NO_x) in determining stratospheric ozone concentrations. In 1974, Ralph Cicerone and Richard Stolarski suggested the potential for chlorine to affect ozone catalytically.

In the early 1950s, it was established that the health symptoms and plant damage observed in the vicinity of urban centres like Los Angeles during pollution events were due to high levels of atmospheric ozone and other oxidants. Arie Jan Haagen-Smit suggested in 1951 that ozone episodes were due to solar ultraviolet light acting on hydrocarbons and nitrogen oxides emitted by motor vehicles and industry. In 1973, Paul Crutzen suggested that similar "smog reactions" involving the oxidation of methane and carbon monoxide in the presence of nitrogen oxides should lead to a substantial production of ozone in remote regions of the troposphere. A few years earlier, in 1970, Hiram Levy II had suggested that the hydroxyl (OH) radical, which provides the dominant oxidation mechanism in the troposphere, was formed in the background atmosphere by the mechanism that had been recognised previously as occurring in polluted air, namely the reaction of water vapour with the electronically excited oxygen atom $O(^1D)$, itself produced by the photolysis of ozone.

Halocarbons, such as CFC-11 ($CFCl_3$), were first detected following the invention of the electron capture detector by James Lovelock in 1971. In 1974, a study by Mario Molina and F. Sherwood Rowland suggested that the presence of ozone-destroying chlorine in the stratosphere resulted to a large extent from the photolysis by solar ultraviolet radiation of industrially manufactured chlorofluorocarbons. The large seasonal change in ozone over Antarctica, referred to as the ozone "hole", was discovered by Joseph Farman and colleagues in 1985. A year later, Susan Solomon and co-workers proposed a connection between chlorine chemistry and polar stratospheric clouds to explain this ozone "hole". The fundamental photochemical mechanism leading to the ozone destruction that creates the hole was discovered by Mario and Luisa Molina in 1987. Paul Crutzen, Mario Molina, and F. Sherwood Rowland were awarded the 1995 Nobel Prize in chemistry for their accomplishments.

selves to environmental management to more ominous and more remote prospects of catastrophic and irreversible change.

It is important to understand the mechanisms and rates of emissions of trace gases and fine particles, and their subsequent transformation and dispersion; the mechanisms for their eventual removal from the atmosphere and associated deposition fluxes are of importance as well. These processes extend over a wide range of spatial scales, from the microscopic to local, regional, and global, with the global scale being affected by all others. Understanding the complicated interplay of these processes at work is an important goal of the atmospheric chemistry community.

Another key goal is to detect and track widespread and long-term changes in the chemical content and radiative properties of the atmosphere and to predict their potential impacts. The community also aspires to address the complex challenge of identifying which atmospheric emissions need to be controlled (and to what levels) in order to prevent unacceptable atmospheric degradation, undesirable climate modification, and irreversible harm to the habitability of our planet.

The field of atmospheric chemistry and air pollution has a long history (see Box 1.1). It is only during the 20th century, however, as more sensitive instruments have been developed, that systematic progress has been made. In the last 30 years, the importance of atmospheric trace substances (e.g. ozone, greenhouse gases) in global environmental issues has been stressed. The purpose of this first chapter is to provide a framework for evaluating recent advances in our understanding of the role of atmospheric chemistry in the Earth system. It is also designed to introduce major uncertainties or gaps in our knowledge that limit our understanding of the present state of the atmosphere and prevent us from fully predicting future developments.

We begin this by outlining a number of key environmental issues related to the observed changes in atmospheric composition and their impact on ecosystems and

---

1 Much of the content of this box is based on discussion in the 1999 book, *Atmospheric Chemistry and Global Change*, edited by G. P. Brasseur, J. J. Orlando, and G. S. Tyndall.

Box 1.2. Definitions and units

Number density: Number of particles (e.g. atoms, molecules or aerosols) of substance per unit volume of air expressed in particles m^{-3} or particles cm^{-3}.

Mass density: Mass of substance per unit volume of air expressed in $kg\,m^{-3}$ or in $g\,m^{-3}$.

Mole fraction: Number density of substance divided by number density of air (dimensionless variable). Abundances expressed in $cmol\,mol^{-1}$, $mmol\,mol^{-1}$, $\mu mol\,mol^{-1}$, $nmol\,mol^{-1}$, and $pmol\,mol^{-1}$ correspond to volume mixing ratios of 10^{-2}, 10^{-3}, 10^{-6}, 10^{-9}, and 10^{-12}, respectively. The American literature commonly adopts percent, per mille, ppmv, ppbv, and pptv for the corresponding mole fractions.

Mass fraction: Mass density of substance divided by mass density of air (dimensionless variable). The pertinent unit for mass mixing ratio is $kg\,kg^{-1}$, but often is expressed as $g\,kg^{-1}$ or $g\,g^{-1}$. Abundances expressed in $\mu g\,g^{-1}$, $ng\,g^{-1}$, and $pg\,g^{-1}$ correspond to mass mixing ratios of 10^{-6}, 10^{-9}, and 10^{-12}, respectively.

Column abundance: Vertically integrated number density of substance (expressed in particles m^{-2} or particles cm^{-2}). In the case of ozone, the column abundance is often expressed in Dobson units (DU). One Dobson unit corresponds to the height (in 10^{-3} cm) of an ozone column if the gas were at standard temperature and pressure. It is equivalent to 2.687×10^{16} molecules cm^{-2}. The unit of $mmol\,m^{-2}$ is the recommended unit, but is currently not widely used in the literature.

Particle or mass flux: Number of particles or mass of substance crossing a unit surface per unit time. Expressed in particles $cm^{-2}\,s^{-1}$ or $g\,cm^{-2}\,s^{-1}$. Total mass flux of a substance (emitted for example at Earth's surface) is often expressed in $Tg\,yr^{-1}$ ($10^{12}\,g\,yr^{-1}$).

Atmospheric lifetime: Average time that a substance spends in the atmosphere. Atmospheric residence time is a synonymous concept. Expressed in days, months, or years. If the removal of a substance is directly proportional to its concentration, the lifetime is also a measure of the time that would be needed to remove 1/e (63%) of its mass if its source were discontinued.

Turnover time: Atmospheric burden of a substance at equilibrium divided by the globally integrated removal rate. Under steady-state conditions the turnover time equals the lifetime (residence time).

Greenhouse gases: Gases (water vapour, carbon dioxide, methane, ozone, nitrous oxide, halocarbons) whose interactions with terrestrial radiation leads to warming of Earth's surface.

Aerosols: An assembly of liquid or solid particles suspended in a gaseous medium (e.g. air) long enough to enable observation or measurement. Although water droplets and ice crystals in clouds fit this definition formally, by convention they are often considered as separate categories.

Radiative or climate forcing: According to the IPCC terminology, the term "climate forcing" denotes the change in the net irradiance (in $W\,m^{-2}$) at the tropopause (after allowing for stratospheric temperatures to re-adjust to radiative equilibrium, but with the surface and tropospheric temperature and moisture held fixed) created by changes imposed on the climate system by natural (solar radiation changes, volcanic emissions of SO_2, and particulate matter) or human-induced processes (greenhouse gases, aerosols, changes in surface albedo). The climate forcing concept provides a very useful means of quantitatively comparing the importance of the different factors forcing climate change. It avoids the difficult problem of how the climate system will actually respond (in terms of temperature, precipitation, winds, etc.) to the imposed changes. This latter problem, which involves an intricate web of interactions and feedbacks within the atmosphere and between the atmosphere and the underlying surface, has to be addressed using complex models of the climate system.

human society. These issues have been a motivating force for atmospheric chemistry research during the past few decades. We then examine past variations in atmospheric composition, the causes of atmospheric change, and the potential impacts resulting from atmospheric change. A few definitions of frequently used terms, which may be useful to the reader, are provided in Box 1.2.

1.1.2 Environmental Issues and Atmospheric Chemistry

Many major environmental issues comprise strong atmospheric chemistry components, and human activities affect the atmosphere's chemical composition in important ways. Chemical species impact climate, key ecosystems, and much of life on Earth. The concentrations of chemical compounds and their effects need to be known across a wide range of temporal (sub-second to decadal) and spatial (molecular to global) scales. The challenge to atmospheric chemists is to develop the research tools and scientific infrastructure necessary to understand and resolve the environmental issues facing humanity.

1.1.2.1 *Greenhouse Gases, Aerosols, and Climate Forcing*

As mentioned earlier, increasing concentrations of CO_2, CH_4, N_2O, and several other gases that absorb in the infrared part of the radiative spectrum affect the earth's climate. The man-made contribution of these gases to the so-called greenhouse effect is likely to result in a significant warming of the earth's surface and troposphere, and a cooling of the stratosphere and mesosphere. The potential for greenhouse gas effects on climate was first studied quantitatively by Arrhenius in 1896. Anthropogenic greenhouse gases have provided an additional radiative energy input to the troposphere and surface part of the climate system of about 2.4 $W\,m^{-2}$ since the beginning of the agro-industrial era (IPCC 1996, 2001).

An increasing load of aerosol particles in the troposphere (sulphates, organics, black carbon, etc. – see Chap. 4) caused by both primary particle and secondary particle precursor emissions from industrial processes and from biomass burning are also contributing to climate change. IPCC (2001) estimates the direct radiative forcing due to aerosols to be −0.4 $W\,m^{-2}$ for sulphate, $-0.2\ W\,m^{-2}$ for biomass burning aerosols, $-0.1\ W\,m^{-2}$ for fossil fuel organic carbon, and $+0.2\ W\,m^{-2}$ for fossil fuel black carbon. Uncertainties in these numbers are large. Indirect aerosol radiative forcing, in which aerosols affect cloud reflectivity and lifetime, could cause substantial but poorly quantified additional cooling (IPCC 2001). This net cooling due to aerosols may have counteracted a substantial part of the warming expected from greenhouse gases during past decades. However, because of their relatively short atmospheric lifetimes (days to weeks), the actual cooling or warming effects from aerosols are largely regional and depend on the actual particle distributions in the given region. The large uncertainty in the quantitative estimate of aerosol effects on climate is due to the lack of data on the actual change in atmospheric composition of aerosols, and the lack of understanding of how the increasing aerosol mass concentration has affected the number concentration of cloud-forming particles and, thereby, the reflectivity and lifetime of clouds (IPCC 1996).

Estimates of the radiative forcing by different types of aerosols have been made by IPCC (2001), as noted above, although uncertainties in these estimates are also large. For example, the radiative forcing due to the direct effect of sulphate aerosols, about which relatively much is known, has been estimated to be about -0.2 to $-0.8\ W\,m^{-2}$, a range of a factor of four. Looking to the future, the IPCC (2001) projects for year 2100 a radiative forcing of 4.1 to $9.2\ W\,m^{-2}$ relative to pre-industrial times. IPCC (2001) also projects a surface temperature increase of 1.4 to 5.8 °C above 1990 values: "the projected rate of warming is much larger than the observed changes during the 20th century and is very likely to be without precedent during at least the last 10 000 years, based on paleoclimate data."

1.1.2.2 *Acidification and Eutrophication*

Increasing deposition of acidifying sulphur and nitrogen compounds has led to widespread damage to terrestrial and limnic ecosystems in some parts of the world – e.g. northern Europe and north-eastern North America – near and downwind of large industrial emission regions, following deposition on poorly buffered soils (e.g. McNeely et al. 1995; Mannion 1992; Rodhe et al. 1995). Whereas sulphur emissions are now declining in these particular regions, emissions are increasing in some other regions (China, India, and Southeast Asia) raising concern about more widespread acidification problems in the future.

Fixed nitrogen (NO_3^-, NH_4^+) is a limiting nutrient in many ecosystems. The increased nitrogen deposition in and around many industrial parts of the world has consequently caused eutrophication (increased primary production caused by the enrichment of an ecosystem with nutrients) of terrestrial, fresh-water and marine ecosystems with consequences for plant and animal species composition and also for the atmospheric carbon balance (e.g. McNeely et al. 1995; Mannion 1992).

1.1.2.3 *Enhanced Tropospheric Ozone*

Observations at the earth's surface at mountain sites in Europe suggest that tropospheric ozone (O_3) has increased by a factor of two or more during the past century (WMO 1998). Increases in ozone concentrations since the pre-industrial period are evident in Europe and East Asia, and appear to be associated with large increases in gaseous precursors, reactive oxides of nitrogen (NO_x), hydrocarbons, and CO, which interact with solar radiation to produce O_3 (see Chap. 3) (Crutzen and Zimmermann 1991). Recent measurements suggest little change in ozone in the last two decades (Logan et al. 1999), but there are not enough observations to assess thoroughly present and past changes in tropospheric ozone.

The elevated concentration of ozone causes concern because of its importance as a greenhouse gas and because of its adverse effects on plants, animals, and human health through contact exposure. Recent model studies (see, e.g. IPCC 2001) suggest that estimated increases in tropospheric ozone have produced a positive radiative forcing of approximately $0.35\ W\,m^{-2}$ since pre-industrial times.

In addition, photolysis of tropospheric O_3 is the primary source of the hydroxyl radical (OH), an extremely strong oxidant. Oxidation by OH is the main sink for a large number of environmentally important atmospheric chemical species including CO, greenhouse gases such as CH_4, and some gases responsible for stratospheric O_3 depletion such as methyl bromide (CH_3Br). Thus, tropospheric O_3 plays a critical role in determining the oxidising and cleansing efficiency of the atmosphere. The rise in O_3 and NO_x since pre-industrial times has increased the source of OH, but this increase has been compensated for by more rapid removal of OH due to the rise in other gases (particularly CO and CH_4). Atmospheric models predict little change in the global mean OH concentration since pre-industrial times (less than 10% in most models), reflecting these compensating factors. Model studies (e.g. Krol et al. 1998; Karlsdottir and Isaksen 2000) suggest that the tropospheric OH concentration has increased in response to increasing concentrations of CH_4, CO, and other gases. Recent evaluation of the 1978–2000 record of methyl chloroform, whose primary sink is reaction with OH, shows, in comparison, that the global average concentration of OH increased between 1978 and 1988 and decreased between 1988 and 2000 (Prinn et al. 2001). Issues related to tropospheric ozone and other oxidants are addressed in Chap. 3.

1.1.2.4 *Depletion of Stratospheric Ozone*

One of the most dramatic changes in atmospheric composition observed during the past decades is the depletion of ozone in the stratosphere. Stratospheric ozone has decreased globally by more than 5% since 1970. This change, primarily in extra-tropical regions, is most pronounced at high southern latitudes during the spring months where the ozone destruction has approached 60%, but significant decreases in ozone are also occurring in the Northern Hemisphere. There is now general consensus that the ozone depletion is largely caused by chlorine- and bromine-containing gases primarily of industrial origin. This decrease in stratospheric ozone leads to an increasing flux of UV radiation into the troposphere with potential impacts on human health, ecosystems, and tropospheric photochemistry. Since there is approximately ten times more ozone in the stratosphere than in the troposphere, the rise in tropospheric ozone has not compensated for the larger loss of stratospheric ozone. Furthermore, stratospheric ozone depletion provides a negative net contribution to the tropospheric energy balance (-0.15 ± 0.1 W m^{-2}; IPCC 2001).

1.1.2.5 *Transport of Toxic Substances – Metals, Organic Compounds, and Radionuclides*

Elevated levels of toxic substances (gaseous or bound to particles) in the atmosphere represent an issue of a different character. These toxic compounds – for example, mercury (Hg), cadmium (Cd), polycyclic aromatic hydrocarbons (PAH), polychlorinated biphenyls (PCB), dioxins, and anthropogenic radionuclides – generally occur in very low concentrations and are not of any importance for the balance of the major atmospheric components. However, the atmosphere serves as the major transport medium for these types of toxic compounds and evaluation of their effects on human health and ecosystems requires that processes of emission, transport, transformation, and deposition be understood.

1.1.2.6 *Transport of Nutrients*

Ecological systems, including those supporting economically important activities such as agriculture and forestry, can be profoundly impacted by the wet and dry deposition of atmospheric substances such as nitrogen- and sulphur-containing compounds. Although the coupling of the atmosphere and biosphere through the exchange of gases and aerosols occurs naturally, this coupling is being affected dramatically by human activities such as agriculture and fossil fuel burning. Forests are being damaged by the combined effects of ozone exposure and pollutant deposition, i.e. nutrient overloads and acidification. Fisheries are being affected by runoff of nutrients from agricultural activities as well as toxic substances such as pesticides.

1.1.2.7 *Policy Considerations*

The issues described above have received growing attention not only from the members of the scientific community, but also from decision-makers in governments and industries. With society's increased recognition of the importance and value of the environment, the relation between atmospheric chemistry research and environmental policy design has been growing substantially over recent decades. In some cases, international treaties to reduce emissions have been enacted and actions to protect the global environment have been taken. Major challenges remain, however. Although substantial advances have been made in understanding fundamental processes in the chemical system of the atmosphere, our predictive capability remains limited in spite of its importance for informed decision making. New and challenging problems at the chemistry-weather, chemistry-climate, and chemistry-ecology interfaces are emerging and will require much attention in the future. The availability of space instrumentation anticipated in the near future to observe chemical species in the troposphere will provide unprecedented information on the global distribution and evolution of key biogenic and anthropogenic compounds (see Chap. 5). These observations will require associated ground-based and airborne measurements to be of greatest value. Laboratory experiments and new developments in numerical modelling of associated processes will also be required (see Chap. 6).

1.1.3 The Atmosphere in the Earth System

The atmosphere is composed primarily (>99.999%) of molecular nitrogen (N_2), molecular oxygen (O_2), water (H_2O), argon (Ar), and carbon dioxide (CO_2). Many environmental problems are linked to changes in the atmospheric abundance of trace chemical compounds whose mole fractions vary from a few µmol mol^{-1} to well below a pmol mol^{-1}.

How can such seemingly small changes in the trace gas and fine particle content of Earth's atmosphere have such large effects on climate and on the biosphere? The answer lies in the complex, non-linear interconnections among the major systems that define Earth as we know her. These include the atmosphere's central role in mediating the energy flux between the sun and the planet's surface, as well as its major role in conveying fluxes of energy and materials between other components of the Earth system. A schematic view of the key processes that

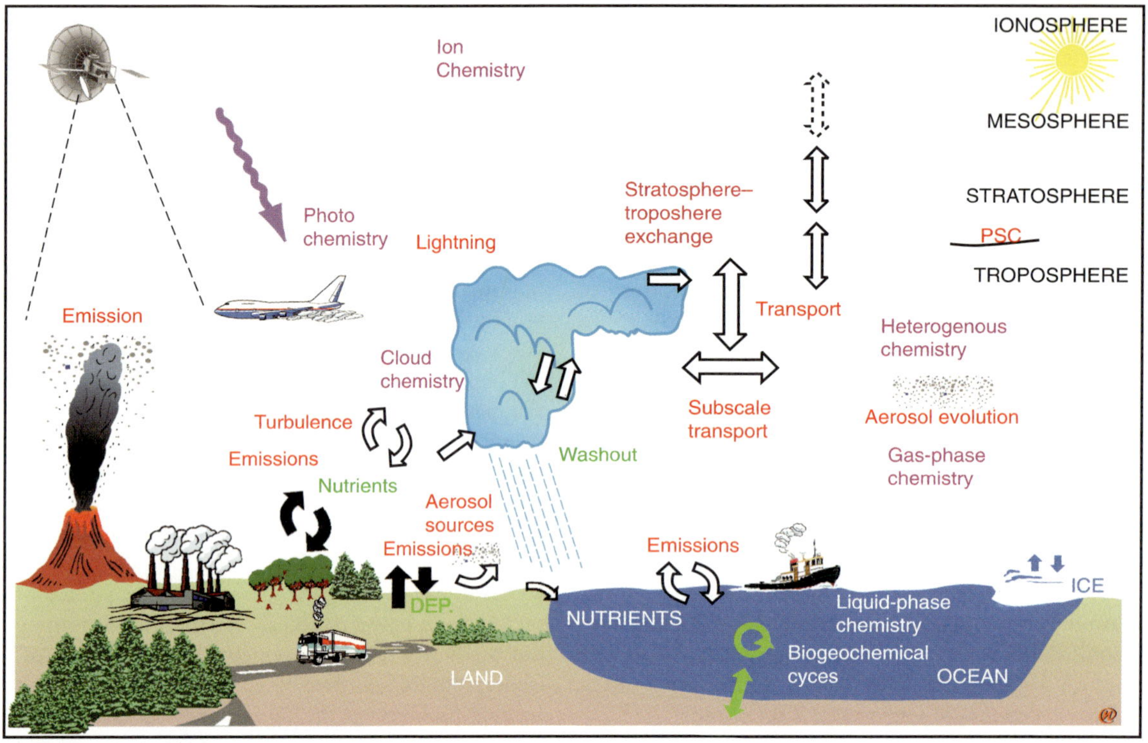

Fig. 1.1. Schematic representation of physical and chemical processes affecting chemical compounds in the atmosphere, including natural and anthropogenic emissions, large-scale transport, subscale transport including convective exchanges, gas-phase and aqueous-phase chemical and photochemical reactions, heterogeneous chemical conversion, stratosphere-troposphere exchanges, wet scavenging, and surface deposition (courtesy of M. Schultz, Max Planck Institute for Meteorology, Hamburg, Germany)

affect the formation and fate of chemical compounds in the troposphere is presented in Fig. 1.1.

A particular property of Earth's atmosphere is that many of the chemical compounds present in it, including the most abundant gases, N_2 and O_2, are constantly taken up and regenerated by biological processes (microbial activity in soils, photosynthesis and respiration, foliage emissions, etc.). Reduced chemical compounds that are released to the atmosphere by terrestrial and oceanic ecosystems and as anthropogenic emissions are often oxidised in the atmosphere and the resulting products are removed by deposition to the earth's surface. Subsequent assimilation and reduction in the biosphere closes the atmosphere-biosphere cycle. The circulation of chemical elements in the Earth system is often described in terms of global biogeochemical cycles (see Chap. 2). One of the challenges for the scientific community is to quantify the global budget (burden and turnover time in each component of the geosphere and transfer rates among them) of key chemical elements or compounds. These cycles produce important feedback mechanisms for the Earth system. They have frequently and dramatically been perturbed by human activities including agricultural practices, industrialisation, and urbanisation. Changes in land use including massive biomass burning, particularly in the Tropics, and energy consumption, mostly fossil fuel burning, have been the sources of substantial changes in the chemical composition of the atmosphere and in the deposition rate of acids and other toxic products on Earth's surface. An important task for the scientific community is to "close" the biogeochemical cycles and to "balance" the chemical budgets in the Earth system in order to understand, and eventually manage, the impact of human activities (see Box 1.3).

The global distribution of chemical compounds in the atmosphere depends on a variety of chemical, biological, and physical processes. Transport is the process by which atmospheric motions carry physical or chemical properties from one region of the atmosphere to another. It enables different chemical compounds, with different sources, to interact. Without transport, local production and destruction would tend to balance and some equilibrium concentration would be reached. It is transport that drives the atmosphere away from these equilibrium conditions and, hence, determines which regions are net sources and which are net sinks for chemical substances. Atmospheric motions are often categorised according to their spatial scales: planetary, global, synoptic (of the order of 1 000 km), mesoscale (10–500 km), and small scale (below 10 km). Many compounds are released in the planetary boundary layer (PBL), where the horizontal winds are weak but often strongly turbulent. This dynamically variable layer

Box 1.3. Biosphere-atmosphere interactions and human activities

Exchange between the biosphere and the atmosphere has been modified substantially by anthropogenic activities, in particular land use changes, such as deforestation, urbanisation, and agricultural activities. These perturbations have severe ecological and socio-economic consequences that make management and mitigation strategies necessary for avoiding or limiting environmental problems. Defining efficient and cost effective management and mitigation strategies requires a detailed understanding of the actual gas exchange among terrestrial, aquatic, and atmospheric systems, and of the biological, physical, and chemical processes that control the fluxes across the biosphere-atmosphere interface. This information is also urgently needed to improve the forecasts of climate change using global climate models.

Relationships between fluxes and concentration changes derived from increasingly sophisticated models of atmospheric transport and chemistry are being employed to evaluate and interpret fluxes estimated from atmospheric observations. The increasing knowledge in this area indicates that biosphere-atmosphere interactions are more complex than thought several years ago. Continued development of numerical models and algorithms is needed to understand the non-linear behaviour of terrestrial/marine systems and the related fluxes of constituents across the biosphere-atmosphere interface. Further interdisciplinary research with an integrated approach needs be done to cover the remaining gaps and to provide policy makers with a basis for making informed decisions on mitigation strategies on regional and global scales.

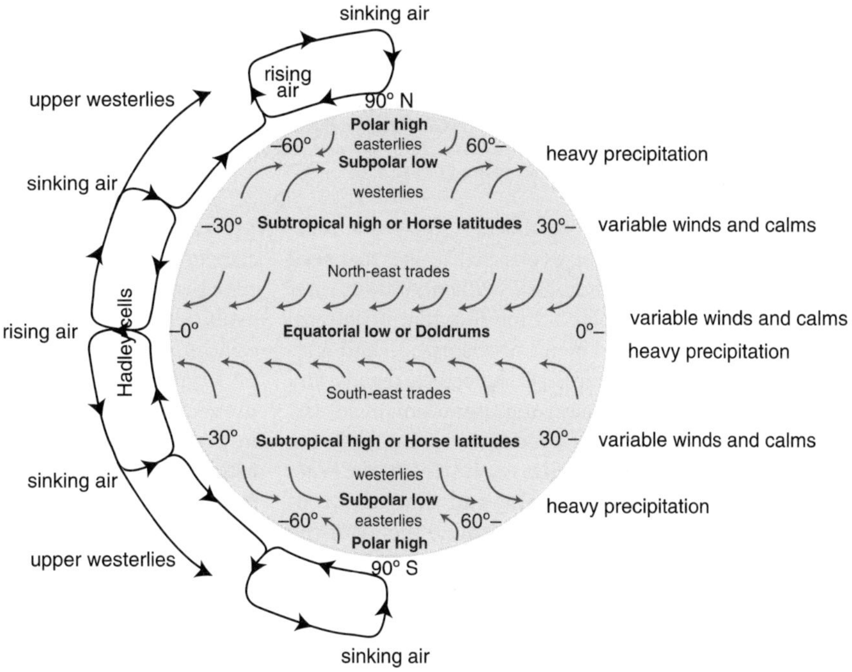

Fig. 1.2.
A schematic presentation of the wind system for a hypothetical water-covered earth, showing major winds and zones of low and high pressure. Vertical air movements and circulation are shown in exaggerated profile on the left of the diagram; characteristic surface conditions are given on the right. The two north-south cells on either side of the Equator make up the Hadley circulation (Brown et al. 1989)

(shallow during the night and morning, deeper during the afternoon and evening) is generally characterised by high humidity and high concentrations of compounds emitted at the surface (e.g. pollutants). In the presence of shallow or deep convection, chemical compounds are injected from the boundary layer to the free troposphere, but the rate of this boundary layer ventilation remains poorly quantified. Transport by deep convective cells and frontal systems is an important mechanism for moving air (and chemical compounds) vertically throughout the free troposphere. As air moves upward and cools, condensation takes place and precipitation can occur, which removes soluble gases and airborne particles from the atmosphere. Large-scale quasi-horizontal advection leads to intercontinental transport of relatively long lived chemical compounds. This process is responsible for trans-boundary air pollution transport. At midlatitudes, flow in the free troposphere is predominantly directed eastward in all seasons, and is considerably stronger during winter. The surface winds are highly variable, with different regimes separated by fronts. Equatorward of 30°, the weather regime is characterised by fast rising motions and heavy precipitation in the Inter-Tropical Convergence Zone (ITCZ) near the equator and slow descending motions in the subtropics. Regional seasonal patterns such as the Asian monsoon or temporal variations on longer time scales, such as the El Niño phenomenon, have large impacts on the budget of chemical compounds. Mass exchanges through the tropopause also affect significantly the distribution of several chemical compounds in the troposphere. This is particularly the case for ozone, as discussed in Chap. 3. Because the air is continuously stirred by weather patterns and convective activity, long lived compounds (lifetime of several months or more) tend to be quasi-uniformly mixed in the troposphere. The complexity of the tropospheric circulation is illustrated schematically in Fig. 1.2 and 1.3.

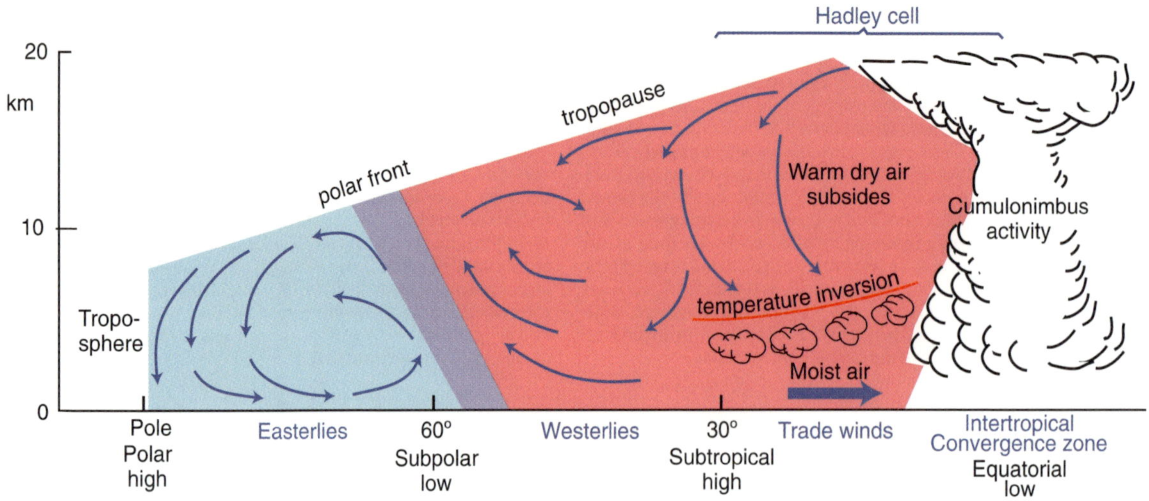

Fig. 1.3. Schematic presentation of the atmospheric circulation below 20 km altitude, from the Equator to the pole. This cross section shows the Inter-Tropical Convergence Zone (ITCZ) near the Equator, where strong upward motions take place, tropical clouds form, and intense thunderstorm activity occurs. Notice the return flow along the tropopause and the slow air subsidence in the subtropics. The polar front and the polar circulation are also shown (Brown et al. 1989)

Chemical transformations through gas-phase reactions or multi-phase processes proceed at rates that need to be measured in the laboratory, while photolysis rates are a function of the solar actinic flux that penetrates into the atmosphere, and hence of absorption and scattering processes by atmospheric molecules, clouds, and aerosol particles. A schematic representation of several of the important chemical processes occurring in the troposphere is presented in Fig. 1.4. It stresses the reactions that control the formation and destruction of ozone and hydroxyl, and that determine the oxidising efficiency of the atmosphere (see also Box 1.4). It also highlights the complexity of the tropospheric chemical system.

Due to the non-linearity of these processes, the response of the tropospheric chemical system to human-induced perturbations is very complex, and exhibits different regimes with abrupt transitions. For example, the existence of thresholds (such as a temperature threshold associated with the formation of particles) contributes to the non-linearity of the chemical system. As a result, the behaviour of the atmosphere in response to human-induced forcings is difficult to predict and can produce surprises.

1.2 Global Atmospheric Chemistry and the IGAC Project

During recent decades the field of atmospheric chemistry has evolved and grown significantly in response to the global and regional concerns discussed in this book. However, in contrast with the situation prevailing in meteorology, where continuous monitoring efforts provide detailed global observations on the physical state of the atmosphere, only limited observational data have been available on the global distributions, seasonal evolution, and trends of chemical compounds in the atmosphere. A major effort of the atmospheric chemistry community has therefore focussed on the measurement of chemical compounds (gases and aerosols) in the troposphere and stratosphere. Many projects, often sponsored by the International Global Atmospheric Chemistry (IGAC) Project, have used different types of instrumentation (e.g. *in situ* or remote sensing techniques) on different platforms (aircraft, balloons, spacecraft, ground-based, and satellites) (see Chap. 5). Field campaigns performed over limited time periods (see Appendix A.3) have been organised to investigate specific photochemical processes at different locations. Other observational projects have provided information on large-scale distributions of specific chemical constituents (see Appendix A.3). Periodic soundings of the atmosphere (balloon-borne, lidar) have revealed the structure of the vertical distribution for a limited number of atmospheric chemical species at a few locations. Future efforts will focus on the miniaturisation of instrumentation (so that it can be used routinely on commercial aircraft, for example), and on the development of new platforms (such as unmanned aircraft). Satellite-based measurements have played an important role in determining concentrations of a variety of gases and some types of particles and in getting data sets that are truly global, particularly for the stratosphere. While relatively little information exists now, satellite measurements will become an important future source of data on the distributions of tropospheric gases and aerosols. Even with these efforts by the international scientific

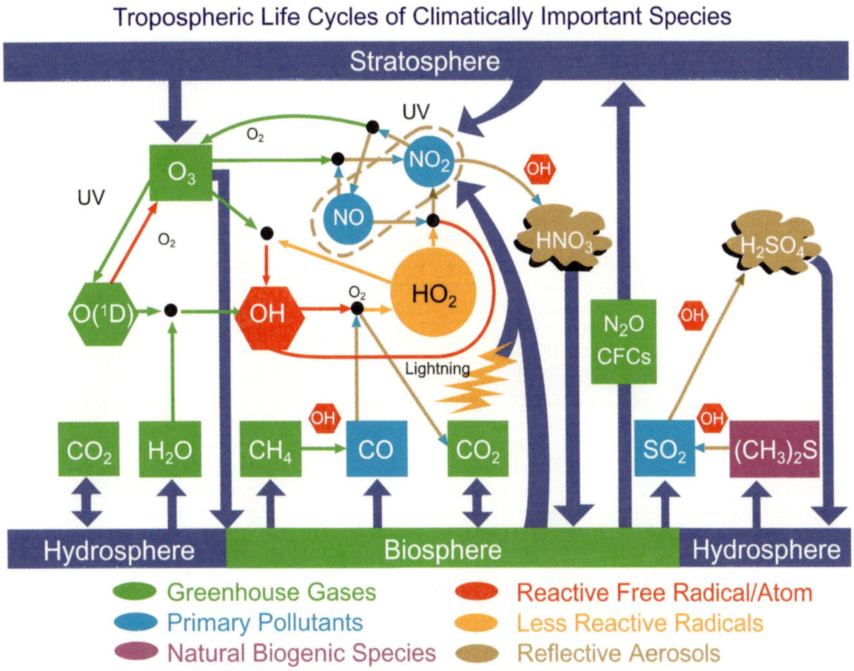

Fig. 1.4. Schematic representation of important chemical processes occurring in the troposphere (adapted from Prinn 1994)

Box 1.4. Oxidising power and atmospheric lifetime

The average time for which a gas molecule or aerosol particle remains in the atmosphere is determined by a number of atmospheric chemical and physical processes. For example, many chemical compounds are destroyed by oxidation processes and the resulting products, when they are soluble in water, are scavenged in precipitation. Other compounds that are chemically stable and not very soluble in water (such as the chlorofluorocarbons) may remain in the atmosphere for years or even decades. In this case, they are transported to the middle atmosphere and may affect stratospheric chemistry.

The atmospheric lifetimes of many atmospheric gases are determined mainly by their reactions with the OH radical. Other oxidation processes include reactions with ozone (O_3), hydrogen peroxide (H_2O_2), the nitrogen trioxide free radical (NO_3), and halogen species. Hydrogen peroxide, for example, converts sulphur dioxide to sulphuric acid in cloud and rain droplets, and hence contributes to the formation of acid precipitation. The NO_3 radical plays a significant role at night, primarily in polluted areas. The oxidation by OH and the other gases mentioned above is referred to as the oxidising power or the oxidising efficiency of the atmosphere.

community, a satisfactory global climatology of tropospheric compounds is far from a reality. A strategy to gather more information on the spatial and temporal distributions of the major chemical species needs to be established.

Laboratory investigations have provided measurements of reaction rate constants and absorption cross sections for many trace atmospheric chemical species and have provided the fundamental knowledge necessary to understand atmospheric photochemistry. While much progress has been made regarding gas-phase reactions, significant uncertainties remain in our understanding of multi-phase processes occurring through cloud droplet, ice surface, and aerosol interactions with trace gases. Regularly updated evaluations of both gas phase and heterogeneous experimental kinetics data provide inputs for atmospheric chemistry models and highlight poorly characterised kinetic processes that require further study (e.g. NASA JPL and IUPAC compilations).

Numerical models of atmospheric chemistry and physics (see Chap. 6) have been used to assess the consequences of newly measured chemical kinetics data, to analyse observations made during field campaigns, and to determine the sensitivity of calculated concentrations to external forcings. This has made it possible to derive (for example, through inverse modelling) the magnitude and location of the surface emissions that can explain these observations and to simulate the evolution of the chemical composition in response to natural or human-induced perturbations. Progress has often resulted from a combined use of several of these tools.

As one might expect for a new field such as global atmospheric chemistry, progress has been largely *ad hoc*, so that the level of international coordination and integration that is required to address such large-scale problems is not yet adequate. By contrast, for example, the international meteorological community, which represents a far more mature and operational discipline, has achieved a state of coordination and inte-

Box 1.5. IGAC and the atmospheric science community

Atmospheric chemistry is a central process in Earth system dynamics. Thus, research in atmospheric chemistry connects to many other efforts in global change research, assessment, and observation. Tackling global environmental problems requires a collaborative international suite of observational, research, and assessment activities.

IGAC is the leading international research project on tropospheric chemistry. It is a Core Project of the International Geosphere-Biosphere Programme (IGBP), one of a triad of international research programmes in global environmental change. The other two are the World Climate Research Programme (WCRP) and the International Human Dimensions Programme on Global Environmental Change (IHDP). IGAC interacts with many other elements of IGBP, WCRP, and IHDP, and in assessment and observation. The goals of IGAC are to develop a fundamental understanding of the processes that determine atmospheric composition, to understand the interactions among atmospheric chemical composition and physical, biospheric, and climatic processes, and to predict the impact of natural and anthropogenic forcings on the chemical composition of the atmosphere.

To achieve its goals, IGAC carries out collaborative research with other elements of IGBP:

- JGOFS, LOICZ and SOLAS (Joint Global Ocean Flux Study, Land Ocean Interactions in the Coastal Zone, and Surface Ocean Lower Atmosphere Study). These programmes study the chemical, physical, and biological processes that control the exchange of trace gases between the ocean and the atmosphere (note: SOLAS is currently in its planning stage).
- GCTE (Global Change and Terrestrial Ecosystems) studies the ecological processes that influence trace gas production from the land surface and the impacts of changes in atmospheric composition on terrestrial ecosystems.
- LUCC (Land Use and Cover Change) investigates the causes, patterns and projected rates of land-use and land-cover change, that affect rates of terrestrial emissions of trace gases.
- PAGES (Past Global Changes) investigates the composition of the atmosphere in the past, from ice core records and other paleo techniques.
- GAIM (Global Analysis, Integration and Modelling) undertakes intercomparisons and facilitates the development of atmospheric transport models and of Earth system models.
- In addition, IGAC interacts with other projects within the Programmes. Of particular relevance are:
  - SPARC (Stratospheric Processes and their Role in Climate) studies the physical and chemical dynamics of the stratosphere, especially the impact of human activities.
  - The WCRP modelling group facilitates the development of models of the physical climate system, drawing on atmospheric composition as a driver of changing climate.

IGAC also has major connection to long-term observational programs via GAW (Global Atmospheric Watch of the World Meteorological Organisation, WMO), which monitors the long-term evolution of atmospheric composition on a global and regional scale. GAW is a component of the Global Climate Observing System (GCOS). IGAC research has also contributed significantly to assessments of climate and atmospheric change. These include assessments of climate change by the Intergovernmental Panel on Climate Change (IPCC), and those on stratospheric ozone sponsored by the World Meteorological Organisation and the United Nations Environment Programme (UNEP).

As highlighted here, many organisations have strong connections to atmospheric chemistry in addition to IGAC. Because of ongoing activities within these other organisations, this document primarily focusses on chemistry in the global troposphere and provides only a cursory discussion of research activities relating to the carbon cycle, to stratospheric ozone, and to local and regional air quality.

gration which provides a sound basis for studying the physical state of the global atmosphere. Atmospheric chemistry requires a similarly sound basis for carrying out fundamental long-term studies of the chemical state of the global atmosphere. The field currently suffers from inadequate spatial and temporal observational capabilities and, especially, from the lack of sufficiently strong international efforts to verify observational methods, to improve analytical precision, and to provide the accurate calibration standards upon which any study of long-term change fundamentally depends.

IGAC, which has established strong links with other international initiatives (see Box 1.5), has sponsored a number of significant experimental and interpretative research programmes, as noted throughout this volume.

Many of the results of the IGAC efforts in the last ten years are summarised within the larger context of global and regional atmospheric chemistry in this and the following chapters of this volume. Where appropriate, research opportunities and priorities are suggested. As an introduction to the other chapters, we present in the next sections an overview of past changes in the chemical composition of the atmosphere, summarise the plausible causes of these changes, and assess possible impacts.

1.3 Past Changes in Atmospheric Chemical Composition

1.3.1 Long-term Variability: Evidence of Feedbacks

Quantitative records of past atmospheric composition are available in polar ice caps on a time scale spanning several ice ages. Gases are encapsulated in air bubbles and aerosols species are entrapped in the ice matrix. Such data allow us to deduce limits to the natural variability of atmospheric composition, document feedbacks, and offer the possibility of testing particular hypotheses and evaluating the capabilities of atmospheric models. These paleo-records clearly demonstrate that past climatic and environmental changes on Earth are highly correlated with atmospheric composition, which suggests that the nature of biogeochemical feedbacks is key to understanding global change. These data sets also suggest a number of questions, such as: How important were biogeochemical cycle feedbacks in the flip from one climate mode to another? How strong are these feedbacks on climate relative to other purely physical responses such as polar ice-albedo effects? Have biological feedbacks contributed to the unprecedented rate of greenhouse gas change observed today?

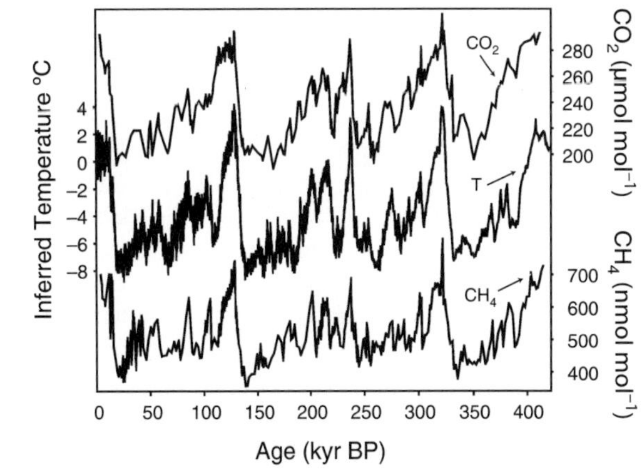

Fig. 1.5. Concentration changes of carbon dioxide, methane, and dust along the Vostok ice core (Antarctica) for the last 400 000 years. Isotopic temperature (from $\delta^{18}O$) is also shown as climate proxy (data from Petit et al. 1999)

Detailed information for the past 400 000 years has been obtained from the Vostok (Antarctica) ice core for two major greenhouse gases CO_2 and CH_4 (Fig. 1.5). A remarkable overall correlation is found between the concentrations of these gases and temperature changes. The lowest levels of greenhouse gases are observed under full glacial conditions (190 µmol mol^{-1} and 0.35 µmol mol^{-1} for CO_2 and CH_4, respectively, during the last glacial maximum, 20 kyr B.P.). All the available records obtained from Antarctic and Greenland glacial ice cores show that the rises during deglaciation are about 80 and 0.3 µmol mol^{-1} for CO_2 and CH_4, respectively. While the data shown in Fig. 1.5 suggest that both greenhouse gases are strikingly in phase with the climate record, some differences emerge when the data are examined in detail. For example, as a glaciation begins, CH_4 concentration decreases in phase with Antarctic cooling, whereas CO_2 lags markedly behind the temperature change (Raynaud et al. 1993). Other data show that the abrupt climatic changes that affected the North Atlantic region on the decadal to centennial time scale during the last glacial-interglacial transition and during the last ice age are remarkably reflected in the CH_4 records. Ice core measurements (Leuenberger and Siegenthaler 1992) also indicate a depletion of the concentrations of N_2O during ice ages in relation, probably, with reduced emissions of this greenhouse gas by soils under drier climatic conditions.

Ice core data also reveal that during ice ages the terrestrial and marine contributions to Antarctic aerosol deposition were increased significantly due to stronger winds and dryer environmental conditions. Concentrations of calcium and sodium (crustal and marine primary aerosol reference elements, respectively) in Vostok ice were higher by factors of about 30 and four, respectively, relative to Holocene (i.e. the past 10 000 years) values. Greenland ice core data suggest that the dustiness of the glacial atmosphere was most likely a global phenomenon (Hammer et al. 1985; Hansson 1994).

Despite higher secondary aerosol (sulphate, nitrate, and methanesulphonate) concentrations, the atmosphere (at least as revealed by polar studies) was not more acidic during glacial periods. The present day atmosphere has become more acidic as a result of anthropogenic emissions. The cause of the higher secondary aerosol levels in the glacial Antarctic atmosphere is not fully understood.

Finally, ice cores from Greenland and Antarctica have recorded numerous large volcanic eruptions (Zielinski et al. 1996; Clausen et al. 1997). Sulphate and chloride concentrations are available for volcanic periods in numerous polar ice cores. Some of these events were powerful enough to have markedly disturbed global atmospheric composition for up to three years after eruption.

In conclusion, strong correlations have been observed between the temperature fluctuations through the glacial cycles and the concentrations of several trace gases and aerosol types. These correlations indicate strong couplings of these components through biogeochemical and physical climate processes. Elucidation of these coupling processes is one of the most exciting challenges facing global change scientists, including atmospheric chemists, during the years ahead.

1.3.2 Changing Atmospheric Composition During the Last Few Centuries

The concentrations of many atmospheric gases have increased since the pre-industrial period to reach levels probably unprecedented in the last 400 000 years. A combination of ice core data and direct monitoring data reveals the globally-averaged changes in atmospheric concentration for several key gases over the last few centuries (see Fig. 1.6). The gap between polar ice core data and the first direct atmospheric monitoring is now entirely filled by firn air data for the three major long lived greenhouse gases: CO_2 (Etheridge et al. 1996), CH_4 (Etheridge et al. 1992; Blunier et al. 1993; Etheridge et al. 1998), and N_2O (Machida et al. 1995; Battle et al. 1996). Prior to the 17th century, CO_2 concentrations fluctuated naturally in the range from 275–285 µmol mol^{-1} except during ice ages. The pre-industrial level of methane in the Antarctic was 670 ±70 nmol mol^{-1}. The increase in CH_4 concentration started during the second half of the 18th century, reaching concentrations greater than 1700 nmol mol^{-1} in the current atmosphere. Increases in CH_4 emissions appear to be the primary factor in explaining the observed changes. In addition, carbon isotope ratio ($^{13}C/^{12}C$ and $^{14}C/^{12}C$) determinations in CH_4 now available for the Southern Hemisphere since 1978 from air archives and Antarctic firn air (Francey et al. 1999) constrain estimates of the natural and an-

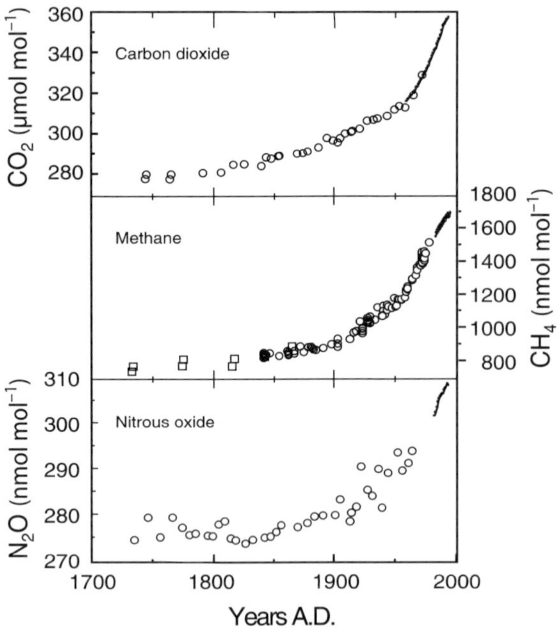

Fig. 1.6. Atmospheric concentration of major greenhouse gases CO_2, CH_4, and N_2O over the last 300 years, as reconstructed from Greenland and Antarctic ice core studies. *Continuous line:* atmospheric record (Etheridge et al. 1992, 1996; Blunier et al. 1993; Machida et al. 1995; see also IPCC 2001)

thropogenic global sources and sinks of methane. Nitrous oxide concentrations have increased from 285 to more than 310 nmol mol^{-1} since the pre-industrial era.

Sulphates from fossil fuel burning and smelting, as well as nitrates from fossil fuel burning (including automotive emissions), have increased in polar snow and ice since the very beginning of the 20th century; however, their impacts are only noticeable in the Northern Hemisphere (Mayewski et al. 1990; Legrand et al. 1997). In comparison to a pre-industrial sulphate level of about 25 ng g^{-1}, concentrations peaked at 110 ng g^{-1} at the end of the 1960s. Thereafter, decreasing anthropogenic sulphur emissions in Europe and North America are clearly reflected in Greenland sulphate profiles, which exhibit a steady downward trend since about 1980.

Nitrate concentration increases in Greenland ice from a pre-industrial mean value of about 70 to 120 ng g^{-1} for the period 1950–1989. This trend, valid all over Greenland, starts markedly around 1950, i.e. much later than the sulphate trend. A decline of nitrate is evident only for recent years. The nitrate trend observed in Greenland firn must be considered with some caution due to the reversibility of gaseous HNO_3 deposition onto snow, even in Greenland (Fischer et al. 1998). Moreover, it must be kept in mind that the nitrogen-bearing chemical species deposited in Greenland remain unidentified. They may have been different in the pre-industrial and present atmospheres.

From pre-industrial (1755–1890) to modern times (1950–1975), carbonaceous particle concentrations at a high-Alpine glacier site have increased by factors of 3.7, 3.0, 2.5, and 2.6, for black carbon, elemental carbon, organic carbon, and total carbon, respectively (Lavanchy et al. 1999).

Only a few data are presently available on CO concentrations in the past. An increase of about 20 nmol mol^{-1} is apparent between 1800 and 1950 for this compound at Summit, Greenland, where the pre-industrial level is 90 nmol mol^{-1} (and 55–60 nmol mol^{-1} in the Antarctic) (Haan et al. 1996). These figures need to be confirmed because glaciological artefacts in the measurements are possible (Haan and Raynaud 1998).

Formaldehyde measurements in Greenland snow are scarce and have to be viewed with caution because the deposition of this compound onto snow is reversible. At Summit, Staffelbach et al. (1991) found a doubling of pre-industrial concentration (about 2 ng g^{-1}) in recent years, with an acceleration of the increase in recent decades. Post-depositional chemical transformations taking place in the surface snow pack can even be a source of compounds such as formaldehyde (Sumner and Shepson 1999), nitrogen oxides (Honrath et al. 1999), and CO to the atmosphere. A task ahead is to determine the role played by the surface snow pack in lower atmospheric composition and in influencing the transfer function of a chemical constituent between atmosphere and glacial firn and ice.

As indicated above, changes in the tropospheric concentration of ozone and its precursors (NO_x, CO, and hydrocarbons) have been substantial since the pre-industrial era. Very few observational data are available to quantify these changes, so that estimates rely primarily on model simulations. This issue will be discussed further in Chap. 3 and 7.

1.3.3 Future Changes in Atmospheric Composition

Although considerable progress has been made in the past decade in understanding the biogeochemical cycles of greenhouse gases, oxidants, and aerosols, large gaps in knowledge still remain. Furthermore, uncertainties in the climate feedback response of these processes are also large. Nonetheless, many analyses suggest that concentrations of key greenhouse gases and other atmospheric constituents will continue to change in the future as a result of human activities. Concentrations of important greenhouse gases are expected to continue to increase in the coming decades. The concentration of CO_2 could double or even triple its pre-industrial concentration by 2100 (IPCC 1996, 2001). On the other hand, atmospheric concentrations of CFCs and several other halocarbons should decline over the coming decades as a result of global controls to protect the stratospheric ozone layer.

Future projections of CH_4 are affected by uncertainties in the temperature dependence of wetland sources and permafrost melting, and in atmospheric OH radical concentrations (the major sink). The evolution of future N_2O levels in the atmosphere is also uncertain because of poorly constrained sources and sinks. Agricultural inputs of large and growing amounts of nitrogen fertilisers lead to questions as to where the nitrogen is going and how much N_2O may be released in the future from fixed nitrogen accumulations in environmental subsystems (soils, waters). Future emissions of SO_2 and NO_x will depend mainly on the use of fossil fuels and the extent to which measures are taken to control these emissions.

1.4 Causes of Atmospheric Changes

The rapid changes in atmospheric composition observed over the past century have been driven largely by emissions from fossil fuel combustion, industry, biomass burning, and intensive agriculture. A sample of the many long lived gases whose atmospheric concentrations are being significantly affected by human activities is shown in Table 1.1. Estimates of the growth rates and lifetimes remain somewhat uncertain for several of the compounds listed. Current global inventories of human-related atmospheric emissions for some of the environmentally important gases discussed in previous sections are presented in Table 1.2 along with comparisons to the magnitudes of natural sources (biosphere, oceans, volcanoes, lightning). Human related emissions are important, and often dominant, contributors to the abundance of many chemical species in the current atmosphere. Also included in Table 1.2 are forecast ranges of anthropogenic emissions for 2100 presented by IPCC (2001) for different potential socio-economic scenarios. These forecasts suggest that human perturbations to atmospheric composition in general are likely to increase over the coming century. In addition, future climate change may affect the natural (and human related) emissions of radiatively active gases and aerosols, resulting in complex chemistry-aerosol-climate feedbacks.

1.4.1 Fossil Fuel Combustion and Industry

Our technological society is largely powered by the combustion of fossil fuels. Fossil fuel combustion has been the main cause of the rise in atmospheric CO_2 over the past 200 years from ~280 µmol mol^{-1} (natural background) to ~370 µmol mol^{-1} in the year 2001. In addition to CO_2, combustion processes emit a large number of pollutants including CO, SO_2, NO_x, soot particles, trace metals, and hydrocarbons (including toxic organic compounds). For many of these chemical species, the global source from fossil fuel combustion exceeds natural sources (Table 1.3). Other emissions related to fossil fuel use include CH_4 from natural gas, coal, and oil production and distribution, and SO_2 from petroleum refining. Motor vehicle traffic is a major source of CO and NO_x.

Industrial activities other than fossil fuel combustion also represent large sources of pollutants to the atmosphere. Metal smelters and the paper industry are major sulphur emitters. Carbon steel manufacturing is an important source of CO. Many organic compounds are emitted by the chemical industry and their environmental impacts in terms of toxicity, climate forcing, and tropospheric ozone are poorly understood.

1.4.2 Biomass Burning

Biomass burning is associated with deforestation, agricultural fires, wood and dung fuel use, and natural fires. Most of the world's biomass burning takes place in the Tropics and is highly seasonal, peaking at the end of the dry season (January–April in the northern Tropics, August–October in the southern Tropics). The largest contribution to this tropical source is agricultural burning of savannas; deforestation is also an important source in some regions. Natural fires are important mainly in boreal forests at high northern latitudes, where their frequency and intensity peak during the summer months.

Table 1.1. A sample of long lived important atmospheric gases that are affected by human activities (data based on IPCC 2001)

	CO_2 (carbon dioxide)	CH_4 (methane)	N_2O (nitrous oxide)	$CFCl_3$ (CFC-11)	CHF_3 (HFC-23)	CF_4 (perfluoromethane)
Pre-industrial concentration	~280 µmol mol^{-1}	~700 nmol mol^{-1}	~270 nmol mol^{-1}	0	0	40 pmol mol^{-1}
Concentration in 1998	365 µmol mol^{-1}	1745 nmol mol^{-1}	314 nmol mol^{-1}	268 pmol mol^{-1}	14 pmol mol^{-1}	80 pmol mol^{-1}
Rate of concentration change	+1.5 µmol mol^{-1} yr^{-1}	+8.4 nmol mol^{-1} yr^{-1}	+0.8 nmol mol^{-1} yr^{-1}	−1.4 pmol mol^{-1} yr^{-1}	+0.55 pmol mol^{-1} yr^{-1}	+1 pmol mol^{-1} yr^{-1}
Atmospheric lifetime	50–200 yr[a]	12 yr[b]	114 yr[b]	45 yr	257 yr	50000 yr

[a] No single lifetime can be defined for carbon dioxide because of the different rates of uptake by different removal processes.
[b] This lifetime is an adjustment time that accounts for the indirect effect of the gas on its own residence time.

Table 1.2. Current global inventories of human related emissions for environmentally important gases compared to the magnitudes of natural sources (e.g., from the biosphere, oceans, volcanoes, and lightning) mostly adapted from IPCC (1995, 1996), with some values based on IPCC (2001). Also included are forecast ranges of anthropogenic emissions for year 2100 presented by the IPCC (2000) Special Report on Emissions Scenarios (SRES) for different possible socio-economic assumptions

	CO_2 Pg C yr^{-1}	CH_4 Tg CH_4 yr^{-1}	N_2O Tg N yr^{-1}	Sulphur Tg S yr^{-1}	NO_x Tg N yr^{-1}	CO Tg CO yr^{-1}
Natural	150[a]	160	10.2	18	11	400
Human-related						
Fossil fuel and industry	6.8	100	1.3	65	22	400
Biomass burning	1.6[a]	40	0.5	2	12	500
Agriculture		170	3.9		1	
Landfills/sewage		65				
Other						550
Human total, present	7.6	375	5.7	67	35	1 450
Human-related, 2100[b]	3.3–36.8	236–1 069	4.8–19.3	11–69	16–110	363–3 766

[a] Annual terrestrial biosphere and ocean fluxes for carbon dioxide are generally balanced by fluxes back to biosphere and oceans.
[b] Based on the harmonised 26 scenarios in IPCC (2000) SRES.

Table 1.3. Combined effect estimates of daily mean particulate pollution (after American Thoracic Society)

% change in health indicator per each 10 µg m^{-3} increase in PM_{10}	
Increase in daily mortality	
Total deaths	1.0
Respiratory deaths	3.4
Cardiovascular deaths	1.4
Increase in hospital usage (all respiratory diagnoses)	
Admissions	1.4
Emergency department visits	0.9
Exacerbation of asthma	
Asthmatic attacks	3.0
Bronchodilator use	12.2
Emergency department visits[a]	3.4
Hospital admissions	1.9
Increases in respiratory symptom reports	
Lower respiratory	3.0
Upper respiratory	0.7
Cough	2.5
Decrease in lung function	
Forced expiratory volume	0.15
Peak expiratory flow	0.08

[a] One study only.

Biomass burning emits the same general suite of chemical species to the atmosphere as fossil fuel combustion (CO_2, CO, VOCs, NO_x, etc.), but often in very different proportions; biomass burning is less efficient than fossil fuel combustion and the combustion temperatures are lower.

There is little knowledge of historical trends in biomass burning, and IPCC (2001) does not venture to forecast trends for the next century. Greenland ice core records of biomass burning tracers over the past 1 000 years identify several historical periods of enhanced biomass burning at high northern latitudes but indicate a decrease since the 1930s. Fire records maintained by governmental agencies in North America and Europe over the past 50 years indicate a general increase in the number of fires due to human negligence, but not much change in area burned because of improved fire-fighting ability. In the Tropics, where most of global biomass burning takes place, there is essentially no information on historical trends. One would expect a rise in biomass burning from increasing agriculture and deforestation over the past century, but this effect could have been offset by better fire control in response to the growing rural population. The issue of biomass burning is discussed in detail in Chap. 2.

1.4.3 Land-use Changes

Major land-use changes over the past 100 years have included urbanisation driven by population pressure, increase in cultivated land in the Tropics, reforestation of formerly cultivated land in northern midlatitudes regions such as the eastern United States, and desertification in some areas such as northern Africa. Tropical deforestation now makes a significant contribution to the rise in CO_2 (Table 1.1). However, this contribution is expected to decrease over the next century due to decimation of primary forests and it may actually be overcome by the atmospheric CO_2 sink from reforestation and fertilisation at northern midlatitudes (IPCC 2001). Agriculture is thought to be the principal cause for the rises of CH_4 and N_2O over the past 100 years (Table 1.2). Methane is emitted by anaerobes in ruminants and rice paddies (which are effectively human-generated wetlands). Nitrogen fertiliser application to crops stimulates microbial emission of N_2O; it also stimulates emission of NO_x, but the resulting source is small compared to that from fossil fuel combustion (Table 1.3). The impact of land-use changes on the emissions of chemical compounds, as well as the question of dust (iron and other metals) emission associated with desertification and deposition to the ocean are addressed in Chap. 2.

1.4.4 Climate Changes

Changes in the climate system could substantially affect the chemical composition of the atmosphere. Such changes are related to interannual variability in the dynamics of the atmosphere, to quasi-periodic oscillations in the atmosphere-ocean system (such as the El Niño and La Niña phenomena in the tropical Pacific), and to longer time scale climatic trends. Long-term trends are believed to be due to external forcing including human-induced emissions of greenhouse gases.

Although the mechanisms governing Earth's climate are not fully understood, the effects of climate changes on the chemical composition of the atmosphere need to be identified because they provide potential positive or negative feedback mechanisms in the climate system. Issues to be considered are the expected warming resulting from enhanced concentrations of carbon dioxide, methane, and other greenhouse gases, possible increases in convective activity and lightning frequency, changes in precipitation rates and in soil moisture, and changes in surface emissions of temperature-sensitive biogenic compounds, among others.

1.5 Impacts of Changes in Atmospheric Composition

Impacts resulting from changes in atmospheric composition can be substantial. In this section, we briefly highlight three major categories of impacts associated with changes in atmospheric composition: climate change, effects on ecosystems, and effects on human health. There is also some discussion of composition change impacts on corrosion, agriculture, and water resources.

Because changes in the chemical composition of the atmosphere are expected to be the main drivers of human-induced climate change, atmospheric chemists have been deeply engaged in the climate change issue, and have established strong links with the more physically oriented climate researchers. They have also developed links with ecologists and agriculture experts who study the effects on ecosystems and crops of increased exposure to various atmospheric gases and aerosols, and to the increased levels of UV radiation associated with decreases in stratospheric ozone. Connections to the community of scientists studying the effects of atmospheric changes on human health have been considerably weaker.

1.5.1 Climate Change

The issue of climate change, in general, and of the impact of human activities on climate, in particular, is the subject of extensive analysis and assessment by the scientific community. At the international level, these assessments are led by the Intergovernmental Panel on Climate Change (IPCC 1990, 1992, 1995, 1996, 2001). Its recent analysis (IPCC 2001) of the global average climate forcing during the past 250 years due to recognised human-related and natural processes is shown in Fig. 1.7. The dominant positive forcing (heating tendency) results from the increased concentration of rather well mixed greenhouse gases (CO_2, N_2O, CH_4, CFCs, etc.). Tropospheric and stratospheric ozone changes are treated separately mainly because this greenhouse gas is not well mixed in the atmosphere and the observed changes have distinct horizontal and vertical spatial patterns. The largest negative forcing (cooling tendency) is due to aerosol particles of human origin. Because of the limited tropospheric lifetime of aerosol particles (days to a few weeks) they are also not well mixed and their forcing patterns are less uniform than that of the greenhouse gases. As shown in Fig. 1.7, evaluation of the radiative forcing from all of the different sources since pre-industrial times indicates that the net globally-averaged radiative forcing on climate is positive.

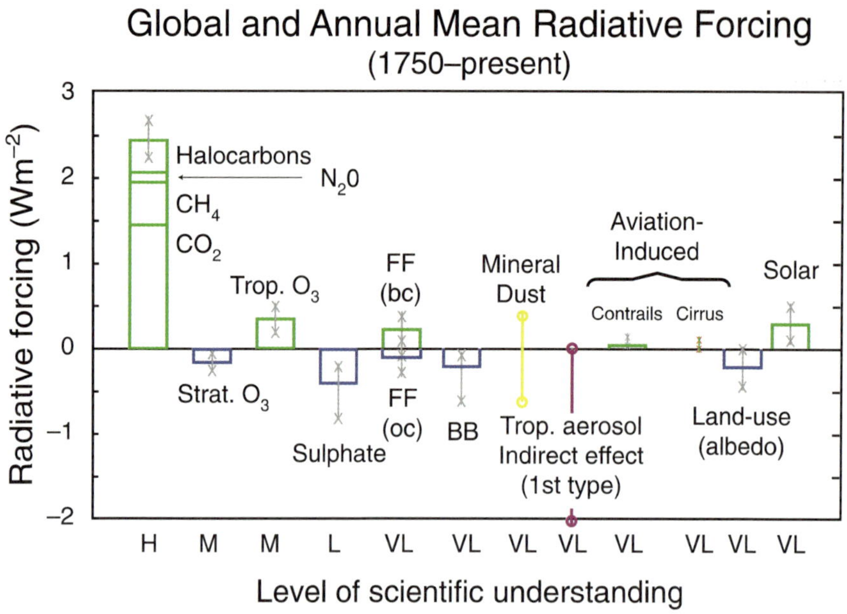

Fig. 1.7. Globally-averaged climate forcing (in W m^{-2}) during the past century due to recognized human related and natural processes (IPCC 2001). *Solid bars* represent best estimates and *vertical lines* uncertainty ranges (*H* = high, *M* = medium, *L* = low, *VL* = very low, *FF* = fossil fuel, *BB* = biomass burning, *bc* = black carbon, *oc* = organic carbon)

1.5.2 Impacts on Ecosystems

Ecosystems are affected by atmospheric changes in several different ways:

- Direct deposition of gases such as O_3, SO_2, and NO_2;
- Impact of increased levels of UV radiation caused by stratospheric ozone depletion; and
- Indirect impacts caused by changes in the chemical balance of soils, lakes, and surface ocean water following the deposition of acidifying and eutrophying substances, such as sulphates, ammonium, nitrates, etc.

The impacts may include anything from elimination of sensitive species and altered productivity to subtle shifts in the relative abundance of plant and animal species. These issues will be addressed in Chap. 2.

1.5.3 Corrosion

Acid deposition and high concentrations of gaseous pollutants (SO_2, NO_2, and O_3) are known to cause damage to buildings and cultural and historical monuments mainly in urban and suburban areas. The economic loss due to such man-made corrosion in Western Europe and North America has been substantial, where, mainly because of reductions in SO_2 concentrations, the corrosion rates are now decreasing significantly. However, in many other areas, especially in Eastern Europe and some developing countries, corrosion problems remain severe or are becoming more acute. There are indications that in warm climates NO_2 and O_3 alone (even without SO_2) may cause serious corrosion. Modern communication and information devices are also subject to atmospherically induced corrosion. Telecommunication switching gear, magnetic recording and data storage media, and metallic connections in semiconductor structures are all susceptible to atmospheric corrosion processes.

1.5.4 Health Effects

Air pollutants have long been recognised as health hazards. Well known examples include smog episodes in London – the one in December 1952 is estimated to have caused 4 000 excess deaths – and the photochemical pollution (O_3, PAN, etc.) in Los Angeles, which causes lung damage and eye irritation, among other effects. During recent years more subtle effects, some of them believed to occur not only in urban and suburban areas but also over larger regions, have been identified. Here we list some of the important health issues related to specific air pollutants.

Halogenated hydrocarbons (including the CFCs), while not harmful as such, contribute to depletion of the stratospheric ozone layer. The resulting *increase in UV-B radiation* at the earth's surface is estimated to cause adverse health effects including higher incidences of eye cancer and cataract, skin cancer (in fair skinned populations), and a depression of the immune system for certain tumours and infectious diseases (UNEP 1998).

Elevated levels of O_3, as low as 75 nmol mol^{-1} in surface air, have been demonstrated to have adverse health effects, even at short-term exposures. Today, such concentrations are commonly observed during the warm seasons over large regions in Europe and North America and probably also in other parts of the world where emissions of ozone precursors (NO_x, hydrocarbons, CO) are large.

Epidemiological studies show that short-term exposure to current ambient levels of *aerosols* (i.e. 24-hour average concentrations of 30 to 150 µg m^{-3} of particles <10 µm diameter (PM_{10})) are associated with adverse health effects ranging from reduced lung function to mortality (Table 1.3).

The *acidification of air and precipitation*, in addition to its adverse effects on terrestrial and limnic ecosystems, also has consequences for human health. Direct health effects of elevated levels of SO_2 and NO_x have been well established (Lübkert-Alcamo and Krzyzanowski 1995). Indirect health effects due to increased dissolution of metals in soils (e.g. Al, Cd) and in water pipes (Cu) are also suspected.

Mercury (Hg) is an interesting example of a contaminant that has been mobilised by man and dispersed globally through the atmosphere. The current deposition of Hg is estimated to exceed the pre-industrial value by at least a factor of two world-wide and up to a factor of five to ten in northern Europe and North America (Bergan et al. 1999). The elevated deposition of Hg over large regions, aggravated by the increased leaching of Hg from acidified soils and a subsequent bioaccumulation of organic Hg in fish, has resulted in toxic concentration in pike in many lakes in Scandinavia and Canada (Lindqvist 1991).

1.6 Some Important Questions

The objective of this book is to synthesise and integrate the results of research conducted by the atmospheric chemistry community over the last 10 to 15 years and, specifically, to highlight key findings of projects facilitated by IGAC. This work provides the scientific basis for answering questions that are important to society:

- What determines the chemical composition of the natural atmosphere?
- How have human activities altered atmospheric composition?
- How have human activities changed the global atmospheric budgets of carbon, nitrogen, and sulphur?
- What controls tropospheric ozone?
- Is the "cleansing efficiency" of the atmosphere changing?
- How does atmospheric chemistry affect the biosphere and food production?
- How does atmospheric chemistry affect human health?
- What is the connection between atmospheric composition and climate?
- How might chemical composition evolve in the future?
- Are there risks of abrupt changes and/or irreversible changes in atmospheric composition?
- What should the research strategy be to address unresolved questions?

Although no definitive answer to some of these questions can yet be given, many of the projects completed by the scientific community in the last decade have provided information that will help address many of them. Some of the important scientific research being conducted by the international atmospheric chemistry community in order to provide the answers to these questions is summarised in Chap. 2 through 4. The importance of the biosphere in controlling the chemical composition of the atmosphere, and an assessment of the impact on the biosphere of changes in the chemical composition of the atmosphere are considered in Chap. 2. The role of chemical transformations of gas-phase compounds, with emphasis on tropospheric ozone and other photooxidants, is examined in Chap. 3. The physical and chemical processes governing the formation of aerosols, plus their evolution and removal from the atmosphere, are discussed in Chap. 4. The advances that have been made in instrumentation and modelling are described in Chap. 5 and 6, respectively. Finally, an integrated view of the causes and impacts of atmospheric chemical changes in the form of responses to the questions listed above is provided in Chap. 7.

Chapter 2

Biosphere-Atmosphere Interactions

Lead authors: Mary C. Scholes · Patricia A. Matrai · Meinrat O. Andreae · Keith A. Smith · Martin R. Manning
Co-authors: Paulo Artaxo · Leonard A. Barrie · Timothy S. Bates · James H. Butler · Paolo Ciccioli · Stanislaw A. Cieslik Robert J. Delmas · Frank J. Dentener · Robert A. Duce · David J. Erickson III · Ian E. Galbally · Alex B. Guenther Ruprecht Jaenicke · Bernd Jähne · Anthony J. Kettle · Ronald P. Kiene · Jean-Pierre Lacaux · Peter S. Liss · G. Malin Pamela A. Matson · Arvin R. Mosier · Heinz-Ulrich Neue · Hans W. Paerl · Ulrich F. Platt · Patricia K. Quinn Wolfgang Seiler · Ray F. Weiss

2.1 Introduction

The contemporary atmosphere was created as a result of biological activity some two billion years ago. To this day, its natural composition is supported and modified, mostly through biological processes of trace gas production and destruction, while also involving physical and chemical degradation processes. The biosphere has a major influence on present environmental conditions, both on a regional and global scale. One of the best-documented and most important indicators of global change is the progressive increase of a number of trace gases in the atmosphere, among them carbon dioxide (CO_2), methane (CH_4), and nitrous oxide (N_2O), all of which are of biospheric origin. There is considerable uncertainty, however, regarding the processes that determine the concentration and distribution of trace gases and aerosols in the atmosphere and the causes and consequences of atmospheric change (Andreae and Schimel 1989). To improve our understanding IGAC created an environment for multi-disciplinary collaboration among biologists, chemists, and atmospheric scientists. This was essential to develop analytical methods, to characterise ecosystems, to investigate physiological controls, to develop and validate micrometeorological theory, and to design and develop diagnostic and predictive models (Matson and Ojima 1990).

Interactions between the biosphere and the atmosphere are part of a complex, interconnected system. The emission and uptake of atmospheric constituents by the biota influence chemical and physical climate through interactions with atmospheric photochemistry and Earth's radiation budget. Comparatively small amounts of CH_4 and N_2O present in the atmosphere make substantial contributions to the global greenhouse effect. In addition, emissions of hydrocarbons and nitrogen oxides from biomass burning in the Tropics result in the photochemical production of large amounts of ozone (O_3) and acidity in the tropical atmosphere. In turn, climate change and atmospheric pollution alter the rates and sometimes even the direction of chemical exchange between the biosphere and atmosphere through influences at both individual organism and ecosystem levels. Recent and expected future changes in land use and land management practices provide further impetus for closely examining climate-gas flux interactions. Anthropogenic influences, e.g. tropical deforestation and the widespread implementation of agricultural technologies, have and will continue to make significant alterations in the sources and sinks for the various trace gases.

Ten years ago, at the beginning of IGAC, researchers sought to establish the source and sink strength of gases in different kinds of ecosystems, in different areas of the world. Specific goals of the programme, related to the biosphere included:

- to understand the interactions between atmospheric chemical composition and biological and climatic processes;
- to predict the impact of natural and anthropogenic forcings on the chemical composition of the atmosphere; and
- to provide the necessary knowledge for the proper maintenance of the biosphere and climate.

Earlier extrapolations of gas fluxes over space and time were often based on a single, or very small, set of measurements, and researchers sought for "representative" sites at which to make those crucial measurements. IGAC brought a new focus to the variability among ecosystems and regions of the world, in order to understand better the factors controlling fluxes (Galbally 1989). For example, studies of CH_4 flux from wetlands and rice paddies of N_2O flux from natural and managed ecosystems, and of dimethylsulphide (DMS) emissions from oceans, consciously spanned gradients of temperature, hydrological characteristics, soil types, marine systems, management regimes, and nitrogen deposition. One result of this strategy has been the recognition that the same basic processes were responsible for gas fluxes across regions, latitudinal zones, and environments. This chapter gives a general overview of the progress that has been made in the field as a whole within the last decade, with emphasis on research activities stimulated, initiated, and/or endorsed by the IGAC community. It is not our intent to provide current assessments of all trace gas source and sink strengths, as those budgets have been compiled and published (with considerable contributions by IGAC researchers) in recent Intergovernmental Panel on Climate Change (IPCC) documents. Examples of research not conducted within the IGAC framework but relevant to the topic are CH_4 from landfills, ruminant livestock, and termites; information on these topics can be found in IPCC (1996, 1999).

Exchanges of biogenic trace gases between surfaces and the atmosphere depend on the production and consumption of gases by microbial and plant processes, on physical transport through soils, sediments, and water, and on flux across the surface-air boundaries. Thus, to understand and predict fluxes, studies of whole ecosystems are required. The goals of research over the past decade have been to develop an understanding of the factors that control flux, organise the measurements so that they are useful for regional and global scale budgets, and use the knowledge to predict how fluxes are likely to change in the future.

The IGAC Project focussed on issues of specific interest over a number of different geographical regions of Earth. A variety of projects have been conducted over the last ten years, many of which addressed issues related to exchange between the biosphere and the atmosphere. Several field campaigns, using a combination of measurement and modelling techniques, have been conducted very successfully under the IGAC umbrella, e.g. in southern Africa (SAFARI 1992 and 2000) and in various oceanic regions (ACE-1, ACE-2, and ACE-Asia) (see A.5).

Why certain trace gases were studied together and why various scientific approaches were adopted to study them is described in this chapter. Research findings specifically related to the exchange of trace gases and aerosols between the atmosphere and the terrestrial and marine biospheres will be given. In the terrestrial section, special attention is given to biomass burning and wet deposition in the Tropics, because of the significant contribution made by IGAC to these programmes. We also consider some of the anthropogenic activities that alter biosphere-atmosphere exchange and discuss potential feedbacks related to climate change, regional level air pollution, and deposition. In the marine section, emphasis is on the biogeochemistry of DMS, given that the greatest advances were made on this topic. The chapter concludes by summarising the major accomplishments of the last decade and highlighting some of the remaining research challenges.

2.2 Key Biogenic Gases or Families and their Relevance to Atmospheric Chemistry

The study of atmospheric composition has largely focussed on trace compounds that affect either the radiative properties of the atmosphere, or the biosphere as nutrients or toxins, or play a key role in atmospheric chemistry. The trace gases that are important in this regard have been summarised in the preceding chapter. This chapter considers the role of the biosphere in emission or removal of such compounds.

Although CO_2 and water (H_2O) are both greenhouse gases which are strongly affected by the biosphere, studies of these compounds have generally been conducted in parallel scientific communities, and IGAC has maintained a focus on the chemically reactive greenhouse gases. Thus no attempt is made here to cover the large body of research on the global carbon cycle and the interactions of CO_2 with the biosphere.

2.2.1 The Carbon Family of Gases: CH_4, Volatile Organic Carbon Compounds (VOCs), and Carbon Monoxide (CO)

Methane (CH_4) is a greenhouse gas with a lifetime in the atmosphere of about nine years. Its atmospheric concentration is largely controlled by the biosphere, with 70% or more of current emissions and virtually all of pre-industrial emissions being biogenic (Fig. 2.1; Milich 1999). The dominant biogenic production process for CH_4 is microbial breakdown of organic compounds in

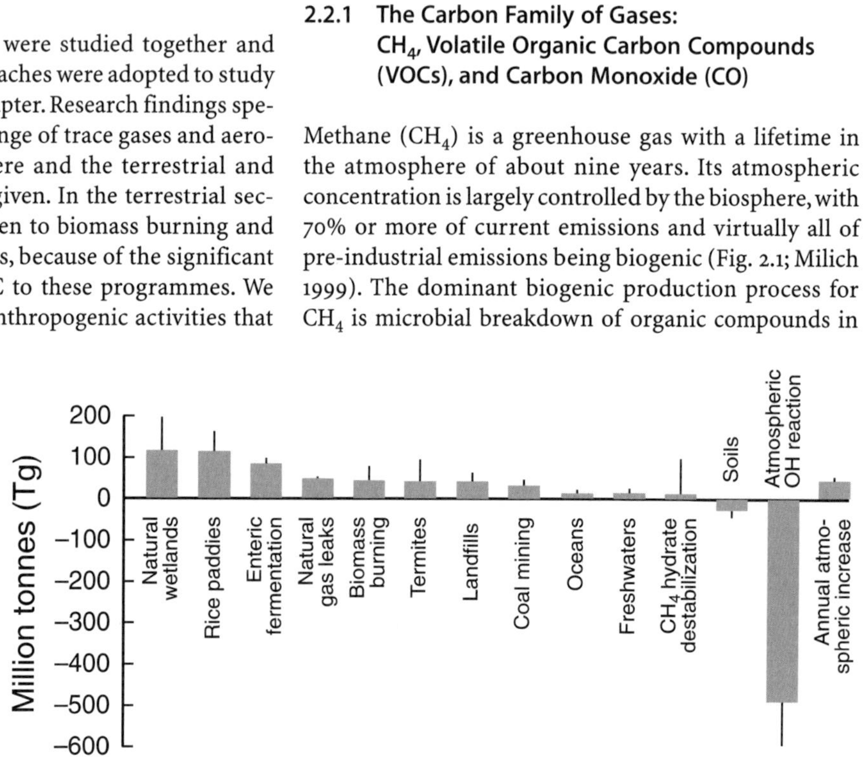

Fig. 2.1. Estimated annual anthropogenic and natural sources and sinks of methane (*thick bars*) in millions of tons, and uncertainty ranges (*thin lines*) (Milich 1999)

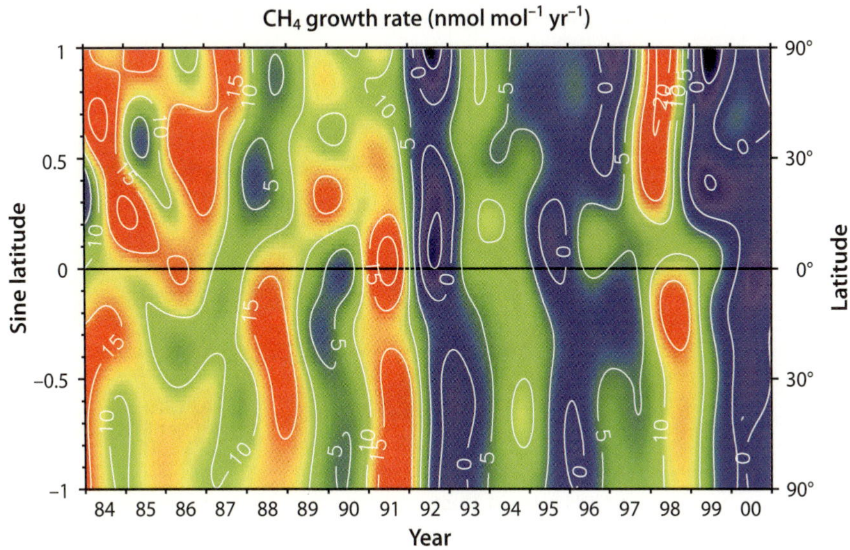

Fig. 2.2. Methane growth rates (figure courtesy of the National Oceanic and Atmospheric Administration (NOAA), Climate Monitoring and Diagnostics Laboratory (CMDL), and Carbon Cycle-Greenhouse Gases (CCGG))

anaerobic conditions. This occurs in flooded soils such as natural wetlands and rice paddies, in the rumen of animals such as cattle, in landfills, and in anoxic layers in the marine water column and sediments. Methane is also emitted directly to the atmosphere from burning vegetation as a product of pyrolytic breakdown of organic material.

Changes in land use, particularly increases in numbers of domestic ruminants, the extent of rice paddies, and biomass burning, have more than doubled biogenic CH_4 emissions since the pre-industrial era (Milich 1999; Ehhalt et al. 2001). Fossil fuel related emissions and a decrease in atmospheric oxidation rates (Thompson 1992) have further increased CH_4 concentrations but those aspects fall outside the scope of this chapter.

The atmospheric concentration of CH_4 is now about 1745 nmol mol^{-1}, compared to pre-industrial levels of about 700 nmol mol^{-1}. Growth rates have been observed directly in the atmosphere since the 1950s (Rinsland et al. 1985; Zander et al. 1989) and on an increasingly systematic basis since 1978 (Blake and Rowland 1988; Dlugokencky et al. 1994). Concentrations were increasing at about 20 nmol mol^{-1} yr^{-1} in the 1970s, but that rate has generally declined to an average of 5 nmol mol^{-1} yr^{-1} over the period 1992 to 1998. High growth rates of about 15 nmol mol^{-1} yr^{-1} occurred in 1991 and 1998 (Fig. 2.2, Dlugokencky et al. 1998; Ehhalt et al. 2001) and appear to be caused by climate related increases in wetland and/or biomass burning emissions (Dlugokencky et al. 2000; Walter and Matthews 2000). IPCC (2001) estimates its rate of increase at 8.4 nmol mol^{-1} yr^{-1}.

The evolution of the CH_4 budget since the pre-industrial era provides a good example of interactions between land use and atmospheric change. A schematic of the change in total emissions from the 18th century to the present is shown in Fig. 2.3 (based on Stern and Kaufmann 1996, Lelieveld et al. 1998, and Houweling et al. 1999). The removal rates that are required to balance the source-sink budget at pre-industrial and present concentrations imply an increase in the methane lifetime. This is consistent with independent estimates of a decrease in atmospheric oxidation rates inferred from chemistry models.

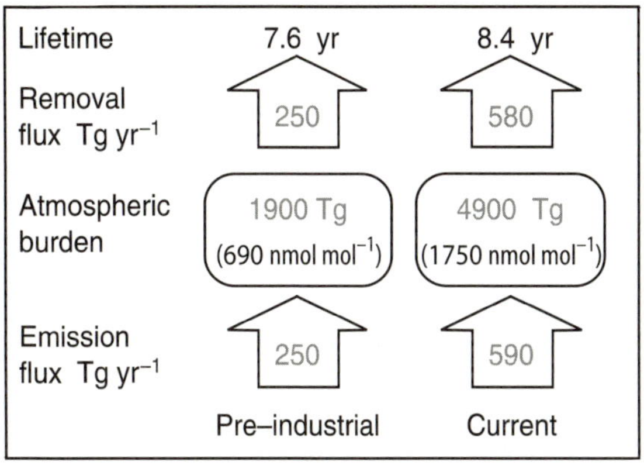

Fig. 2.3. Schematic of the pre-industrial Holocene and current (1990s) atmospheric methane budget. The mean lifetime derived from the ratio of atmospheric burden to removal rate has increased by ca. 10%, which is broadly consistent with estimates of the relative decrease in OH from atmospheric chemistry models (based on Stern and Kaufmann (1996), Lelieveld et al. (1998) and Houweling et al. (1999) (see also Fig. 7.4.)

Plants emit a range of volatile organic carbon (VOCs) compounds, which include hydrocarbons, alcohols, carbonyls, fatty acids, and esters, together with organic sulphur compounds, halocarbons, nitric oxide (NO), CO, and organic particles. Estimates of anthropogenic emissions for 1990 are shown in Fig. 2.4. According to current estimates plants emit up to 1200 Tg C yr^{-1} as VOCs (Guenther et al. 1995). The amount of carbon released from the biosphere this way may be up to 30% of net ecosystem productivity (NEP), i.e. the annual accumulation of carbon in an ecosystem before taking account

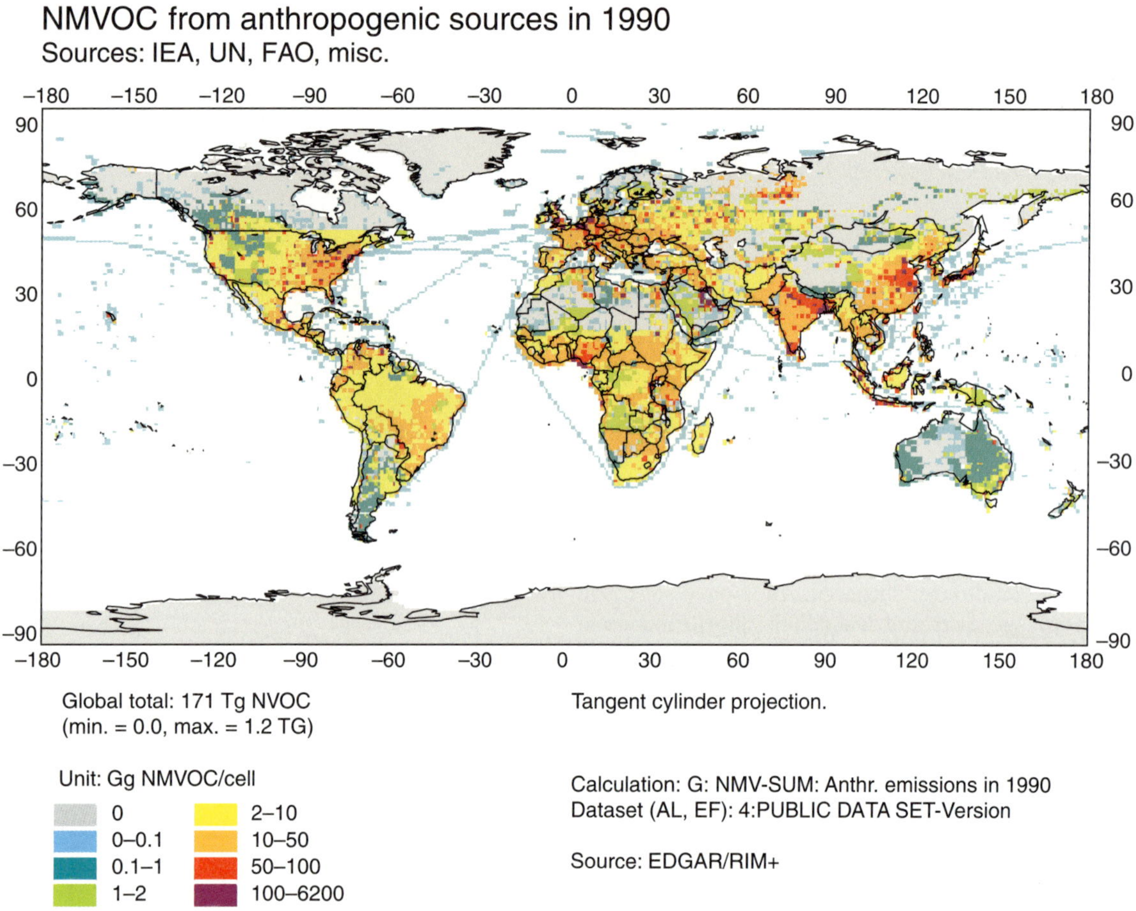

Fig. 2.4. Anthropogenic yearly non-methane VOC emissions in 1990 from the EDGAR (Emission Database for Global Atmospheric Research) database (Olivier et al. 1996)

of ecosystem disturbance (e.g. Valentini et al. 1997; Kesselmeier et al. 1998; Crutzen et al. 1999). Neglect of VOC and CO terrestrial emissions may cause significant errors in estimates of NEP and changes in carbon storage for some ecosystems.

While the oceans are supersaturated with CO and surface production of VOCs is widespread, the ocean-atmosphere fluxes are small, but less well studied, compared with terrestrial emission estimates. VOCs show a wide range of reactivities in the troposphere, with lifetimes ranging from minutes (e.g. β-caryophyllene) to two weeks (e.g. methanol) (Atkinson and Arey 1998). Many are emitted at very low rates, and in some cases are offset by plant uptake, thus having a negligible impact on atmospheric chemistry; others impact ozone production (see Chap. 3), aerosol production (see Chap. 4), and the global CO budget.

Primary pollutants emitted mainly as a result of human activity include hydrocarbons, CO, and nitrogen oxides. About half the terrestrial surface emissions of CO are due to direct emissions from vegetation and biomass burning. In addition about 45% of the total CO source to the atmosphere is due to oxidation of methane and other organics in the atmosphere, which themselves are predominantly biogenic compounds. Because CO is the end product in the methane oxidation chain the two budgets are closely linked; in addition, CO also originates from the breakdown of VOCs. The concentrations of CO are temporally and spatially highly variable due to the short lifetime of CO and the nature of its discontinuous land based sources. Estimates of anthropogenic CO emissions for 1990 are shown in Fig. 2.5.

2.2.2 The Nitrogen Family of Gases: Ammonia (NH_3), N_2O, and NO

Despite its importance for particle formation and climate, relatively little effort has been spent on understanding the sources and removal processes of NH_3. Most work on atmospheric ammonia has been performed with respect to eutrophication and acidification close to the terrestrial sources; large scale transport and chemistry of NH_3 and ammonium (NH_4^+) have received much less attention, especially over remote marine regions. The global source strength of ammonia is about

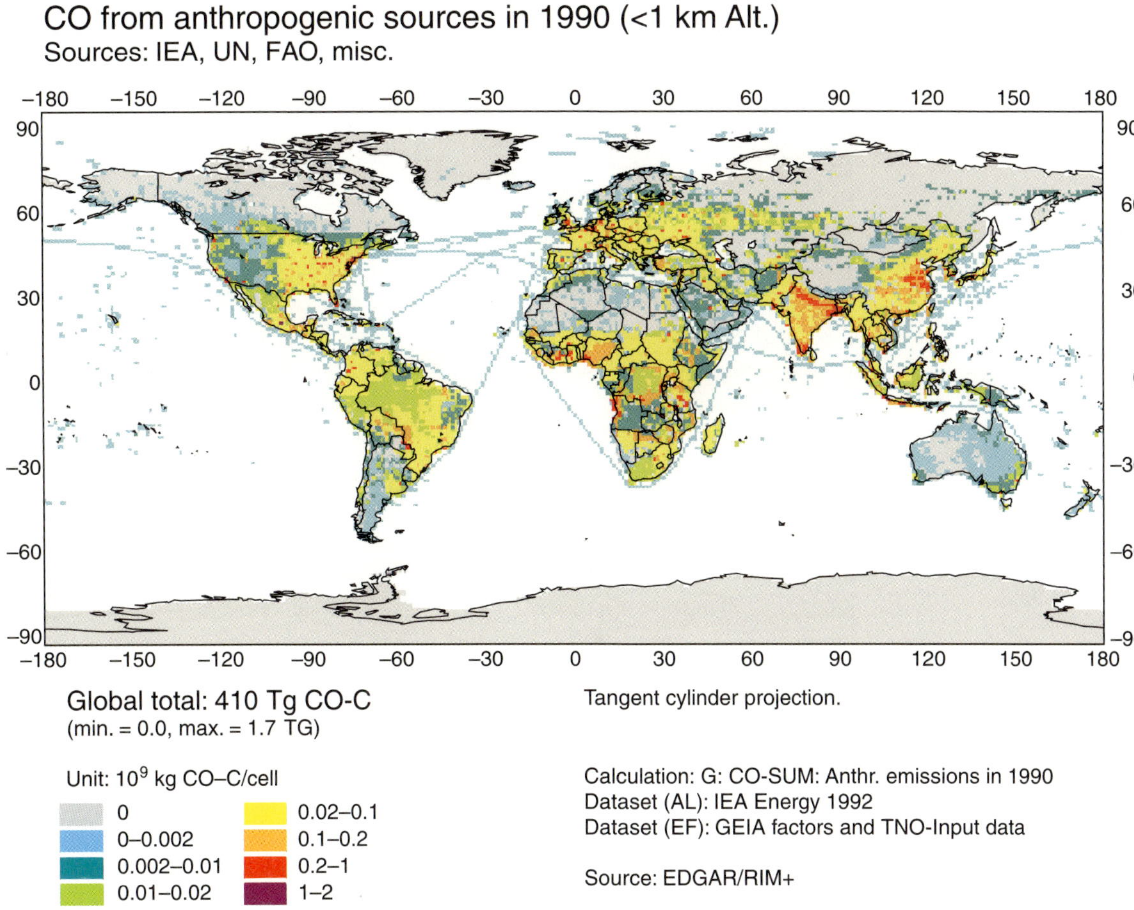

Fig. 2.5. Anthropogenic annual CO emissions in 1990 from the EDGAR database (Olivier et al. 1996)

55 Tg N yr^{-1}, which is of similar magnitude to global NO_x-N emission (Bouwman et al. 1997).

The most recent estimate for global NH_3 emissions (Bouwman et al. 1997) from animals relied on constant emission factors and amounted to 21.7 Tg N yr^{-1}, which is of similar magnitude as fossil fuel related global NO_x-N emissions. The second most important emission category is N-containing synthetic fertiliser. Again, huge differences in agricultural practice and environmental conditions cause a large variation of emissions factors. Overall global emission of ammonia derived from nitrogen fertiliser was estimated to be 9 Tg yr^{-1}, which is 10% of the amount applied. Interestingly, ammonia losses from application of urea fertiliser to rice paddies seem to contribute strongly to this. Other anthropogenic sources, such as biomass burning, cropland, and humans additionally emit about 10 Tg yr^{-1}. Natural sources, such as soils, vegetation, and oceans, emit about 10–20 Tg yr^{-1} (Bouwman et al. 1997; Schlesinger and Hartley 1992) and are highly uncertain.

Nitrous oxide is an important greenhouse gas with a lifetime of about 120 years. The largest production process for N_2O is "leakage" during microbial nitrification and dentrification processes in soil and aquatic systems. Significant emissions also occur from decomposition of animal waste, oxidation of ammonia (NH_3), and biomass burning. Biogenic sources of N_2O have increased with expansion of food production systems, intensification of agriculture, and anthropogenic modification of the global nitrogen cycle.

The concentration of N_2O has increased from about 270 nmol mol^{-1} in pre-industrial times (Kroeze et al. 1999) to 314 nmol mol^{-1} today (CMDL 2001). There is some evidence for small variations in growth rates in the early 1990s, but during the period of precise *in situ* measurements growth rates have remained near constant at around 0.8 nmol mol^{-1} yr^{-1} in both hemispheres (CMDL 2001). The global N_2O flux from the ocean to the atmosphere has been calculated based on more than 60 000 field measurements of the partial pressure of N_2O in surface water (Fig. 2.6). These data were extrapolated globally and coupled with air-sea gas transfer coefficients estimated on a daily basis (Nevison et al. 1995). A global ocean source of about 4 (1.2–6.8) Tg N yr^{-1} was determined and latitudinal bands of varying emission were delimited.

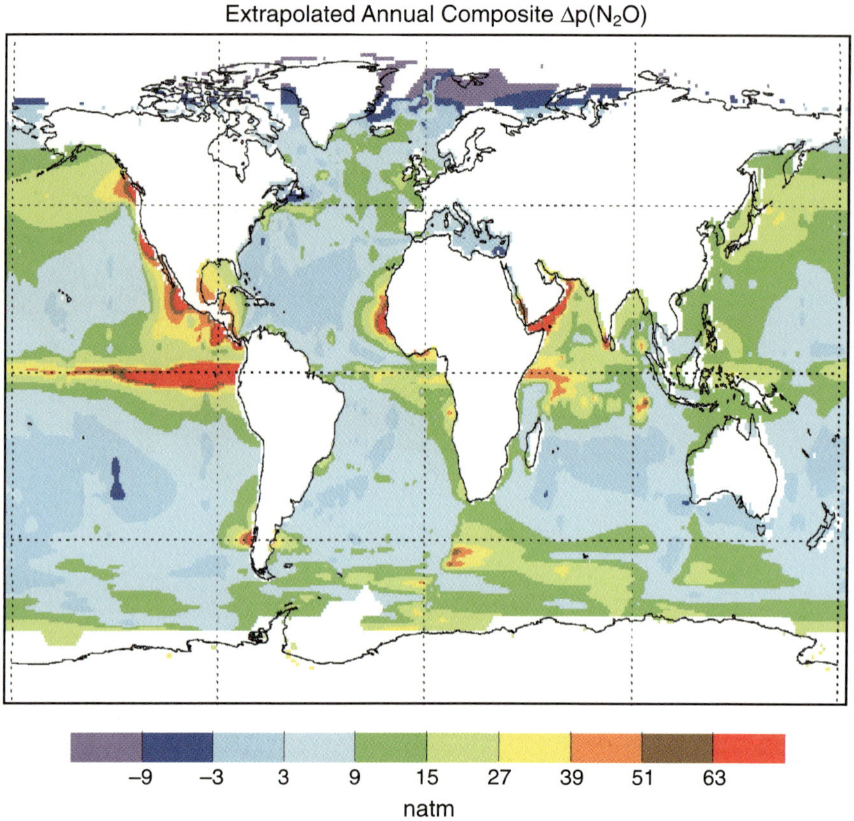

Fig. 2.6. Annual composite surface $\Delta p(N_2O)$ (10^{-9} atmospheres) (Nevison et al. 1995)

Quantifying the wide range of N_2O sources has proved difficult and upper and lower bound estimates for specific source types can differ by a factor of ten (Ehhalt et al. 2001). However, progress has been made in balancing the source-sink budget and its recent evolution has been reviewed by Kroeze et al. (1999).

The radiative forcing of climate due to increases in CH_4 (see above) and N_2O during the industrial era is about 25% of the total due to all well mixed greenhouse gases (Ramaswamy et al. 2001). In addition both gases play a significant role in atmospheric chemistry. Increases in CH_4 tend to decrease atmospheric oxidation rates (e.g. Thompson and Cicerone 1986), but increase O_3 and stratospheric H_2O levels. The result of these indirect effects is to amplify the radiative forcing due to CH_4 emissions by around 70%. Changes in concentrations of N_2O over time have tended to decrease stratospheric O_3 (Crutzen 1979) but this effect is small (see Chap. 3).

Nitric oxide has a short lifetime (approximately one day) in the atmosphere and takes part there in a complex cycle of reactions with CO and hydrocarbons to form tropospheric ozone. Total emissions, both natural and anthropogenic, range from 37 to 59 Tg N yr^{-1} (Graedel and Crutzen 1993). Estimates in the 1980s of global annual emissions of NO from soils, its largest natural source, were ca. 8 Tg N yr^{-1}; these have been revised using an extended data set and are now estimated to be as large as 21 Tg NO-N yr^{-1} with an error term of at least ±4 and perhaps as large as ±10 Tg N yr^{-1} (see Sect. 2.7.1.3). The available data confirm that the soil source of NO is similar in magnitude to fossil fuel emissions of NO_x (Davidson and Kingerlee 1997; Skiba et al. 1997). Minor sources include lightning, transport from the stratosphere, biomass burning, and aircraft emissions.

2.2.3 The Sulphur Family: Dimethylsulphide and Carbonyl Sulphide

Sulphur containing gases are major participants in gas to particle conversion (see Chap. 4). Anthropogenic sulphur emissions from fossil fuel oxidised to sulphate particles can act, in addition to sea salt particles, as condensation nuclei for marine clouds (see Chap. 4). Natural biological emissions of sulphur are predominantly marine in origin, with minor emissions from volcanoes. Dimethylsulphide (DMS), which is produced by microbial processes in the ocean, is emitted at the rate of 15–30 Tg S yr^{-1} (Bates et al. 1992). A recent global inventory of DMS emissions to the atmosphere has been created using the data from more than 16 000 observations of surface ocean DMS concentrations (Kettle et al. 1999) (Fig. 2.7). The estimates of DMS emitted from the ocean to the atmosphere are constrained largely due to the increased number of field observations and mass bal-

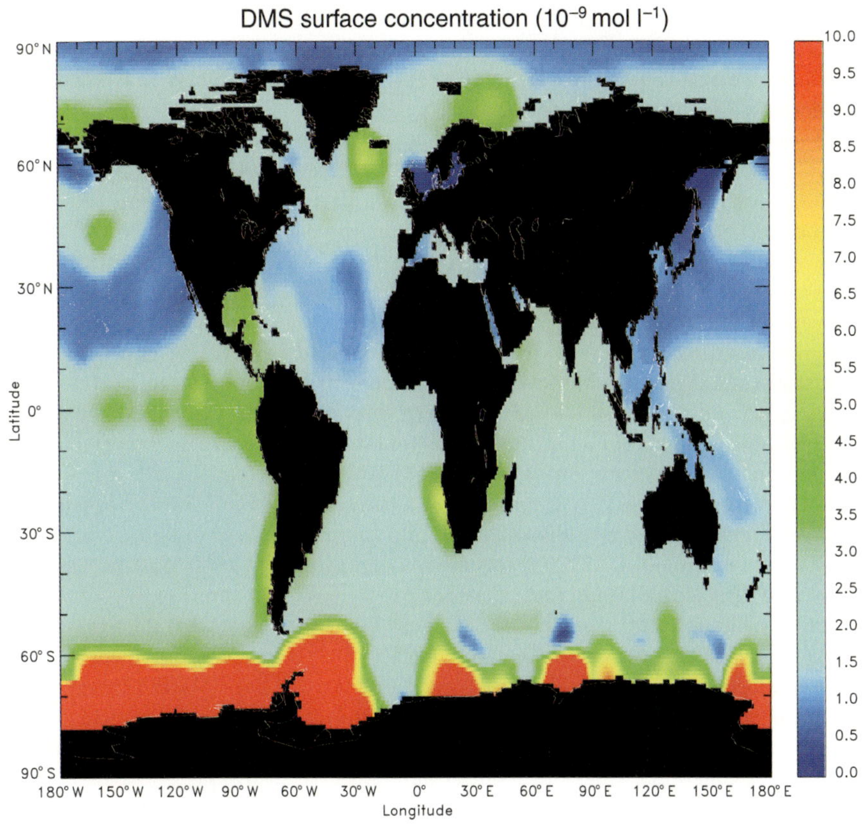

Fig. 2.7. Smoothed field of January mean DMS sea surface concentration (10^{-9} mol l^{-1}). The original field was smoothed with an 11-point unweighted filter to remove discontinuities between biogeochemical provinces (Kettle et al. 1999)

ance of the sulphur budget in the marine boundary layer (Chen et al. 1999; Davis et al. 1999).

Carbonyl sulphide (COS) in the atmosphere originates predominantly from the outgassing of the upper ocean (30%), atmospheric oxidation of carbon disulphide (unknown), and biomass burning (20%), with a total emission of about 1 Tg S yr^{-1} (Andreae and Crutzen 1997; Chin and Davis 1993). With the longest tropospheric lifetime of all atmospheric sulphur compounds, COS can reach the stratosphere where it is oxidised to sulphate particles, which may impact the radiation budget of Earth's surface (Crutzen 1976) and influence the stratospheric ozone cycle.

2.3 A Paleoclimatic Perspective on CH_4 and DMS

Information on past concentrations of several trace gases is preserved in air bubbles trapped when snow is progressively buried and compacted to form ice in areas of Greenland and the Antarctic where temperatures are cold enough to prevent surface melting. The archived air preserved in this way has provided reliable estimates of changes in atmospheric CH_4 and N_2O for up to 400 000 years in the past.

Methane concentration changes are now well depicted in both hemispheres and vary from about 350 nmol mol^{-1} for glacial to about 700 nmol mol^{-1} for interglacial climatic conditions (Stauffer et al. 1988; Raynaud et al. 1988; Chappellaz et al. 1990). Significant rapid CH_4 changes are associated with nearly all abrupt climatic changes that affected the northern hemisphere over the last ice age (Chappellaz et al. 1990, 1993; Brook et al. 1996), indicating a very tight response of the natural CH_4 cycle to climate fluctuations.

The Holocene record (11 500 B.P. to present) provides the natural atmospheric CH_4 variability in relatively stable climatic conditions (Blunier et al. 1995; Chappellaz et al. 1997). The early Holocene (11 500–9 000 B.P.) is a period of relatively high concentrations (720 nmol mol^{-1}), with a lower mean value (570 nmol mol^{-1}) centred around 5 000 B.P. and marked drops of 200-year duration around 11 300, 9 700, and 8 200 B.P. The mean inter-hemispheric difference of concentrations, which is mainly a function of the latitudinal distribution of sources and sinks, has been found to be 45 ±3 nmol mol^{-1}, i.e. markedly lower than the present-day difference of ca. 140 nmol mol^{-1} (Dlugokencky et al. 1994).

A high precision record for CH_4 in the Antarctic (Etheridge et al. 1998), shows mixing ratios increasing from about 670 nmol mol^{-1} 1 000 yr ago with an anthropogenic increase evident from the second half of the 18th century. Similar information is available from Greenland ice (Blunier et al. 1993). Over the pre-industrial period, natural variability is about 70 nmol mol^{-1} around the mean level.

For the last 50 years both concentration and isotopic data ($^{13}C/^{12}C$ and $^{14}C/^{12}C$) for CH_4 are now becoming available from analyses of firn air samples (e.g. Francey et al. 1999). The concentration data indicate a pause in the increase of anthropogenic emissions during the period 1920–1945, probably due to a stabilisation of fossil fuel emissions at that time, while the isotopic data have placed constraints on the relative role of natural and anthropogenic sources and sinks in the 1978 to 1995 period.

Paleo data from ice core studies have had a strong impact on our understanding of the global CH_4 cycle, in particular the latitudinal distribution of wetland emissions. Changes in monsoon patterns (Chappellaz et al. 1990) and the distribution of northern mid- and highlatitude wetlands (Chappellaz et al. 1993) have been considered. More recently Brook et al. (1996) favoured a boreal control on the CH_4 global budget. Changes in methane removal rates must also be taken into account, and model calculations (Thompson 1992; Thompson et al. 1993; Crutzen and Brühl 1993; Martinerie et al. 1995) generally, though not unanimously, suggest that hydroxyl radical (OH) concentrations were higher in glacial conditions than today. The consequent increased removal rate explains at most 30% of the reduction in concentration, implying that the larger effect is that due to lowered emissions.

Ice core data do not support a sudden release to the atmosphere of large amounts of CH_4 from clathrate (hydrate) decomposition at the last deglaciation (Thorpe et al. 1996), as proposed by several authors (e.g. Paull et al. 1991; Nisbet 1992). However, more gradual release of CH_4 from clathrates cannot be discounted as a potentially significant factor and there is some isotopic evidence for clathrate methane releases synchronous with reorganisation of ocean circulation (Kennett et al. 2000).

Global DMS emissions may be modulated by climatic conditions. Could global warming trigger a change of marine biogenic activity and consequently of DMS emissions? Human-induced atmospheric changes could also disturb the oxidation processes of DMS and modify the branching ratio between methanesulphonic acid (MSA) and non-sea salt (nss) SO_4 formation. Ice core studies may help to elucidate these questions, provided that DMS or at least a DMS-related compound is recorded in polar ice. In this regard, MSA has been considered as the most promising parameter to determine in polar ice cores. Over the last decade, a few firn and ice cores have been analysed in detail for MSA and $nssSO_4$, in the hope of finding a correlation between concentrations in ice and climate fluctuations on various time scales. Some interesting results have been obtained, but glaciological phenomena have been pointed out recently that obscure the interpretation of the data.

At Antarctic locations where accumulation is relatively high (>20 g cm^{-1} yr^{-1}), MSA concentration records seem to be reliable and decadal variations can be seen in shallow firn cores. In the Weddell Sea area, Pasteur et al. (1995) found from an ice core covering the last three centuries that MSA marine production increases at warmer temperatures, in relation probably to the amount of broken sea ice where phytoplankton can develop favourably. MSA concentration in coastal Antarctic snow seems to be linked with sea-ice extent (Welch et al. 1993). On the other hand, the validity of MSA ice records is questionable inland. A marked decreasing trend of MSA concentration was found in upper firn layers (the first 6 m) at Vostok (Wagnon et al. 1999). It is suggested that MSA scavenged in the snow crystals is progressively released from the solid phase by snow metamorphism. Part of the initially deposited MSA probably escapes back to the atmosphere. The profile obtained at Dome F (Dome-F Ice Core Research Group 1998) shows very low MSA concentrations between about 30 and 70 m depth, thereafter a rise from about 70 m up to 110 m. The effect can be attributed tentatively to the trapping of interstitial gaseous MSA in the air bubbles at the firn-ice transition (pore close-off). These observations, corroborated by MSA measurements at Byrd Station (West Antarctica) (Langway et al. 1994), lead to the conclusion that MSA concentration depth profiles from central Antarctica are most probably strongly affected by post-deposition phenomena. Sulphate records are not perturbed.

At Amundsen Scott Station (the South Pole), some decreasing trend of MSA concentrations with depth is observable in the firn layers, but it is less steep than at Vostok, probably related to the higher snow accumulation rate. Interestingly, Legrand and Feniet-Saigne (1991) detected marked spikes of MSA concentration in the upper 12 m of firn (i.e. over the last 60 years) at this site. These were attributed to the impact of El Niño events on the production rate of MSA in the sub-Antarctic marine areas or on its transport to inner Antarctica. The changes are superimposed on the general decreasing trend of MSA profiles found in the upper firn layers.

MSA records in Greenland firn cores over the last 200 years, on the other hand, show a rise starting from surface layers and lasting several decades (Whung et al. 1994; Legrand et al. 1997). This surprising trend, opposite to what is found at the South Pole, could be attributed to a change in DMS marine productivity during this period or to the marked increase of atmospheric acidity caused by anthropogenic sulphur emissions. In the latter case, the amount of MSA remaining in the snow could depend on the pH of the atmosphere or of the snow.

Long-term changes in DMS-derived compounds can be seen in both Antarctica and Greenland records. The covariance of MSA and $nssSO_4$ concentrations observed in the Vostok core suggests that both compounds are mainly derived from marine DMS emissions. MSA and $nssSO_4$ concentrations are both higher in glacial conditions, with higher values of the ratio MSA / $nssSO_4$ found for ice ages. An increase of marine biogenic productivity has been put forward to explain this observation (Legrand et al. 1988, 1991, 1992), but the glaciological artefacts reported above for MSA records in central Antarctic firn layers cast some doubt on the proposition. Clearly more work has to be done on the understanding of chemical composition changes of ice on the scale of several glaciations, all the more since Greenland data are contradictory to Antarctic observations. In the Renland ice core (East Greenland), MSA concentration and the MSA / $nssSO_4$ ratio are markedly lower for cold than for warm climatic stages (Hansson and Saltzman 1993). For the two deep cores recovered at Summit (GRIP and GISP 2), conclusions are similar (Saltzman et al. 1997; Legrand et al. 1997). These observations suggest that, for the sulphur cycle, the cases of the northern and the southern hemispheres have to be discussed differently. In particular, the interaction of the primary aerosol (continental dust, sea salt) with acid sulphur compounds has to be investigated.

2.4 Atmospheric Compounds as Nutrients or Toxins

Deposition of atmospheric trace compounds can act as a significant source of nutrients or toxic substances to ecosystems, and their effects on these systems may in turn affect other trace atmospheric constituents. An example is natural fertilisation of the oceans by dust deposition, which leads to increased biological productivity, hence increased uptake of atmospheric CO_2 and release of DMS. The effect of dust deposition on community structure in certain marine systems is currently a key research topic among oceanographers.

Natural biogenic aerosol particles emitted by plants play an important role in nutrient cycling in tropical ecosystems. Many tropical systems are limited by nitrogen and phosphorus and depend on atmospheric input of certain mineral nutrients to maintain productivity (Vitousek and Sanford 1986). Work conducted in the Okavango Delta in southern Africa showed that in channel fringes water is the dominant source of nutrients but that in backswamps aerosols may provide as much as 50% of the phosphorus requirement of the ecosystem (Garstang et al. 1998). Sulphur emissions have been studied since the 1970s when their role in acid rain and forest die-back became key environmental issues (see, e.g. reviews by Sehmel 1980; Hosker and Lindenberg 1982; Voldner et al. 1986). Other acids (e.g. nitric acid) or anhydrides (e.g. sulphur dioxide) can also be deposited in gaseous form.

Ozone is a significant greenhouse gas and in addition plays a major role in the atmospheric chemistry of both the troposphere and stratosphere (see Chap. 3). In the stratosphere its role in removing biologically damaging UV radiation has received considerable attention. In the troposphere this gas is associated with negative impacts to human health and plant physiology and it can have significant negative impacts on plant productivity in polluted regions. Ozone damage occurs in most crop plants at concentrations of 0.05 to 0.3 $\mu mol\ mol^{-1}$, with some more sensitive plants being affected at 0.01 $\mu mol\ mol^{-1}$. Ozone directly affects the photosynthetic processes, which results in decreases in plant yield (Tingey and Taylor 1982). As O_3 has a short lifetime and is produced and consumed in the atmosphere, its concentration is highly variable both spatially and temporally. This makes accurate estimates of the total atmospheric burden difficult and estimates of global scale trends even more so. Surface O_3 measurements from background stations have shown both positive and negative trends of less than or about 1% yr^{-1} (e.g. Oltmans et al. 1998; Logan 1999). This complex picture may reflect real re-distribution of O_3 abundance due to changes in the emissions of precursors.

2.5 Approaches for Studying Exchange

A basic organising principle for understanding the fluxes of trace gases to and from the atmosphere is that of a source-sink budget. For each compound, there is a mass balance between the fluxes into the atmosphere (sources), removals from the atmosphere (sinks), including chemical conversions and changes in the atmospheric burden. Budgets provide the conceptual framework for bringing together a process-based understanding of surface exchange fluxes and atmospheric chemistry through demonstration of balanced source-sink budgets.

Exchanges of biogenic trace gases and particles between surfaces and the atmosphere are typically driven by the production and consumption of gases by plant, microbial, and chemical processes, and influenced by physical transport through soils, sediments, water, or across gas-liquid boundaries.

For many chemical compounds, demonstrating a balanced budget based on process models of these fluxes remains a goal rather than a reality. However, substantial progress has been made in the last decade through collaborations between a number of disciplines, including atmospheric chemistry, ecology, biogeochemistry, geochemistry, microbiology, soil science, meteorology, hydrology, and oceanography. One of the hallmarks and great successes of IGAC research has been the integration of knowledge from such relevant disciplines toward the understanding of trace gas sources and sinks.

Understanding the source-sink budget for a trace gas involves establishing and validating process models across a range of scales. Most terrestrial process studies of trace gas fluxes are carried out at small spatial scales, e.g. of the order of 1 m, in order to control the relevant environmental factors. Validation at this scale typically uses flux measurements derived from chamber studies. However, process models are also increasingly used as extrapolation tools to derive landscape, regional, and even global scale flux estimates. Most models can account for short term changes (minutes to hours) of some compounds but are limited in their ability to predict longer term (days to years) variations (Otter et al. 1999).

This up-scaling provides flux inventories that are relevant for environmental management, but requires estimation of the key inputs to the process model such as marine plankton speciation, soil or vegetation type, land cover and management, and climatic, radiation, hydrological, and marine parameters. Validation of these scaled-up inventories requires measurement of average fluxes at the corresponding scale. These may be determined by direct flux measurements near the surface, e.g. using eddy-covariance or relaxed eddy accumulation techniques, inferred from vertical gradients in the atmospheric boundary layer, or derived from regional or global scale transport models used in an inverse mode to calculate the flux distribution that reproduces observed concentration distributions. Coupled land-ocean-atmosphere models are only available for CO_2 and H_2O with little attention being paid to other chemical compounds of biogenic origin. A few modelling studies have included the effects of anthropogenic sulphur (Erickson et al. 1995; Meehl et al. 1996; Haywood et al. 1997) for example, on climate and plant growth, but much more research is required to include a very detailed treatment of sulphur and other aerosol dynamics in on-line climate simulations. Few global climate models have examined the climate response of DMS emissions from the oceans or variability thereof (Bopp et al. 2000). Similar modelling work is required for the emissions of many other compounds as well as for deposition to the surface.

Additional validation of budgets or constraints on individual source and sink terms can be derived from dual-tracer studies. For example co-variation of ^{222}Rn and CH_4 or N_2O concentrations has been used to determine regional scale terrestrial fluxes of the latter where the corresponding ^{222}Rn fluxes are better known (Wilson et al. 1997; Schmidt et al. 1996). A special case of dual-tracer studies is the use of isotope ratio measurements in trace gases. Where different source types emit a trace gas with different isotopic ratios, measurement of those ratios in the atmosphere provides a means of separating the influence of each source. Typical examples of this are the use of the ^{13}C fraction in methane to place constraints on the biogenic source fraction (e.g. Sugawara et al. 1996; Conny and Curie 1996; Hein et al. 1997; Bergamaschi et al. 1998; Lassey et al. 2000) and the ^{14}C fraction to place constraints on the fossil fuel source (e.g. Lowe et al. 1988; Wahlen et al. 1989; Manning et al. 1990; Quay et al. 1999). Isotopic studies in marine regions are currently used in the parameterisation of air-sea exchange. Methodological difficulties still prevent this approach from fully extending into marine process studies.

Process models, both diagnostic and prognostic, require large data sets for initialisation and validation. Compilation of trace gas emission inventories has been carried out by the Global Emissions Inventory Activity (GEIA) (*http://weather.engin.umich.edu/geia*). This component of IGAC was created in 1990 to develop and distribute scientifically sound and policy-relevant inventories of gases and aerosols emitted into the atmosphere from natural and anthropogenic sources. Most GEIA inventories currently available are for emissions from anthropogenic sources. Current inventories for natural sources include emissions of N_2O, NO_x, VOCs, and organic halogens. Inventories are in progress for natural sources of methane, reduced sulphur compounds, and some source-specific emissions such as biomass burning. There is still uncertainty, however, associated with all global emission inventories. The extrapolation of space- and time-limited observations to regional and global scales invites many venues for error. For example, coastal regions typically have higher concentrations than open ocean regions but the patterns are very local; in addition, marine measurements are more biased towards spring and summer than terrestrial measurements but annual scaling frequently takes place.

2.6 Terrestrial Highlights

2.6.1 Exchange of Trace Gases and Aerosols from Terrestrial Ecosystems

A Dahlem workshop was held in 1989 where the delegates focussed on research needs in the area of exchange of trace gases between terrestrial ecosystems and the atmosphere (Andreae and Schimel 1989). They focussed on five priority areas for research:

- To determine what processes are involved in production of CH_4, N_2O, and NO in different ecosystems, and if they are constant or change with time, and why different ecosystems have evolved different production pathways.
- To describe characteristics of soils that influence the area and depth distributions of production-consumption reactions modulating trace gas emissions.
- To develop mechanistic models that include microbiological and physical-chemical processes applicable at the scale of trace gas exchange experiments and to test these models with field and laboratory experiments.

- To develop ecosystem scale models for biogenic trace gas fluxes.
- To assess what quantitative changes in CH_4, N_2O, and NO fluxes can be expected in response to physical and chemical climate changes.

A large number of studies has been conducted in the last ten years attempting to address these questions. Substantial progress has been made in both expanding the databases by conducting more measurements, and improving markedly the level of sophistication with which these measurements have been carried out (see Chap. 5), together with the way they are linked to auxiliary data, e.g. isotopic data. Not only do we now have better databases but we also understand better the mechanistic processes and controlling factors regulating the fluxes. This has enabled adequate models to be formulated, although many of them are very limited in their applicability (see Chap. 6). A significant part of the effort has come via the TRAGNET trace gas network, developed in the US with strong European participation, and the BATGE trace gas exchange programme centred on the Tropics. The following section summarises the progress made in the last decade in the quantification of the terrestrial sources and sinks of methane, volatile organic carbon compounds, and nitrous and nitric oxides, and advances in the understanding of the processes controlling their fluxes. Additional sections follow on biomass burning and wet deposition in the Tropics.

2.7 Background: Emissions and Deposition

Biogenic emissions from and deposition to vegetation and soils occur in a more or less continuous way over the year with the magnitude of the exchange controlled by a complex interaction of biotic and abiotic factors. On the other hand, biomass burning releases large amounts of emissions in pulses varying in frequency depending on the geographic location, the biome, and the management. The natural biogenic aerosol comprises many different types of particles, including pollen, spores, bacteria, algae, protozoa, fungi, fragments of leaves and insects, and excrement. The mechanisms of particle emission are still not well understood, but probably include mechanical abrasion by wind, biological activity of microorganisms on plant surfaces and forest litter, and plant physiological processes such as transpiration and guttation. Vegetation has long been recognised as an important source of both primary and secondary aerosol particles. Forest vegetation is the principal global source of atmospheric organic particles (Cachier et al. 1985) and tropical forests make a major contribution to airborne particle concentrations (Andreae and Crutzen 1997). However, only a few studies of natural biogenic aerosols from vegetation in tropical rain forests have been undertaken (Artaxo et al. 1988, 1990, 1994; Echalar et al. 1998).

Gaseous or particulate matter may be removed from the atmosphere and transferred to Earth's surface by various mechanisms, known under the generic terms of "dry deposition" and "wet deposition". Research findings related to the latter in tropical systems are addressed specifically in Sect. 2.7.3. Dry deposition is the removal of particles or gases from the atmosphere through the delivery of mass to the surface by non-precipitation atmospheric processes and the subsequent chemical reaction with, or physical attachment to, vegetation, soil, or the built environment (Dolske and Gatz 1985). Dry deposition is best described by the surface flux, F, corresponding to an amount of matter crossing a unit surface area per unit time. In most modelling work, another quantity, called deposition velocity, $v_d = F / C$ (flux divided by concentration), is preferred for practical numerical reasons, because its time variations are smoother. Deposition velocities are also easier to parameterise and most data on dry deposition are actually expressed as deposition velocities, usually in cm s^{-1}. A powerful parameterisation of dry deposition is the resistance analogy (Chamberlain and Chadwick 1953), where the difference between concentrations in the air and at the surface (C_s) is equal to the product of the flux and a resistance R, an empirical quantity to be parameterised. Through parameterisation of resistances, deposition velocities are readily derived. Further, this scheme may be extended and adapted to the degree of complexity of the surface, e.g. as in the case of a forest canopy, by using a greater number of resistances, in series or in parallel, according to the rules of an electric circuit. The most powerful mechanism by which deposition occurs over a canopy is penetration into plant tissues through the stomata.

Although most pollutants undergo deposition only (downward flux), some of them show bidirectional fluxes. An illustration of such behaviour is the case of nitrogen oxides NO and NO_2, as shown by Delany et al. (1986) and Wesely et al. (1989). Nitric oxide is emitted by soils (Williams et al. 1992; Wildt et al. 1997). Once emitted, it can readily be oxidised to nitrogen dioxide, with a resulting upward flux of the latter. If the concentration of nitrogen dioxide is high, as in the case of polluted air, its flux can be directed downwards.

Contrary to nitrogen oxides, ozone undergoes deposition only, since there is no known process which could produce ozone at the surface. The deposition velocity of ozone depends mostly on the nature of the surface. If vegetation is present, ozone is deposited ("taken up" would be a more appropriate term) by penetration into plant tissues through the stomatal cavities present on leaves. This process is likely to cause damage, and, in extreme cases, decreases in crop yields. Ozone uptake by vegetation has been put forward to explain ozone downward fluxes by Rich et al. (1970), and Turner et al. (1974), and subsequently by many other authors. Another ozone deposition mechanism occurs on bare soils, where ozone molecules are destroyed by a process similar to wall reactions observed in the laboratory, e.g. in a glass vessel. On mixed surfaces, both processes occur. Dry deposition of ozone has been extensively studied over the last forty years (Regener 1957; Galbally 1971; Galbally and Roy 1980; Wesely et al. 1978, 1982; Delany et al. 1986; Guesten and Heinrich 1996; Labatut 1997; Cieslik 1998).

Resistance analysis has been applied to the interpretation of ozone flux observations (Massman 1993; Padro 1996; Cieslik and Labatut 1996, 1997; Sun and Massman 1999), in particular to discriminate between the relative contributions of stomatal uptake and direct deposition on soil to the overall process. Most authors used an approach in which stomatal resistance for ozone uptake was deduced from stomatal resistance for evaporation, since both processes depend on stomatal aperture. Combining direct ozone deposition measurements and the inferred ozone stomatal resistance, its partial resistance for deposition on the soil was deduced as a residual.

The diurnal pattern of ozone deposition is governed by both turbulence and physiological activity of the vegetation. At night, ozone deposition is close to zero. It increases during the morning hours, both because air turbulence increases, bringing more molecules into contact with the surface, and because stomata are open for transpiration and carbon assimilation. The noon maximum value of deposition velocity ranges between 0.2 and 0.8 cm s^{-1}, depending on the intensity of turbulence and on the state of vegetation: the more active the vegetation, the more ozone is taken up. The daily variation in ozone deposition generally follows the pattern of the surface heat flux. For example, rapid deposition of ozone was observed in the lowest layers of a tropical forest canopy in Brazil, with an average flux of $-5.6 \pm 2.5 \times 10^{11}$ molecules cm^{-2} s^{-1}. This co-occurred with a large NO flux of $5.2 \pm 1.7 \times 10^{10}$ molecules cm^{-2} s^{-1}, which was about three times larger than the flux of N_2O. The rapid destruction of O_3 in the forest environment was also manifested by a pronounced ozone deficit in the atmospheric boundary layer. Rapid removal by the forest clearly plays a role in the regional ozone balance, and, potentially, in the global ozone balance. The location of strong NO sources and sinks in the humid Tropics makes these ecosystems pivotal in the chemistry of the atmosphere (Kaplan et al. 1988).

2.7.1 Production and Consumption of CH_4

The state of understanding of the CH_4 budget in 1990 was well summarised by Fung et al. (1991) who showed that the observed seasonal cycles at sites remote from sources could be reproduced using estimates of sources and removal rates consistent with the literature at that time. However, large uncertainties in individual components of the budget were evident and the atmospheric global observation network did not provide sufficiently strong constraints to reduce these uncertainties.

Since 1990 considerable progress has been made, particularly through studies of CH_4 emissions from wetlands and rice paddies, but also through improved estimates of oxidation rates, better data on animal and landfill emissions, and extension of the observational network. One significant outcome of these studies has been to decrease estimates of rice emissions and increase estimates of natural wetland emissions. At the regional scale there has been a reduction in the uncertainty of some type of emissions, e.g. from ruminant animals, with some studies being prompted by requirements to report national greenhouse gas emission inventories under the United Nations Framework Convention for Climate Change (UNFCCC).

Early successes of IGAC included a systematic characterisation of CH_4 fluxes from wetlands obtained from field programs in the ABLE, BOREAS, and related projects. This area has received considerable attention by many groups during the last decade; although a comprehensive literature review is beyond the scope of this chapter, a partial summary follows. A summary of parallel studies of CH_4 from rice paddies is given separately (see Box 2.1). Dependencies of CH_4 production rates in wetlands and closely related systems on water table depth, temperature, and precipitation, were examined and used to develop regression-based explanatory models (e.g. Wahlen and Reeburgh 1992; Roulet et al. 1993; Frolking and Crill 1994). Consumption by methanotrophic communities, which may intercept a substantial fraction of below ground production, was also quantified in a variety of situations and related to environmental variables (e.g. Wahlen et al. 1992; Koschorreck and Conrad 1993; Bender and Conrad 1994).

As the available data grew, the value of organising them in terms of latitudinal transects and of using consistent methodologies and reporting formats was recognised. The US Trace Gas network (TRAGNET), a component of the IGAC BATREX project, was established to meet this need (Ojima et al. 2000) and has created a database of flux measurements covering 29 sites ranging largely, but not exclusively, from 10° N to 68° N on the American continent (see *http://www.nrel.colostate.edu/projects/tragnet/*). These flux data along with site and climate characteristics are stimulating the development and validation of more sophisticated models. Recent wetland CH_4 models have improved their ability to simulate observations by explicit treatment of net primary productivity as an underlying driver of production (Cao et al. 1996; Walter and Heimann 2000). More sophisticated models of soil oxidation processes have also been developed (Del Grosso et al. 2000) and comparison of models across a

Box 2.1. Case study: Methane emissions from rice

IGAC researchers have been very active in studying methane emissions from rice paddies and considering mitigation options (RICE Activity). This is particularly relevant given projections that rice production will increase from 520 million t today up to 1 billion t during this century. Rice agriculture is subdividable into dryland, rainfed, deepwater, and wet-paddy production. The latter three categories have land continuously under water at some time of the year, creating anoxic conditions. They comprise some 50% of the rice crop area and contribute 70% of total rice production (Minami and Neue 1994).

Emissions from rice fields are influenced by many factors, of which the most important are water management, the amount of decomposable organic matter (e.g. rice straw) incorporated into the soil, and the cultivar of rice grown (Neue 1997). Other factors such as temperature, soil redox potential, soil pH, and the type and amount of mineral fertiliser applied also affect the emission, which reflects a net balance between gross production and microbial oxidation in the rhizosphere.

Substantial CH_4 emissions occur only during those parts of the cultivation period when rice paddies are flooded, although a delay of typically two weeks occurs after flooding. The main control of CH_4 production is the availability of degradable organic substrates (Yao and Conrad 1999). Readily mineralisable carbon, e.g. in rice straw or green manure, produces more CH_4 per unit carbon than humified substrates like compost (Van der Gon and Neue 1995). Higher soil temperature also speeds up the initiation of CH_4 formation but not necessarily the total emitted over a growing season.

Earlier estimates that up to 80–90% of the CH_4 produced in a paddy field is oxidised (e.g. Sass and Fisher 1995) may be too high. The use of a novel gaseous inhibitor, difluoromethane, which is specific for CH_4 oxidising bacteria in rice fields and which does not affect the CH_4-producing bacteria, showed that CH_4 oxidation was important only during a rather short period of time at the beginning of the season, when ca. 40% of the CH_4 produced was oxidised before it could enter the atmosphere. This fraction then decreased rapidly and for most of the season the CH_4 oxidation was only of minor importance (Krüger et al. 2000). There is now evidence of a nitrogen limitation of the oxidation process (Bodelier et al. 2000). There is also evidence of systematic changes during the rice-growing season in the $\delta^{13}C$ value of emitted CH_4 due to changes in production, transport, and oxidation (Tyler et al. 1994; Bergamaschi 1997). This may have an impact on the $\delta^{13}C$ signal of atmospheric CH_4, which is relevant for inverse modelling of methane sources.

Up to 90% of the CH_4 emitted from rice fields passes through the rice plant. Well-developed intracellular air spaces (aerenchyma) in leaf blades, leaf sheaths, culm, and roots provide a transport system for the conduction of CH_4 from the bulk soil into the atmosphere (Nouchi et al. 1990). Modern cultivars emit generally less per plant than traditional varieties because the improved harvest index often results in less unproductive tiller, root biomass, and root exudates (Neue 1997). Work in China (Lin 1993) and the US (Huang et al. 1997) has demonstrated a two-fold difference in emission rates between rice varieties grown under similar conditions. However, under field conditions, a comparison of cultivars is more complex because farmers adjust planting densities or seed rates to achieve an optimum canopy and tiller density.

Existing model approaches are still crude, with low resolution, but they provide good regional estimates within the range of observed and extrapolated fluxes. The best estimate of the global emission of CH_4 from rice fields is likely to be in the range of 30–70 Tg (Neue 1997). Recently Matthews et al. (2000) developed a simulation model describing the main processes involved in CH_4 emission from flooded rice fields by linking an existing crop simulation model (CERES-Rice) to a model describing the steady-state concentrations of CH_4 and oxygen (O_2) in soils. Experimental field and laboratory data from five Asian countries participating in the Inter-regional Research Program were used to develop, parameterise, and test the model. Field measurements of CH_4 emissions were extrapolated to national levels for various crop management scenarios using spatial databases of required inputs on a province-district level. Lack of geographic information on required inputs at appropriate scales limits application of this model in determining current, and predicting future, source strengths.

Promising candidates for mitigation of rice emissions are changes in water management, organic amendments, fertilisation, cultural practices, and rice cultivars (Neue et al. 1998). However, while present knowledge of processes controlling fluxes allows the development of mitigation technologies, information is still lacking on trade-offs and socio-economic feasibilities. Climate change will tend to extend rice production northwards, especially in Japan and China. Elevated CO_2 concentrations will enhance the production of rice yields, but also increase carbon exudation from roots, enhancing CH_4 emissions. Breeding of new rice cultivars will be the most effective strategy for dealing with this issue (Milich 1999). However, enhanced temperatures are likely to limit the potential increases.

range of soil sources (wetland, rice paddy, and landfills) now suggests an ability to explain variations over orders of magnitude in the net emission resulting from production and oxidation processes involving both natural and anthropogenic factors (Bogner et al. 2000).

As understanding of the CH_4 budget has improved, attention has turned to explaining interannual variability and, in particular, the high growth rates observed in 1991 and 1998, which appear to be associated with anomalous climatic conditions. A key factor in this respect has been the development of better process models for wetland emissions outlined above. An important factor in both the contemporary and pre-industrial global CH_4 budgets is the relative role of tropical vs. temperate and boreal wetlands. Recent estimates of current total wetland emissions cover a wide range from about 100 Tg yr^{-1} to over 200 Tg yr^{-1} (Hein et al. 1997; Cao et al. 1996; Houweling et al. 1999; Walter and Matthews 2000). Inversion methods tend to favour the upper half of this range and both inverse estimates and process-model estimates now suggest that tropical emissions dominate over those at higher latitudes. The best determined biogenic source of CH_4, based on the consistency of different estimates, is that from ruminant animals. Total emissions from this source are estimated in the range 80 to 115 Tg yr^{-1} (Mosier et al. 1998; Lelieveld et al. 1998). In recent years, several studies have produced a large amount of data on emission factors per animal or per unit of production and related these to models of rumen physiology (e.g. Lassey and Ulyatt 2000). Emissions from rice paddies and biomass burning are covered separately below, and other biogenic sources such as termites and marine methanogenesis are relatively minor in significance.

Most CH_4 removal occurs through atmospheric oxidation by OH; however, consumption by methanotrophic bacterial communities in soils is estimated to be responsible for 3 to 6% of total removals. This process is also important as it is responsible for reducing the net emissions from soils, e.g. those from rice paddies, landfills, and natural wetlands, through consumption in aerobic conditions near the surface. Thus, changes in water table can shift the balance between CH_4 production and consumption in soils.

At the outset of the IGAC programme, there were few data available on the oxidation of atmospheric CH_4 in soils and the total sink was estimated at 30 (range 15–45) Tg CH_4 yr^{-1} (Watson et al. 1990). Now there are many more flux measurements available (including some from studies lasting more than one year), there is more information on the impact of land use change, and the relationships between oxidation rates and soil parameters have been modelled. Flux values for different ecosystems show consistent median values of 10–20 Mg CH_4 m^{-2} yr^{-1}, but with skewed (log-normal) distributions (Smith et al. 2000). A major reason for the similarities between different ecosystems is that the effect of temperature on oxidation rate is small, as the organisms responsible are substrate-limited due to diffusion resistance and low atmospheric concentration. This analysis gives a global sink of 29 Tg CH_4 yr^{-1}, and a $\pm 1\sigma$ range from a ¼ to 4 times this value (Smith et al. 2000). Thus, the best estimate is essentially unchanged but the uncertainty is increased. Global estimates from models range from 17–23 Tg CH_4 yr^{-1} (Potter et al. 1996a) to 38 (range 20–51) Tg CH_4 yr^{-1} (Ridgwell et al. 1999).

Changes in land use between natural grassland, pasture or arable land, and forestry can produce a large relative difference in CH_4 removal rates in soils (Smith et al. 2000; Del Grosso et al. 2000). Recent data have shown that methane-oxidising bacteria associated with the roots of rice are stimulated by fertilisation rather than inhibited, as had been generally believed; these data will make a re-evaluation of the link between fertiliser use and methane emissions necessary. The impact of disturbance on oxidation rate can be long lasting, e.g. it may take 100 years or more to recover (Priemé et al. 1997; Fig. 2.8), but nothing is known about the ecological reasons for this. There is evidence that the microorganisms principally responsible for CH_4 oxidation differ from those responsible for CH_4 oxidation in environments such as landfill cover soils, wetland hummocks, termite mounds, and oxidised zones within rice paddy soils, where much higher gas concentrations are the norm (Conrad 1996).

Ongoing climate change is expected to increase temperatures and thaw depth in tundra ecosystems, which will tend to increase methane emissions. On the other hand, increased evaporation at the surface may create an oxygenated zone, increasing methanotrophic activity. However, climate models also indicate an increase in precipitation for northern latitudes (Vourlitis et al. 1993) suggesting that wet tundra soils will continue to be waterlogged and that the temperature effect will dominate.

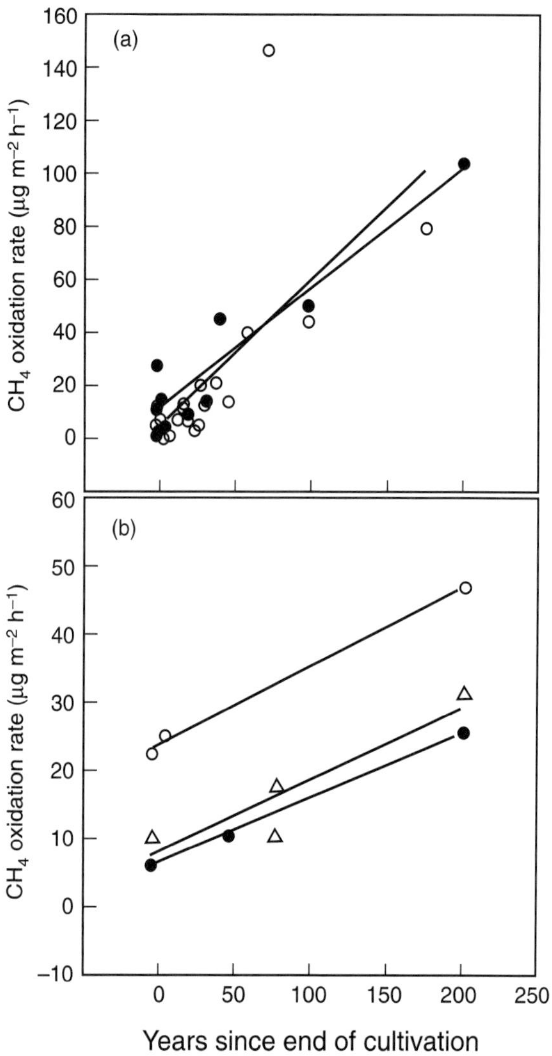

Fig. 2.8. Increase in rate of soil oxidation of atmospheric methane with time after reversion of former agricultural land to forest-woodland or to grassland. **a** European forest-woodland: Denmark (*filled circles*) and Scotland (*open circles*); **b** North American grassland (*circles*) and forest (*triangles*) (Smith et al. 2000)

An indication of the potential for climate feedbacks on CH_4 wetland emissions is given by recent analyses suggesting that higher CH_4 growth rate in the atmosphere during 1998 can be explained by temperature and precipitation effects on wetlands (Dlugokencky et al. 2000; Walter and Matthews 2000). Extrapolation of these results has suggested that a global mean warming of 1 °C would lead to an increase in wetland emissions of 20 to 40 Tg yr^{-1} (B. Walter, private communication 2000).

Dlugokencky et al. (1998) showed that the overall decline in CH_4 growth rates during the 1990s was broadly consistent with a constant total emissions and removal rate and that CH_4 concentrations would stabilise at a level 4% higher than observed in 1996 if this situation continued. Alternatively, CH_4 could be stabilised at 1996 concentrations if the total emissions were reduced by 4%. However, there is some evidence that removal rates have been increasing during the last decade (Krol et al. 1998; Karlsdottir and Isaksen 2000) at ca. 0.5% yr^{-1}. This would imply that total sources were increasing at about the same rate and is consistent with an analysis of trends in the $^{13}C/^{12}C$ ratios in CH_4 (Francey et al. 1999).

Longer term scenarios for CH_4 emissions have not been studied in as much detail as those for CO_2 emissions. One perspective is that emissions will generally follow human population because of the connection to agriculture, sewage, and landfill. However, recent trends indicate a decoupling of emissions from population (D. Etheridge, private communication 2001) and several authors have noted that anthropogenic CH_4 emissions are generally associated with inadvertent losses of energy for both animals and fossil fuel use. These lead to an alternative view that abatement of current CH_4 emissions may be possible at low or negative cost.

2.7.1.1 *Production of Volatile Organic Carbon Compounds from Vegetation*

Plant growth involves the uptake of CO_2, H_2O, and nutrients and the release of particles, water vapour, O_2, and reduced carbon compounds to the atmosphere. These reduced carbon compounds are usually described as VOCs and consist of a range of short chain organic compounds including hydrocarbons, alcohols, carbonyls, fatty acids, and esters. Recent reviews of biogenic VOC research have been published in journals of the biological (Sharkey 1996; Lerdau et al. 1997; Harley et al. 1999), chemical (Atkinson and Arey 1998), and atmospheric (Kesselmeier et al. 1998; Guenther et al. 2000) science communities as well as in a book (Helas et al. 1997) and several book chapters (e.g. Fall 1999; Guenther 1999; Steinbrecher and Ziegler 1997). The achievements of IGAC and associated research activities during the last decade on VOC emissions from plants and the currently identified research gaps are discussed in the following section.

2.7.1.2 *New Emission Measurements*

The advances in measurement techniques described in Chap. 5 have greatly increased capabilities for investigating biogenic VOC fluxes at multiple spatial and temporal scales. The resulting data have provided a more complete and accurate picture of biogenic VOC emissions. New analytical methods have extended the range of chemical compounds that can be investigated. Enclosures coupled with environmental control systems have been used to characterise the environmental and genetic controls over emissions, while above-canopy flux measurements provide an integrated measurement of landscape-level trace gas exchange. Tower-based flux measurements are particularly useful for investigating diurnal and seasonal variations without disturbing the emission source. Aircraft and tethered balloon measurement systems can be used to characterise fluxes over scales similar to those used in regional models. These regional measurements are especially useful in tropical landscapes with high plant species diversity.

The measurement database that can be used to characterise biogenic emission processes and distributions has been greatly increased by large international field programs including EXPRESSO, LBA, SAFARI, NARSTO/SOS, EC-BEMA, EC-BIPHOREP, EC-EUSTACH (see Appendix A.3). Over a thousand plant species have been investigated, for at least a few VOCs, by these studies. Equally important has been the large number of landscapes that have been studied. IGAC-endorsed research has been particularly important for advancing measurements in tropical regions. A number of these studies have investigated emissions on multiple scales resulting in measurements that can be used to evaluate biogenic VOC emission model estimates.

Investigations of emission mechanisms, often conducted under controlled laboratory conditions, have also added to our understanding of biogenic VOC emissions. These measurements have been used to relate emissions to both environmental and genetic controls. Although these measurements have not revealed distinct taxonomic relationships, some patterns have emerged (Harley et al. 1999; Csiky and Seufert 1999).

2.7.1.3 *Newly Identified Compounds*

A substantial improvement has been achieved in the last ten years in the identification and accurate quantification of VOCs emitted by terrestrial ecosystems. The number of components reported as biogenic VOC emissions has increased from seven (ethylene, isoprene, α-pinene, β-pinene, limonene, Δ-3-carene, and p-cymene) to more than 50, belonging to ten different classes. The list of detected compounds is reported in Table 2.1 together with information on:

- hypothesised biological production pathways occurring inside and outside the chloroplast;
- numerical algorithms adopted for describing emission variations;
- relative abundance in vegetation emission; and
- degree of removal by OH and O_3 attack under certain atmospheric conditions.

Table 2.1. List of VOC so far identified and quantified in terrestrial vegetation emissions

a The existence of a possible biosynthetic pathway

- 0 = not yet hypothesized
- 1 = only hypothesized
- 2 = hypothesised and partly supported by enzyme isolation or carbon labeling
- 3 = hypothesised and partly supported by both enzyme isolation and carbon labeling

b The type of algorithm followed

- 1 = light and temperature dependent
- 2 = temperature dependent

c The relative abundance in the emission

- 4 = high abundant
- 3 = abundant
- 2 = moderately abundant
- 1 = present at trace level
- * = episodic emission due to injuries

d The production or losses occurring by within-canopy processes when levels of O_3 and OH radicals in air exceed 60 nmol mol^{-1} and 106 molecules cm^{-3} of OH radicals, respectively

- −1 = partial losses
- −2 = complete losses
- +1 = moderate production
- +2 = strong production

VOC	a	b	c	d
Alkanes				
n-hexane			1	
Alkenes				
Ethylene	2		2	
a C5 isoprenoids				
Isoprene	3	1	4	
b C10 isoprenoids (monoterpenes)				
Camphene	1	2	1	
p-cymene	1	1,2	2	
Δ3-carene	1	1,2	2	
Fenchene	1	1,2	1	?
d-limonene	3	1,2	3	−1
Myrcene	3	1,2	2	−1
t-*β*-ocimene	2	1,2	2	−1
c-*β*-ocimene	2	1,2	1	−1
α-phellandrene	2	1,2	1	−1
β-phellandrene	2	1,2	1	−1
α-pinene	3	1,2	3	
β-pinene	3	1,2	2	
Sabinene	3	1,2	2	
α-terpinene	2	1,2	1	−1
γ-terpinene	2	1,2	1	−1
Terpinolene	2	1,2	1	−1
Thujene	2	1.2	1	?
Tricyclene	1	1,2	1	
c C15 isoprenoids (sesquiterpenes)				
t-*α*-bisabolene	2	2	1	?
α-copaene	1	2	2	?
β-caryophyllene	1	2	2	−2
α-humulene	1	2	1	−2
Longifolene	1	2	1	
Valencene	1	2	1	?
Arenes				
Toluene		1	1	
Aldehydes				
Formaldehyde	1	2	2	+2
Acetaldehyde	1	2	1	+2
n-hexanal			1	
n-nonanal			1	
n-decanal			1	
t-2-hexenal	2		3*	
Benzaldehyde			1	
Ketones				
Acetone	1		2	+2
Camphor	1	1,2	2	
6-methyl-5-heptene-2-one (MHO)			1	+1,−1
Butanone			1	
2,3-butadione			1	
2-pentanone			1	
4-methyl-2-pentanone			1	
Ethers				
1,8-cineole	2	1,2	2	
Esters				
Acetyl salicilate			1	
Bornyl acetate	1	2	1	
Endobornyl acetate	1	2	1	
3-hexenyl acetate	2		3*	
Lynalyl acetate	2	1,2	1	
3-methyl-3-butenyl acetate		1	1	
Alcohols				
Methanol	2		2	
2-methyl-3-buten-2-ol (MBO)	2	1	2	
3-methyl-3-buten-1-ol	2	1	1	
cis-3-hexen-1-ol	2		3*	
Linalool	3	1,2	2	
α-terpineol	2	1,2	1	?
γ-terpineol	2	1,2	1	?
Acids				
Formic acid	1	1	1	
Acetic acid	1	1	1	
Oxides				
c-linalool oxide	2	1,2	1	
t-linalool oxide	2	1,2	1	

The increased knowledge of VOC composition can be attributed to the recent measurement campaigns and analytical advances described in the previous section. Some compounds were not previously observed because of their reactivity with O_3 during sampling and analysis. The use of suitable O_3 scrubbers has reduced this problem for some compounds.

Although impressive, the list reported in Table 2.1 cannot be considered exhaustive of the potential VOCs that could be released by vegetation because of the influence that mechanical injuries, pathogen attack, ozone exposure, and natural decomposition can have on emissions. Such effects can induce the release of organic compounds that are not normally emitted by the plant. In addition, some compounds have a limited distribution within the plant kingdom and may be produced by plant species that have not yet been investigated.

2.7.1.3.1 *Mechanisms and Pathways Controlling Production and Emission*

At least seven different biogenic VOC production-emission categories have been identified: chloroplast, metabolic by-products, decaying and drying vegetation, specialised defence, unspecialised defence, plant growth hormones, and floral scents. Recent studies have shown that some compounds can be produced by more than one pathway (Table 2.1).There have been significant advancements in elucidating the biochemical pathways responsible for VOC production in chloroplasts. These include the identification of the enzymes associated with the synthesis of these VOCs, the chemical precursors, the production sites, and the demonstration that compounds other than isoprene (e.g. 3-methyl-3-buten-1-ol and α-pinene) are emitted in this manner (Silver and Fall 1995; Loreto et al. 1996). It has also been observed that these VOCs penetrate the intercellular space of the leaf and exit the plant via the stomata; yet emissions are not directly controlled by stomatal conductance. Instead, emissions depend on the rate of synthesis in the plant, which is coupled to the availability of precursors.

Some oxygenated VOCs (e.g. acetone, methanol, formaldehyde) could be produced as metabolic by-products and, although there has been little experimental investigation, plausible pathways have been proposed (Fall 1999). The production pathways associated with the remaining categories (floral scents, growth hormones, and specialised and unspecialised defence) have been studied primarily due to their biological importance. The production mechanisms for VOC emissions associated with these four categories have been described and are summarised by Guenther et al. (2000).

2.7.1.3.2 *Environmental Influences on Emissions*

A quantitative description of the environmental factors controlling biogenic emissions is needed for predicting regional emissions and how they might change. The progress made for each individual VOC is shown in Table 2.1. However, since each VOC can be emitted by more than one process, it is more convenient to discuss the environmental controls associated with the seven emission categories mentioned above.

IGAC-related research has resulted in significant advancements in descriptions of the factors controlling VOC emission, particularly isoprene, from chloroplasts. This has included the development of numerical algorithms that accurately describe short-term variations in emissions (Guenther et al. 1993). A general description of longer-term variations has been developed but

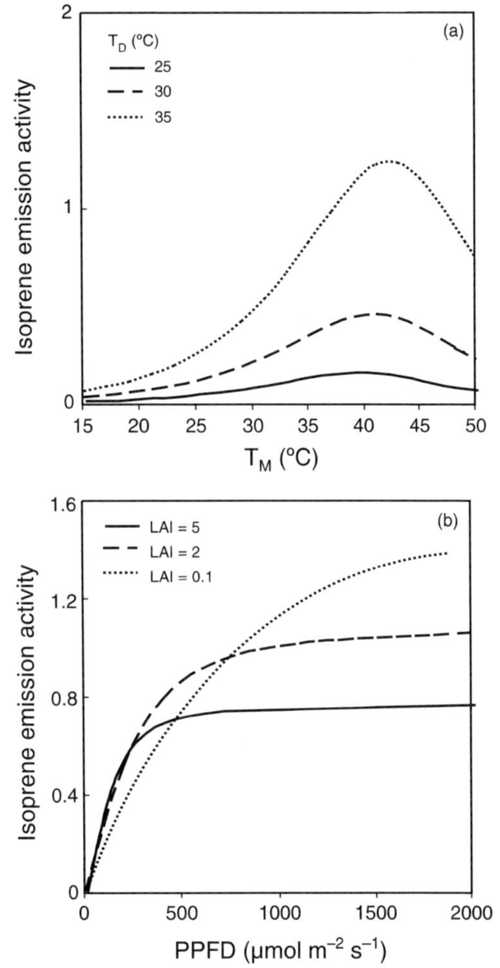

Fig. 2.9. Influence of **a** temperature, and **b** photosynthetic photon flux density (*PPFD*) on isoprene emission activity factors predicted by the algorithm described by Guenther et al. (2000). Temperature during the past 15 min (T_M) and temperature during the past 15 days (T_D) both influence isoprene emission. *LAI* is the cumulative leaf area index above a point in the canopy

an improved physiological and biochemical understanding is needed (Schnitzler et al. 1997; Guenther et al. 2000). It is known that the light response of shade leaves differs considerably from sun leaves and that the temperature that the plant has been exposed to in the past can influence its temperature dependence. This general understanding is illustrated by the response curves shown in Fig. 2.9.

Emission of stored monoterpenes from specialised defence structures results from the diffusion of compounds through the cell barrier around the resin vessels or ducts. The amount released by this process at a given temperature is dependent on the nature of the compound, resistance properties of the cell layers, and other transport resistances within and outside of the leaf. As cell resistances and vapour pressure of the compound are temperature dependent, the emission source strength is strongly dependent on the temperature of the leaf. Recent advancements have been made in describing how the exponential increase in emissions with temperature is dependent on VOC compounds and the resistance properties of the plant.

The environmental controls over metabolic by-products, decaying and drying vegetation, plant hormones, floral scents, and unspecialised defence have not been characterised in a manner that is useful for emission modelling. Some of the primary controlling factors are known but there are no quantitative algorithms for simulating emission variations (Fall 1999; Kirstine et al. 1998; Guenther 1999). A significant obstacle to regional-scale extrapolation of these emissions is the need for databases of driving variables, e.g. temperature and irradiance intensity, which are currently unavailable.

2.7.1.3.3 *Production and Loss Mechanisms in the Plant Canopy*

Field experiments carried out within the frame of the EC-BEMA project (Valentini et al. 1997; Ciccioli et al. 1999) have shown that within-canopy losses are significant for VOCs that have atmospheric lifetimes comparable to the transport time from the canopy to the atmospheric boundary layer (ABL). Compounds with atmospheric lifetimes ranging between one and three minutes (ranked as –2 in Table 2.1) never reach the ABL whereas severe losses (>50%) are observed for compounds characterised by atmospheric lifetimes ranging between three and ten minutes (ranked as –1 in Table 2.1) (Ciccioli et al. 1999).

In addition to gas-phase reactions, adsorption and partition processes can also play an important role in removing emitted components inside the forest canopy. These effects can be particularly important in the case of polar compounds (such as alcohols and carboxylic acids) that are three orders of magnitude more soluble than isoprenoids in water droplets and stick on particles and surfaces. They have been invoked to explain the reduced flux of linalool from pine-oak forests and orange orchards.

Degradation of VOCs in the canopy may lead to the formation of secondary organic aerosols (SOA), as mentioned earlier (see also Chap. 4), or gaseous products (mainly very volatile carbonyls) that can diffuse in the ABL. Photochemical degradation of VOCs has been suspected to be the main source of the huge fluxes of acetaldehyde, formaldehyde, and, partly, acetone from orange orchards. A substantial contribution to carbonyl fluxes can also arise from heterogeneous ozonolysis of lipids covering the leaf surface (Fruekilde et al. 1998), which can produce acetone, 6-methyl-5-hepten-2-one, and geranyl acetone as a function of the levels of ozone in air. Mesoscale modelling studies applied to a region north of Valencia, Spain (Thunis and Cuvelier 2000) have shown that secondary products formed by within-canopy reactions accounted for more than 70% of the ozone formed by biogenic emission from orange orchards. The complexity of within-canopy processes occurring in certain ecosystems can only be assessed by incorporating chemical processes into models describing the transport of VOCs into the ABL. At the present time, development of such models is made difficult by the fact that degradation pathways of primary products formed by photochemical reactions of mono- and sesquiterpenes are still unknown and it is not clear to what extent and in which conditions they can possibly nucleate to form SOA.

2.7.1.3.4 *Models of Emissions*

The IGAC-GEIA natural VOC project has compiled and synthesised the available information on biogenic VOC emissions and their driving variables into a global model that has been used to generate inputs for regional and global chemistry and transport models. The initial effort described by Guenther et al. (1995) provided monthly emissions of isoprene and three VOC categories (monoterpenes, other reactive VOC, less reactive VOC) with a spatial resolution of 0.5 degree of latitude/longitude. The global distribution of isoprene emissions predicted by this model for the month of July is illustrated in Fig. 2.10.

The model was constructed using the following information. Emission factors were based on the results of 20 studies that were primarily located at temperate forest field sites. Two emission types were utilised: isoprene emissions were estimated using current light and temperature conditions while all other emissions were assumed to be dependent on current temperature. Landcover characteristics were primarily based on values assigned to landscape types and a global database of current landcover distributions. Satellite (AVHRR) measurements were used to estimate monthly foliar

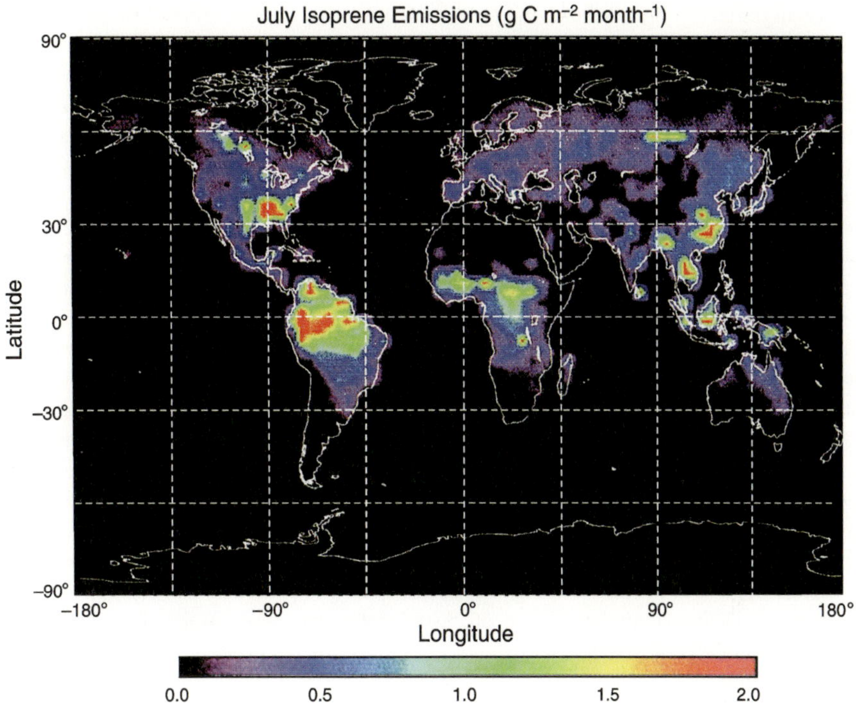

Fig. 2.10. Global distribution of isoprene emission rate estimates (g C m^{-2} $month^{-1}$) for July (Guenther et al. 1995)

variations. The model predicted annual emissions of about 500 Tg C of isoprene, 130 Tg C of monoterpenes, and slightly over 500 Tg C of other VOCs. These estimates are higher for isoprene and lower for monoterpenes in comparison with previous estimates. The differences are primarily due to an improved and expanded emission factor database.

Recent biogenic VOC observations have been incorporated into regional emission models for Europe (Simpson et al. 1999), Africa (Guenther et al. 1996, 1999) and North America (Guenther et al. 2000), and the IGAC-GEIA project is updating the global model. The regional totals tend to be within a factor of two of the Guenther et al. (1995) global model estimates but emissions for a particular location and time differ by more than a factor of five. Model improvements include a better chemical speciation, additional emission mechanisms, improved and expanded emission factors, better landcover characteristics and plant species distribution databases, improved canopy environment model, and higher spatial (1 km) and temporal (hourly) resolution.

2.7.1.3.5 *VOCs and the Carbon Cycle*

The emissions of VOCs are one pathway in the flow of carbon through plants and ecosystems. These emissions must be accounted for if the carbon balance of the system is to be accurately determined. The carbon balance of some systems is currently determined by the direct measurement of the incoming and outgoing CO_2. For ecosystems, the balance determined by such measurements is known as the net ecosystem productivity and reflects the change in carbon storage of the ecosystem. However this approach neglects the loss of carbon as VOCs. Estimates of the fraction that VOC emissions make up of the net carbon assimilated are 2 to 4% (Valentini et al. 1997; Kesselmeier et al. 1998; Crutzen et al. 1999). Given that the net ecosystem productivity is often only a small fraction (10%) of net carbon assimilation, the neglect of VOC emissions can cause errors of the order of 20 to 40% in the estimate of net ecosystem productivity and carbon storage. Fortunately, VOC emissions are being measured in some systems and more accurate values of net ecosystem productivity are being determined.

2.7.1.3.6 *Global Change and the Ecology of Emissions*

Three related aspects of global change have the potential to dramatically affect biogenic hydrocarbon emissions: increases in atmospheric levels of CO_2, increases in surface temperatures, and changes in precipitation patterns. Increases in temperature and decreases in rainfall would both be expected to increase VOC emissions; however, there is little understanding of the magnitude of the change or whether acclimation would take place (Fuentes et al. 2000; Guenther et al. 2000). Plant species compositional change due either to climate change or land use change may result in changed VOC emissions. The direction of this change will be controlled by the phenology of the vegetation and the geographic region. Savannah areas that may undergo bush encroachment would produce higher VOC emissions with a different chemical signature.

Changes in air temperature, the length of the growing season, precipitation, and atmospheric CO_2 concentrations could lead to large changes in the VOC emissions from temperate latitude ecosystems. In tropical regions, little change may take place because plant leaves already may be near their optimal temperature for isoprene emissions. There are some suggestions that plants adapt to changing temperature regimes and that VOC emissions would also rise in the Tropics. It is also unclear if cultivating increased areas of genetically modified plants could alter the nature of VOC emissions significantly.

Changes in vegetation type can lead to large changes in VOC emissions. For instance, replacing C3 grass species with C4 types could change the direct emissions and emissions from the plant decay process. Current knowledge is inadequate, however, to quantify these changes. Woody plants (shrubs and sun tolerant trees) tend to have much higher isoprene and monoterpene emissions rates, compared to annual crops and grasses; therefore deforestation involving conversion of closed forest to grassland could greatly reduce biogenic VOC emissions. However, there is a tendency for higher emissions from the woody plants that replace a closed canopy forest (Klinger et al. 1998); therefore, the effect of conversion of closed forest to open woodland is unknown. Fire dominated systems in the more arid areas have suppressed woody plants and kept emissions low, but agricultural practices of grazing and fire suppression have allowed shrublands to spread with resultant increased emissions. The chestnut blight of the late nineteenth and early twentieth centuries on the US East Coast lowland forests caused massive change in forest composition with oak replacing the chestnut. Unlike oak, chestnut does not emit isoprene. The chestnut blight has thus resulted in an approximate doubling of the biomass of isoprene-emitting species (Lerdau et al. 1997).

2.7.1.4 *Ammonia Emissions and Interactions with Particles*

Terrestrial emissions of NH_3 are associated with animal waste, fertilisers, biomass combustion, soils, vegetation, and some other minor sources (Bouwman et al. 1997). The microbial breakdown of urea and uric acid present in animal waste produces ammonium, which subsequently partly volatilises as NH_3. The overall emission of NH_3 from waste is dependent on the specific N-excretion per animal, and the NH_3 loss during housing, storage of waste outside the stable, grazing, and application of manure on grassland or arable land. Further important properties influencing NH_3 volatilisation involve soil pH and moisture, and temperature. There is a pH-dependent equilibrium between NH_3 and NH_4^+, with NH_3 being emitted from soils when they are alkaline. Some northern European countries have measured and calculated country- and animal-specific emission factors. The applications of such emission factors to calculate animal related emissions elsewhere may be quite problematic, since agricultural practice and climatic factors may differ substantially from those in northern Europe. In addition, the number of animals per country may fluctuate strongly from year to year.

Dry deposition and reactions with acidic particles and particle precursor gases are the main removal mechanisms for NH_3 (see also Sect. 2.7.3). Oxidation chemistry of NH_3 is thought to play a relatively minor role (Dentener and Crutzen 1994). Because NH_3 emissions occur almost exclusively close to Earth's surface, and because plants utilise nitrogen in their metabolism, dry deposition is a very efficient process that may remove 40–60% of all emissions. Ammonia that has reacted with sulphuric or nitric acid to form NH_4^+ is removed mainly by wet deposition and much less efficiently by dry deposition. It has an average residence time in the atmosphere of up to one week, in contrast to the much shorter residence time of gas phase ammonia, which is less than one day.

The understanding of the atmospheric NH_3 cycle is still limited, because:

- NH_3 emissions are estimated for most countries, rather than measured because of the difficulty of making such measurements (Fehsenfeld 1995).
- Emissions of NH_3 show a large spatial and temporal variability (e.g. farm-scale, winter-summer). Transport models of NH_3 and NH_4^+ utilise larger grid scales and the temporal variability of NH_3 emissions is not accounted for in such models.
- Most models of NH_3 chemistry and transport are highly simplified and parameterised and may therefore produce results that may be spurious.
- Measurements of particulate NH_4^+ are common, although their quality is frequently suspect; gas phase NH_3 data are scarcely available.

The comparison of models with such measurements is difficult since the measurements may not be representative for the model grid scale. This is especially the case for gas phase NH_3, which may have an atmospheric lifetime of hours. In addition, there may be substantial instrumental problems, e.g. the evaporation of ammonium nitrate from filter packs, which makes it difficult to interpret routinely performed aerosol measurements.

Wet deposition measurements of NH_4^+ are relatively straightforward, and there are, at least in Europe and the US, substantial data sets available. However, these measurements may again not be representative for a larger model region. Also, discrepancies of model results and measurements may be due to a host of reasons, such as a poor representation of dry deposition, vertical mixing, and in- and below-cloud scavenging or emissions. An evaluation of global model deposition with measurements in Europe and the US by Holland et al. (1999) indeed indicated substantial discrepancies but no single process could be identified as the cause of the problem.

Despite all uncertainties involved, several studies (e.g. Galloway et al. 1995) have indicated the significance of terrestrial NH_3 emissions for the global nitrogen cycle. Recent modelling studies (Adams et al. 1999; Metzger et al. 1999) indicated the potential significance of NH_4^+ and nitrate (NO_3^-) for aerosol burden and composition. These studies were performed using thermodynamic equilibrium models, developed originally for urban smog conditions. Substantial effort has been spent on extending these schemes to global modelling (e.g. Nenes et al. 1998).

A major challenge is presented by the need to increase the resolution of the models and develop sub-grid parameterisations that represent the variability of ammonia emissions and the resulting effect on particle composition. Long-term, representative, and reliable measurements of NH_3, NH_4^+, SO_4^{2-}, and NO_3^- are needed in conjunction with deposition measurements to constrain the NH_4^+ budget further.

2.7.1.5 *Production and Consumption of N_2O and NO in Soils*

Nitrogen oxides are produced in soils as obligate intermediates or by-products of the microbially mediated processes of nitrification and denitrification (Conrad 1996). The same environmental factors of soil temperature, nitrogen availability, and soil moisture affect the production of both nitrogen oxides. The pathways and enzymatic mechanisms of these processes were not well understood in the 1980s. The development of chemiluminescence instruments for NO measurement and reports in the late 1970s that increased use of nitrogen fertilisers could be one of the main causes of accumulation of nitrous oxide in the atmosphere, thus contributing to global warming, stimulated scientists to research the mechanisms involved (Firestone and Davidson 1989; Davidson and Kingerlee 1997). The conceptual model offered by Firestone and Davidson in 1989, which has since become known as the "hole-in-the-pipe" (HIP) model, synthesised the information known at that time about the microbiological and ecological factors influencing soil emissions of NO and N_2O. The HIP model linked the two gases through their common processes of microbial production and consumption. It was a breakthrough in understanding factors controlling emissions and the model has stood the test of time. Ongoing testing of the model over the last decade with numerous data sets from temperate and tropical agroecosystems shows that it provides a sound ecological and mechanistic basis for interpreting temporal and spatial variation at all scales of study by neatly encapsulating into two functions – nitrogen availability and soil water content – a large fraction of the variability caused by numerous environmental factors that influence the production and consumption of NO and N_2O by nitrifying and denitrifying bacteria (Davidson et al. 2000b).

Matson et al. (1989) stated that empirical models that are based on correlation analysis involving easily measured soil variables (e.g. temperature, moisture, texture, and organic carbon) often predict trace gas fluxes quite well. As data sets became more available, this set of variables has been further reduced to moisture and temperature with some corrections needed to take account of texture differences. Empirical relationships have been established for a number of different ecosystems around the world for both NO and N_2O emissions with water-filled pore space values of approximately 35% being the switch from NO to N_2O emissions. The magnitude of the emissions varies with substrate availability; the use of ^{15}N labelling techniques to measure the turnover of the soil ammonium and nitrate pools has greatly enhanced our capacity to partition nitrogen gas production among NO, N_2O, and N_2. Trace gases are produced and consumed by defined reactions in individual microorganisms and control must be exerted at this level initially. To date, empirical models based on various physical and chemical parameters have been successful without considering the structure of the microbial community; even those models that differentiate between nitrification and denitrification neglect microbial community structure. It is still unknown whether microbial species diversity is an important factor especially if one considers changes associated with land use (Conrad 1996).

Soil mineral N, resulting from additions of synthetic N fertilisers and N from animal manure, crop residues, etc., and the mineralisation of soil organic matter and deposition from the atmosphere, is recognised as a major driver of these emissions. Much work has gone into establishing the relationships between the fluxes of N_2O and NO and the other key drivers, soil moisture and temperature. Although some questions remain to be answered, significant developments in this direction have been achieved. The logarithmic relationship between N_2O flux and soil water-filled pore space (WFPS) in a tropical forest soil is illustrated in Fig. 2.11a (Keller and Reiners 1994). A remarkably similar relationship has been found for temperate fertilised grassland, when mineral N was not limiting (Dobbie et al. 1999). These data can be contrasted with NO flux data (Fig. 2.11b) obtained from a semi-arid southern African savannah, where temperature and soil moisture are the major controlling factors on emissions. Multiple regression analyses reveal the following sequence of importance of environmental factors on NO flux: soil temperature > water-filled pore space > soil nitrate concentrations > soil ammonium concentrations (Otter et al. 1999) (see Box 2.2).

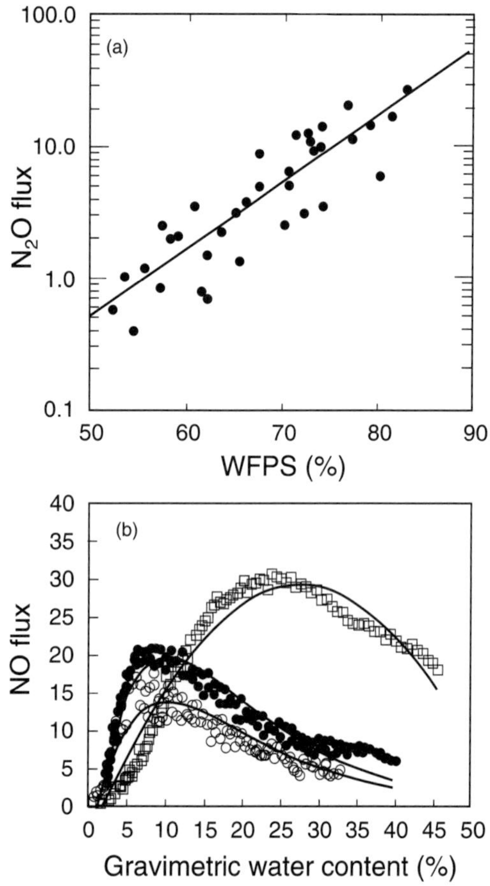

Fig. 2.11. a N_2O flux (ng N m^{-2} s^{-1}) versus soil water filled pore space (*WFPS*), old-growth forest, Atlantic lowlands, and Costa Rica (Keller and Reiners 1994); **b** NO flux rate (ng N m^{-2} s^{-1}) increases with gravimetric soil moisture to a maximum around field capacity (*FC*) for nutrient-poor (*FC* = 7.5%) (*white circles*), nutrient-rich (*FC* = 10.6%) (*black circles*), and floodplain (*FC* = 27.1%) (*squares*) soils, and declines thereafter (Otter et al. 1999)

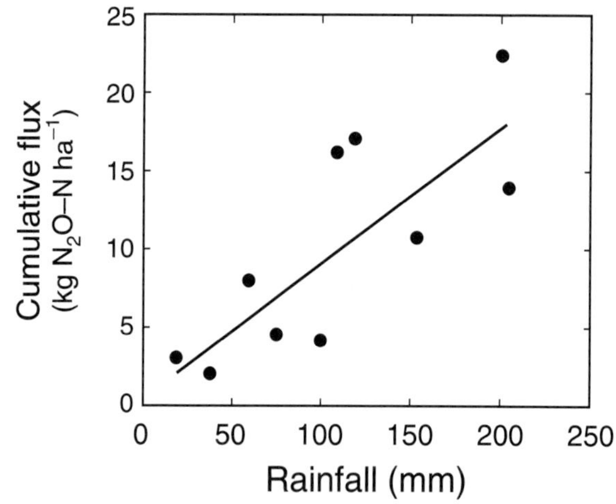

Fig. 2.12. Annual N_2O fluxes from intensively managed grassland in Scotland, UK, as a function of summer rainfall around times of N fertiliser application (Groffman et al. 2000, based on data in Dobbie et al. 1999)

Continuous flux measurements are few and far between, with many more studies of N_2O than NO, mostly driven by the global warming community rather than the broader atmospheric community. Among the significant observations that have emerged from continuous N_2O flux studies over extended periods are: (*a*) the great variation in annual flux (from 20 to >200 Mg N_2O-N m^{-2} yr^{-1}) that can occur from fertilised temperate grassland as a result of variations in the timing and amount of rainfall (Fig. 2.12); (*b*) the large proportion of the annual N_2O that can be derived from soil freezing/thawing events in winter, both in agricultural soils (Flessa et al. 1995; Kaiser et al. 1998) and in forest soils (Papen and Butterbach-Bahl 1999); and (*c*) that nitrification can be as important as dentrification in producing N_2O (Davidson et al. 1993; Panek et al. 2000). In hot dry environments, variations in the time elapsed between fertilisation, sowing, and irrigation of cereal crops have been shown to be almost as important (Matson et al. 1998).

Soil emissions contribute 70% and 20% of the global budgets of N_2O and NO respectively, with humid tropical forests accounting for 20–50% of all the global sources of atmospheric N_2O (Keller et al. 1986; Potter et al. 1996b; Verchot et al. 1999). Despite the different roles of N_2O and NO in the atmosphere and the many different reasons why scientists from several disciplines study one or the other gas, combining studies of the two gases and linking them mechanistically in conceptual and empirical models makes good biological, ecological, and practical sense (Davidson et al. 2000b). At the outset of the IGAC project, tropical forest soils were considered to be the major contributor (3.2–7.7 Tg N yr^{-1}) of N_2O, with agricultural emissions being much smaller and more uncertain (0.03–3.0 Tg N yr^{-1}) (IPCC 1992). Rapid land use change in the Tropics was expected to result in markedly increased emissions. Early measurements of the flux of nitrous oxide from recently formed pastures in the Amazon basin showed a threefold increase relative to the flux from the original forest soil (Luizao et al. 1989). However, more recent work by Keller and Reiners (1994) shows that very large increases of soil N_2O emissions are only observed in young pastures compared with forests but that the periods of high emissions are limited to only about a decade following clearing. Thereafter, N_2O emissions from pastures fall below forest levels, probably as a result of the depletion of available nitrogen. In addition, data show that old-growth tropical forests have high fluxes, and young successional forests highly variable ones (Davidson et al. 2000b). However, Hall and Matson (1999) showed that in the presence of N deposition phosphorus-limited tropical forests exhibited enhanced emissions of N_2O; in addition, enhanced nitrogen inputs and irrigation in an intensively managed Mexican wheat system resulted in large emissions (Panek et al. 2000). There is now solid

Box 2.2. Impact of nitrogen fertiliser and deposition on nitrogen trace gas emissions

Worldwide consumption of synthetic N fertilisers has increased 20-fold since 1950 to about 82 Tg N yr^{-1} in 1996 (Bouwman 1998). About 40% of current consumption is in the Tropics and subtropics, particularly in Asia, and is expected to rise to 60% by 2020 (Hall and Matson 1999). Animal wastes used as fertiliser supplied an estimated additional 65 Tg N yr^{-1} in 1996, compared with 37 Tg N yr^{-1} in 1950 (Mosier and Kroeze 1999), and this input is likely to increase in the future. The expected effect of the additional N use is a further major increase in N_2O from agricultural sources. These increases in N fertiliser use are also expected to raise the agricultural contribution to soil NO emissions to over 50% (Yienger and Levy 1995). In the IPCC (1999) assessment, direct emissions of N_2O from agricultural soils were taken to be 1.25 ±1% of the N applied (Bouwman 1996), but a re-evaluation indicates that the observed emission factors are strongly skewed, giving an uncertainty range from one-fifth to five times the mean value, i.e., from 0.25% to about 6% of the N applied, and suggesting that the mean flux from this source may be even higher or lower than presently accepted. Recent data (Fig. 2.13) on N_2O and NO emissions from N fertilised and N-saturated systems in temperate regions give indications as to how global change and changing land management practices may be enhancing emissions. However, Hall and Matson (1999), raise the possibility that increasing nitrogen deposition in tropical regions is likely to have very different effects than nitrogen deposition in the temperate zone, with much greater feedbacks to the atmosphere (see Sect. 2.7.3).

Figure 2.13a and b show a remarkable similarity in graphs of N_2O emission vs. deposition and N_2O emission (nmol mol^{-1}) vs. fertiliser application for sites in Scotland. In a three-year continuous record study of nitrogen trace gas fluxes from untreated and limed soils of a N-saturated spruce and beech forest ecosystem in Germany (Fig. 2.13c and d) there was a significant positive correlation between the amount of *in situ* N input by wet deposition and magnitude of *in situ* N_2O and NO emissions (Papen and Butterbach-Bahl 1999; Gasche and Papen 1999). At the beech site, 10% of the actual N input was released from the soil in the form of N_2O whereas at the spruce site the fraction was 0.5%, indicating that forest type itself is an important modulator of N_2O release from soil. However, there is a marked similarity in the N_2O data obtained from Scotland and from Germany. Approximately 15% and 7% of the actual N input was lost as NO from the German soils stocked with spruce and beech, respectively. Liming resulted in 49% reduction of NO emissions as compared to an unlimed spruce control site. The results indicate that the reduction in NO emission was due to an increase in NO consumption within the limed soil. Liming of a spruce site resulted in a significant increase in ammonification, nitrification, and N_2O emissions as compared with an untreated spruce control site. On the basis of these results it was concluded that the importance of temperate and boreal forests for the global N_2O source strength may have been significantly underestimated in the past and that these forests, in which N deposition is high, most likely contribute in excess of 1.0 Tg N_2O-N and 0.3 Tg NO_x-N yr^{-1} (Papen and Butterbach-Bahl 1999; Gasche and Papen 1999).

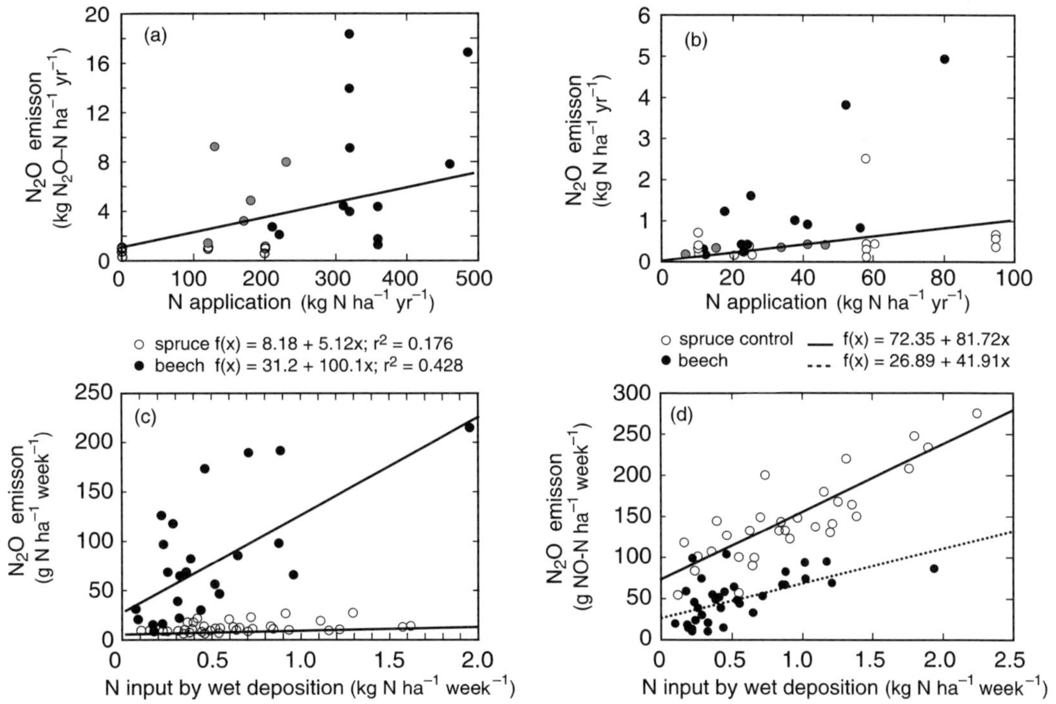

Fig. 2.13. Deviations of N-induced emissions of N_2O in Scotland from the IPCC "default" values. **a** Mineral N-induced emissions from grass (*solid black circles*), arable land with cereal crops (*open circles*), and arable land with non-cereal crops (*grey-shaded circles*), together with the IPCC default emission factor (1.25% of the N applied) (*solid line*) (Dobbie et al. 1999; Skiba and Smith 2000); **b** N deposition-induced emissions from forest and moorland soils downwind of poultry and pig farms (*solid black circles*), in large-scale acid mist experiments (*open circles*) and in upland areas (*grey-shaded circles*), together with the IPCC default emission factor (1% of the N applied) (*solid line*) (Skiba and Smith 1999); **c** linear regression analysis between *in situ* N input by wet deposition and *in situ* N_2O emission rates from soil of spruce and beech control sites. For correlation analysis, data for mean weekly N_2O emission rates and mean weekly N input (data from Huber 1997) by wet deposition measured in the throughfall were used (1994–1995) (Papen and Butterbach-Bahl 1999); and, **d** linear regression analysis between weekly mean NO emission rates and weekly amounts of N input (sum of ammonium and nitrate) by wet deposition (measured in the throughfall) (Huber 1997) at the spruce control and beech sites of the Höglwald forest; data obtained in the time period January 1994 to December 1994 (spruce sites) and September 1994 to June 1995 (beech sites), respectively, were used for correlation analysis (Gasche and Papen 1999)

evidence that agricultural emissions of N_2O exceed those from the tropical forests (2.1 ±1 Tg N yr^{-1} directly from agricultural soils, plus another 4.2 ±2.1 Tg from animal production and from N leached from agricultural soils into surface and ground waters (Mosier et al. 1998)). However, management and environmental conditions play a major role in determining emissions from tropical and temperate systems. For example, in tropical pastures with a fertile loamy soil developed from volcanic ash, the fraction of fertiliser nitrogen lost as N_2O (6.8%) and NO (1.3%) was much higher than the loss percentages generally observed in temperate agricultural lands (about 1% and 0.5% for N_2O and NO respectively). On a banana plantation with similar soil, the loss percentages for N_2O and NO were 1.3–2.9% and 5.1–5.7%, respectively. In a fertilised sugar cane field in Hawaii and fertilised pastures with well-aerated soils in Puerto Rico, however, the percentage (1.1–2.5%) of fertiliser lost as N_2O came close to the loss rates generally observed in temperate regions (Bouwman 1996; Matson et al. 1996). There is still no clear cut answer; while some evidence indicates that emissions from land use change in the Tropics may be lower than originally expected, but increased fertiliser usage and nitrogen deposition may, yet again, lead to expectations of increased emissions.

Nitric oxide emissions are greatest from savannah areas of the world (Table 2.2). The greatest uncertainties lie with tropical grasslands and agricultural systems due to the difficulty in estimating land area and the extent of land use change. Emissions in sub-tropical regions are pulsed by rainfall, making annual integrations difficult. Data sets from natural regions converted to a variety of agro-ecosystems are few and there are indications that as data become available emission estimates will increase as shown in Table 2.2 when estimates published in 1995 and 1997 are compared (Davidson and Kingerlee 1997).

Adsorption of NO_x onto plant canopy surfaces may reduce emissions to the atmosphere to as low as 13 Tg N yr^{-1}, although the absorption effect is probably smaller than this. Measurements of NO_x exchange are complicated by the partial oxidation of NO to NO_2, the presence of NO_2 as an atmospheric pollutant from other sources, and the deposition of NO_2 onto plant leaf and soil surfaces. Nitric oxide undergoes oxidation, mostly by reactions with ozone, to form NO_2. This process begins immediately following NO emission from the soil. During upward transport through the plant canopy the NO_2 product may be partly deposited, resulting in a reduced NO_2 flux from the canopy. The presence of NO_2 from other sources can result in a net deposition, rather than emission, of this compound; this makes field flux measurements hard to interpret (e.g. Delany et al. 1986). Several attempts have been made to resolve this problem by modelling (Kramm et al. 1991; Vila-Gereau et al. 1995; Galmarini et al. 1997; Kirstensen et al. 1997), in particular by parameterising the vertical transfer of NO in terms of resistances (Chamberlain and Chadwick 1953) (see Sect. 2.6.1).

Several process-oriented models have been developed over recent years that simulate N trace gas emissions as part of more general simulations of C and N biogeochemical transformations in terrestrial ecosystems. Key examples are: DNDC (Li et al. 1992); Century-NGAS (Parton et al. 1996), ExpertN (Engel and Priesack 1993); NLOSS (Riley and Matson 2000); CASA (Potter et al. 1996b); and DAYCENT (Parton et al. 2001). These and other models have been applied with varying degrees of success to the simulation of N_2O and NO emission data from contrasting temperate and tropical re-

Table 2.2. Comparison of NO estimates by biome (Tg NO-N yr^{-1}) (Davidson and Kingerlee 1997)

Biome	Y and L[a] (no canopy)	D and K[b] (no canopy)	Y and L (W/canopy effect)	D and K (w/Y and L's canopy effect)
Tundra	0.02	0.1	0.02	0.0
Temperate grassland	0.52	1.1	0.34	0.7
Temperate woodland	0.09	4.7	0.05	2.9
Temperate forest	0.07	4.0	0.04	0.2
Temperate agriculture	1.82	1.8	1.33	1.0
Tropical grassland	2.50	7.4	1.60	4.3
Tropical woodland	0.39	5.0	0.22	0.3
Tropical dry forest	0.11	2.0	0.06	1.0
Tropical rainforest	3.4	1.1	0.85	0.3
Tropical agriculture	1.16	3.6	0.92	2.9
Deserts and semi-deserts	–	5.0	–	0.5
Total	10.2	21.0	5.45	13

[a] Yienger and Levy.
[b] Davidson and Kingerlee.

gions. Uptake and emissions may occur simultaneously and models using unidirectional fluxes are no longer appropriate, yet they are still used widely. A modification of one of the models (Century), using a different trace gas production module, has also been used to simulate the important changes in N_2O and NO emissions that are associated with conversion of tropical rain forests to pastures. Modelled net increases in gas fluxes over the first 15 years after conversion were over 5 Gg N m^{-2} (a mean of 340 Gg N m^{-2} yr^{-1}), of which N_2O accounted for 90% (Liu et al. 1999). The DNDC model has been applied in a modified form to N-limited pastures in Australia (Wang et al. 1997). The model estimates of annual N_2O emissions and N transformations agreed well with observations, for conditions where the N flows are an order of magnitude smaller than those in Northern Hemisphere midlatitude systems already modelled by DNDC (Li et al. 1992; Frolking et al. 1998). N_2O fluxes from N fertilised grassland in the UK (and those for cereal-growing sites in the same region) have been successfully predicted by a summary model based on relationships between flux and soil water-filled pore space, mineral N content, and temperature (Conen et al. 2000). The NLOSS model has been used to simulate nitrification and denitrification sources of N gases and has been tested against ^{15}N data from irrigated wheat systems in Mexico (Riley and Matson 2000). At the global scale, the CASA model uses satellite estimates of absorbed infrared radiation in combination with soils and climate databases to estimate N_2O fluxes as a proportion of N mineralisation in the soil (Potter et al. 1996b).

Groffman et al. (2000) have explored the relationships between annual, rather than hourly or daily, fluxes and ecosystem-scale variables such as plant community and soil type, and annual climate rather than field-scale variables such as soil moisture and temperature. They concluded that there are indeed coherent patterns in annual N_2O flux at the ecosystem scale in forest, cropland, and rangeland ecosystems, although these patterns vary by region and only emerge with continuous (daily or weekly) flux measurements over several years. An ecosystem approach to evaluating N_2O fluxes is useful for regional and global modelling and for computation of national N_2O flux inventories for regulatory purposes, but only if measurement programmes are comprehensive and continuous.

2.7.2 Biomass Burning

2.7.2.1 *Introduction and History*

Fire and its impact on Earth's atmosphere have been present ever since the evolution of land plants, some 350 to 400 million years ago. Before the advent of humans, fires were ignited naturally by lightning strikes, especially during dry periods. Today, however, fire is almost exclusively the result of human activities, such as the burning of forested areas for land clearing, of natural grasslands and savannas to sustain nomadic agriculture, of agricultural residues, and of biomass as fuel for cooking and heating. Even wildfires are frequently caused by human activities, e.g. camp fires, cigarettes, or sparks from engines. Natural wildfires play a significant role only in the boreal and savannah regions of the world. The return frequency of wildfires varies widely across the biomes of the world; for example, in savannas it is typically three to five years, whereas in boreal regions fire may recur only once every 500 years.

As a result of the increasing human impact on our planet, it is likely that the amount of biomass burned annually has strongly increased (by some 30–50%) over the past century, especially because of increasing tropical deforestation and domestic biofuel use. In some regions, such as Southeast Asia and Brazil, smoke from deforestation fires has been so intense in recent years as to cause serious health concerns. In principle, the fact that at present most vegetation fires are the result of human activities would imply the capacity to control and manage emissions from biomass burning better. Unfortunately, this has seldom been translated into government policy and even less often implemented effectively.

The first pioneering papers on the impact of biomass burning on the chemistry of the atmosphere were published in the late 1970s and early 1980s (e.g. Radke et al. 1978; Crutzen et al. 1979). Scientific interest in this topic grew when early estimates suggested that pyrogenic (i.e. fire-related) emissions of some atmospheric pollutants could rival or exceed those from fossil fuel burning (Crutzen and Andreae 1990). Further impetus to study biomass burning came from the discovery that pollution from pyrogenic emissions could affect large areas of the world as a consequence of long-range transport (Andreae 1983; Fishman et al. 1990; Reichle et al. 1986). The investigation of the role of biomass burning in atmospheric chemistry was therefore seen as a high priority when the objectives of IGAC were formulated in 1988.

Since the IGAC Biomass Burning Experiment (BIBEX) became active in 1990, research activity in this field has increased rapidly, and, over the last decade, fire has been recognised widely as a major source of important trace gases and aerosol particles to the global atmosphere. Following well-publicised large fire catastrophes in recent years and intensive scientific efforts over the last decade, the general public as well as the scientific community is now aware that emissions from biomass burning represent a large perturbation to global atmospheric chemistry, especially in the Tropics. Satellite and airborne observations have shown elevated levels of O_3, CO, and other trace gases over vast areas of Africa, South America, the tropical Atlantic, the Indian Ocean, and the

Pacific Ocean. There is now also strong evidence that smoke aerosols perturb climate by scattering and absorbing sunlight and by influencing cloud microphysical processes.

We have also learned that the effects of burning are not limited to the emissions from the fires themselves, but that vegetation fires have pronounced effects on trace gas emissions from plants and soils. In the case of CO_2, NO, and N_2O, post-fire emissions may be more significant than the immediate pyrogenic release. Fire also alters the long-term dynamics of the cycling and storage of elements within terrestrial ecosystems, thereby changing their potential as sources or sinks of various trace gases. Finally, deposition of pyrogenic compounds onto tropical ecosystems may affect their composition and dynamics. In the following sections, we review some of the results and attempt to put them into the larger context of global change research.

2.7.2.2 *Scientific Approach*

Since the early 1990s, BIBEX has designed and carried out a number of biomass burning experiments in various ecosystems throughout the world, often in collaboration with other international programmes, particularly with other IGBP Core Projects. These experiments have produced extensive local-scale data on vegetation fire characteristics, emissions, and ecology, while simultaneous regional-scale measurements, using remote sensing and aircraft sampling platforms, have provided a capability to scale results up. Typically, these experiments have involved ground measurements on individual fires, airborne sampling and analysis of smoke plumes, and remote sensing of regional and global fire activity. Emphasis to date has been placed on tropical ecosystems, but an increasing number of experiments are now being organised in the boreal zone in response to climate change concerns.

STARE (Southern Tropical Atlantic Regional Experiment), with its two components SAFARI (Southern Africa Fire-Atmosphere Research Initiative) (see Box 2.3), and TRACE-A (Transport and Chemistry near the Equator) was the first large experiment coordinated by BIBEX. Conducted in 1992, STARE brought together scientists from many countries to investigate the chemical composition, transport, and fate of fire emissions originating from South America and southern Africa.

2.7.2.3 *Land-Use Fires, Wildfires, and Domestic Biomass Burning: General Trends, Uncertainties, and Possible Changes*

In the regional and global research activities on fire ecology and atmospheric chemistry, key questions have been addressed: What is the current state of vegetation fires at the global scale? Are there quantitative and qualitative changes of vegetation fires compared to historic times? The baseload of natural fires and anthropogenic fires during evolutionary time scales has been determined by several factors: climate and vegetation changes, changes of land occupation, and cultural practices. The magnitude of historic and prehistoric vegetation burning remains largely unknown, however, because only fragmentary data obtained by case studies are available (summarised in Clark et al. (1997)). BIBEX research and other observations reveal uncertainties, recent changes, and new insights of fire occurrence in the following main vegetation zones.

Tropical evergreen forest. Deforestation statistics by the FAO and others in many studies have provided the baseline data for calculation of pyrogenic emissions due to land use change. While these numbers are useful for estimating the net release of carbon to the atmosphere, they do not reflect the entire spectrum of fire activities. Recurrent fires following the initial deforestation burns not only present additional emission pulses but also lead to impoverishment of forest ecosystems resulting in reduced above- and below-ground phytomass (Goldammer 1999a; Nepstad et al. 1999). Extreme climate variability such as the ENSO-related droughts of 1982–1983 and 1997–1998 favour the application of fire for land use change and maintenance of agricultural systems and facilitate the spread of uncontrolled fires (wildfires) in humid tropical ecosystems that under average climate conditions are subjected to less fire. The area burned by wildfire in the Indonesian and Malaysian provinces on Borneo Island in 1982–1983 covered ca. 5×10^6 ha, and in 1997–1998 land use fires and wildfires combined burned ca. 8–9×10^6 ha in Indonesia alone.

Tropical savannas and open seasonal forests. Assessments made in the early 1990s on the average annual amount of savannah phytomass burned were in the range of 3–4 Pg yr^{-1} (Andreae 1993). Model predictions on the savannah area annually burned ranged between 750×10^{10} m^{-2} yr^{-1} (Hao et al. 1990) and 1500×10^{10} m^{-2} yr^{-1} (Goldammer 1993). More detailed studies on fire regimes and fuel loads in Africa point towards lesser amounts of regional and global combustion of savannah phytomass (Menaut et al. 1991; Scholes et al. 1996). Recent and ongoing growth of rural populations and intensity of land use involves landscape fragmentation and competitive utilisation of phytomass for grazing and domestic burning (biofuel use) and may represent a reason for a decrease of fire activities in tropical savannas and open forests; desertification in the sub-Saharan Sahel zone of Africa and other regions leads to a reduction and discontinuity of fuel loads and wildfire occurrence.

Box 2.3. Case study: SAFARI-92

The following example from the SAFARI-92 campaign (Lindesay et al. 1996) highlights the scientific approaches used to test hypotheses and validate models related to biogenic and biomass burning emissions and depositions. It is approaches like these that have allowed for an integrated understanding of the magnitude and controllers of sources, sinks, and exchange processes. During SAFARI-92, experimental vegetation fires were set and studied in the Kruger National Park, South Africa, and at some sites in Zambia and Swaziland. These experiments provided a broad set of data on trace gases and aerosol emissions, from which emission factors for fires in dry savannas and related biomes could be derived. The relationships between fuel characteristics, burning conditions, and fire behaviour were elucidated.

Regional studies on atmospheric chemistry and air mass transport showed that savannah fires in southern Africa account for a substantial amount of photochemical oxidants and haze over the subcontinent. These studies also showed that the export of smoke-laden air masses contributes strongly to the burden of ozone and other trace gases and aerosols over the tropical ocean surrounding Africa. However, results also showed that biogenic soil emissions severely impacted atmospheric chemistry. Investigations on the relationships among fire, soil moisture status, and soil trace gas emissions showed that soil moisture played a crucial role but that fire history also had an important influence on the emission of several trace gases. Figure 2.14a shows the relationship among daily NO emissions and nitrate concentrations plotted against water-filled pore space. Figure 2.14b describes the relationship between NO emission rate and nitrification rate in areas where fire has been excluded and in areas where the vegetation has been burned every two years (Parsons et al. 1996). These relationships were later incorporated into a simulation model to predict NO emissions from semi-arid savannas thereby reducing the large uncertainty associated with the magnitude of previous savannah measurements (Otter et al. 1999).

Remote sensing studies confirmed that Advanced Very High Resolution Radiometery/Land Aerial Cover (AVHRR/LAC, 1 km) imagery was a useful tool for fire monitoring in the region. In combination with biomass models, the remote sensing data could be used for the estimation of the seasonal and geographical distribution of pyrogenic emissions. The results from SAFARI-92 confirmed that it is justified to consider biomass burning as a significant contributor to the overall increase in greenhouse gases that has occurred over the last 150 years, accounting for some 10–25% of current estimates (Andreae 1993).

In order to establish accurately the global budgets of trace gases, reliable source strength and distribution estimates are needed. At present, the uncertainties associated with budget calculations are necessarily large, owing to the often-inadequate quantification of individual sources and the problems associated with extrapolating from a number of poorly known sources to achieve a global estimate. The contribution of vegetation fires in the savannah regions of southern Africa has been such a poorly quantified source, despite the fact that savannas are recognised as one of the most significant biomes in terms of global biomass burning emissions (Andreae 1993) and that a large portion of the savannah burns each year. It will now be possible to refine these estimates on the basis of results obtained from SAFARI-92. Modelling studies incorporating the emission data, meteorological information, and the chemical measurement data obtained during these campaigns indicate that the fires on the African and South American continents are indeed a major source of the gaseous and particulate pollutants, particularly ozone, found in the troposphere over the study regions (Thompson et al. 1996a). Data from airborne observations (Fig. 2.15) aboard a DC-3 using a combination of spectrometers and chemiluminescence instruments, showed that episodic pyrogenic emissions were not adequate to account for the buildup of tropospheric ozone in the region but that the continuous production of biogenic NO_x emissions and especially the amounts produced at the start of the rainy seasons have important consequences for regional scale ozone formation (Harris et al. 1996). The vertical distribution of NO_2 and NO as well as that of CO_2 showed markedly different characteristics. All three compounds have a strong gradient toward higher values near the ground, and the CO_2 and NO_x mixing ratios correlated linearly. The anticorrelation of the profiles of these compounds with that of CO rules out biomass burning as a source of the observed NO_x and CO_2 near the ground, supporting the field evidence of no active fires in the region. It was concluded that the source of the elevated NO_x mixing ratios near the surface was biogenic emission from the soil (Harris et al. 1996).

SAFARI-92 was an innovative project in many ways. In addition to being the largest international, interdisciplinary investigation of biomass burning and its atmospheric emissions, it also represented the first time that a large-scale fire emission measurement campaign included, as integral components: the characteristics of the biomass, the fire ecology, the fire dynamics in the area, the biogenic emissions, and the long-range transport of the aerosols and particulates.

As a follow-up to SAFARI-92, a much smaller experiment, SAFARI-94, was organised by BIBEX to investigate the composition of trace gases in the troposphere over Africa outside the burning season. EXPRESSO (Experiment for Regional Sources and Sinks of Oxidants), designed primarily to investigate the exchange fluxes of trace gases between the tropical biosphere and atmosphere, took place in the Congo Basin in 1996–1997 (Delmas et al. 1999). In 1997, AFARI-97 (African Fire-Atmosphere Research Initiative) was carried out in Kenya, investigating the atmospheric effects of fires occurring in the fertile savannas of East Africa. At the same time, an experiment designed to quantify aerosol and trace gas fluxes from the Miombo woodlands of southern Africa was initiated: ZIBBEE (The Zambian International Biomass Burning Emissions Experiment) began in 1997 and is ongoing. At the present time, BIBEX is heavily involved in the planning of two large tropical fire-atmosphere experiments: SAFARI-2000 is studying the transport and climatic effect of biogenic, pyrogenic, and anthropogenic emissions in southern Africa, while LBA (The Large Scale Biosphere-Atmosphere Experiment in Amazonia) is investigating the climatological, ecological, biogeochemical, and hydrological functioning of Amazonia, and the sustainability of development in this region.

In the boreal zone, BIBEX has been involved in the development of research programs addressing the role of fire in boreal ecosystems and its consequences for the global atmosphere and climate. FIRESCAN (Fire Research Campaign Asia-North) conducted the first joint Russian-western experimental fire in central Siberia in 1993, and continues with the planning of further such under the auspices of the IGBP Northern Eurasia Study (FIRESCAN Science Team 1996). In addition, BIBEX is active in ICFME (The International Crown Fire Modeling Experiment), a series of high-intensity experimental crown fires carried out in the Canadian Northwest Territories during the 1997–2000 period for the purpose of developing a physical model of crown fire initiation and propagation.

Areas of Mediterranean and temperate vegetation. Mediterranean forest and shrub vegetation, including Californian chaparral and South African fynbos, are increasingly converted to suburban residential use. The consequent suppression of natural and human-caused wildfires results in a buildup of fuels that often cannot be burned by prescribed fires. High-intensity wildfires are an inevitable consequence of fire suppression in these ecosystems. However, there is no indication of change in the average area burned in the recent decade. In the industrial countries of the temperate region the application of fire in non-forest land use systems has

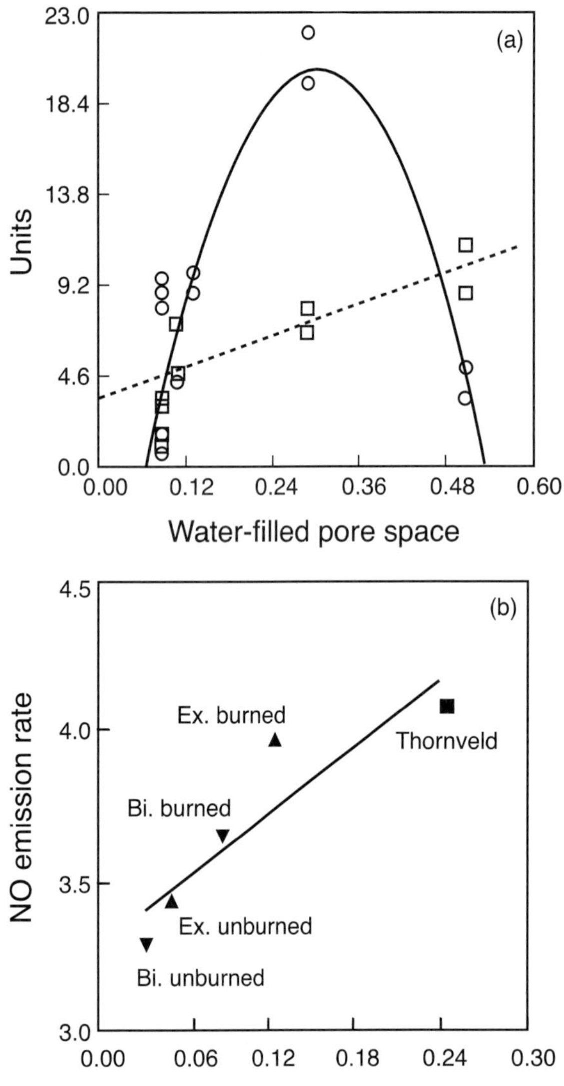

Fig. 2.14. a Mean daily NO emissions (ng N-NO $m^{-2} s^{-1}$, *circles*) and NO_3^- concentrations (μg N-NO_3^- g^{-1} dry soil, *squares*) in the fire exclusion plots, plotted against water-filled pore space simulated using the HotWet model. *Solid* and *dashed lines* represent fitted functions to the NO emissions and NO_3^- concentrations, respectively; **b** Mean NO emissions (ng N-NO $m^{-2} s^{-1}$) measured by Parsons et al. (1996) and Levine et al. (1996) plotted against mean nitrification rate (mg N kg^{-1} dry soil) measured in the corresponding plots. Linear function: NO emission rate = 0.04 (nitrification rate) + 0.003, $r^2 = 0.911$, $P < 0.030$, $n = 5$

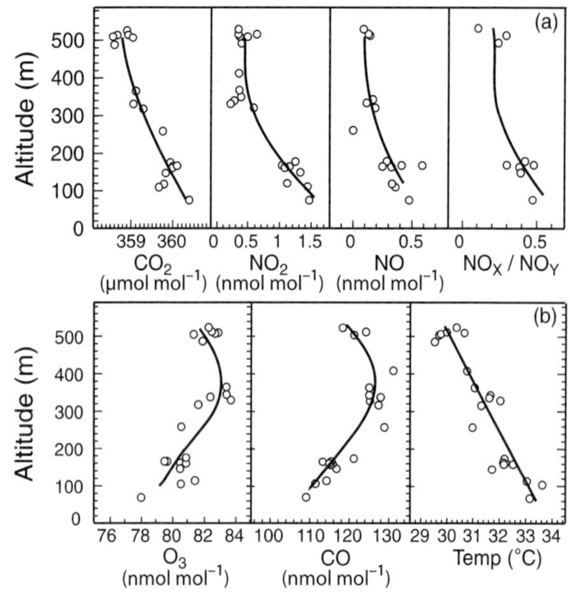

Fig. 2.15. a Vertical profiles of CO_2, NO_2, and NO and the ratio NO_X/NO_Y measured in the SAF11 profile during SAFARI 92 (Harris et al. 1996); **b** Vertical profiles of ozone, CO, and temperature measured during SAFARI on September 28, 1992 (Harris et al. 1996)

been eliminated (e.g. in Europe) or is subject to legal restrictions due to air pollution and traffic risks (e.g. in North America). Natural and human-caused wildfires in temperate forests are usually suppressed. Prescribed burning in forestry has been receiving more attention in the US where it is envisaged to expand the prescribed burned area under the jurisdiction of the USDA Forest Service to 1.2×10^{10} m^{-2} by 2010 (Haines et al. 1998).

Temperate-boreal steppe-forest ecotones. A typical region representing the steppe-forest fire environment is central Asia. Recent remotely sensed data from Mongolia indicate that in the past years political and socio-economic changes in the country were responsible for a sharp increase in the area burned by wildfires. In 1996 and 1997 more than 10×10^{10} m^{-2} and 12×10^{10} m^{-2} burned in the grass steppes and adjoining coniferous forests (Goldammer 1999b).

Boreal forest. More than 70% of the global boreal forest area is located in Russia. Fire statistics published after the dissolution of the USSR indicate that more than 650 000 ha of forests were burned annually. This number most likely is still an underestimation. In the period 1990–1996, burn areas totalling more than 1.12×10^{10} $m^{-2} yr^{-1}$ were recorded (Stocks et al. 1999, 2000). Satellite imagery revealed that a large area was burning in central Siberia during the 1987 fire season, totalling ca. 10×10^{10} m^{-2} (Cahoon et al. 1994). While the fire exclusion policy of the USSR reduced the area burned by natural fires, the scale of human-caused fires increased over the same period. Current economic problems resulting in a weakening of the fire control system in Russia are responsible for a recent increase in area burned.

In Canada, detailed forest fire statistics have been archived since 1920 and, within limits, this extensive record permits a general analysis of trends in this country (Stocks et al. 1999). Fire occurrence has increased rather steadily from approximately 6 000 fires annually in the 1930–1960 period, to almost 10 000 fires during the 1980s and 1990s. This is due to a growing population and increased forest use, but also reflects an expanded fire detection capability. During the 1981–1996 period an average of 9 246 fires annually burned over an average of 2.5×10^{10} m^{-2} in Canada, with the annual area burned fluctuating by an order of magnitude ($0.76–7.28 \times 10^{10}$ m^{-2}). Lightning accounts for 35% of Canada's fires, yet these fires result in 85% of the total area burned, due to the fact that lightning fires occur randomly and therefore present access problems usually not associated with human-caused fires, with the end result that lightning fires generally grow larger, and detection and control efforts are often delayed. In addition, the practice of "modified" or "selective" protection in remote regions of Canada results in many large fires in low-priority areas being allowed to perform their natural function.

Domestic biofuel use. Plant biomass provides about 14% of the world's demand of primary energy. Half of the global population meets an average of 35% of its energy needs by domestic biomass burning. In Africa, for example, the biomass contribution to the total energy use typically ranges from 90–100% in poor, 55–65% in middle, and 30–40% in high income groups. Unlike free-burning vegetation fires, which are usually restricted to a few months during the dry season, domestic biofuel combustion takes place during the whole year (Marufu et al. 2000).

Summary assessment of trends in global vegetation fire occurrence. The trends of changing fire occurrence and fire regimes are not uniform. Qualitative and quantitative data on fire occurrence and fire effects are still insufficient to determine reliably the amount of phytomass burned in all eco- and land-use systems worldwide. However, improved remote sensing capabilities and rigorous fire detection algorithms now provide regional fuel load and burning estimates within a much narrower range of uncertainty. Fire in boreal and tropical forests and the resulting ecological effects play a potentially critical role in determining the rate of global climate change (Goldammer and Price 1998; Stocks et al. 2000; Nepstad et al. 1999). Changes in the carbon balance of these two forest biomes could strongly influence global warming through impacts on atmospheric CO_2. The implications of regional circumpolar changes of climate and fire regimes on boreal ecosystem properties, permafrost changes, and the release of gas and carbon stored in organic terrain and ice must be further addressed by research.

2.7.2.4 *Characterisation of Emissions*

A central objective of BIBEX was to characterise and quantify the production of chemically and radiatively important gases and aerosol compounds from biomass burning. To meet this goal, the BIBEX scientific community has produced a large set of measurements that describe qualitatively and quantitatively the pyrogenic emission of gases and aerosols. The results show that the composition of fire emissions is mainly determined by two factors: the elemental composition (carbon, nitrogen, sulphur, halogen, minerals, etc.) of the biomass fuel, and the relative contribution of flaming and smouldering combustion in the vegetation fires.

Heating of vegetation fuels produces combustible gases by pyrolysis and volatilisation of waxes, oils, etc. Sustained flaming conditions are obtained when the vegetation reaches a temperature of about 600 K. Smouldering combustion involves heterogeneous reactions of atmospheric oxygen with solid fuel. The combustion type dominating in a given fire is influenced by the water content, the density and structure of the fuel, the oxygen availability in intense flaming, and the meteorological conditions prevailing during the fire.

Generally, emissions from fires occurring in natural vegetation are a mixture of compounds from flaming and smouldering combustion, with different proportions being typical of the various types of fires. The major part of savannah and domestic fuels is consumed in the flaming stage, while charcoal making is a purely smouldering process; forest biomass is combusted about equally by both processes. Lobert et al. (1991) summarised the composition of emissions released during the different burning stages. Relatively oxidised compounds, such as CO_2, NO, NO_2, SO_2, N_2O, as well as N_2 and elemental carbon particles are emitted during the flaming stage of a fire. The emission of more reduced compounds (CO, CH_4, nonmethane hydrocarbons, PAH, NH_3, HCN, CH_3CN, amines, CH_3Cl, H_2S, COS, DMS, and organic particles) occurs during the smouldering stage (e.g. Lobert et al. 1991; Yokelson et al. 1997).

2.7.2.5 *Emission Ratios and Emission Factors for Different Chemical Compounds from Fires in Various Ecological Systems or Vegetation Types*

To express the emission of trace gases and aerosols from fires quantitatively, we use the concept of emission ratios and emission factors. These parameters relate the emission of a particular compound of interest to that of a reference compound, such as CO_2 or CO (emission ratio), or to the amount of fuel burned (emission factor). Emission ratios are obtained by dividing the excess trace compound concentrations measured in a fire plume by the excess concentration of a simultaneously measured reference gas, such as CO_2 or CO. To obtain "excess" concentrations, the ambient background concentrations must be subtracted from the values measured in the smoke. For example the emission ratio of methyl chloride, CH_3Cl, relative to CO is:

$$ER_{CH_3Cl/CO} = \frac{\Delta CH_3Cl}{\Delta CO} = \frac{(CH_3Cl)_{Smoke} - (CH_3Cl)_{Ambient}}{(CO)_{Smoke} - (CO)_{Ambient}}$$

The various techniques for these calculations and the associated errors are discussed in Le Canut et al. (1996). For gases, the results are expressed in terms of molar ratios. For aerosols, emission ratios are usually given in units of mass of aerosol per kg carbon in the form of CO_2 (g kg^{-1} $C(CO_2)$). In this chapter, all results will be presented after conversion to emission factors (see below).

The selection of CO or CO_2 as reference gas is determined by the ultimate objective of the analysis and on the fire phase (flaming or smouldering) during which the compound is preferentially released. For compounds emitted predominantly in the smouldering stage of fires, CO is a suitable reference gas as it is also emitted predominantly during this stage. Close correlations between derived-derived gases and CO can usually be obtained, which allows fairly accurate estimation of trace gas emissions from fires for which the CO emission is known. For compounds containing nitrogen or halogen elements, the emission ratio relative to CO is also dependent on fuel composition, i.e. the nitrogen or halogen element content of the fuel.

Flaming-derived compounds correlate well with CO_2, while correlations of derived-derived gases with CO_2 tend to be relatively poor because the variation in the relative proportion of flaming versus smouldering combustion between different fires or even different parts of the same fire results in variable trace gas to CO_2 ratios. On the other hand, the emission ratio relative to CO_2 permits the estimation of trace gas emission from fires based on the amount of biomass burned, because most of the biomass carbon is released as CO_2. Therefore this ratio is the most suitable for regional or global estimations; however, it is worth noting that when multiple ratios are used to estimate emissions, errors are propagated making overall estimates quite uncertain.

Another parameter frequently used to characterise emissions from fires is the emission factor, which is defined as the amount of a compound released per amount of fuel consumed (g kg^{-1} dm; *dm:* dry matter). Calculation of this parameter requires knowledge of the carbon content of the biomass burned and the carbon budget of the fire (usually expressed as combustion efficiency, see Ward et al. 1996); both parameters are difficult to establish in the field as opposed to laboratory experiments where they are readily determined. Where fuel and residue data at the ground are not available, a fuel carbon content of 45% is usually assumed in order to derive emission factors from emission ratios.

During the various BIBEX field experiments, and in other studies over the last decade, a large number of emission ratios/factors have been determined. Recently, these data have been compiled into a coherent set of recommended emission factors (Andreae and Merlet 2001). In Table 2.3 we present emission data from this compilation for selected gaseous and particulate emission products and for the most important types of fire regimes (savannas and grasslands, tropical forest, ex-tratropical forest, domestic biofuel burning, charcoal combustion, and agricultural waste burning). These emission factors are based on an analysis of some 130 publications, a large fraction of which were produced as a result of BIBEX campaigns. The values given are means and standard deviations wherever possible; when only two values for an emission factor are available in the literature, these two values are given as a range, and where only a single measurement is available, it is given without an uncertainty estimate. It is evident from this compilation that only for savannah fires do we have adequate data for most compounds, whereas for other fire types only the emissions of some key compounds have been satisfactorily determined. The release of compounds for which data are missing for a given type of fire can, however, be estimated by scaling emissions to CO or CO_2.

2.7.2.6 *Emissions from Global Biomass Burning*

Estimation of the amounts of trace substances emitted from biomass burning requires knowledge of both the emission factors (i.e. the amount of trace substance per amount of fuel combusted) and the actual amount of fuel burned. We have shown above that the emission factors for many important compounds, such as CO and CH_4, are now fairly well known, with a typical uncertainty of about 20–30%. In spite of this progress, large uncertainties persist for regional and global fire emissions because of the difficulties inherent in estimating the amount of biomass burned. In particular, there are differences of as much as an order of magnitude in regional estimates based on estimates of typical fire frequencies in the various vegetation types, and those based on actual fire counts obtained from remote sensing. These issues will be discussed in more detail in Sect. 2.6.2.7. Table 2.4 provides a summary of estimates made over the last decade using the former approach. In the following paragraphs, we discuss global emissions for key compounds based on the emissions factors in Table 2.3 and the biomass burning estimates of Logan and Yevich (R. Yevich, personal communication 2001) given in Table 2.4 (also see Andreae and Merlet (2001) for further information).

Carbon compounds. CO_2 is the major carbon compound emitted and accounts on average for about 80–90% of the mass of carbon burned, ca. 3 650 Tg $C(CO_2)$ yr^{-1}. CO represents around five to eight percent of the carbon burned, ca. 300 Tg C(CO) yr^{-1}. Hydrocarbon (methane

Table 2.3. Emission factors for selected gases and aerosols for different forms of biomass burning (in g species per kg dry fuel burned)

Species	Savanna and grassland	Tropical forest	Extratropical forest	Biofuel burning	Charcoal burning	Agricultural residues
CO_2	1613 ±95	1580 ±90	1569 ±131	1550 ±95	2611 ±241	1515 ±177
CO	65 ±20	104 ±20	107 ±37	78 ±31	200 ±38	92 ±84
CH_4	2.3 ±0.9	6.8 ±2.0	4.7 ±1.9	6.1 ±2.2	6.2 ±3.3	2.7
NMHC	3.4 ±1.0	8.1 ±3.0	5.7 ±4.6	7.3 ±4.7	2.7 ±1.9	–
C_2H_2	0.29 ±0.27	0.21–0.59	0.27 ±0.09	0.51–0.90	0.05–0.13	–
C_2H_4	0.79 ±0.56	1.0–2.9	1.12 ±0.55	1.8 ±0.6	0.46 ±0.33	–
C_2H_6	0.32 ±0.16	0.5–1.9	0.60 ±0.15	1.2 ±0.6	0.53 ±0.48	–
C_3H_4	0.022 ±0.014	0.013	0.04–0.06	–	–	–
C_3H_6	0.26 ±0.14	0.55	0.59 ±0.16	0.5–1.9	0.13–0.56	–
C_3H_8	0.09 ±0.03	0.15	0.25 ±0.11	0.2–0.8	0.07–0.30	–
1-butene	0.09 ±0.06	0.13	0.09–0.16	0.1–0.5	0.02–0.20	–
i-butene	0.030 ±0.012	0.11	0.05–0.11	0.1–0.5	0.01–0.16	–
n-butane	0.019 ±0.09	0.041	0.069 ±0.038	0.03–0.13	0.02–0.10	–
Isoprene	0.020 ±0.012	0.016	0.10	0.15–0.42	0.017	–
Benzene	0.23 ±0.11	0.39–0.41	0.49 ±0.08	1.9 ±1.0	0.3–1.7	0.14
Toluene	0.13 ±0.06	0.21–0.29	0.40 ±0.10	1.1 ±0.7	0.08–0.61	0.026
Methanol	–	–	2.0 ±1.4	–	–	–
Formaldehyde	0.26–0.44	–	2.2 ±0.5	0.13 ±0.05	–	–
Acetaldehyde	0.50 ±0.39	–	0.48–0.52	0.14 ±0.05	–	–
Acetone	0.25–0.62	–	0.52–0.59	0.01–0.04	–	–
Benzaldehyde	0.029	0.027	–	0.02–0.03	–	0.009
Furan	0.095	–	0.40–0.45	–	–	–
2-methyl-furan	0.044–0.048	0.17	0.47	–	–	0.012
Furfural	–	–	0.29–0.63	0.22	–	–
Acetonitrile	0.11	–	0.19	–	–	–
Formic acid	–	–	2.9 ±2.4	0.13	–	0.22
Acetic acid	–	–	3.8 ±1.8	0.4–1.4	–	0.8
H_2	0.97 ±0.38	3.6–4.0	1.8 ±0.5	–	–	–
NO_x (as NO)	3.9 ±2.4	1.6 ±0.7	3.0 ±1.4	1.1 ±0.6	3.9	2.5 ±1.0
N_2O	0.21 ±0.10	–	0.26 ±0.07	0.06	–	0.07
NH_3	0.6–1.5	–	1.4 ±0.8	–	–	–
HCN	0.025–0.031	–	–	–	–	–
SO_2	0.35 ±0.16	0.57 ±0.23	1.0	0.27 ±0.30	–	–
COS	0.015 ±0.009	–	0.030–0.036	–	–	0.065 ±0.077
CH_3Cl	0.075 ±0.029	0.02–0.18	0.050 ±0.032	0.04–0.07	0.012	0.24 ±0.14
CH_3Br	0.0021 ±0.0010	0.0078 ±0.0035	0.0032 ±0.0012	–	–	–
CH_3I	0.0005 ±0.0002	0.0068	0.0006	–	–	–
$PM_{2.5}$	5.4 ±1.5	9.1 ±1.5	13.0 ±7.0	7.2 ±2.3	–	3.9
TPM	8.3 ±3.2	6.5–10.5	17.6 ±6.4	9.4 ±6.0	–	13
TC	3.7 ±1.3	6.6 ±1.5	6.1–10.4	5.2 ±1.1	6.3	4.0
OC	3.4 ±1.4	5.2 ±1.5	8.6–9.7	4.0 ±1.2	4.8	3.3
BC	0.48 ±0.18	0.66 ±0.31	0.56 ±0.19	0.59 ±0.37	1.5	0.69 ±0.13
K	0.34 ±0.15	0.29 ±0.22	0.08–0.41	0.05 ±0.01	0.40	0.13–0.43

Table 2.4. Biomass burning estimates in Tg dry mass per year

	Crutzen and Andreae 1990	Andreae 1993	Hao and Liu 1994	Liousse et al. 1996	Logan and Yevich 2001
Tropical forest	1 560 – 3 800	1 260	1 820	1 260	1 330
Extratropical forest	–	1 150	–	–	640
Savanna and grassland	670 – 3 600	3 690	2 670	2 680	3 160
Domestic biomass fuel	670 – 1 300	1 960	620	1 380	2 900
Agricultural waste	1 100 – 1 800	850	280	300	540
Total	4 000 – 10 400	8 910	5 390	(5 620)	8 600

+ non-methane hydrocarbons) emissions range around 66 Tg C. Carbonaceous particulate matter emissions are estimated to be around 42 Tg C in organic compounds and 4.8 Tg of black carbon per year.

Nitrogen compounds. Emissions of nitrogen compounds are closely related to fuel composition and combustion type. Linear relationships have been found between the emissions of N_2O (Lobert et al. 1991) and NO_x (Lacaux et al. 1993) and the fuel nitrogen content. The release of NO, N_2O, and molecular N_2 occurs predominantly during flaming combustion, while NH_3, amines, and nitriles are related to smouldering combustion. NO_x emissions from biomass burning are estimated at 9.7 Tg N(NO_x) yr^{-1}. The emission of N_2, which accounts for about one-third of the fuel nitrogen released, represents a conversion of fixed nitrogen to atmospheric N_2, and consequently a loss of nitrogen available as a nutrient ("pyrodenitrification", Kuhlbusch et al. 1991). Little information is available on organic nitrogen released or the organic carbon:nitrogen ratios.

Sulphur compounds. Because of the relatively low sulphur content of biomass compared to fossil fuels, SO_2 emissions from biomass burning make only a small contribution to the atmospheric sulphur budget (ca. 1.7 Tg S(SO_2) yr^{-1}). On the other hand, the COS emitted from biomass fires (ca. 0.2 Tg S(COS) yr^{-1}), makes up about 20% of the source of this trace gas (Andreae and Crutzen 1997).

Emissions from domestic biofuel use. To assess the emissions from domestic biofuel use, the concentrations of CO_2, CO, NO, and some organic compounds and aerosols have been determined in the smoke (Brocard and Lacaux 1998; Ludwig et al. 2002). The emission figures are combined with biofuel consumption rates obtained from surveys of per capita consumption and appropriate demographic information. These rates may vary considerably as they depend on many factors, among them biofuel availability, traditional habits for cooking and heating, prevailing temperatures, etc. At present, uncertainties in emissions from domestic fuel use are thought to stem mainly from insufficient knowledge of the consumption rates.

In a recent study by Marufu et al. (2000) global emissions of CO_2 from domestic biofuel use were estimated at roughly 17% of total sources (1 420 of 8 350 Tg C yr^{-1}). For CO, the contribution from biofuel combustion makes up 16% of total sources (80 of 480 Tg C yr^{-1}). For CH_4 and VOCs much of the respective two and four percent of total are derived, since biogenic sources dominate in these cases, while three and a half percent of the total (1.4 of 40 Tg N yr^{-1}) was calculated for NO_x from domestic biofuel combustion. More recent assessments indicate that emissions from biomass burning may be even higher (R. Yevich, personal communication 2001). All currently available estimates thus agree that biofuel use is a significant source for many atmospheric trace compounds, especially because the emissions occur predominantly within the chemically very active tropical atmosphere and because these gases (except CO_2) contribute to ozone formation. Recent measurements in the INDOEX campaign have confirmed the large impact of biofuel use on atmospheric chemistry and climate in the Asian region (Lelieveld et al. 2001).

The available database, in particular the biofuel consumption figures, for biofuel emissions should be improved further, even if extrapolations from spot assessments will remain necessary. Most measurements have been made in Africa. However, cooking and heating habits vary considerably between different developing regions of the world. Additionally, agricultural waste burning and, even more, smouldering dump sites are not yet characterised adequately, but are expected to contribute significantly to global emissions.

2.7.2.7 *Detection of Fires and Burned Area by Remote Sensing*

Reporting of national estimates of anthropogenic trace gas emissions, including those from biomass burning, are a requirement of the Framework Convention on Climate Change, and the IPCC provides guidelines for these emissions calculations (Callander 1995). For many parts of the world however, national emissions estimates from biomass burning are based largely on expert opinion or summary statistics, and the resulting accuracies are largely unknown. Synoptic fire information derived from satellites provides a source of information for augmenting available national fire statistics. Satellite detection of active fire occurrence has been used to identify the timing and location of fires, and has been used in emission product transport studies, for example. Polar orbiting and geostationary satellite systems have been used to provide fire information (Elvidge et al. 1996; French et al. 1996; Prins et al. 1998). The first global data set of annual satellite fire distributions was developed directly as a contribution to BIBEX (Stroppiana et al. 2000).

Automated algorithms for direct estimation of burned area are currently under development with the intent of providing direct input to emissions modelling (Roy et al. 1999). Satellite based techniques for direct estimation of emitted energy, fire intensity, atmospheric aerosol loading, and vegetation recovery are also being developed. Since in most cases the data products are to be used in numerical modelling, there is a need to provide a quantitative assessment of their accuracy. For satellite products, validation using independent data sources needs to be undertaken to determine product accuracy.

New satellite systems are planned that will improve our current fire monitoring capability (e.g. Kaufman et al. 1998b). The requirements for these systems come in part from the experience gained from BIBEX. The satellite fire research community is working to secure the necessary long-term fire observations from the next generation of operational satellite systems, such as the US National Polar Orbiting Environmental Satellite System (NPOESS).

With the operational availability of satellite-derived information on the location and timing of fires and on the area burned, it will be feasible to run an improved class of models to estimate emissions on an annual basis. These improved models will require ground-based estimates of emission factors and modelled estimates of fuel load and fuel consumed for a given year, rather than representative values for a given vegetation type. As new satellite information becomes available on fire intensity, emitted energy, and fuel moisture content, these first order emissions estimates can be improved. Providing robust models that can be used for operational generation of annual emissions estimates and developing approaches to validate them provide the next challenge for the fire and global change research community.

2.7.2.8 *Impacts of Burning on Trace Gas Exchange from Soils*

The process of biomass burning represents a vast reallocation of nutrients in cleared tropical forest and savannah systems. Large proportions of system carbon, nitrogen, and sulphur are volatilised. Soils are affected by changes in nutrient levels, pH, and temperature, with associated changes in microbial communities. Studies conducted during the SAFARI 92 campaign showed that the mean NO emissions increased after burning, reaching 15 ng N m^{-2} s^{-1} from dry sites and exceeding 60 ng N m^{-2} s^{-1} from the wetted sites (Levine et al. 1996). The long-term effect of excluding fire from a savannah is to increase the soil nitrogen content through increased litter inputs, which in turn increases nitrification rates and soil NO emissions (Parsons et al. 1996). Soil emissions of CO_2 and CO were increased by an order of magnitude after burning, whereas exchange of CH_4 was not affected. In all cases the increases were short lived and dropped back to pre-burn levels within a few days (Zepp et al. 1996). Studies on the impact of burning on soil carbon pools showed that annual burning in a semi-arid savannah reduced the light-fraction carbon markedly but did not impact the intermediate or passive carbon pools. This has implications for the amount of soil carbon that can be readily metabolised by the soil microorganisms. Burning the savannas at longer time intervals had no effect on the pool size or the turnover rates of the various soil carbon pools (Otter 1992).

2.7.2.9 *Importance to Atmospheric Chemistry and Climate*

We have already pointed out that biomass burning is a significant source of several greenhouse gases, among them CO_2, CH_4, and, to a much lesser extent, N_2O. It also makes important contributions to the budget of several gases of stratospheric significance, such as methyl chloride and methyl bromide, N_2O, and COS. Of particular importance to the chemistry and radiative characteristics of the atmosphere are the emissions of ozone precursors, particularly NO_x, VOC, CO, and CH_4. Because vegetation fires in tropical regions can occur only when the vegetation is dry enough to burn, fires are most abundant in the dry season, when the trade wind inversion with its large-scale subsidence and suppression of rain-forming convection prevails over the region. Because this inversion prevents convection to heights of more than a few kilometres, it was initially thought that the linkage between dry conditions and subsidence more or less precluded the transport of pyrogenic ozone precursors to the middle and upper troposphere. Recent work has shown, however, that large amounts of smoke can get swept by low-level circulation, e.g. the trade winds, towards convergent regions over the continents or the Inter-Tropical Convergence Zone, and there become subject to deep convection (Andreae et al. 2000; Chatfield et al. 1996; Thompson et al. 1996a). This transport pattern can explain the abundance of fire-related O_3 and O_3-precursors in the middle and upper troposphere as observed by remote sensing and *in situ* measurements (Browell et al. 1996a; Connors et al. 1996; Olson et al. 1996).

The aerosols from biomass fires, the most obvious and visible sign of pyrogenic air pollution, may have an important impact on climate. Biomass burning is the second largest source of anthropogenic sub-micrometer aerosol (after sulphates from fossil fuel combustion), and possibly the largest source of black carbon particles. These aerosols influence climate and the hydrological cycle by scattering and absorbing solar radiation, and by changing the properties of clouds in ways that are just now being elucidated (Hobbs et al. 1997; Kaufman et al. 1998a; Ramanathan et al. 2000). Further characterisation of the radiative and cloud-nucleating properties of pyrogenic aerosols and their effect on regional and global climate remains a major challenge to the scientific community.

Whether the impact of biomass burning will grow in the future depends both on climate change and on human factors. The amount of fuel available for burning at a given place and time is a function of ecological factors, e.g. soil fertility, precipitation, and temperature. It also depends on land use, i.e. if the area has been burned previously, is used for grazing or agriculture, and so on. If climatic variations become more extreme, as climate models have suggested, we can expect a more frequent occurrence of drought years following very wet years. This would result in large amounts of fuel ready to burn in the fire season. Furthermore, in a warmer and drier climate, fire frequency is likely to increase, which would reduce biomass carbon storage by changing the age class structure of vegetation, as well as causing increased emissions of ozone precursors. To monitor the regional and global evolution of pyrogenic emissions, it would be very useful to develop unique tracers for biomass burning, and to set up continuous measurements of these tracers at selected sites.

Human activities are of central importance to the frequency and severity of biomass fires. If large parts of the humid Tropics are deforested further, they will be transformed from a biome essentially free of fires (the tropical rainforest) to biomes with much more frequent fires (grazing lands, agricultural lands, and wastelands). With a higher human population density, the frequency of ignition will go up as well. And finally, the amount of biomass burned for cooking and domestic heating, already a major source of emissions in tropical countries, will increase further. To follow these changes, we will need to develop further and validate techniques to determine the spatial and temporal distribution of biomass burning and the amounts of biomass burned in the various fire regimes.

2.7.3 Wet Deposition in the Tropics

Wet deposition plays an essential role in controlling the concentrations of trace gases and aerosol particles in the atmosphere and in providing the essential nutrients for the biological functioning of ecosystems. Wet and dry deposition affect the budgets of key nutrients and trace gases in both terrestrial and marine ecosystems, as described in other sections of this chapter.

The Tropics are a particularly important region regarding global atmospheric chemistry. Due to intense ultraviolet radiation and high water vapour concentrations, high OH concentrations oxidise inorganic and organic gases, and induce an efficient removal from the atmosphere of the oxidised products. Strong convection in the tropical regions results in huge volumes of air being drawn out of the sub-cloud layer with the resultant chemical composition of the precipitation coming from the capture of gases and small particles by the liquid phases of cloud and rain. Knowledge of the chemical composition of wet deposition allows one to track seasonal emissions from various ecosystems.

In the 1990s, due to the lack of information on wet deposition in the Tropics, a cooperative programme was undertaken, involving the Global Atmosphere Watch (GAW) of the World Meteorological Organisation and the Deposition of Biogeochemically Important Trace Species (DEBITS) Activity of IGAC (see A.5), mostly in Asia. It was followed by the Composition and Acidity of the Asian Precipitation (CAAP) programme, later expanded into Africa and South America (Lacaux 1999).

In some tropical areas, however, dry deposition is at least as important as wet deposition and must be considered in the calculation of total deposition. Dry deposition of acidic gases impacts soil and plants, as indicated in Sect. 2.4 and 2.6.1. High concentrations of sulphur and nitrogen oxides and nitric and sulphuric acids may increase acidification processes. In many arid or semi-arid regions, transport and deposition of alkaline soil particles to adjacent ecosystems are also important. The deposition of alkaline particles partly mitigates the effects of wet and dry deposition of acidic compounds.

In addition to acidity, the N content of wet deposition may strongly affect ecosystem properties such as C storage, trace gas exchange, cation leaching, biodiversity, and estuarine eutrophication. This has been shown for temperate regions with altered N inputs (e.g. Howarth et al. 1996; Mansfield et al. 1998 and papers therein). Now, however, 40% of global applications of industrial N fertiliser takes place in the Tropics and subtropics, and over two-thirds is expected to occur in now-developing regions by 2020 (Matthews 1993; Bouwman 1998). Similarly, fossil fuel combustion is increasing dramatically in less economically developed regions, including much of the tropical and subtropical regions. Galloway et al. (1994) estimated that by 2020 nearly two-thirds of Earth's energy-related N inputs would take place in the Tropics and subtropics. In addition, N emissions associated with biomass burning are heavily concentrated in the Tropics, and will likely remain so for decades (Andreae 1993) (see Sect. 2.6.2).

2.7.3.1 *Precipitation Chemistry in Equatorial Forests*

To illustrate work undertaken by DEBITS, data for precipitation chemistry and associated wet deposition from several sampling sites located in equatorial forests are presented in Table 2.5. Hydrogen ion is abundant at all sites on an annual mean basis, indicating the generally acidic character of equatorial precipitation. This acidity is due to a mixture of mineral acids (HNO_3, H_2SO_4, etc.) and organic acids (formic, acetic, propionic, and others) (Andreae et al. 1990; Ayers and Gillet 1988; Galloway et al. 1982; Lacaux et al. 1991; Williams et al. 1997). In equatorial African forest precipitation, the acidity contributed by organic acids (40–60%) is equivalent to that contributed by mineral acids (ca. 40%). In Amazonia the composition of precipitation is very different, with organic acids accounting for 80–90% of the total acidity. In the rainwater collected at several remote locations in the Northern Territory of Australia, Gillet et al. (1990) found a volume-weighted mean (vwm) pH for all samples of 4.89, with organic acids contributing about 50% of the free acidity, the remainder being H_2SO_4 and HNO_3.

During the dry season, biomass burning has a drastic influence on rainwater composition. The chemical content of rainwater from Amazonia (ABLE-2A, wet season) and African equatorial sites (Dimonika, Congo and Zoétélé, Cameroon) can be compared to get a rough estimate of the contribution of some of the chemical compounds from the vegetation fires (Table 2.6). In the case of Amazonia, it was assumed that the precipitation chemistry reflected the biogenic emissions of soils and vegetation, with little influence of biomass burning emissions. Therefore, the mean contribution of the vegetation-burning source in the African sites was estimated to be about 60 to 70% of the NO_3^-, NH_4^+, and acidity contents. On the other hand, the African sites, located on opposite sides of the Equator, are alternately affected by savannah burning sources from the Southern (June to October) and Northern (November to February) Hemispheres, as shown by the ubiquitous presence of high concentrations of NO_3^-, NH_4^+, and H^+ in rainwater samples. The gases and particles produced by savannah burning in the Northern and Southern Hemispheres are transported by the north-east and south-

Table 2.5. Weighed volume mean concentrations in µeq l^{-1} and wet deposition in meq m^{-2} yr^{-1} for precipitation collected in sites located in equatorial forests

	Location	Reference	pH	H^+	Na^+	K^+	NH_4^+	CA^{2+}	Mg^{2+}	NO_3^-	Cl^-	SO_4^{2-}	HCO_T	CH_3COO_T
Africa	Dimonika (Congo)	Lacaux et al. 1992	4.74	18.1	11.1	2.0	6.4	9.3	–	8.6	13.3	10.5	6.6	3.0
	Zoetélé	Lacaux 1999	4.83	14.8 (23.7)	1.7 (2.7)	3.6 (6.5)	9.6 (15.4)	6.3 (10.2)	1.9 (3.2)	7.0 (11.6)	2.2 (3.7)	5.0 (8.4)	8.1 (13.1)	5.3 (8.4)
South America	Central Amazon	Andreae et al. 1990	4.91	12.2 (29.4)	4.0 (9.5)	1.8 (4.2)	4.0 (9.6)	4.2 (10.0)	– –	2.1 (5.0)	4.7 (11.3)	3.5 (8.2)	6.3 (15.1)	4.3 (10.3)
		Williams et al. 1997	4.70	17.0 (46.8)	2.4 (6.7)	0.8 (2.3)	3.0 (8.2)	2.4 13.2	0.9 (5.0)	4.2 (11.5)	4.6 (12.6)	2.0 (10.8)	2.9 (8.0)	9.3 (25.6)

Table 2.6. Biomass burning contribution to chemical precipitation in African rainforests, assuming biogenic inputs equivalent to Amazonia

	Amazonia	Equatorial Africa			
	ABLE-2A Biogenic reference	South Equator Dimonika, Congo		North Equator Zoétélé, Cameroon	
	µeq l^{-1}	µeq l^{-1}	Biomass burning contribution (%)	µeq l^{-1}	Biomass burning contribution (%)
H^+	5.8	18.1	68	14.8	61
pH	5.24	4.74		4.83	
NO_3^-	1.1	8.6	87	7.0	84
Organic acids	5.2	7.4	30	13.4	61
NH_4^+	1.9	6.4	70	9.6	80

east Trade Winds, respectively, to the equatorial forests and progressively scavenged by convective clouds.

The wet deposition measured in the semi-arid and humid savannas surrounding the forested ecosystems presents a source of high potential acidity, which may not result in final strong acidity of the deposition. For example, Galy and Modi (1998) have shown that the precipitation from arid savannas is characterised by a weak acidity ($H^+ = 2$ μeq l^{-1}) in spite of a high potential acidity (nitrate + dissociated formate + dissociated acetate = 22 μeq l^{-1}). This result is explained by heterogeneous interactions occurring between alkaline soil dust particles and acidic gases. Many of these mineral dust particles are able to entirely neutralise gaseous nitric acid. These gas-particle interactions occur before the incorporation of these particles into cloud droplets or raindrops. Furthermore, high concentrations of organic aerosols (from biomass burning and condensation of biogenic hydrocarbons) and mineral dust (from deserts and arid areas) could also promote intense heterogeneous atmospheric chemistry (e.g. Dentener et al. 1996; Carmichael et al. 1996, 1997) (also see Chap. 4). These processes may affect the cycles of nitrogen, sulphur, and atmospheric oxidants significantly.

2.7.3.2 *Effect of Wet Deposition on Tropical Terrestrial Ecosystems*

Acidification effects are mainly due to deposition of mineral sulphur and nitrogen compounds. In tropical regions, organic acid deposition may contribute as much as 80%; however, these acids are oxidised in soils and will not participate directly in soil acidification. Soil particles exchange alkaline cations with H^+ and the concentration of alkaline ions determines the soil base saturation. When base saturation is low, acids may release aluminium ions from soil particles. In spite of its limitations, the "critical load" concept, characterising ecosystem sensitivity to acidic deposition, has been adopted as a tool for estimating potential impacts on ecosystems. In order to facilitate the development of strategies to control pollution in tropical countries, the Stockholm Environment Institute (SEI) has recently proposed a global assessment of terrestrial ecosystem sensitivity to acidic deposition that uses soil buffering capacity as a key indicator (Cinderby et al. 1998). This assessment depends on two factors: the buffering capacity of the base layer to identify soils that have high weathering rate, and the cation exchange capacity to quantify the capacity of a soil to buffer acidity.

A global map prepared by SEI (see Fig. 2.16), shows five classes of sensitivity to acidic deposition, from a critical load of 200 meq m^{-2} yr^{-1} for the insensitive class to a critical load of 25 meq m^{-2} yr^{-1} for the most sensitive class. Some selected wet deposition measurements of non-sea salt sulphate, nitrate, and organic acids, mainly obtained by the DEBITS Activity, are also included. These combined measurements provide an overall view of tropical regions where the potential risk of acidification is important. All the equatorial rainforests of South America, Africa, and Asia are classified in the most sensitive classes. For South American soils, which have a level of mineral acidity deposition of about 10–20 meq m^{-2} yr^{-1}, future acidification problems may become severe if further land use change and industrial activities occur in these regions. For tropical Africa, due to the high contribution of mineral acidity from wet deposition from biomass burning sources, the critical load is nearly exceeded in many parts of western Africa. For Asia, in some parts of China, Japan, and other industrialised and populated zones, the critical load has already been exceeded (Fig. 2.16).

Work by IGBP researchers suggests that there is substantial, although mostly indirect, evidence that the supply of N may not limit plant production in some tropical forests (Hall and Matson 1999). Thus, additions of N may have little direct effect on plant production and carbon storage, but may substantially affect rate and timing of N losses. As indicated above, tropical forest soils are highly acidic; additions of anthropogenic N may increase that acidity, leading to increased losses of cations and decreased availability of phosphorus and other nutrients, ultimately limiting plant production and other ecosystem functions. Moreover, N additions to tropical soils may result in immediate and relatively large proportional losses of N in trace gas forms, as discussed above. On-going work strives to identify the direct and indirect effects of wet deposition on tropical agro-ecosystems, and to determine its implications for ecosystem functioning and feed-backs to the atmosphere locally, regionally, and globally.

2.8 Marine Highlights

From its inception, IGAC stimulated and sponsored research on marine aerosol and gas exchange of compounds of biological origin through the Marine Aerosol and Gas Exchange (MAGE) Activity (see A.5), examples of which are given below. Similar to terrestrial biosphere-atmosphere research, integrated field campaigns in marine regions, such as the ACE-1, ACE-2 and ASGA/MAGE experiments (see A.5) have been an IGAC hallmark. These field research efforts have linked studies of emissions, transformations, and transport in the marine boundary layer. The study of pertinent marine biogeochemical cycles resulting in sea-air fluxes, however, has yet to be fully integrated into these field campaigns. While research on biosphere-atmosphere interactions in marine regions has progressed significantly in the last decade, it remains less advanced than

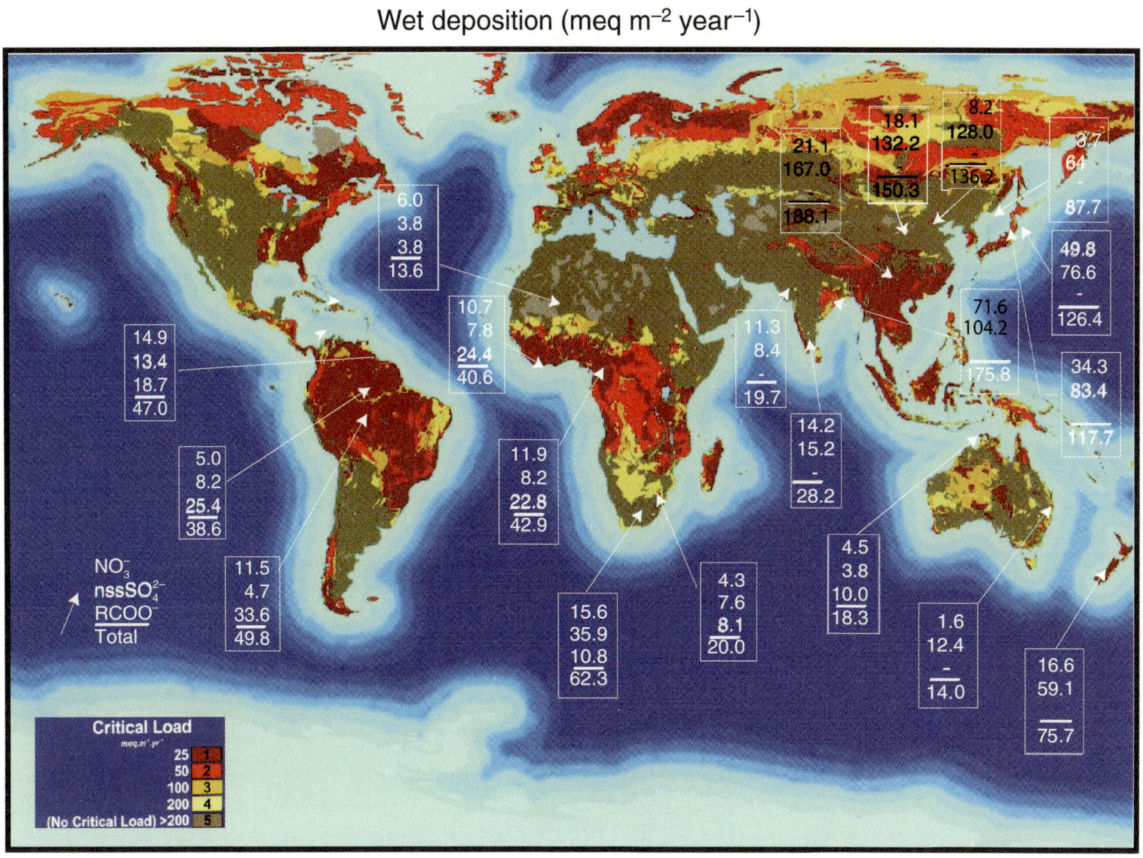

Fig. 2.16. Wet deposition (meq m^{-2} yr^{-1}) of nitrate, non-sea-salt sulphate and organic acids compared with "critical load", a measure of ecosystem sensitivity to acidic deposition. Asia and Oceania (Ayers et al. 1996a); Africa (Galy and Modi 1998; Turner et al. 1996c); Amazonia (Andreae et al. 1990; Williams et al. 1997); and ecosystem sensitivity to acidic deposition (Cinderby et al. 1998)

that for terrestrial regions due to the higher demands of field logistics as well as, perhaps, a lesser recognition of the interaction by the whole of the research communities involved. In the last decade, the greatest advances occurred in DMS biogeochemistry while the marine cycling of other compounds (e.g. organohalogens) is less well, or not at all, understood. Currently, it is still necessary to estimate air-sea fluxes of most gases given existing measurements of the mixing ratios in the atmosphere and sea surface waters. The recent establishment of a new IGBP programme element, Surface Ocean – Lower Atmosphere Study (SOLAS), hopefully will inspire accelerated progress in these various research areas.

2.8.1 Air-Water Gas Exchange Parameterisation

The exchange of inert and sparingly soluble gases including CO_2, O_2, CH_4, and DMS between the atmosphere and oceans is controlled by a thin (20–200 µm) boundary layer at the top of the ocean. Laboratory and field measurements show that wind waves significantly increase the gas transfer rate and that it may be significantly influenced by the presence of surfactants. The mechanisms are still understood only marginally. Empirical gas transfer rate/wind speed relations imply an uncertainty of between 50 and 100%.

The transfer across the boundary layer at an interface shows characteristic mean properties that can be described by a transfer velocity, k, a boundary layer thickness, z, and time constant, t (Jähne and Haußecke 1998). The flux density divided by the concentration difference between the surface and the bulk at some reference level is defined as the transfer velocity, k (also known as the piston velocity transfer coefficient). The equilibrium partitioning between air and water (as measured by the Henry's law constant, H) is another key parameter of air-water gas transfer. A strong partitioning in favour of the water phase shifts control of the transfer process to the gas-phase boundary layer, and a partitioning in favour of the air phase moves control to the aqueous layer. The value of H for a transition at which the control shifts from one phase to the other depends on the ratio of the transfer velocities. For all sparingly soluble gases only the water-side controlled process is relevant (Fig. 2.17). Some environmentally important compounds (e.g. polychlorinated benzenes and some pesticides) lie in a transition zone where it is

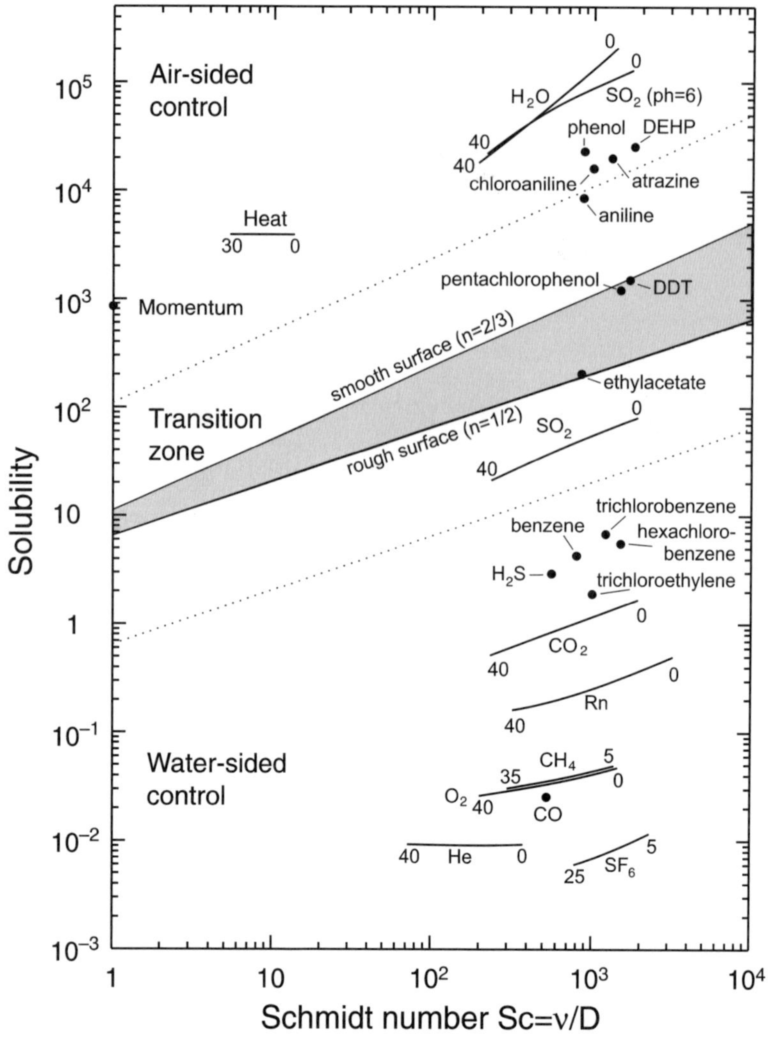

Fig. 2.17.
Schmidt number/solubility diagram including various volatile tracers, momentum, and heat for temperature ranges (°C) as indicated. *Filled circles* refer to only a temperature of 20 °C. Regions for air-sided, mixed, and water-sided control of transfer process between the gas and liquid phase are marked. At the *solid lines* the transfer resistance is equal in both phases. The following dimensional transfer resistances were used: $r_a = 31$, $r_w = 12Sc^{2/3}$ (smooth), $r_w = 6.5Sc^{1/2}$ (wavy surface) with $r_a = R_a u^*_a$ and $r_w = R_w u^*_w$ (after Jähne 1982, and Jähne and Haußecke 1998)

required to consider both transport processes. For reactive gases, as for SO_2, the high transfer resistance in the water is shortcut by its very fast hydration reaction, and the transfer of SO_2 is controlled, as water vapour, on the air side (compare SO_2 only physically dissolved at a low pH with SO_2 at pH = 6 in Fig. 2.17).

The intensity of turbulence determines the transfer resistance: the more intense the turbulence, the thinner the boundary layers. At the scales of the viscous boundary layer, turbulence is strongly attenuated by viscous forces. Thus, the turbulent diffusivity must decrease much faster to zero at the interface than the linear decrease found in the turbulent layer. A free water surface is, however, not solid, nor is it smooth as soon as short wind waves are generated. On a free water surface, velocity fluctuations are possible that make convergence or divergence zones at the surface possible. A film on the water surface, however, creates pressure that works against the contraction of surface elements. This is the point at which the physicochemical structure of the surface influences the structure of the near-surface turbulence as well as the generation of waves. As at a rigid wall, a strong film pressure at the surface maintains two-dimensional continuity at the interface.

A significant influence of surfactants from oceanic conditions has been found by Goldman et al. (1988) and Frew et al. (1990), although contrary results have recently been presented by Nightingale et al. (2000b). The effect of surface films on the boundary layer processes is also discussed in detail in Liss and Duce (1997) and Frew (1997).

Given the lack of knowledge, all theories about the enhancement of gas transfer by waves are rather speculative (for a recent review, see Jähne and Haußecke 1998). Even worse, by just measuring the transfer rates and the wave parameters at the current state of the art it is impossible to verify one of these models conclusively. At high wind speeds, wave breaking with the entrainment of bubbles enhances gas transfer further. The uncertainties of this phenomenon are also large; less soluble gases are affected most (Keeling 1993; Woolf 1993).

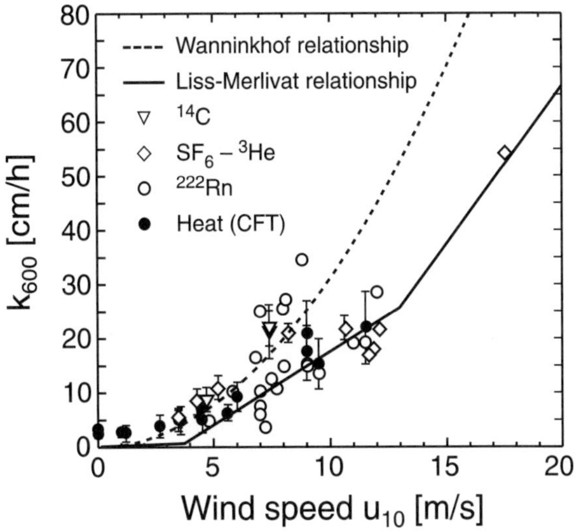

Fig. 2.18. Summary of gas exchange coefficients normalised to a Schmidt number of 600 and plotted versus wind speed (10 m above sea level), plus two empirical relationships (Jähne and Haussecker 1998)

A collection of field data is shown in Fig. 2.18. Although the data show a clear increase of the transfer velocity with wind speed, there is significant scatter in the data that can only partly be attributed to uncertainties and systematic errors in the measurements. The gas transfer velocity is not simply a function of the wind speed. The scatter mainly reflects the additional influence of the wind wave field, which may vary with all parameters that modify the microturbulence in the boundary layer such as the viscoelastic properties of the surface films present and the wind wave field.

Part of the data shown in Fig. 2.18 is based on geochemical tracers such as the ^{14}C, ^{3}He / T, or ^{222}Rn / ^{226}Ra methods. The transfer velocities obtained in this way are only mean values. Thus a parameterisation is only possible under steady state conditions over extended periods; it is questionable under changing conditions. The changes of the parameters (e.g. wind speed) are some orders of magnitude faster. Thus mass balance methods are not suitable for a study of the mechanisms of air-water gas transfer. This is also true for the tracer injection techniques pioneered by Wanninkhof et al. (1985, 1987) in lakes, and Watson et al. (1991a) and Nightingale et al. (2000a) in oceans. Progress in better understanding the mechanisms of air-water gas exchange has been hindered by inadequate measuring technology. Promising new techniques are now available (Jähne and Haußecke 1998; McGillis et al. 1999) but there are currently too few measurements using them for definite conclusions to be drawn. Thus, empirical gas exchange-wind speed relationships (see Fig. 2.17) must still be applied with caution since they have an uncertainty of up to a factor of two.

2.8.2 Marine Biogenic Emissions: A Few Examples

2.8.2.1 *Methane*

The ocean is a small source of methane to the atmosphere. Open Pacific Ocean saturation ratios (ratio of seawater CH_4 partial pressure to the overlying atmospheric CH_4 partial pressure) range from 0.95 to 1.17. Large areas of the Pacific Ocean are undersaturated with respect to atmospheric CH_4 partial pressures during the fall and winter. On a seasonal time scale, the driving force controlling saturation ratios outside the Tropics appears to be the change in sea surface temperature. Saturation ratios in the equatorial region have always been positive and appear to be driven by the strength of the equatorial upwelling. Extrapolating the Pacific data globally and regionally into ten zones, the calculated average flux of CH_4 to the atmosphere is 0.4 Tg yr^{-1} (0.2–0.6 Tg yr^{-1}) (Bates et al. 1996). This is approximately an order of magnitude less than previous estimates, which lacked fall and winter data. Thus the open ocean is a very minor source of methane to the atmosphere (<0.1%) compared with other sources (IPCC 1996). However, the coastal ocean and marginal seas appear to be a much larger source (Owens et al. 1991; Kvenvolden et al. 1993; Bange et al. 1994; Lammers et al. 1995; Scranton and McShane 1991) due to CH_4 emissions from bottom sediments; this definitely warrants further investigation.

2.8.2.2 *Carbon Monoxide*

The ocean is ubiquitously supersaturated with CO with respect to the atmosphere, resulting in a net flux to the atmosphere ranging seasonally and regionally from 0.25 to 13 μmol m^{-2} d^{-1}. However, the total annual emission to the atmosphere (13 Tg; see Table 3.1) is small compared to current estimates from both terrestrial natural and anthropogenic sources (1150 Tg yr^{-1}) (Bates et al. 1995; WMO 1999). Even in the Southern Hemisphere, which accounts for two-thirds of the oceanic emissions, the ocean source is relatively small (<1%), since both methane oxidation and biomass burning are large sources of CO (Bates et al. 1995).

2.8.2.3 *Volatile Organic Carbon Compounds*

Volatile organic carbon (VOC) compounds, or non-methane hydrocarbons, are produced in surface seawater possibly by photochemical mechanisms, phytoplankton activity, and/or microbial breakdown of organic matter (Plass-Dülmer et al. 1995; Ratte et al. 1995; Broadgate et al. 1997). Oceanic concentrations show a strong seasonal cycle (Broadgate et al. 1997). The ocean-atmosphere flux is dominated by alkenes and is small compared to terrestrial emission estimates (<1%). However, the emissions may be significant on local scales considering the short lifetimes of the unsaturated compounds (Donahue and Prinn 1993; Pszenny et al. 1999). Additional seasonal measurements of isoprene, ethene, and propene in particular are needed in different oceanic regions.

2.8.2.4 *Ammonia*

Ammonia is the dominant gas phase basic compound in the marine atmosphere and, as such, has a unique influence on marine multi-phase atmospheric chemistry. Ammonia exists in seawater as both ionised ammonium, $NH_4^+(s)$, and dissolved ammonia, $NH_3(s)$. Dissolved ammonia makes up about ten percent of the total seawater ammonium concentration, $NH_x(s)$, at a pH of 8.2 and a temperature of 25 °C (Quinn et al. 1996a) and is the compound that is emitted across the air-sea interface. $NH_x(s)$ is produced in the upper ocean from the degradation of organic nitrogen-containing compounds and excretion from zooplankton. It also is released from bottom sediments to overlying waters. Loss processes for $NH_x(s)$ include bacterial nitrification, uptake by phytoplankton and bacteria, and emission across the air-sea interface. There will be a net flux of ammonia from the ocean to the atmosphere if the atmospheric $NH_3(g)$ concentration is less than the gas phase concentration in equilibrium with $NH_3(s)$. Alternatively, there will be a net flux into the ocean if the atmospheric $NH_3(g)$ concentration is greater. In either direction, the magnitude of the flux depends on the concentration difference and the transfer velocity.

Attempts to estimate the air-sea flux of ammonia have been hindered by a lack of techniques with sufficient sensitivity and by difficulties in avoiding sample contamination (Williams et al. 1992). As a result, the contribution of NH_3 to the oceanic biogeochemical cycling of N is poorly understood. The few estimates of the air-sea flux of NH_3 that have been reported and that are based on measurements of ammonia in the gas, particle, and/or seawater phases are summarised below.

The first estimates of the flux for the Pacific Ocean were based on filter collection of $NH_3(g)$ and $NH_3(s)$ (Quinn et al. 1988, 1990). These measurements indicated a net flux of ammonia from the ocean to the atmosphere in the northeastern and central Pacific ranging between 1.8 and 16 $\mu mol\ m^{-2}\ d^{-1}$. Clarke and Porter (1993) used measurements of aerosol volatility (which indicate the degree of neutralisation of sulphate aerosol by ammonia) to infer an efflux of ammonia from the ocean to the atmosphere of about 10 $\mu mol\ m^{-2}\ d^{-1}$ over the equatorial Pacific. Similar results have been reported for the Atlantic Ocean and the Arabian Sea. Based on aircraft measurements of aerosol ammonium during a Lagrangian experiment near the Azores, Zhuang and Huebert (1996) estimated a flux of NH_3 from the ocean to the atmosphere of 26 ±20 $\mu mol\ m^{-2}\ d^{-1}$. Simultaneous measurements of $NH_x(s)$ and $NH_3(g)$ were made in the Arabian Sea (Gibb et al. 1999). It was found that in both coastal and remote oligotrophic regions there was a flux of NH_3 from the ocean to the atmosphere. Hence, to date, measurements over portions of the Pacific and Atlantic Oceans, and the Arabian Sea indicate that the remote ocean serves as a source of NH_3 to the atmosphere even in regions of low nutrient concentrations.

Given the importance of $NH_x(s)$ as an oceanic micronutrient, the loss of ammonia through venting to the atmosphere may seem surprising. However, only a small percentage of $NH_x(s)$ exists as $NH_3(s)$ so that this efflux most likely represents a relatively minor loss of NH_3 (Gibb et al. 1999). In addition, this loss can be episodically compensated for through the wet and dry deposition of ammonium-containing aerosol particles. For example, Quinn et al. (1988) estimated that, over the northeastern Pacific, the transfer of $NH_3(g)$ from the ocean to the atmosphere was balanced by wet and dry deposition processes. In certain regions, such as the Southern Bight of the North Sea, there is a flux of ammonia from the atmosphere to the ocean due to the advection of high concentrations of ammonia from adjacent land (Asman et al. 1994). The extent and impact of the deposition of continentally derived ammonia to marine regions is unknown but may be significant. Model results suggest that about six percent of the global continental emissions of ammonia are deposited to the North Atlantic and Caribbean (Prospero et al. 1996). The deposition would be greatest in coastal waters.

It is clear that ammonia, as an oceanic micronutrient and the dominant atmospheric gas phase compound, plays a unique role in both the ocean and the atmosphere. The flux of ammonia from the ocean to the atmosphere affects aerosol chemical composition, pH, and hygroscopicity. The reverse flux, of ammonia plus ammonium in particles and rain from the atmosphere, to the ocean, may affect biological productivity. Simultaneous measurements of ammonia in the atmospheric gas and particle phases, in seawater, and in rainwater are needed to improve our understanding of the multi-phase marine ammonia system in general and the air-sea exchange of ammonia in particular. It is interesting to note that ammonia, due to its strong partitioning into the water phase, is the only gas discussed in this chapter whose transfer velocity is under the control of air-side transfer processes.

2.8.2.5 *Nitrous Oxide*

The world oceans represent a significant natural source of N_2O to the atmosphere (e.g. Seitzinger et al. 2000). The surface waters of many oceanic regions are supersaturated in N_2O with respect to solubility equilibrium with the atmosphere, giving rise to a net air-sea gas exchange flux of N_2O to the atmosphere. Nitrous oxide is produced in subsurface waters both as an intermediate of denitrification, the reduction of nitrate ion (NO_3^-) to nitrogen (N_2), and as a trace by-product of nitrification, the oxidation of NH_4^+ to NO_3^-. Denitrification occurs under suboxic to anoxic conditions, and is thought to take place mainly in restricted low-oxygen regions such as the eastern tropical Pacific and the Arabian Sea, and in the sediments of continental shelves. The distribution of nitrification is thought to be more widespread, since it occurs under aerobic conditions in association with the internal recycling of fixed nitrogen.

During the past decade there have been significant improvements in our understanding of oceanic processes of N_2O production, of the distribution of N_2O in the surface and subsurface ocean, and of the magnitude of the oceanic N_2O source to the global atmosphere. The relative roles of nitrification and denitrification processes have been addressed by measuring nitrogen stable isotopes and their fractionation between N_2O and other dissolved nitrogen-bearing compounds. The interpretation of these difficult measurements is complicated by the likelihood that both nitrification and denitrification are coupled in many oceanic systems, and no clear picture has yet emerged. There have also been recent advances in the study of air-sea gas exchange processes, as indicated in Sect. 2.8.1, which will lead to improvements in the quantification of exchange coefficients as a function of wind speed.

Finally, our understanding of the large-scale distribution of N_2O in the oceans has been improved through a number of shipboard measurement programs, such as those associated with the World Ocean Circulation Experiment (WOCE) and Joint Global Ocean Flux Study (JGOFS) programs. These have generally reinforced our view that open ocean upwelling regions along eastern ocean boundaries and in equatorial and coastal regions, represent major sources of atmospheric N_2O. By contrast, the great subtropical gyres, which represent a large portion of the surface area of the oceans, are relatively close to atmospheric equilibrium for N_2O. In recent years, some extremely high N_2O concentrations have been found in the eastern Arabian Sea, in suboxic waters over the Indian Shelf (Naqvi et al. 2000). Since anthropogenic impingements on the coastal ocean may cause an increase in hypoxia, suboxia, and anoxia in some areas, these recent observations from the Arabian Sea are provocative. By modelling these distributions together with the wind field (e.g. Nevison et al. 1995), we have come to believe that the global oceans constitute a net source to the atmosphere of about 4–5 Tg of N_2O, or about one third of the global natural source strength. This value may increase as more is learned about the diverse distribution of N_2O in coastal waters (e.g. Seitzinger and Kroeze 1998).

2.8.2.6 *Dimethylsulphide*

2.8.2.6.1 *Introduction*

Dimethylsulphide was discovered in ocean waters some 30 years ago by Lovelock et al. (1972). However, it remained a compound of marginal scientific interest for about a decade, until it was established that DMS is the main volatile sulphur compound emanating from the oceans and therefore plays a major role in the atmospheric sulphur cycle (e.g. Nguyen et al. 1978; Leck and Rodhe 1991). Interest in the biogeochemical cycle of DMS increased sharply again in the late 1980s, when Charlson et al. (1987) proposed a hypothesis linking biogenic DMS emission and global climate. In short, this hypothesis states that DMS released by marine phytoplankton enters the troposphere and is oxidised there to sulphate particles, which then act as cloud condensation nuclei (CCN) for marine clouds (see Box 4.3, however, regarding the utility of the CCN concept). Changes in CCN concentration affect the cloud droplet number concentration, which influences cloud albedo and consequently climate. Large-scale climate change, in turn, affects the phytoplankton number and speciation in the oceans and thereby completes, but does not necessarily close, a feedback loop. Recent assessments of the DMS-climate link can be found in Watson and Liss (1998) and Bigg and Leck (2001) (also see Chap. 4).

In the years since publication of the DMS-CCN-climate hypothesis, almost 1 000 papers have been published discussing the distribution and biogeochemistry of DMS (and its precursors) and its link to climate. Several IGAC-inspired studies have addressed aspects of the DMS-aerosol-climate connection, most prominently among them ASTEX/MAGE (e.g. Huebert et al. 1996), ACE-1 (e.g. Bates et al. 1998), and AOE-91, 96 (e.g. Leck et al. 1996, 2001). As a result of these projects, and the large number of independently conducted studies related to the DMS-climate hypothesis, we now understand many of the details of DMS production in the oceans, its transfer to the atmosphere, and the atmospheric oxidation processes (see Chap. 3) that lead to the formation of aerosols (see Chap. 4) that can act as CCN. However, in spite of this progress, fundamental gaps remain in our understanding of key issues in this biosphere-climate interaction, in particular with regard to the processes that regulate the concentration of DMS in seawater. While the basic processes have been identified, and even quantified in specific locations (e.g. Bates et al. 1994; Simo and Pedro-Alio 1999), generally applicable models of DMS-plankton relationships are still in their infancy (e.g. Gabric et al. 1993; Jodwalis et al. 2000). Therefore, we are still not able to represent the DMS-CCN-climate hypothesis in the form of a process-based, quantitative, and predictive model. Even the overall sign of the feedback cannot be deduced with certainty, since it is not known yet if a warming climate would result in an increase or decrease of DMS emissions. Glacial-to-interglacial changes in the amounts of DMS oxidation products in polar ice cores cannot answer this question unambiguously, as they may reflect variations in atmospheric transport patterns as much as differences in DMS production (e.g. Whung et al. 1994), as is discussed in detail in Sect. 2.3.

Early, limited data sets had suggested a possible correlation between DMS and phytoplankton concentration (e.g. Andreae and Barnard 1984). This correlation is particularly evident in vertical profiles of DMS and chlorophyll *a*, which in most instances show a sharp drop of both parameters around the depth corresponding to a light penetration of one percent of the surface light flux. Close correlations between DMS and phytoplankton densities were also found in situations where a single species accounted for much of the DMS production or phytoplankton biomass (e.g. Barnard et al. 1984; Matrai and Keller 1994). These findings led to the hope that global DMS distributions could be estimated from chlorophyll concentrations obtained by remote sensing, but experimental investigations of this proposal were not encouraging (e.g. Matrai et al. 1993), except in frontal regions (e.g. Holligan et al. 1993; Belviso et al. 2000). Furthermore, a statistical analysis of almost 16 000 measurements of DMS in surface seawater failed to show any useful correlations between DMS and chlorophyll or other chemical or physical parameters (Kettle et al. 1999). One reason for the absence of a general correlation between plankton biomass and DMS is that the intracellular concentration of its metabolic precursor, dimethylsulphoniopropionate (DMSP), varies between different phytoplankton species over a range of five orders of magnitude. While it is clear that some taxonomic groups typically contain higher amounts of DMSP, these relationships are by no means clear-cut (e.g. Keller et al. 1989). At least as important, however, are the complexities of DMS cycling by biological and abiotic processes in the surface ocean, which will be addressed below.

2.8.2.6.2 *Physiological and Ecological Controls of DMS Production*

The pathways of DMSP biosynthesis in phytoplankton have been studied and have shed light on potential regulating mechanisms such as nitrogen nutrition (e.g. Gröne and Kirst 1992; Keller et al. 1999a,b), temperature (Baumann et al. 1994), and light (Vetter and Sharp 1993; Matrai et al. 1995). While DMSP, and sometimes DMS, is directly released by phytoplankton, zooplankton also play a role by grazing, or avoiding, DMSP-rich cells (e.g. Dacey and Wakeham 1986; Wolfe et al. 1997; Tang 2000).

Very high concentrations of DMS and dissolved DMSP have been reported from several coastal and/or high latitude areas, especially where blooms of DMSP-producing phytoplankton such as the coccolithophore *Emiliania huxleyi* and the prymnesiophyte *Phaeocystis pouchetii* occur (e.g. Malin et al. 1993; Barnard et al. 1984). In this context, it is interesting to note that Kettle et al.'s (1999) DMS database revealed that high DMS regions corresponded roughly to the coccolithophorid bloom areas derived by Brown and Yoder (1994) from remotely sensed ocean colour data. Prymnesiophytes (including coccolithophores) and dinoflagellates are phytoplankton groups that tend to have high DMSP cell quotas (Keller et al. 1989) and, not surprisingly, DMS is often relatively high when these groups dominate the phytoplankton assemblage. Diatoms, on the other hand, tend to have low intracellular DMSP concentrations and it is generally observed that diatoms are less important DMSP producers in the field (e.g. Keller et al. 1989). Predicting DMS concentrations from the algal assemblage is not straightforward, however. For example, Matrai and Vernet (1997) reported that DMS concentrations were as high in diatom-dominated, Arctic waters as they were in those dominated by *Phaeocystis* sp. It is now recognised that some phytoplankton species not only produce high intracellular concentrations of DMSP, but they also have cell-surface (Stefels and Dijkhuizen 1996) and intracellular (Steinke et al. 1996) DMSP lyase enzymes that may be involved actively in DMS production, thereby contributing further to the elevated DMS concentrations associated with these organisms. The ecological roles of these lyase enzymes are not well understood but several recent studies have pointed to very interesting functions such as in grazing deterrence, carbon acquisition, and bacterial inhibition (Noordkamp et al. 1998; Wolfe and Steinke 1996; Wolfe et al. 1997).

Blooms of marine phytoplankton provide convenient natural "laboratories" for investigating the production of DMS in relation to phytoplankton community dynamics and species succession and associated processes, including grazing and bacterial turnover. However, this apparent focus on "hotspots" of DMS production in relatively nutrient rich areas can be criticised in that oligotrophic areas of the oceans, which generally have relatively low levels of DMS and DMSP throughout the year, make up a large fraction of the total ocean area and so must contribute significantly to the total global flux of DMS (Bates et al. 1992). These pioneering studies established the link between phytoplankton and DMS levels, but failed to account for a large part of the natural variability in DMS concentrations. There have been rather few actual DMS time-series studies (Leck et al. 1990; Turner et al. 1996a; Dacey et al. 1998), all of which noted seasonal periods of elevated DMS concentrations.

We now realise that bacterial processes are also very important in the overall DMS cycle. More isolates of bacteria are available with which to study biochemical pathways and physiology of DMSP and DMS metabolism (e.g. Ledyard and Dacey 1994; Yoch et al. 1997). New methods, including use of ^{35}S tracers, improved inhibitors, and molecular genetics techniques have allowed ever more sensitive analyses of DMSP-DMS cycling rates and fates, and have permitted more detailed examination of the complex microbial communities involved (e.g. González et al. 1999; Wolfe and Kiene 1993). The potential for DMS production from dissolved DMSP is quite large (e.g. Kiene 1996b; van Duyl et al. 1998), but recent studies indicate that most of the DMSP in the sea is not converted to DMS. A demethylation-demethiolation pathway leading to production of methanethiol (MeSH) can account for 70–95% of DMSP metabolism *in situ* thereby diverting sulphur away from DMS (Kiene 1996a). The predominance of this non-DMS producing demethylation-demethiolation pathway is explained by the fact that bacteria use it to assimilate the sulphur from DMSP into protein amino acids (Kiene et al. 1999). Further understanding of this DMSP-DMS-MeSH-bacteria interaction is critical because a relatively small change in the yield of DMS from DMSP could have a major impact on the gross production of DMS, which would then be available for sea-air exchange.

Removal of DMS from the water column by biological and photochemical mechanisms also exerts a great influence on the net accumulations of DMS in surface waters. Slow biological degradation of DMS may partially explain the rise in DMS concentrations observed at the peak and initial decline phases of phytoplankton blooms (e.g. Matrai and Keller 1993; Nguyen et al. 1988). Net consumption of DMS appears to occur in the later stages of blooms after DMS-consuming bacteria have had time to develop (Kwint et al. 1996; van Duyl et al. 1998). The photochemistry of DMS in seawater remains poorly understood, despite the fact that it has been identified as a major removal mechanism under some circumstances (e.g. Kieber et al. 1996; Sakka et al. 1997; Brugger et al. 1998). DMS photooxidation appears to depend on photosensitisers in seawater, which are most likely part of the coloured dissolved organic matter (CDOM) (Dacey et al. 1998). In the open ocean CDOM originates from autochthonous primary productivity and food web processes so the interaction with DMS is probably complex. Add to this the fact that DMS producing and consuming bacterial populations are likely to be strongly influenced by UV-B in surface waters, and one can easily see the importance of understanding photophysical effects on the DMS cycle. Recently, it has been shown that viruses are significant agents in the control of bacteria and phytoplankton. Viral infections can cause a total release of intracellular DMSP (Hill et al. 1998) and viruses are known to infect DMSP-containing bloom organisms such as *Emiliania huxleyi* (Brussaard et al. 1996) and *Phaeocystis* sp. (Malin et al. 1998). It seems clear from studies such as these that the overall food web dynamics, including macro- and microzooplankton grazing, bacterial, and viral activities, as well as the physicochemical dynamics of the upper ocean (e.g. incoming solar radiation, mixing, temperature, air-sea exchange) are important factors governing DMS accumulation.

Modelling efforts have expanded our understanding of DMS production, both for field situations (e.g. Gabric et al. 1999) and laboratory systems (Laroche et al. 1999). However, our current knowledge base is not sufficient to develop and constrain predictive DMS production models for diverse biogeographic regions, in order to allow interpretation of the role of DMS in climate change, for example. Future research will need to focus on (*1*) gaining a full understanding of the processes that control DMS production and allow the prediction of DMS emissions, and (*2*) obtaining much more data concerning spatial, temporal, and interannual variation in the concentration of DMS and related compounds. Emphasis on undersampled areas and seasons would be valuable. For process studies, there is an increasing need to cross disciplinary and international boundaries to bring together experts on different aspects of DMS and related compounds for integrated field campaigns. For analysis of variability, "remote" sampling systems could be considered (such as attempted in ACE-1). It might be possible to develop a buoy-mounted monitoring system whereby samples were stored on a carousel for later analysis. Alternatively, we might follow the example of the pCO_2 measuring community, who have demonstrated that it is feasible to employ unmanned instruments on merchant ships (Cooper et al. 1998a). This would enable the collection of large data sets during long passage routes, covering diverse biogeographic areas, and different seasons, and the chance to investigate interannual variability at relatively low cost. New techniques will be needed to circumvent the present lack of a reliable storage method for DMS samples. In the first instance, it might be more realistic to concentrate on DMSP analyses.

2.8.2.7 *Carbonyl Sulphide*

The oceans represent approximately 30% of the total atmospheric source of COS, and much of the oceanographic work on COS over the last decade has focussed on assessing the spatial and temporal distributions of COS concentration and understanding the processes that control its temporal and spatial distribution. The photochemical source of COS was first recognised by Ferek and Andreae (1984), who demonstrated a clear diurnal cycle in the sea surface concentration of the compound. A mechanism of formation of COS was proposed by Pos et al. (1998) who suggested that the photochemical production of COS and carbon monoxide proceeds along a coupled pathway which first involves the photochemical formation of an acyl radical from coloured dissolved organic matter (CDOM). Flöck et al.

(1997) and Ulshöfer et al. (1996) suggested that cysteine is probably implicated in the reaction mechanism of COS formation as the result of its reactivity and abundance in the oceans. The photochemical COS production in natural seawater is probably not limited by the concentration of a precursor sulphur compound but rather by the concentration of CDOM represented by its ultraviolet attenuation coefficient (Ulshöfer et al. 1996; Uher and Andreae 1997). Zepp and Andreae (1994) and Weiss et al. (1995) quantified the wavelength dependence of COS photoproduction from CDOM and found that quantum efficiency of photoproduction decreases monotonically with increasing wavelength. The dark (or non-photochemical) production of COS has been proposed on the basis of the non-zero COS concentration observed at ocean depths where there is no photochemical production and where there is no mixing from the surface (Radford-Knoery and Cutter 1994; Flöck and Andreae 1996) and also on the basis of careful interpretation of sea surface COS concentration measurements using inverse models (Ulshöfer 1995). COS hydrolysis varies as a function of temperature and pH and has been evaluated several times over the last decade (Elliott et al. 1989; Radford-Knoery and Cutter 1994; Uher and Andreae 1997).

Recent models have used laboratory results for the photoproduction and hydrolysis rate constants to explain COS sea surface measurements obtained during expeditions made in the 1980s and 1990s (see Ulshöfer (1995) for a review of recent sea surface COS concentration measurements). Andreae and Ferek (1992) developed the first chemical box model to explain the diurnal variation of COS in terms of photochemical formation and hydrolysis destruction. Ulshöfer (1995) adopted an optimisation scheme based on the coupled photochemical-mixed layer used by Kettle (1994) to calculate the photoproduction and dark production constants for COS from a series of sea surface measurements made between 1992 and 1994 in the North Atlantic Ocean. von Hobe (1999) extended this work for other models and expedition measurements. Najjar et al. (1995) generalised a simplified coupled physical-chemical model on a global scale to investigate the sensitivity of COS sea surface concentration on ozone reduction and tropospheric increases of carbon dioxide. Kettle and Andreae (1998) and Preiswerk and Najjar (1998) have used existing measurements of the CDOM absorption coefficient of seawater to predict a seasonal variation in the absolute COS concentration, with maximum values at high latitudes in the summer of either hemisphere.

Future work on COS should aim to quantify more accurately the role of the oceans as a source or sink of the gas to the atmosphere. The global application of the photochemical production model for COS is currently limited by the absence of an algorithm to predict the global CDOM absorption coefficient and by the suggestion that the apparent quantum yield of COS formation may vary by more than an order of magnitude in different regions of the ocean. The scarcity of profile measurements of COS concentration has been problematic for modelling efforts which have so far been developed to explain only the surface COS concentration distributions. Finally, the precise quantification of the sea-air flux of all gases produced in the upper ocean (including COS) is currently limited by the absence of an effective gas exchange parameterisation based on wind speed, average wave slope, or other measure of upper ocean turbulence, as already indicated.

2.8.2.8 *The Ocean's Role as Source and Sink of Atmospheric Methyl Bromide and other Methyl Halides*

Methyl halides are produced and consumed biologically (CH_3Br) (Moore and Webb 1996; Baker et al. 1999); (CH_3I) (Moore and Groszko 1999); photochemically (CH_3I) (Happell and Wallace 1996); and in surface ocean waters (CH_3Cl) (Moore et al. 1996). Recent measurements have shown that the flux of CH_3Cl is significantly less than early estimates (Moore et al. 1996) and that the open ocean is a net sink, rather than a source, for CH_3Br (see below).

Methyl bromide (CH_3Br) in the environment began to receive considerable attention in the early 1990s when it was being evaluated as an ozone-depleting gas, along with chlorofluorocarbons, chlorocarbons, and halons. First-order calculations indicated that methyl bromide was likely to be a significant contributor to stratospheric ozone depletion. Before then, only a few studies of CH_3Br in the ocean and atmosphere had been conducted. Lovelock (1975) detected CH_3Br in coastal waters of England and suggested that this gas could have a large natural source. Singh et al. (1983) later reported widespread supersaturations greater than 200% off the west coast of North America, lending support to the ocean as a large natural source of CH_3Br. Khalil et al. (1993) suggested that the open ocean was supersaturated in methyl bromide by 40–80%. However, prompted in part by calculations showing that the ocean simultaneously had to be a large sink for CH_3Br because of reaction with Cl^- in seawater (Elliott and Rowland 1995; Jeffers and Wolfe 1996; King and Saltzman 1997), numerous investigations, using *in situ* mass spectrometry-gas chromatography, demonstrated that the ocean on average was a net sink for atmospheric CH_3Br, with tropical and subtropical waters of the open ocean highly undersaturated and coastal waters often supersaturated in this gas (Lobert et al. 1995, 1996, 1997; Moore and Webb 1996; Groszko and Moore 1998). Certain species of phytoplankton produce CH_3Br, but apparently not at rates sufficient to explain the observed saturation levels (Saemundsdottir and Matrai 1998; Moore et al. 1995;

Scarratt and Moore 1996). Most recently, there have been suggestions that CH_3Br in temperate and coastal waters might undergo a seasonal cycle, with higher concentrations or supersaturations in the spring and early summer and undersaturations the rest of the year (Baker et al. 1999; King et al. 2000). About the same time, it also became clear that chemical and biological removal of CH_3Br in seawater constituted such a large sink for this gas that it would have a profound effect on the lifetime of CH_3Br in the atmosphere, even if the ocean were everywhere a net source (Butler 1994; Yvon et al. 1996b; Yvon-Lewis and Butler 1997). In the latest budget calculations, irreversible loss of atmospheric CH_3Br to the ocean accounts for one-quarter to one-third of the total removal (Kurylo et al. 1999).

These two findings – that the oceanic source was outweighed by its sinks and that the lifetime of atmospheric CH_3Br depended strongly upon its reaction in seawater – necessitated a re-evaluation of the global budget of this gas in the atmosphere. Once the apparently large soil sink was discovered and confirmed (Serça et al. 1998; Shorter et al. 1995; Varner et al. 1999), the calculated budget of atmospheric CH_3Br was no longer in balance. The latest calculations have sinks outweighing sources by 80 Gg yr^{-1}, out of a budget of 205 Gg yr^{-1} (Kurylo et al. 1999). It is unlikely that this additional source will come from the ocean, as the current global coverage of surface measurements, although not complete, is representative of the various oceanic regimes, although with reduced coverage of coastal waters. Currently, a small net sink is calculated for the ocean (3–30 Gg yr^{-1}) which is unlikely to change much, unless, of course, there is some significant global change driving it. Furthermore, recent studies are identifying terrestrial sources from plants and salt marshes that are making the budget gap smaller (Gan et al. 1998; Rhew et al. 2000; Dimmer et al. 1999).

Perhaps one of the most significant things to come out of these intensified studies of methyl bromide in the ocean is that other halogen gases may behave in similar, quantifiable ways. Many of these gases, which may include CH_3I, $CHBr_3$, CH_2Br_2, CH_2BrCl, and C_2H_5Br, among others, also have climatic implications through their chemistry or radiative effects; however, specific studies of them in the past have been limited (e.g. Sturges et al. 1992, 1993; Nightingale et al. 1995). When gases are produced and destroyed in seawater and exchanged with the atmosphere on similar time scales, their exchange with the atmosphere can be controlled in good part by their biogeochemical cycling in seawater. Recent analyses of polar firn air have provided global temporal trends for CH_3Br, while also showing *in situ*,

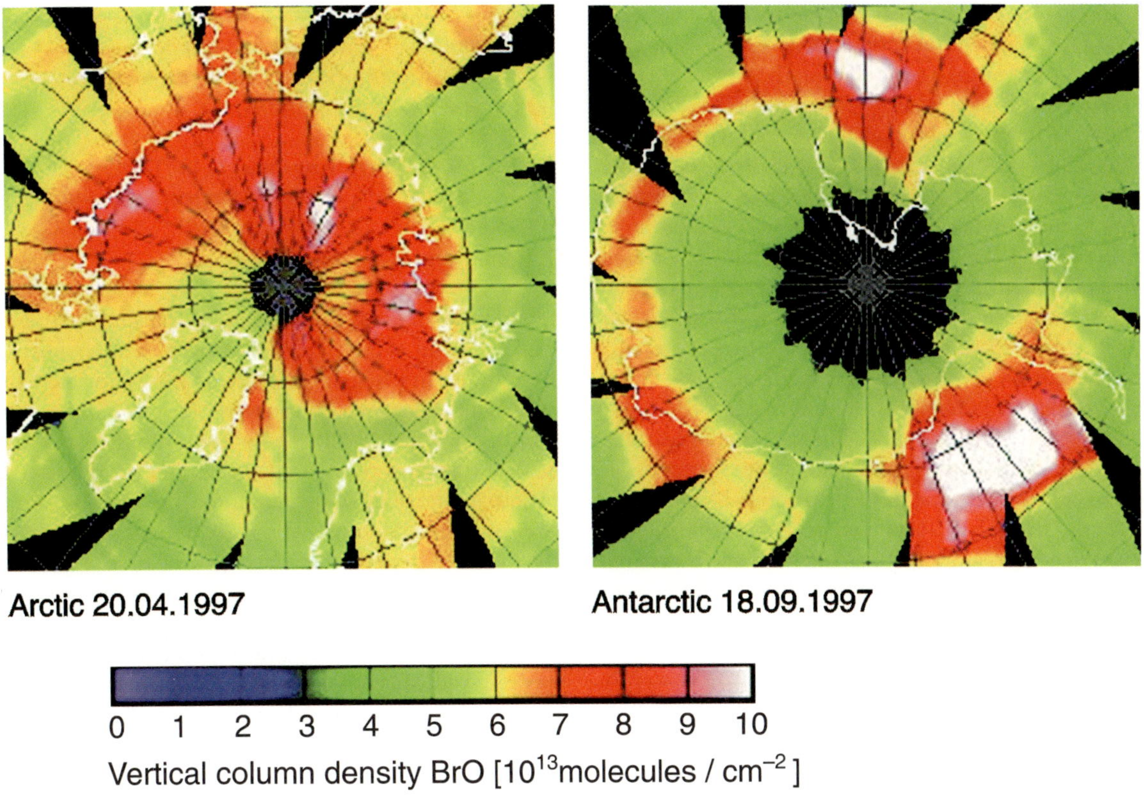

Fig. 2.19. Satellite (Global Ozone Monitoring Experiment, GOME instrument on ERS-2) observations of tropospheric BrO "clouds" in the Arctic and over Antarctica (Wagner et al. 2001). Total columns in the centre of the clouds exceed 10^{14} BrO molecules cm^{-2}. The clouds are visible only in springtime and have a typical lifetime of one to a few days

seasonal production for other organohalogens (Butler et al. 1999; Sturges et al. 2001).

Large amounts of reactive bromine (and smaller amounts of chlorine) are also found in polar regions and near salt pans likely due to oxidation of halides by inorganic reactions (see also Table 3.4). However, the source of iodine compounds in the coastal marine air and of bromine in the free troposphere is much more likely to be the photochemical degradation of organohalogen compounds (like CH_2I_2, or CH_3Br, respectively) of marine biogenic origin, as indicated above.

In field campaigns of the IGAC's Polar Air and Snow Chemistry (PASC) activity, it was discovered in the late 1980s and early 1990s (Barrie et al. 1988, 1992; Barrie and Platt 1997) that surface ozone depletion chemistry occurring in spring over the Arctic Ocean is a tropospheric analogue of stratospheric ozone chemistry, with a difference. It is driven by sea salt halogens from heterogeneous reactions occurring in sunlight on surface snow and ice, rather than by halogens from photolysis of spray can propellants. The existence of BrO and ClO as well as Cl and Br reactions with hydrocarbons was well documented in numerous measurements in air just above the surface of frozen marine areas. In spring, BrO was found in the Arctic (Hausmann and Platt 1994; Tuckermann et al. 1997) and Antarctic marine boundary layer (Kreher et al. 1997) by ground based and satellite observations (Fig. 2.19) (Wagner and Platt 1998; Richter et al. 1998; Hegels et al. 1998; Wagner et al. 2001). In addition, measurements made by chemical amplification (Perner et al. 1999), DOAS (Tuckermann et al. 1997), and by the "hydrocarbon clock" technique (Jobson et al. 1994, Solberg et al. 1996; Ramacher et al. 1999) suggest ClO levels in the pmol mol^{-1} range in the Arctic marine boundary layer. An unexpected link to the mercury cycle was discovered to result in enhanced inputs of mercury to the biosphere in these regions, when long lived elemental mercury is converted to shorter lived particulate and reactive gaseous forms of mercury (Schroeder et al. 1998). Halogen reactions are suspected to be the cause of this conversion of mercury.

As a result of these polar discoveries as well as modelling studies (e.g. Vogt et al. 1996; Sander and Crutzen 1996) (see also Chap. 3), researchers have begun to seek and confirm the occurrence of reactive halogen compounds (IO, BrO, ClO) from air-surface exchange processes in other regions (e.g. remote mid- and lowlatitude marine sites, midlatitudes coastal sites, Dead Sea basin, and the free troposphere).

2.8.2.9 *Primary Marine Aerosols*

Primary aerosols are also emitted directly from the oceans. The work of Blanchard and colleagues (Blanchard 1983) has shown that bubble bursting at the air-water interface injects aerosols into the atmosphere from two sources. One is from fragments of the bubble film (film drops), the other from a jet of water that follows the bubble burst. Bubbles selectively scavenge high molecular weight surface-active compounds (Gershey 1983) and viable particulate material from the water such as bacteria and viruses (Blanchard 1983), leading to a considerable enrichment of these organic components in the aerosol relative to the water. As a result, primary particles in the marine environment will usually contain a wide range of biogenic compounds. Long-chain fatty acids, alcohols, esters, and soluble proteins have all been found in marine aerosols. Proteinaceous material and free amino acids are present in marine rain (Mopper and Zika 1987). In the atmosphere some of these compounds are degraded to form secondary aerosols, such as the fatty acids which may break down to short chain forms such as oxalic acid. Others, like the amino acid L-methionine, are oxidised. Bacteria and remains of organisms have been observed to become separated from the other aerosol components in Arctic conditions (Bigg and Leck 2001). Estimates of the organic components of marine aerosols in relatively unpolluted environments vary widely, the order of magnitude being around 10–20% by number, but this may include secondary aerosols as well as those transported from continents (Mathias-Maser 1998) (see Chap. 4).

2.8.3 Biological and Chemical Impacts of Atmospheric Deposition on Marine and Estuarine Systems

2.8.3.1 *Atmospheric Iron Input to the Ocean and its Role in Marine Biogeochemistry*

2.8.3.1.1 *Introduction*

It is now recognised that a primary transport path for iron found in the ocean is through the atmosphere. Among the first papers to address the importance of atmospherically derived iron were those of Moore et al. (1984) and Duce (1986). These authors calculated the aeolian transport of mineral matter into many areas of the ocean, and pointed out that some fraction of the iron from the mineral matter dissolved into seawater after the dust was deposited to the ocean surface. Duce and Tindale (1991) and, more recently, Jickells and Spokes (2001) have reviewed this topic.

The major reason why atmospheric dust transport has received considerable research effort over the last decade is because of the role iron has been hypothesised to play in controlling marine primary productivity over large areas of the oceans remote from land. Because of their distance from riverine and shelf inputs in these regions (e.g. Southern Oceans, North and Equatorial Pacific) one of the primary ways in which "new" iron gets into the system is via deposition from the atmosphere of terrestrially derived material. The idea of iron being a major control on ocean production is not new. In the early decades of the 20th century it was hypothesised that the reason why large areas of the Southern Oceans contained significant amounts of residual conventional plant nutrients (nitrate and phosphate), when light and other conditions for plant growth were favourable (the HNLC, high nutrient-low chlorophyll, regions), was because of iron deficiency in the water (see, for example, Gran 1931; Harvey 1933; Hart 1934; and the recent review by DeBaar and Boyd 2000). However, it is only in the last decade that analytical techniques for iron and field-going experimental approaches have been good enough to begin to test the hypothesis critically.

2.8.3.1.2 *Sources and Transport of Mineral Aerosol to the Oceans*

The primary sources of mineral aerosol are arid and semi-arid continental regions (e.g. Tegen and Fung 1994; Duce 1995; IPCC 1996) (see also Chap. 4). The atmospheric concentrations of dust and the deposition of dust to the ocean surface are both very episodic and are primarily associated with the transport of aerosol from dust storms or major dust outbreaks. The typical duration of such dust pulses over the ocean may range from one to four days, and the transport and deposition may also vary seasonally. Due to the episodic character of both the atmospheric dust concentrations and local rainfall, the primary removal process for dust (Duce 1995) input to the ocean in a particular region can often occur during a few events covering a relatively short period of time. For example, the results of one multi-year study showed that half of the annual deposition of dust to the ocean at Midway Island in the central Pacific occurred during only two weeks (Prospero et al. 1989). In Bermuda, Arimoto et al. (1992) found that mineral aerosol concentrations ranged over four orders of magnitude, from 0.001 to 11 $\mu g\ m^{-3}$.

We have very few data sets of marine surface dust concentrations collected over long periods of time. In general, the highest atmospheric concentrations of dust in marine areas are found over the North Pacific and the tropical Atlantic. Other high concentration areas are found in the Arabian Sea and the northern Indian Ocean, but there are very limited data in these regions. Accurate estimation or calculation of dust deposition is still quite difficult. An estimate of the geographical distribution of the flux of mineral matter to the global ocean is presented in Fig. 2.20 (Duce et al. 1991). Note that by far the major fraction of mineral dust is deposited in the Northern Hemisphere. The atmospheric deposition has clearly fluctuated significantly in the past, as seen in ice core and deep sea sediment samples (see, for example, Rea 1994; Andersen et al. 1998; Maher and Hounslow 1999). Numerical simulations of the mineral dust cycle are attempting to improve global data sets by linking soil types, particle emissions, gas-particle heterogeneous chemistry, and wind transport in the troposphere with aerosol satellite measurements (e.g. Marticorena and Bergametti 1995; Phadnis and Carmichael 2000) (also see Chap. 6).

2.8.3.1.3 *Iron in Mineral Aerosol over the Oceans*

The atmospheric deposition of iron is associated with the eroded mineral aerosol particles, and the iron is primarily bound in their aluminosilicate matrices. It is thus possible to convert mineral aerosol concentrations or fluxes to an iron concentration or flux by knowing the abundance of iron in the earth's crust. This ranges from ~3 to 5% (Taylor and McClennan 1985). Typically a value of 3.5% is used. With a mineral aerosol flux of 500–2 000 Tg yr^{-1}, the input of iron would be ca. 15–100 Tg yr^{-1}. However, before the iron deposited from the atmosphere can be utilised by phytoplankton, it must be in a form that is available to these organisms. Processes that change the solubility or lability of the iron in the atmosphere will then have potential for influencing

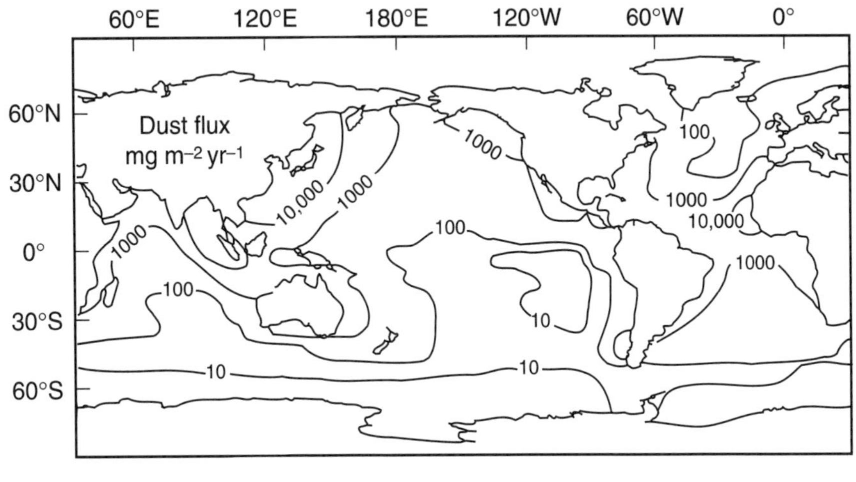

Fig. 2.20. Calculated global fluxes of atmospheric mineral matter to the ocean (Duce et al. 1991)

the availability of the iron when the atmospheric material enters the ocean. Jickells and Spokes (2001) have carefully reviewed the information to date on the mechanisms that may control the distribution of dissolved and/or particulate iron in the material entering the ocean from the atmosphere. Some studies have observed Fe(II) in aerosol iron and its formation is postulated to occur via photochemical reduction of Fe(III) hydroxides. Certain organic compounds such as oxalate, acetate, and formate can facilitate this photoreduction. It has been suggested by several authors that the low pH (0–5) characteristic of the cloud cycling process produces acidic, hygroscopic aerosols. This combined with possible photochemical reactions results in an increase in the lability of crustally derived metals, such as iron, in the atmosphere over that seen in the parent material. In turn, this will play a role in the availability of the iron when the aerosol enters the ocean (e.g. Jickells and Spokes 2001; Zhu et al. 1992). In addition, high ionic strength solutions and alternating wet and dry cycles during cloud formation and evaporation would be common. There are likely to be many such cycles before the particles ultimately enter the ocean by dry deposition or precipitation.

Jickells and Spokes (2001, and references therein) state, in summary, that it is likely that the overall iron solubility of dry deposited mineral aerosol is <1% at a seawater pH of 8, and that a significant proportion of this iron is photoreduced to Fe(II), which is bioavailable. The solubility of iron in marine rains with a pH of 4–7 is generally 14%. Thus the input of soluble atmospheric iron to the oceans is apparently dominated by wet deposition. These estimates, based on laboratory studies, are somewhat lower than those made earlier by other authors. However, Jickells and Spokes (2001) made other oceanographic approaches to estimate the solubility of atmospheric iron. All of these approaches result in low overall iron solubility, probably less than 2%. Their final conclusion is that approximately 0.8 to 2.1% of the total iron deposited in the ocean is soluble. With a total input of 15 to 100 Tg yr^{-1}, this would result in a total soluble iron atmospheric input of from 0.12 to 2.1 Tg yr^{-1}.

2.8.3.1.4 *Iron and Marine Biogeochemistry*

Once the atmospheric iron has entered the oceans by either wet or dry deposition, it is hypothesised to play potentially important roles in the primary productivity of surface waters in substantial areas remote from land. These HNLC regions are estimated to cover ~20–25% of the area of the oceans. An up-to-date and detailed assessment of the chemical form of iron in seawater and how this relates to its uptake by marine organisms is to be found in several chapters in the book edited by Turner and Hunter (2001).

John Martin and his colleagues made some of the first reliable measurements of iron in the oceans and conducted shipboard incubation studies in flasks and carboys of HNLC seawater that had been amended with soluble iron (Martin et al. 1994). The results were promising (e.g. Martin and Fitzwater 1988) and clearly showed that addition of iron (normally added as ferrous sulphate or other simple inorganic salts) could lead to substantial increases in plankton growth, as indicated by increasing chlorophyll concentrations with time in the experimental flasks. An interesting variant on this basic experiment, which is particularly relevant in the present context, was a study conducted in the equatorial Pacific by Johnson et al. (1994). They added the iron in a variety of inorganic and organically complexed forms, but they also used natural Asian dust aerosol particles (collected in Hawaii) and added them to one of the carboys of seawater. In this carboy the rate of plankton growth was found to be the most rapid and attained the highest chlorophyll levels, indicating that the aerosol particles were more effective at promoting growth than artificial iron supplements.

Other avenues have been explored to attack the problem in a more direct way. Young et al. (1991) monitored natural dust inputs to the North Pacific and examined any resulting change in productivity in the receiving water. Several dust deposition events appeared to be correlated with increases in primary productivity measured in on-deck incubators, but with a four day lag between the dust input and the peak in productivity. Although suggestive of a relationship, the results were too few and insufficiently clear-cut to be totally convincing. In addition, interpretation was complicated because productivity change was measured in a deck incubator, not in the ocean itself. Also, when deposition occurred, meteorological conditions changed, with greater stirring of near-surface water, which itself may have changed the productivity. However, this experiment represents a novel and potentially powerful tool since it uses the natural atmospheric input and examines the response of the real oceanic system.

A different approach to testing the iron hypothesis is that of adding inorganic iron ($FeSO_4$) directly to a small patch (of the order of 100 km^2) of the oceans. In order to be able to track the iron enriched patch as it moves in the ocean, the gas sulphur hexafluoride is added along with the iron. Sulphur hexafluoride can be easily measured at tracer (femtomolar) concentrations in almost real time from the research vessel, enabling the enriched patch to be tracked. The principles underlying this approach are outlined in Watson et al. (1991a). It has been utilised three times to date; twice in the equatorial Pacific (IronEx I: Martin et al. (1994) and IronEx II: Coale et al. (1996)) and very recently in the southern oceans (SOIREE: Boyd et al. (2000)). On all three occasions, raising the iron level in the water by a few nanomoles per litre produced a significant enhancement in phytoplankton activity, as measured by chlorophyll concen-

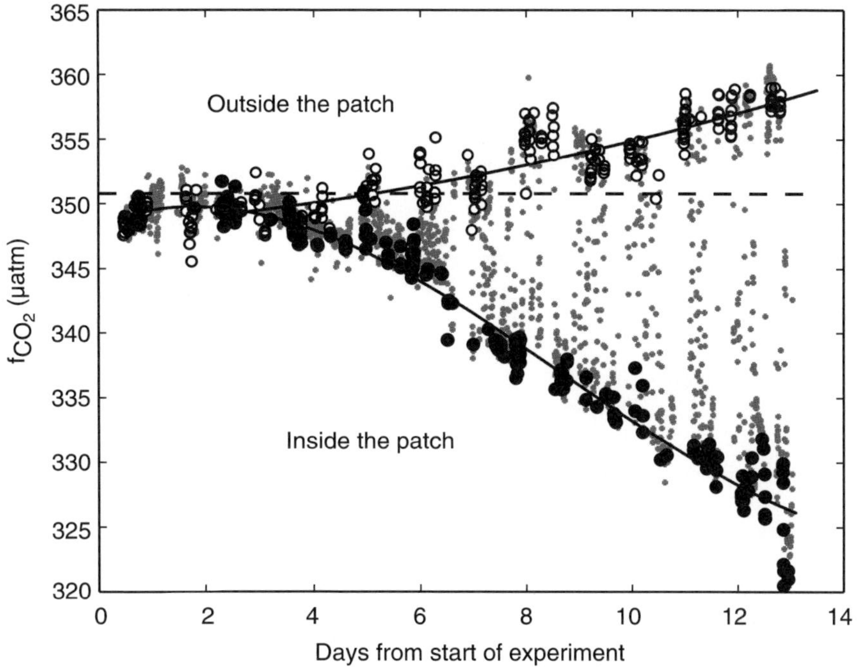

Fig. 2.21. Carbon dioxide changes inside and outside the enriched patch during the course of SOIREE (Boyd et al. 2000)

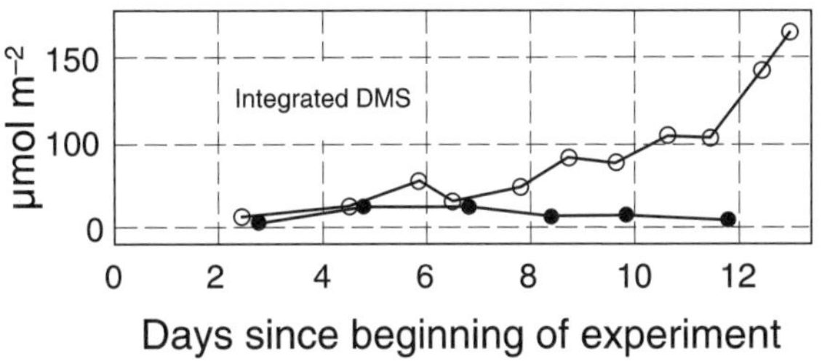

Fig. 2.22. Depth-integrated time evolution of DMS and DMSP inside (*open circles*) and outside (*closed circles*) the enriched patch during the course of SOIREE (Boyd et al. 2000)

tration increase, consistent with the iron fertilisation hypothesis. In the case of IronEx II, the increase was at least an order of magnitude. Smaller organisms were the first to utilise the iron supplement, with the larger plankton (mainly diatoms) benefitting later.

Trace gases measured in these experiments were CO_2 and DMS. The former was drawn down due the enhanced primary production. The extent of CO_2 removal roughly mirrors the increase in chlorophyll, except for IronEx I where it was very small, probably due to rapid recycling of the fixed carbon by grazers. For DMS, three- to five-fold increases occurred in all three studies, with much less variation than for CO_2. Carbon dioxide changes between inside and outside the enriched patch during the course of SOIREE are shown in Fig. 2.21, and the time evolution of DMS and its precursor DMSP integrated over a vertical column are shown in Fig. 2.22. Such a fertilisation experiment is akin to a batch culture perturbation and it is not clear whether long-term Fe enrichment and sustained higher productivity would lead to higher steady-state DMS concentrations. Given

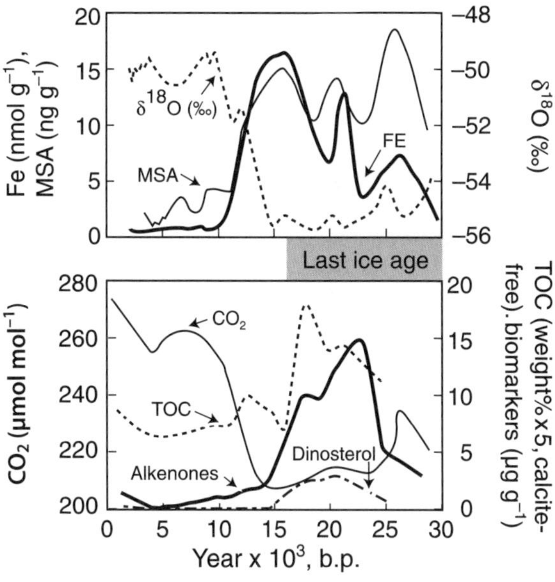

Fig. 2.23. Compendium of results from ice cores for iron, CO_2, MSA, and several other parameters (Turner et al. 1996b)

enough time, shifts in the population structure, or simply growth of the microbial populations, might re-establish low steady-state DMS concentrations with perhaps higher turnover of both DMSP and DMS. Our present understanding of the response of microbial populations to changes in DMSP and DMS supply is insufficient to make confident predictions in this regard. To put these results in a broader time context, a compendium of results from ice cores for iron, CO_2, MSA (an atmospheric oxidation product of DMS), and several other parameters is provided in Fig. 2.23. It is noteworthy that the elevated iron and MSA and lowered CO_2 levels during the last glacial period are consistent with a scenario wherein ocean productivity was higher then due to enhanced atmospheric inputs of iron. For further discussion of the use of ice core records to examine the overall sulphur cycle see Sect. 2.3.

2.8.3.2 *The Input of Atmospheric Nitrogen to the Ocean*

2.8.3.2.1 *Coastal Regions*

There is now widespread evidence that atmospheric fixed nitrogen compounds contribute to enrichment and in some areas probably to coastal and estuarine eutrophication (Jaworski et al. 1997; Howarth et al. 1996). Current estimates of the percentage of total (natural + anthropogenic) new N loading attributed to direct atmospheric deposition at a number of North American and European locations range from 5% to over 50% (Duce 1991; Fisher and Oppenheimer 1991; Valigura et al. 1996; Dennis 1997; Holland et al. 1999). Inputs of N to estuarine systems that result from direct atmospheric deposition by-pass much of the estuarine N "filter" (Kennedy 1983; Paerl 1995, 1997). Thus, atmospheric deposition assumes an increasingly important role as a new N source in lower estuarine and coastal waters below the biological N filtering zone (Fig. 2.24).

Dry and wet atmospheric deposition introduces into estuaries a variety of biologically available inorganic (NO_3^-, NH_4^+, DON) compounds, most of which result from human activities (Likens et al. 1974; Galloway et al. 1994). In addition, organic nitrogen (ON) comprises a significant fraction (from 15 to over 30%) of wet and dry atmospheric deposition in coastal watersheds (Correll and Ford 1982; Scudlark and Church 1993; Peierls and Paerl 1997). Although the composition of atmospheric ON is poorly known, recent work (Peierls and Paerl 1997; Seitzinger and Sanders 1999) indicates that constituents of this pool are biologically utilised and, hence, should be included in eutrophication assessments.

In situ bioassays and field surveys show that enrichment with the major deposition constituents NH_4^+ and NO_3^- at natural dilutions and atmospherically derived dissolved organic nitrogen (DON) results in enhanced phytoplankton primary production and increased biomass (Paerl 1985; Willey and Paerl 1993; Paerl and Fogel 1994; Peierls and Paerl 1997). Atmospheric DON may selectively stimulate growth of specific types of marine phytoplankton (Neilson and Lewin 1974; Antia et al. 1991). These selective phytoplankton responses to specific nitrogen inputs, and changes in stoichiometric C:N ratios resulting from these inputs may induce changes at the zooplankton, invertebrate, herbivorous fish, and higher trophic levels.

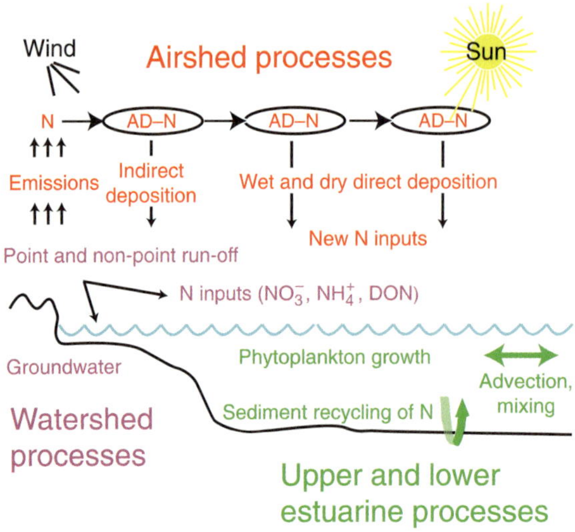

Fig. 2.24. Schematic representation of airshed, upper estuarine, and lower estuarine processes

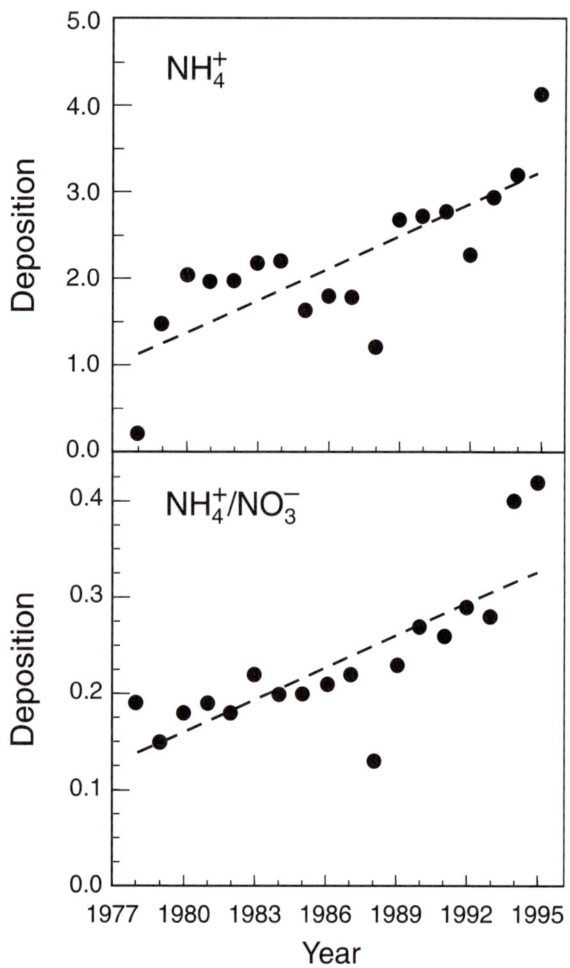

Fig. 2.25. Trends in annual atmospheric deposition (wet deposition as NH_4^+ and NO_3^-, expressed in kg N ha^{-1} yr^{-1}) collected during a 20 year period at the National Atmospheric Deposition Program (NADP) site NC-35 in Sampson County, North Carolina (data re-plotted from NADP information)

Atmospheric nitrogen inputs have been examined recently in North Carolina's Albemarle-Pamlico Sound System (APSS) (Paerl 1995, 1997). Within the APSS, the Neuse River Estuary receives N inputs from a mosaic of upstream and upwind agricultural, urban, and industrial sources. Fossil fuel combustion and agricultural and industrial N emissions represent a significant and growing source of new N to this system (Paerl and Fogel 1994), reflecting a national and world-wide trend (Duce 1986; Luke and Dickerson 1987; Asman 1994; Paerl 1995; Holland et al. 1999). Depending on the relationship between watershed-estuary surface areas, degree of watershed N retention, seasonal rainfall, discharge and flow patterns, and proximity of atmospheric sources, an important fraction of nitrogen from atmospheric deposition is directly deposited on the estuary. In the case of the APSS, recent estimates are on the order of 20% (for its estuarine tributaries) to 40% (for the downstream waters of Pamlico Sound) (Paerl and Fogel 1994; Paerl 1995, 1997) of the N being directly deposited.

Atmospheric N generated from expanding intensive animal farming is of particular concern. Examination of the long-term record of atmospheric NH_4^+ and NO_3^- deposition in Sampson County, eastern North Carolina, shows a nearly three-fold increase in annual NH_4^+ deposition (also relative to NO_3^-) since 1977, with a particularly precipitous rise since the late 1980s (Fig. 2.25). The reason for this may be that unlike human waste, swine waste is stored in open lagoons and remains largely untreated, and substantial amounts (30 to >80%) of N are lost via NH_3 volatilisation alone (O'Halloran 1993).

2.8.3.2.2 *The Open Ocean*

There is also growing concern about the increasing input of human-derived nitrogen compounds to the open ocean. This is especially important in parts of the open ocean where nitrogen is the nutrient that limits biological growth. This is the case in the nutrient-poor waters of the large central oceanic gyres in the North and South Pacific and Atlantic Oceans and the southern Indian Ocean. Current estimates suggest that, at present, atmospheric nitrogen accounts for only a few percent of the total new nitrogen delivered to surface waters in these regions, with upwelling from deep waters being the primary source of new surface nitrogen. It is recognised, however, that the atmospheric input to the ocean is highly episodic, often coming in large pulses extending over a few days. At such times, atmospheric input plays a much more important role as a source for nitrogen in surface waters. A recent estimate of the current input of fixed nitrogen to the global ocean from rivers, the atmosphere, and nitrogen fixation indicates that all three sources are important (Cornell et al. 1995). Paerl and Whitall (1999) estimate that 46–57% of the total human-mobilised nitrogen entering the North Atlantic Ocean is coming via the atmosphere. In addition, the atmospheric organic nitrogen flux may be equal to or perhaps greater than the inorganic (i.e. ammonium and nitrate) nitrogen flux in open ocean regions. The source of the organic nitrogen is not known, but a large fraction of it is likely to be human-derived. This form of atmospheric nitrogen input to the open ocean had not been considered in detail until very recently.

Not only will the input of atmospheric fixed nitrogen to the open ocean increase significantly in the future as a result of increasing human activities, but the geographical locations of much of this input will probably change as well. Galloway et al. (1994, 1995) have evaluated pre-industrial nitrogen fixation (formation of the so-called reactive nitrogen) on the continents; the near-current (1990) reactive nitrogen generated from human activities such as energy production (primarily as nitrogen oxides), fertiliser use, and legume growth; and the estimated reactive nitrogen that will be produced in 2020 as a result of human activities.

The most highly developed regions in the world are predicted to show relatively little increase in the formation of reactive nitrogen, with none of these areas contributing more than a few per cent to the overall global increase. However, other areas will contribute very significantly to increased human-derived reactive nitrogen formation in 2020. For example, it is predicted that Asia will account for ~40% of the global increase in energy-derived reactive nitrogen, while Africa will have a six-fold increase accounting for 15% of the total global increase. It is predicted that production of reactive nitrogen from the use of fertilisers in Asia will account for ~87% of the global increase from this source. Both energy sources (nitrogen oxides and ultimately nitrate) and fertiliser (ammonia, urea) result in the extensive release of reactive nitrogen to the atmosphere. Thus, these predictions indicate very significant potential increases in the atmospheric deposition of nutrient nitrogen compounds to the ocean downwind of such regions as Asia, Central and South America, Africa, and the former Soviet Union (see Fig. 2.26).

The potential problem outlined above was highlighted by a computer modelling study undertaken by Galloway et al. (1994), who generated maps of the recent (1980) and expected (2020) annual deposition of reactive nitrogen compounds from the atmosphere to the global ocean. Figure 2.26 is a map of the projected ratio of the estimated deposition of oxidised forms of nitrogen in 2020 to the values for 1980. It appears that from one and a half to three, and in some limited areas up to four, times the 1980 rate will occur over large areas of the oceans. This increased nitrogen deposition will provide new sources of nutrient nitrogen to some regions of the ocean where biological production is currently limited by nitrogen. There is thus the possibility of important impacts on regional biological production and the marine carbon cycle in these regions of the open ocean.

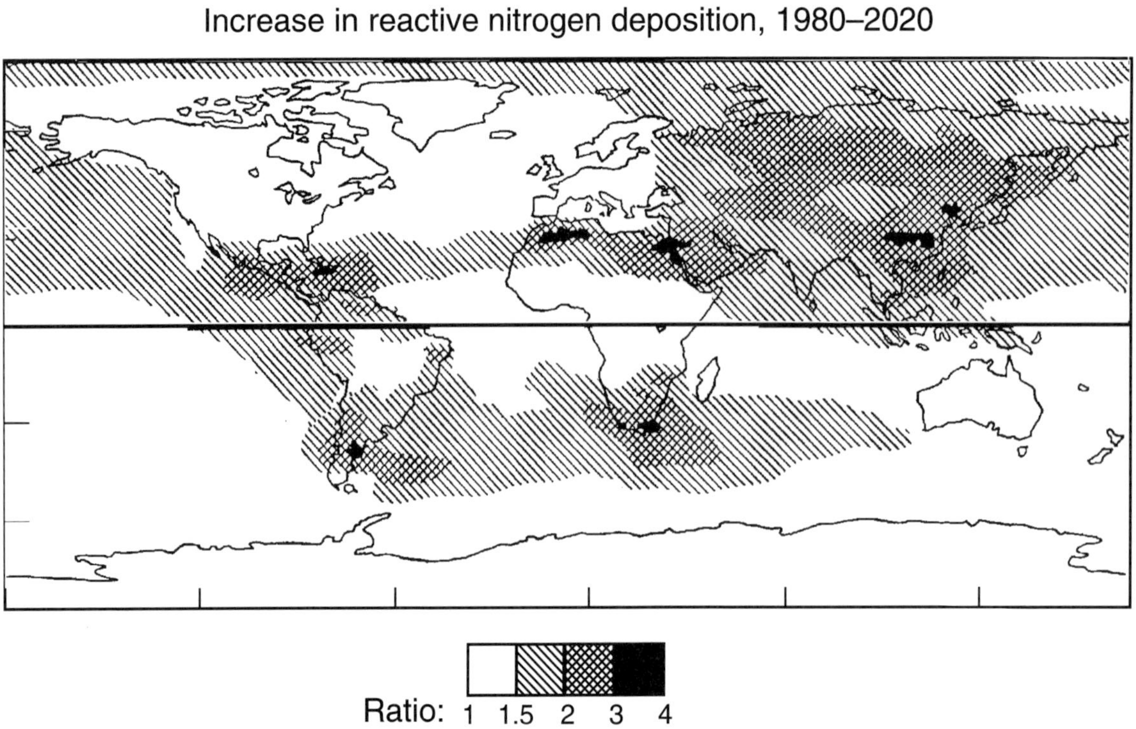

Fig. 2.26. Ratio of the estimated deposition of oxidised forms of nitrogen to ocean and land surfaces in 2020 relative to 1980 (adapted from Galloway et al. 1994, and Watson 1997)

Debate continues about the relative importance of iron, nitrogen, or other compounds as prime determinants of oceanic phytoplankton productivity and, consequently, potential controls of marine gas emissions. On geological time scales, phosphorus is accepted to be the ultimate control. Silica, another component of the same Fe and N-containing dust but with a significantly longer residence time, has also been examined as an altering agent of the species composition of marine phytoplankton in oceanic regions, favouring siliceous organisms (e.g. diatoms) (e.g. Harrison 2000; Treguer and Pondaven 2000). Such organisms would differentially affect total gas emissions but not total primary production. Such a silica hypothesis reinforces the link between marine biogeochemistry and resulting sea-air gas emissions.

2.9 Summary of Achievements and Remaining Research Challenges

Much progress has been achieved over the last decade through technological advances and appropriate scientific approaches. The advances include: remote sensing instrumentation to provide detailed spatial and temporal data; micrometeorological and isotopic techniques for estimating the flux of matter and energy within ecosystems and between ecosystems and the atmosphere, geosphere, and biosphere; techniques for manipulations of local scale selected environmental factors; and, statistical and numerical modelling techniques capable of analysing multi-variate, nonlinear problems.

A developing system-based approach, including Lagrangian studies of air and water masses, comprising all components, e.g. soils, vegetation, and atmosphere, has led to an understanding of biogeochemical cycles of individual chemical compounds and interactions among chemical compounds. The campaign mode of carrying out field measurements has enhanced the understanding of the interconnectedness of systems and the importance of scaling issues.

Highlights of the research include:

- Reduced uncertainties in N_2O, NO, CH_4, DMS, and certain organohalogen emissions, and a better characterisation of local and regional distribution patterns of fluxes together with a mechanistic, but not necessarily integrated, understanding of the surface factors which control these emissions and exchange.
- Effective mitigation strategies have been developed for some CH_4 emissions and a better understanding of how land management practices influence N_2O and CH_4 emissions has been gained.
- Mechanisms and pathways of production and environmental controls have been identified for a large number of VOC compounds emitted from vegetation, including canopy transfer processes. Emission models estimating global emissions have been developed.

- VOC emissions can account for a loss of two to four percent of C taken up by photosynthesis, which has implications for understanding and quantifying the C cycle.
- Improved understanding of atmospheric input of inorganic N and Fe, mainly of anthropogenic and soil origin, into coastal and open ocean environments representing 5–70% of total input in the case of nitrogen and the subsequent impacts on the C uptake of the oceans and the C and S cycles.
- The acidic nature of wet deposition, which differs by source and region, has been characterised.
- Improved understanding of the partitioning of dry deposition (particularly O_3 and NO_2) on leaves and soil surfaces and related physiological mechanisms has been developed.
- Emission ratios for biomass burning are well described for savannas, but less well described for humid forests and biofuels. Broad databases are available of emission factors for a large number of substances.

To achieve more plausible and quantitatively reliable answers, several key issues remain:

- To investigate mechanisms (chemical and biological) responsible for trace gas cycling (emission and deposition) in oceans, soils, and plants, and to establish long-term sites/studies to provide that information, undertaking field experiments to determine, quantify, and discriminate among driving variables.
- To study the exchange of VOCs between vegetation, oceans, and the atmosphere along with the exchange of other trace gases.
- To understand and quantify the effect of soil-released NO and its oxidation product NO_2, under different management practices, to the atmosphere including interaction of these gases both within and above the canopy.
- To determine whether changes in the marine emissions of trace gases and particles are likely to have a significant influence on atmospheric chemistry and vice-versa, resulting from climatic (e.g. rainfall, temperature; perhaps small), elevated CO_2 (perhaps large), and/or land use changes. Key areas may include the greenhouse effect (tropospheric O_3), stratospheric ozone (CH_3Br), radiation and clouds (DMS), VOCs, tropospheric chemistry (dust-FE-DMS-CO_2), and other unpredicted impacts (e.g. the change in marine phytoplankton communities coupled with changes in N and Fe deposition), especially in the Tropics and high latitudes.
- To understand how the hydrological cycle will be affected in various regions with climate change and the subsequent impacts on emissions.
- To improve the parameterisation of air-sea exchange and its links to biogeochemical cycling in surface waters as well as improve Lagrangian studies in water, air, and the combined ocean-atmospheric front, including international participation in order to overcome the intrinsic organisational and logistical difficulties.
- To promote the establishment, wherever possible, of long-term sites for flux measurements, to investigate the magnitude of interannual variation and thus achieve more robust estimates of mean annual fluxes and global budgets.
- To design experiments that will bring synthesis from emission-type studies, regional means of fire detection and prediction, spatially and temporally resolved, and chemical transport models (see Chap. 6) in order to determine the impact of burning on atmospheric chemistry.
- To develop more realistic biological and deposition process-oriented models with interaction and feedback among process-oriented, regional models and global models in order to provide improved estimates of emission and deposition fluxes.

Chapter 3

Atmospheric Photooxidants

Lead authors: Stuart A. Penkett · Kathy S. Law · Tony Cox · Prasad Kasibhatla
Co-authors: Hajime Akimoto · Cyndi Atherton · Elliot Atlas · Carl Brenninkmeijer · John Burrows · Nicola Carslaw Richard G. Derwent · Fred Eisele · Louisa Emmons · Fred Fehsenfeld · Jack Fishman · Claire Granier Dwayne Heard · Øystein Hov · Daniel J. Jacob · Patrick Jöckel · M. Koike · Yutaka Kondo · Jos Lelieveld Jonathan I. Levy · Alain Marenco · Paul S. Monks · Steve Montzka · Jenny Moody · Fiona O'Connor David Parrish · Ken Pickering · John Plane · Alex Pszenny · Geert-Jan Roelofs · Hans Schlager · Paul Seakins Hanwant B. Singh · Andreas Stohl · Anne Thompson

3.1 Introduction

3.1.1 Background

Earth's atmosphere is made up of 78% nitrogen and 21% oxygen. It is therefore highly oxidising but the oxygen only acts as the main source of the more reactive molecules and free radicals that provide the atmosphere's real oxidising power. These are formed almost entirely by photochemistry and many involve coupling between ozone and water vapour, the influence of which on the overall oxidising power is often underestimated. The oxidants to be dealt with here therefore include ozone (O_3), hydroxyl radicals (OH), peroxy radicals (both inorganic (HO_2) and organic (RO_2)), and peroxides (H_2O_2 and RO_2H). Other oxidants include nitrate radicals (NO_3) and halogen atoms; however, these play a subsidiary role and probably are unimportant over large parts of the atmosphere.

The main theme of this chapter is tropospheric ozone and the processes that control its concentration and distribution. These include photochemical processing of precursor molecules that are emitted into the atmosphere, and transport processes, which may be dominant. About 90% of the ozone in the atmosphere resides in the stratosphere as shown in Fig. 3.1. It is formed from the direct photolysis of oxygen with a maximum wavelength of light of 242 nm. Ozone in the stratosphere performs a vital role as a shield against low wavelength ultraviolet (UV) light, which is damaging to many life forms, including our own. Much work has been expended on understanding the chemistry and dynamics of ozone in the stratosphere over the last two decades, mainly brought about by the spectacular losses observed in Polar regions during the spring. For a fuller description of the state of our knowledge on stratospheric ozone and its destruction due to the large-scale production of CFCs and halons, the reader is referred to the excellent reviews produced by WMO/UNEP (WMO 1995, 1999) from where the diagram in Fig. 3.1 was derived. Many current issues concerning stratospheric ozone are described in these reports including long-term global trends and Arctic and Antarctic ozone loss. Also many popular questions about ozone such as: Can chlorofluorocarbons (CFCs) be transported into the stratosphere when they are heavier than air?; and Does the chlorine in the stratosphere mostly come from human or natural sources? are answered in the latest report.

Only about 10% of Earth's ozone is present in the troposphere (Fig. 3.1) and it cannot be made there

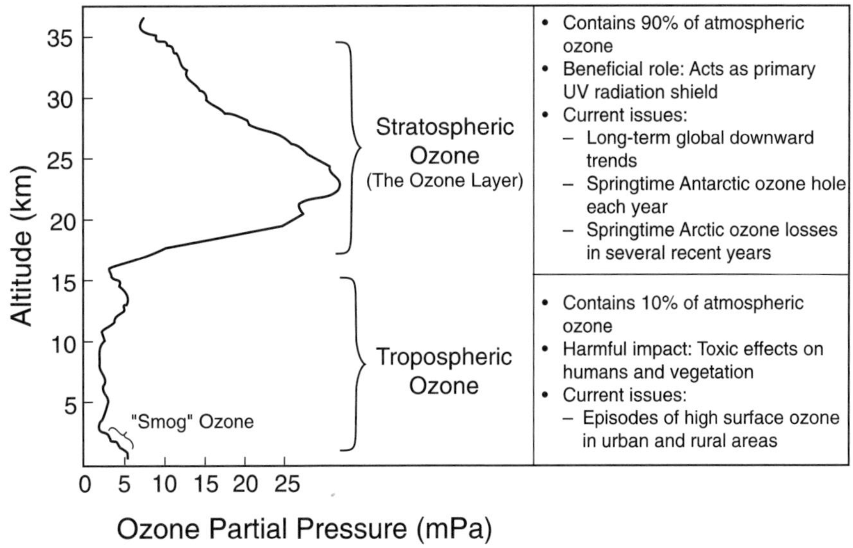

Fig. 3.1. The vertical distribution of ozone in the neutral atmosphere (WMO 1995)

by oxygen photolysis. An obvious source is the large reservoir of stratospheric ozone that enters the troposphere when air is transferred across the tropopause. Earlier theories concerning the origin of ozone in the troposphere (Junge 1962; Danielsen 1968; Fabian and Pruchniewicz 1977) proposed a purely physical picture with the stratosphere as the source and gaseous deposition to the surface as the sink. This explanation with its roots in meteorology is depicted in Fig. 3.2a, which also emphasises the role of transport within the troposphere in distributing ozone by vertical motion and by global circulation. This purely physical picture has now been superseded by a more complex one in which chemistry also plays a major role, partly as a result of advances made by the IGAC Project over the last decade or so. The chemistry involving ozone destruction and production in the troposphere is described in detail later; a summary of the major processes with estimates of their fluxes for comparison with the physical processes is shown in Fig. 3.2b and c for ozone destruction and ozone production respectively. In Fig. 3.2b the process of ozone destruction by sunlight in the presence of water vapour is illustrated; it can easily be shown that much more ozone is destroyed in this way than by deposition. This is indeed the key to showing that chemistry plays a major role in the tropospheric ozone budget. Ozone photolysis produces various free radicals (OH) and peroxy radicals (HO_2), which subsequently become involved in chain reactions contributing to destruction and production of ozone. Figure 3.2c shows the processes that produce ozone in the troposphere which appear to be on a similar scale to the destruction process. Ozone production involves NO_x ($NO + NO_2$) and the switchover from loss to gain of ozone is critically dependent on the concentration of NO at concentrations of the order of 10 pmol mol^{-1}. The chemistry that produces ozone also produces many other photochemical products, some of which are illustrated in Fig. 3.2c under the notation NO_y and NO_z, with the sum NO and NO_2 being NO_x.

The chemistry shown in this figure reveals itself clearly in the troposphere in the form of photochemical smog near the surface as shown in the ozone profile in Fig. 3.1. It can occur anywhere in the troposphere, however, depending on light levels in the near UV region and the concentration of various precursors to ozone, such as hydrocarbons, NO_x, and water vapour. Photochemical smog has been discussed in detail in reports such as those produced by the United States National Research Council (NRC), and IGAC's focus has been on global rather than regional aspects of atmospheric chemistry. Many aspects of the chemistry though are common to the polluted boundary layer and the free troposphere and the difference is more a matter of degree than kind. Also, as will be shown, transport of boundary layer pollution is an important contributor to global ozone.

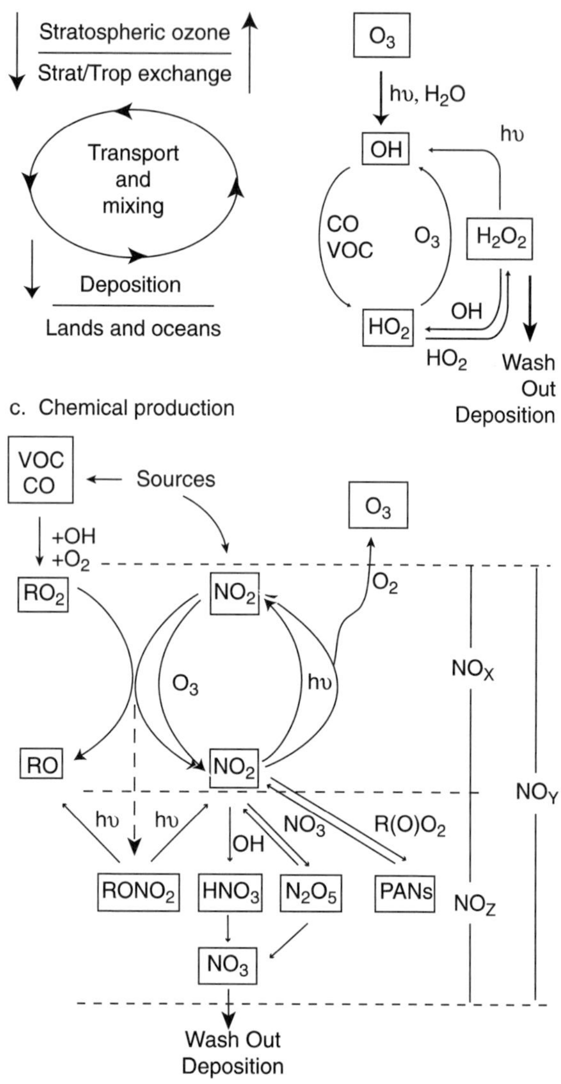

Fig. 3.2. Physical and chemical processes affecting tropospheric ozone (courtesy of Stuart A. Penkett, University of East Anglia, Norwich, England)

Hydroxyl radicals are the main oxidising species in the gas phase troposphere. They react with almost all trace gases containing H, C, N, O, S, and the halogens (with the notable exception of the CFCs and N_2O which are mostly removed by photolysis in the stratosphere) but they do not react with the principal components of the atmosphere such as N_2, O_2, H_2O, and CO_2. The chemistry of the lower atmosphere is thus much concerned with the production of OH radicals from ozone photolysis in the presence of water vapour, their reaction with other trace gases, and their subsequent removal by reaction with oxides of nitrogen to form nitric acid. Other important species are peroxy radicals and peroxides, the former of which are intimately involved in ozone production and loss, and the latter of which are a principal sink for HO_x radicals. Proxides are also involved in reactions with sulphur dioxide (SO_2) in clouds, leading to the wide-scale production of sulphuric acid and of a sulphate aerosol (SO_4). Establishing the extent of tropospheric chemistry occurring has been one of the major achievements in the field of tropospheric science over the last decade; IGAC has played a large role in this venture.

Tropospheric chemistry and tropospheric ozone have until now been treated rather as addenda to other main subject areas in recent reviews of stratospheric ozone by WMO/UNEP (1995, 1999), and of greenhouse gases and their influence on climate by IPCC (1999). Tropospheric chemistry and tropospheric ozone are intrinsically important in their own right. The object of this review is to reveal the relevance of tropospheric chemistry by showing that it largely controls the trace gas composition of the lower atmosphere with all its consequences on different time scales. For example, it is responsible for controlling atmospheric composition over geological time by modifying emissions from the geosphere and the biosphere and allowing them to be removed in soluble forms into the hydrosphere. It is currently responsible for many forms of tropospheric pollution including photochemical smog, sulphate, nitrate, and organic aerosol production, and also for the production of most of the acidity found in rain. Over the last five or ten decades it has almost certainly been responsible for a doubling in ozone concentrations throughout much of the troposphere. For many reasons it must make a large contribution to the extent and nature of the greenhouse effect, particularly in regard to regional effects caused by shorter lived gases and particles. The consequences of its inefficiencies are revealed in a most startling way in the case of emission of CFCs and CO_2, both of which are totally inert in the troposphere and therefore accumulate to cause serious environmental problems.

3.1.2 Chapter Structure

The main theme of this chapter is tropospheric ozone and the processes that determine its production and loss. Many of these processes are common to other gases and particulates in the atmosphere and are discussed in Chap. 1 (see Fig. 1.1). They are described here as they relate to ozone in five main sections, as follows: Sect. 3.1 provides introductory material, while Sect. 3.7 gives a summary of principal achievements over the IGAC period and identifies remaining uncertainties and future research to be undertaken in the study of photooxidants. Section 3.2 describes how gases, which are precursors to tropospheric ozone, particularly nitrogen oxides (NO_x), methane, and carbon monoxide, are emitted from the continents and the oceans. Photochemical processing of precursor species by free radicals leading to ozone production and loss, is the subject of Sect. 3.3, and transport of ozone and its source gases around the troposphere including exchange between the boundary layer and the free troposphere, convection, and synoptic-scale transport, and stratosphere-troposphere exchange, is the subject of Sect. 3.4. A consideration of the impact of all of these processes is presented in Sect. 3.5 along with an up-to-date summary of the global ozone distribution as seen from ground-based measurements, from sonde data, and from new platforms including commercial aircraft and satellites. Section 3.6 highlights results from some recent aircraft experiments carried out principally under the auspices of IGAC, which suggest that long-range transport of ozone formed over the continents is a major contributor to ozone observed throughout the background troposphere. Although this chapter is not comprehensive, it does present a full picture of the current state of knowledge of tropospheric oxidation processes and tropospheric ozone. Much more needs to be done before these two subjects are understood in detail.

3.2 Ozone Precursors

3.2.1 Introduction

As will be shown later ozone (O_3) photochemical production in the troposphere occurs by hydroxyl radical oxidation of carbon monoxide (CO), methane (CH_4), and other hydrocarbons (generally referred to as NMHC) in the presence of nitrogen oxides (NO_x). A quantitative understanding of the budgets and distributions of these O_3 precursors is key to understanding the global budget and distribution of tropospheric O_3, as well as in predicting how its concentrations will likely change in response to future changes in emissions of these precursors. It is also worth noting that the chemical cycles of O_3 precursors are intricately coupled to the chemical cycling of O_3 itself. In this section, we briefly summarise the present state of knowledge of the sources and distribution of these precursors focusing on those known to have significant concentrations in the free troposphere. For a more detailed review of these topics, the reader is referred to Brasseur et al. (1999a) and to Ehhalt (1999).

3.2.2 Primary Emissions of Ozone Precursors

3.2.2.1 *Nitrogen Oxides*

The release of nitrogen oxides into the atmosphere occurs due to the fixation of nitrogen by human activities or by natural phenomena. A breakdown of the esti-

Table 3.1. Estimated primary emissions of O_3 precursors. Best estimates are listed, with ranges given in parentheses (WMO 1998)

Sources	CH_4 (Tg yr^{-1})	CO (Tg yr^{-1})	NMHC (Tg C yr^{-1})	NO_x (Tg N yr^{-1})
Energy use	110 (65–155)	500 (300–900)	70 (60–100)	22 (20–24)
Aircraft				0.5 (0.2–1)
Biomass burning	40 (10–70)	500 (400–700)	40 (30–90)	8 (3–13)
Vegetation		100 (60–160)	400 (230–1 150)	
Soils				7 (5–12)
Lightning				5 (2–20)
Ruminants	85 (60–105)			
Rice paddies	80 (30–120)			
Animal wastes	30 (15–45)			
Landfills	40 (20–60)			
NH_3 oxidation				0.9 (0–1.6)
N_2O breakdown				0.6 (0.4–1)
Domestic sewage	25 (20–30)			
Wetlands	145 (115–175)			
Oceans	10 (5–15)	50 (20–200)	50 (20–150)	
Freshwaters	5 (1–10)			
CH_4 hydrates	10 (5–15)			
Termites	20 (1–40)			
Total	600 (520–680)	1 150 (780–1 960)	560 (340–1 490)	44 (30–73)

[a] NO_y produced in the stratosphere and transported to the troposphere.

mated primary NO_x emissions by source category is provided in Table 3.1 (WMO 1999). The latitudinal variation of the zonally-averaged total column NO_x source is shown in Fig. 3.3. Clearly, the predominant source of NO_x (~22 Tg N yr^{-1}) is surface-based fossil fuel combustion, which accounts for about half of the total global source. This source is concentrated in Northern Hemisphere mid-latitudes. In the tropical boundary layer, the principal sources of NO_x are biomass burning (~8 Tg N yr^{-1}) and emissions from soils (~7 Tg N yr^{-1}). In the free troposphere, production of NO_x by lightning (~5 Tg N yr^{-1}) is the dominant source, while emissions from aircraft and injection from the stratosphere (each about 0.5 Tg N yr^{-1}) make smaller contributions. Table 3.1 also shows that, even from a global perspective, there are considerable uncertainties in the magnitudes of the biomass burning and lightning sources of NO_x. The uncertainty in the biomass-burning source is related to uncertainties in the areal extent of biomass burning and in the associated NO_x emission factors (see Chap. 2 for a more extensive discussion). The uncertainty in the lightning source is related to uncertainties in the characteristics of individual lightning strokes as well as to uncertainties in extrapolating to the global scale. Some recent estimates based on the EULINOX experiment carried out at mid-latitudes over Europe suggest a global NO_x production rate of 3 Tg N yr^{-1} from lightning (Huntrieser et al. 2000), which is in line with other studies at mid-latitudes over the US, based upon the number of moles of NO produced per metre of lightning flash (STEREO-A) (Dye et al. 2000). The subject of aircraft emissions in the context of impact on ozone in the upper troposphere is discussed further in Sect. 3.6 and has been discussed extensively in a recent report by IPCC (1999).

While global estimates of the magnitudes of the individual NO_x sources are useful, there are several issues worth considering in the context of assessing current and future impacts of anthropogenic activities on tropospheric O_3. It is important to note that sources of smaller magnitude can have disproportionately large impacts on NO_x budgets in specific parts of the atmosphere owing to the relatively short tropospheric lifetime of NO_x. It is also worth noting that while the biomass burning and soil sources of NO_x are partly natural in character, they have been strongly impacted by human activities. For example, Crutzen and Zimmerman (1991) estimated that pre-industrial biomass burning emissions were about ten percent of present-day values owing to the lower human population in the Tropics. In a similar vein, Yienger and Levy (1995) estimated that pre-industrial emissions of NO_x from soils were about 35% lower than present-day values. The low pre-industrial emission rate reflects in part the use of fertilisers and in part the conversion of forest with low NO_x emission rates to grasslands and pastures with higher NO_x emission rates. Another issue worth noting is that the ambi-

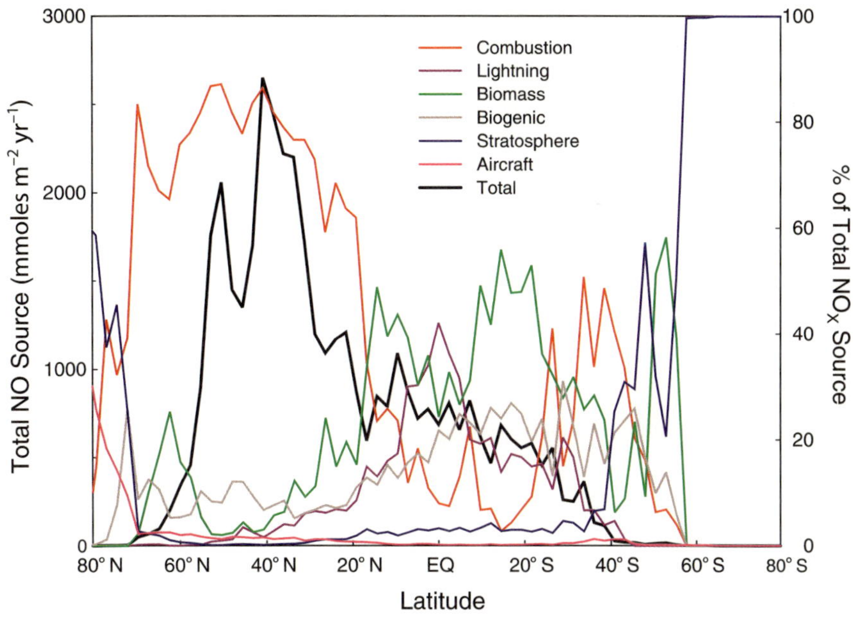

Fig. 3.3.
Latitudinal distribution of zonally-averaged NO_x emissions (courtesy of H. Levy II, NOAA/GFDL)

ent NO_x concentration at certain remote locations during certain times of the year may be largely due to decomposition of PAN and photolysis and oxidation of HNO_3 (Moxim et al. 1996; Schultz et al. 2000) rather than due to direct transport of primary NO_x. In addition, it has been hypothesised that rapid chemical conversion of HNO_3 to NO_x by a heterogeneous pathway can be a significant source of NO_x in the remote troposphere (Chatfield 1994; Hauglustaine et al. 1996; Lary et al. 1997).

3.2.2.2 *Carbon Monoxide, Methane, and Nonmethane Hydrocarbons*

Primary emissions of CO into the atmosphere occur principally due to surface-based fossil fuel combustion and biomass burning (see Table 3.1). Each of these sources is estimated to contribute about 500 Tg CO yr^{-1}. In addition to these primary sources, there are significant secondary sources of CO in the troposphere owing to oxidation of CH_4 and NMHC. According to recent estimates by Wang et al. (1998a), the combined magnitude of these secondary sources (~800 Tg CO yr^{-1} from CH_4 oxidation and ~300 Tg CO yr^{-1} from NMHC oxidation) is comparable to the combined magnitude of the primary sources. The global CH_4 source is estimated to be about 600 Tg yr^{-1}. A significant portion of this total is believed to be associated with anthropogenic activities related to energy use, biomass burning, animal husbandry, waste management, and agriculture. The global NHMC source is also estimated to be 600 Tg C yr^{-1}, with the dominant component being emissions of isoprene and monoterpenes from vegetation and a smaller contribution from anthropogenic activities related to energy use and biomass burning.

As is the case with NO_x sources, there are significant uncertainties in the estimated sources of CO, CH_4, and NMHC. In the case of CO, the uncertainties arise because of poorly known CO emission factors during combustion processes, the areal extent of biomass burning and the associated CO emission factors, and the yield of CO from CH_4 and NMHC oxidation processes. Although oxidation of isoprene is generally considered to be a natural source, the yield of CO depends on the ambient NO_x concentration (Miyoshi et al. 1994) that could have a significant anthropogenic component. In the case of isoprene and monoterpenes, uncertainties in the global emission rates are related to sensitive dependence of emissions on a variety of factors (e.g. light, temperature, tree species, etc.) that make global extrapolation difficult; for a more detailed discussion see Ehhalt (1999).

3.2.3 Global Distribution of Ozone Precursors

3.2.3.1 *Nitrogen Oxides*

Our present understanding of the global distribution of NO_x is derived from sporadic measurements at a variety of locations complemented by results from global-scale model simulations. In general, the spatial NO_x distribution is highly variable as would be expected given its relatively short lifetime and geographically inhomogeneous source distribution. Surface and airborne observations of NO_x have been compiled by Emmons et al. (1997, 2000), Thakur et al. (1999), and Bradshaw et al. (2000).

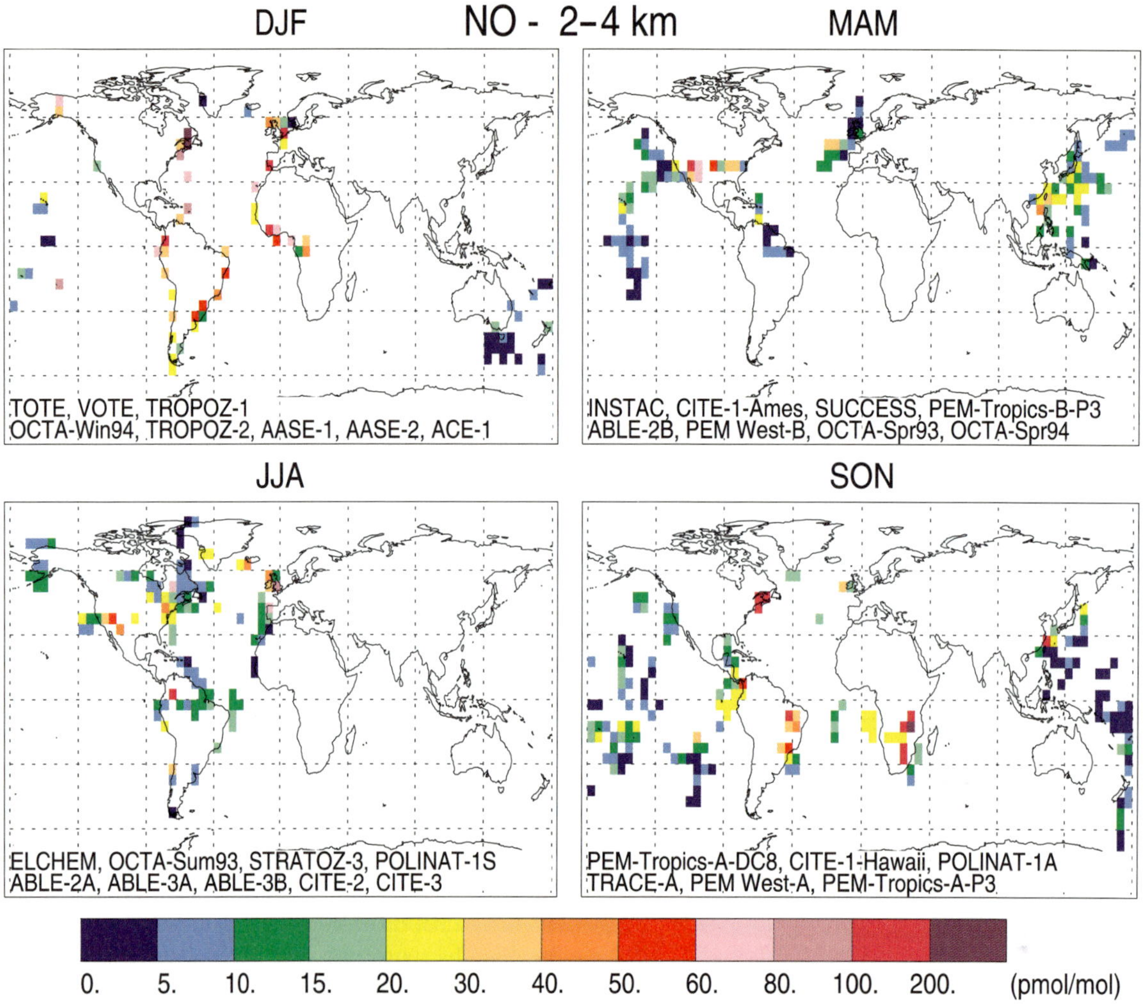

Fig. 3.4 a. Global distribution of NO concentrations at 2–4 km (from *http://aoss.engin.umich.edu/SASSarchive/Species.htm*; L. K. Emmons, NCAR, Oct. 6, 1999)

Ground-based measurements at rural locations in North America and Europe show mean NO_x levels ranging from 1–3 nmol mol^{-1} in polluted regions (Parrish et al. 1993a) to a few tenths of a nmol mol^{-1} in cleaner air masses (e.g., Horowitz et al. 1998). By contrast, measured median summertime NO_x concentrations at Sable Island, off the coast of Nova Scotia, are less than 0.1 nmol mol^{-1} (Wang et al. 1996a) and even lower median NO_x concentrations (<0.05 nmol mol^{-1}) have been reported from shipboard measurements in the eastern North Atlantic (Carsey et al. 1997). Observations over one year at Hawaii (20° N, 3.4 km altitude) were made during MLOPEX 2 and showed median NO_x values of 30–40 pmol mol^{-1} (Atlas and Ridley 1996). At more remote locations, such as in the boundary layer of the equatorial Pacific, NO_x concentrations as low as 5–10 pmol mol^{-1} have been measured (Torres and Thompson 1993; Singh et al. 1996a).

Airborne measurements of NO and NO_2 have also been made during a variety of field campaigns. It is generally accepted that until recently airborne NO observations have been more reliable than NO_2 measurements (Crawford et al. 1996; Bradshaw et al. 1999). Emmons et al. (2000) have recently created a data composite of tropospheric NO measurements from 27 aircraft field campaigns in the 1983–1996 time period, and have developed maps of average NO concentrations at different altitudes based on this data composite. Figure 3.4 shows the NO maps developed by Emmons et al. (2000) for 2–4 km, 6–8 km, and 10–11 km. The aircraft campaigns included in these composites are listed in Table 3.2; many of these were carried out as part of IGAC. In the lower troposphere, high values occur near continental emissions, but NO_x is quickly oxidised to HNO_3 and PAN with a photochemical lifetime of a few hours in mid-latitude summer, and concentrations are quite low in remote regions. In the upper troposphere, NO_x is produced by aircraft and lightning in addition to being transported from the continental surface. Due to these additional sources combined with a longer lifetime (two to seven days) in the upper troposphere, NO_x mixing

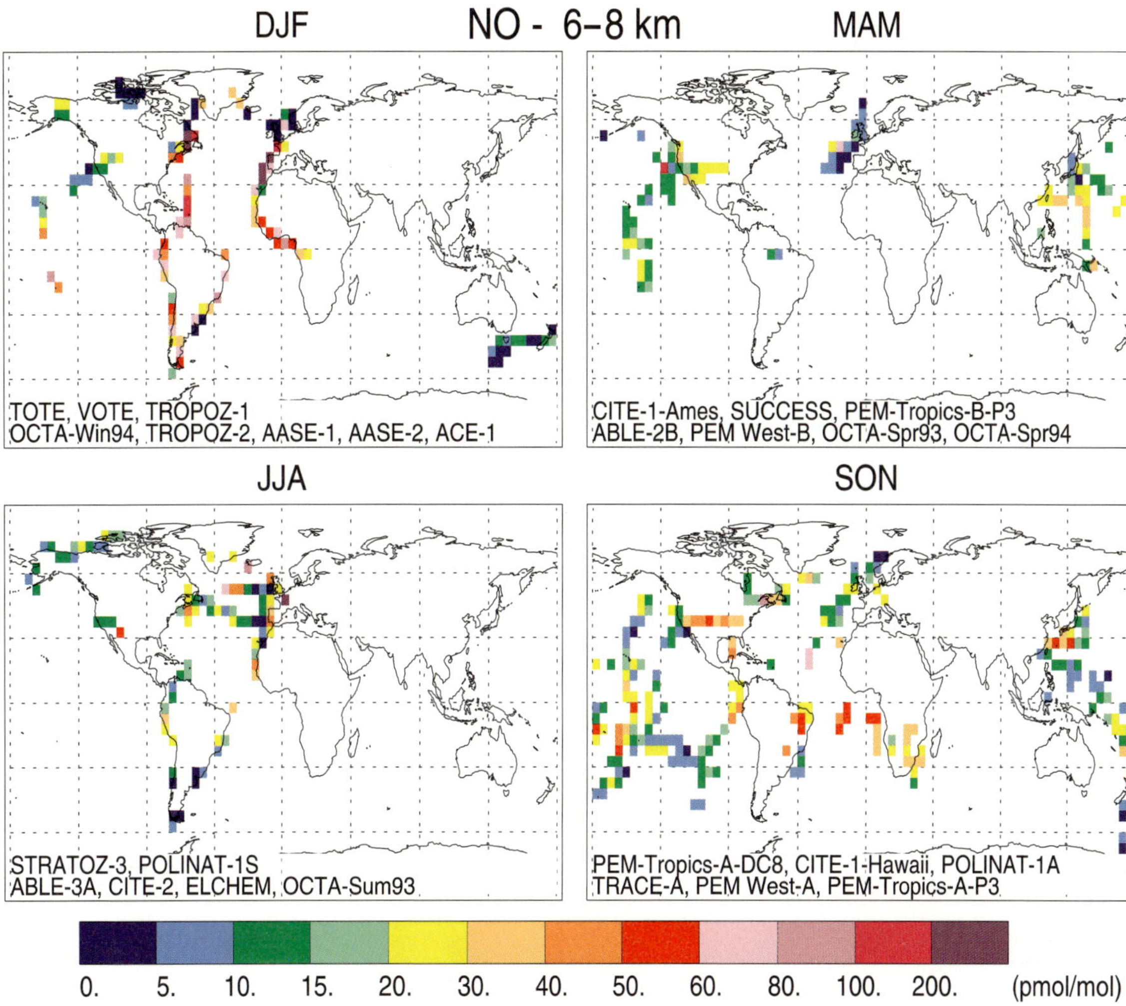

Fig. 3.4 b. Global distribution of NO concentrations at 6–8 km (from *http://aoss.engin.umich.edu/SASSarchive/Species.htm*; L. K. Emmons, NCAR, Oct. 6, 1999)

ratios generally increase with altitude over the ocean, while C-shape vertical profiles are often seen over the polluted continent (Singh et al. 1996a).

In the lower troposphere, average NO concentrations of 20–60 pmol mol^{-1} are evident over the continents. The mean NO concentrations over biomass burning regions in South America and southern Africa during September–November (TRACE-A) are comparable to summertime concentrations over the United States (e.g., CITE-2, ABLE-3B). Evidence for outflow of NO_x from continents to adjacent ocean regions is seen in the measurements off China during March–May (PEM-West-B) and over the tropical south Atlantic during September–November (TRACE-A). Over more remote oceanic regions, NO concentrations in the lower troposphere are typically less than 20 pmol mol^{-1} (ACE-1 during December–February, and PEM-Tropics-A during September–November).

Over South America and Africa the NO_x mixing ratios are lower at 6–8 km than at 2–4 km (TRACE-A) due to its conversion to HNO_3 and PAN. Over the oceans, however, the NO_x values increase with altitude from below 10 pmol mol^{-1} at 2–4 km to above 50 pmol mol^{-1}. During September–November emissions from biomass burning have been convected as NO_x or its reservoirs and then transported to the Atlantic (TRACE-A) and the Pacific (PEM-Tropics-A) (Jacob et al. 1996; Schultz et al. 2000). Lightning-produced NO can also contribute to these high levels.

The maps of NO at 10–12 km (Fig. 3.4c) include observations from commercial aircraft (NOXAR) (see Emmons et al. 2000). Observations at high latitude during winter (AASE, STRATOZ-3) are below 20 pmol mol^{-1} due to the lack of sunlight. Seasonal variation can be seen over North America, Europe, and Asia, with the highest values occurring during June–August, coincident with the maximum in lightning activity. As at 6–8 km, high levels of NO, above 100 pmol mol^{-1}, are observed over the South Pacific and South Atlantic.

The overall features of the observed NO_x concentrations are simulated reasonably well by global-scale

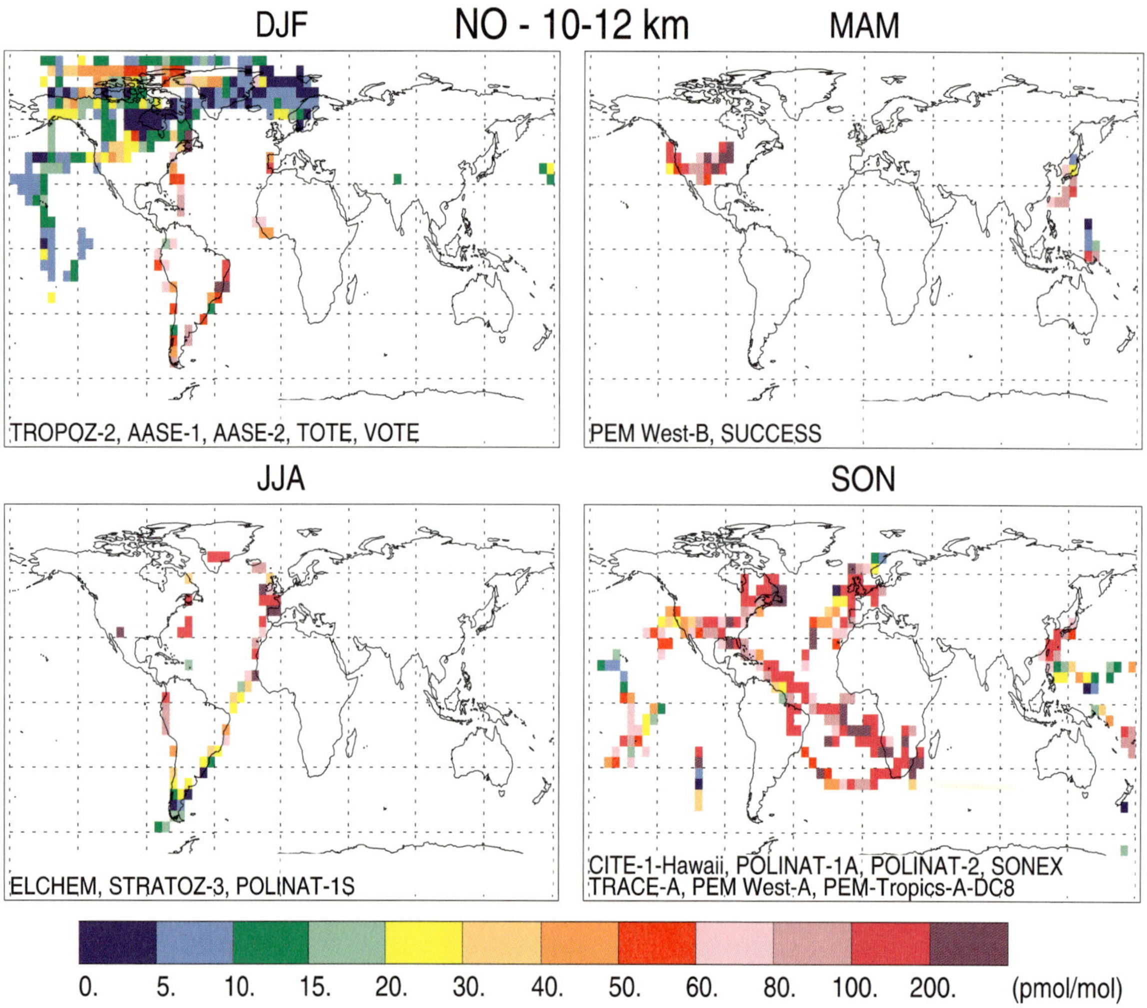

Fig. 3.4 c. Global distribution of NO concentrations at 10–12 km (from *http://aoss.engin.umich.edu/SASSarchive/Species.htm*; L. K. Emmons, NCAR, Oct. 6, 1999)

Table 3.2. Campaigns included in data composites (Emmons et al. 2000)

Campaign	Date	Location
CITE-1-Hawaii	Oct. 19 to Nov. 26, 1983	Hawaii
CITE-1-Ames	April 20 to May 10, 1984	California
STRATOZ-3	June 4–26, 1984	N. and S. Atlantic
STRATOZ-3S	March 11–23, 1985	E. N. Atlantic
ABLE-2A	July 12 to Aug. 13, 1985	Amazon
CITE-2	Aug. 11 to Sept. 5, 1986	N. America
ABLE-2B	April 1 to May 13, 1987	Amazon
TROPOZ-1	Dec. 11–22, 1987	N. and S. Atlantic
ABLE-3A	July 7 to Aug. 17, 1988	Alaska
AASE-1	Jan. 2 to Feb. 15, 1989	Arctic
INSTAC-1	March 5–10, 1989	W. N. Pacific
ELCHEM	July 27 to Aug. 22, 1989	New Mexico
CITE-3	Aug. 22 to Sept. 29, 1989	W. Atlantic
ABLE-3B	July 6 to Aug. 15, 1990	E. Canada
TROPOZ-2	Jan. 9 to Feb. 1, 1991	N. and S. Atlantic
PEM-West-A	Sept. 16 to Oct. 21, 1991	N. Pacific
AASE-2	Jan. 14 to March 20, 1992	Arctic
MLOPEX-2	April 25 to May 11, 1992	Hawaii
TRACE-A	Sept. 21 to Oct. 26, 1992	S. Atlantic
OCTA-1	March 23 to May 21, 1993	N. Atlantic
OCTA-2	Aug. 19 to Sept. 1, 1993	N. Atlantic
OCTA-3	Jan. 12–17, 1994	N. Atlantic
OCTA-4	March 1 to April 26, 1994	N. Atlantic
PEM-West-B	Feb. 7 to March 14, 1994	N. Pacific
ACE-1	Oct. 31 to Dec. 22, 1995	N. and S. Pacific
TOTE	Dec. 6–22, 1995	Tropical N. Pacific
VOTE	Jan. 20 to Feb. 19, 1996	Arctic N. Pacific
SUCCESS	April 15 to May 15, 1996	N. America
PEM-Tropics-A	Aug. 15 to Oct. 5, 1996	Central Pacific

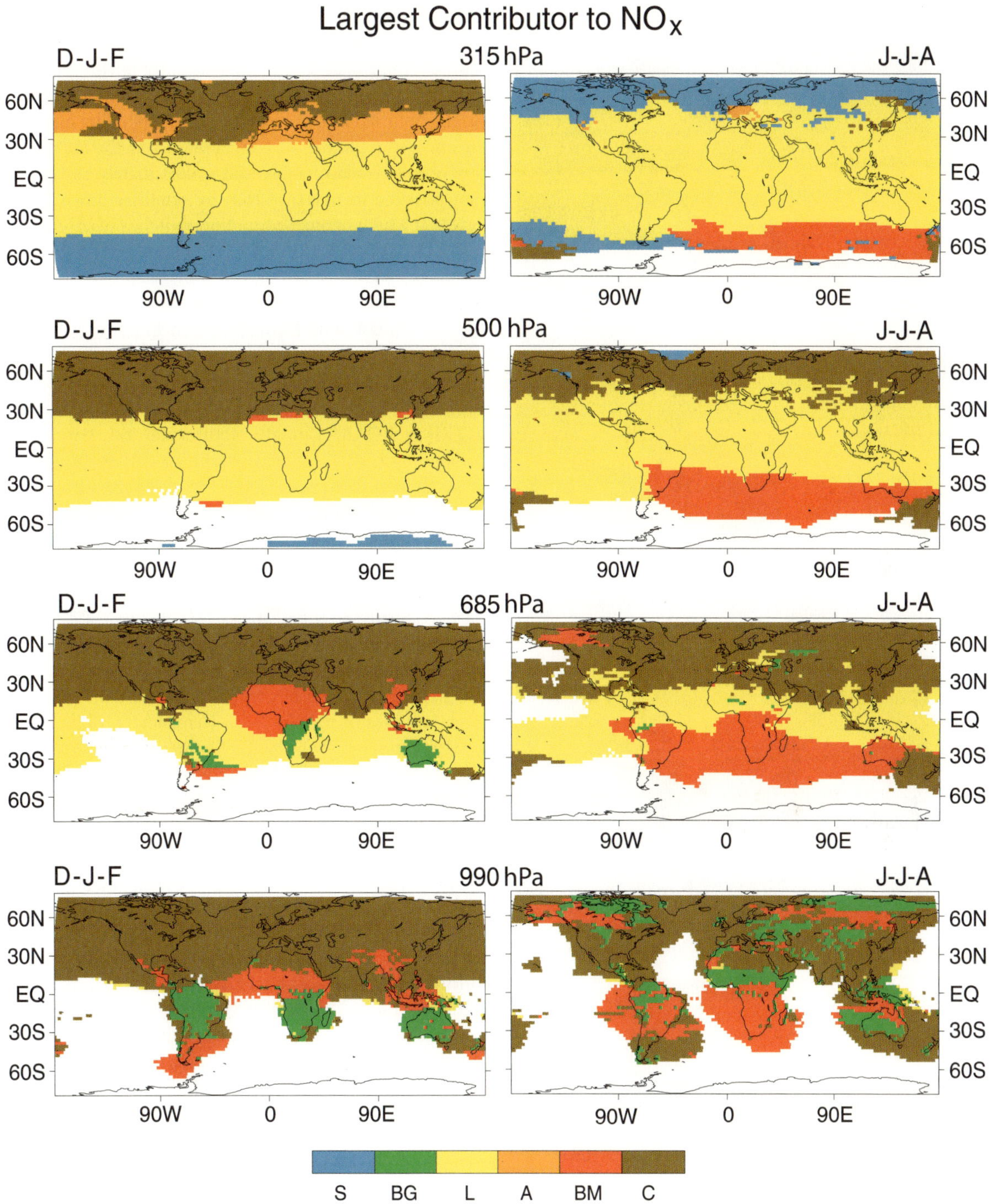

Fig. 3.5. Maps showing largest source contributor to NO_x by season and altitude (*S* = stratosphere; *BG* = biogenic; *L* = lightning; *A* = aircraft; *BM* = biomass burning; *C* = fossil-fuel combustion) (Levy et al. 1999)

models (e.g., Emmons et al. 1997; Wang et al. 1998b; Levy et al. 1999). Based on an analysis of model results, Levy et al. (1999) have delineated the largest source contributing to NO_x in various regions, altitudes, and seasons. Their results (see Fig. 3.5) illustrate the complex controls on tropospheric NO_x as well as the fact that human activities have substantially altered NO_x concentrations in large parts of the atmosphere. It is also worth noting that the model-based analysis of Wang et al. (1998b) provides only limited support for the hypothesis that there is a significant secondary source of NO_x in the remote troposphere associated with a rapid heterogeneous conversion of HNO_3 to NO_x. Schultz et al. (2000) also indicate regions such as the tropical Pacific

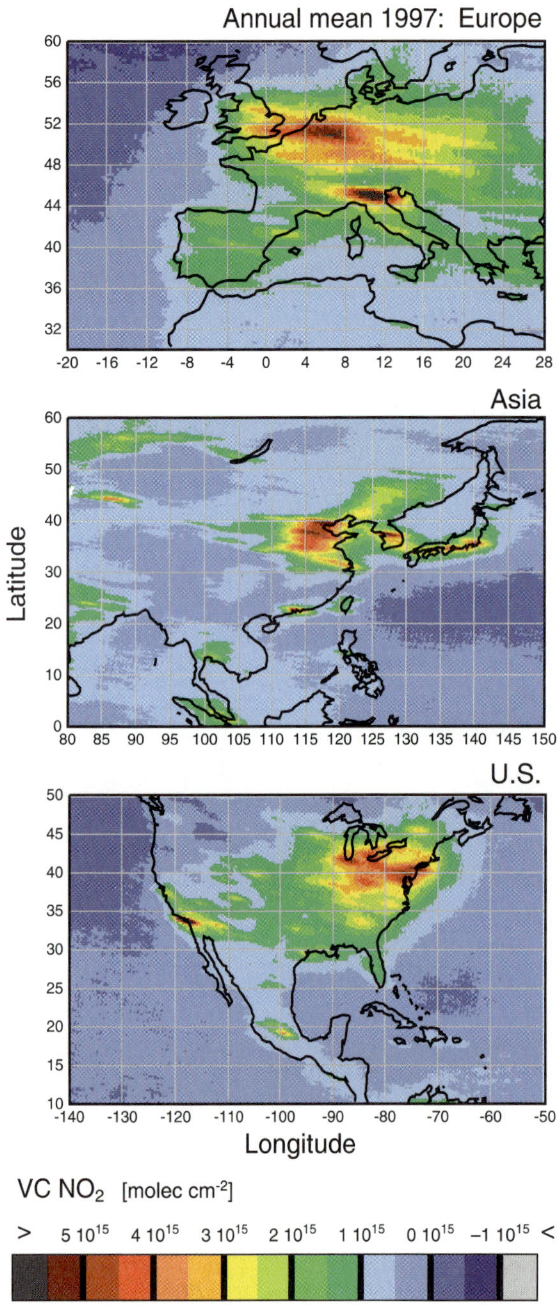

Fig. 3.6 GOME NO_2 distributions over industrialised regions (J. Burrows, University of Bremen, Germany)

will not be conducive to heterogeneous HNO_3 reactions due to the aerosols being dry and fully neutralised.

Since 1995, the Global Ozone Monitoring Experiment (GOME) has been providing satellite measurements with enough spectral resolution that total column nitrogen dioxide (NO_2) and formaldehyde (CH_2O) can be measured. Using a residual technique similar to the methodology described by Fishman et al. (1990), and making the assumption that the stratospheric NO_2 column contribution did not vary as a function of latitude, the depictions shown in Fig. 3.6 have been derived. These three panels show the annual mean distribution of NO_2 over Europe, the US, and southeast Asia. By far the majority of the NO_2 shown in these satellite measurements resides in the boundary layer over the continents. The rapid dilution as the plumes of NO_2 spread over the surrounding oceans is also visible, but unfortunately as yet it is not possible to measure NO_2 by satellites down to the levels at which ozone production switches to ozone loss.

3.2.3.2 *Carbon Monoxide, Methane, and Nonmethane Hydrocarbons*

There exists a well-established monitoring programme for surface CO and CH_4 operated by the Carbon Cycle Group (CCG) at NOAA's Climate Monitoring and Diagnostics Laboratory (CMDL). The geographical distribution of the measurement sites is shown in Fig. 3.7. In addition to the surface sites, this monitoring program makes measurements using tall tower and aircraft platforms at a limited number of locations. The surface measurements reveal distinctive seasonal and latitudinal patterns in CO and CH_4 that are related to the distribution and seasonality of the sources and sinks of these compounds (see Fig. 3.8 and 3.9). As discussed in Novelli et al. (1998), background surface CO mixing ratios are largest (200–225 nmol mol^{-1}) in the high latitudes of the Northern Hemisphere, while lowest mixing ratios are found in the Southern Hemisphere during summer (35–40 nmol mol^{-1}). The absolute seasonal amplitude is also highest in the high northern latitudes and the seasonal cycles in the two hemispheres are out of phase by about six months. CH_4 concentrations generally range from about 1.85 µmol mol^{-1} at high northern latitudes during winter to about 1.68 µmol mol^{-1} at the South Pole during the austral summer.

The CMDL CO and CH_4 measurements also provide information on recent trends in global CO and CH_4 concentrations. The data of Novelli et al. (1998) indicate that background-mixing ratios of CO have shown a downward trend (~2.3 nmol mol^{-1} yr^{-1} on a global- and annual-average basis) over the last decade (see Fig. 3.10). While reasons for this trend are not entirely clear, Novelli et al. (1998) speculate that changes in OH and in sources such as biomass burning have contributed to the overall decrease. While the globally averaged atmospheric CH_4 concentration continues to increase, the rate of increase has decreased (Dlugokencky et al. 1998) (see Fig. 3.11). This decrease in the growth rate has been attributed to a slow approach to a globally averaged steady-state CH_4 concentration of about 1.8 µmol mol^{-1} consistent with relatively constant CH_4 emissions and OH concentrations (Fig. 3.11) although other processes are likely to be operating on shorter timescales.

Unlike the case for CO and CH_4, there is no well-established monitoring network for NMHC. Consequently,

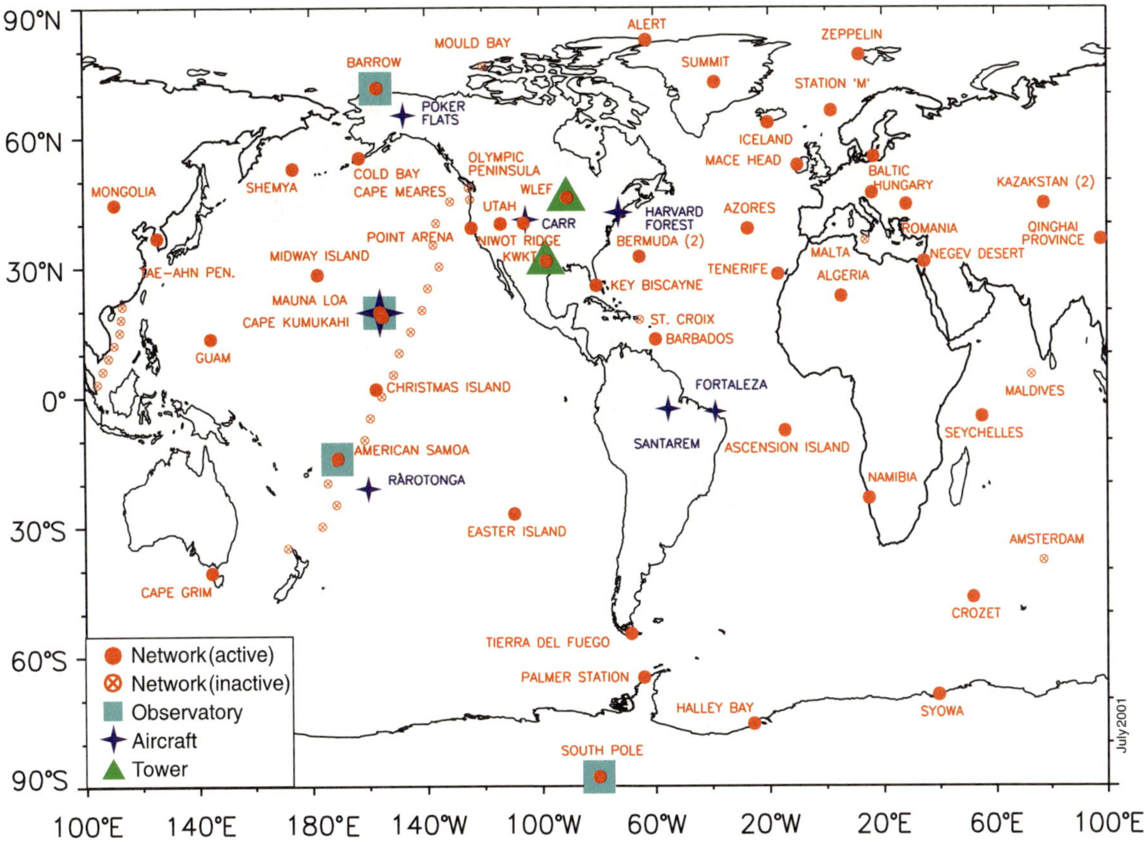

Fig. 3.7. The NOAA/CMDL Carbon Cycle Group Global Air Sampling Network (from *http://www.cmdl.noaa.gov/ccgg/figures/figures.html*)

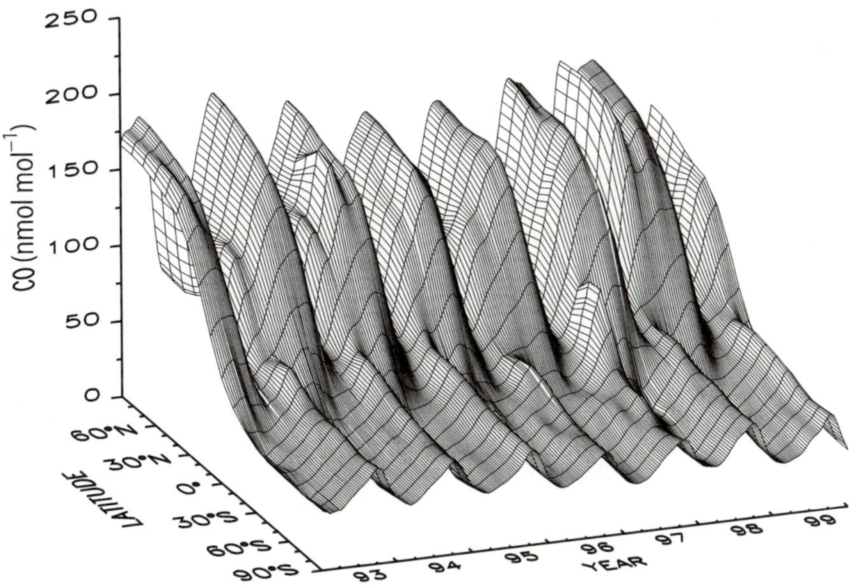

Fig. 3.8. Three dimensional representation of the latitudinal distribution of atmospheric carbon monoxide in the marine boundary layer for the period 1993 through 1997. Data from the NOAA CMDL cooperative air sampling network were used. The surface represents data smoothed in time and latitude (from *http://www.cmdl.noaa.gov/ccgg/figures/figures.html*)

our knowledge of the distribution of NMHC is derived from sporadic field measurements. A summary of a limited set of measurements for selected NMHC compiled by Brasseur et al. (1999a) shows background mixing ratios over continental regions of a few nmol mol^{-1} or less (see Table 3.3). Over oceans, lower concentrations of light NMHC are generally observed though there is evidence of oceanic emissions of NMHC such as ethene and propene (e.g. Rudolph and Ehhalt 1981; Singh and Salas 1982; Rudolph and Johnen 1990; Broadgate et al. 1997). Composites of aircraft observations of NMHC have been made, as for NO discussed above (Emmons et al. 2000).

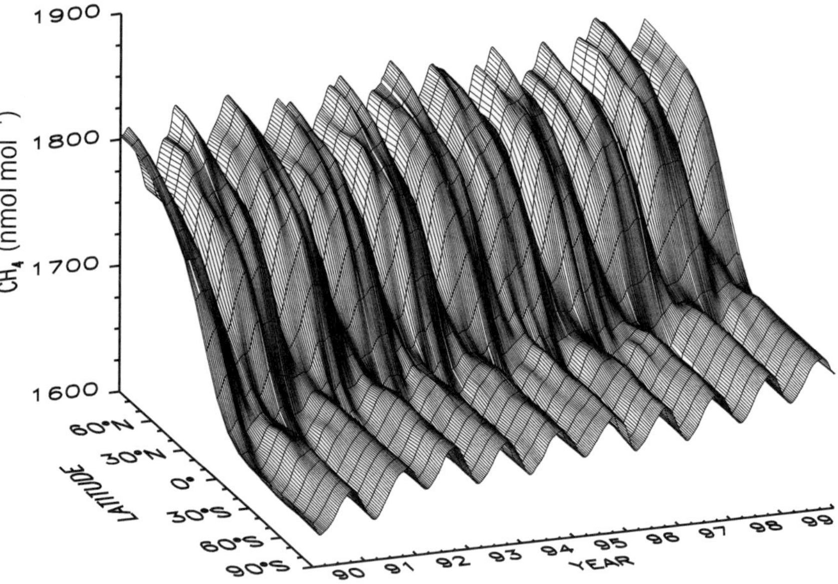

Fig. 3.9. Three dimensional representation of the latitudinal distribution of atmospheric methane in the marine boundary layer for the period 1988 through 1997. Data from the NOAA CMDL cooperative air sampling network were used. The surface represents data smoothed in time and latitude (from *http://www.cmdl.noaa.gov/ccgg/figures/figures.html*)

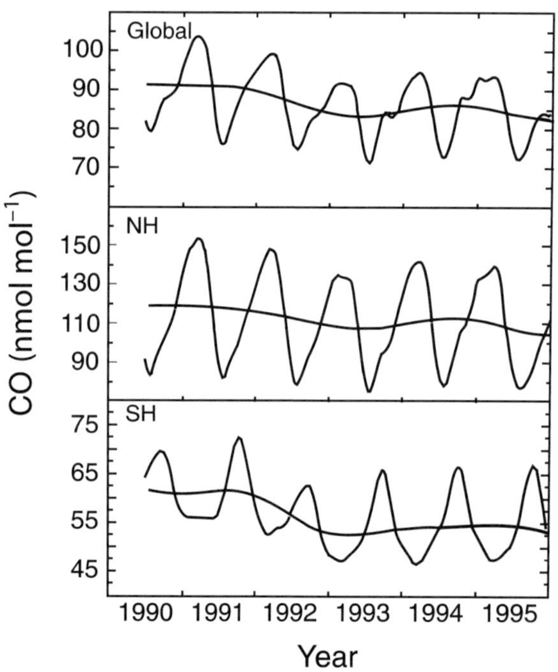

Fig. 3.10. Hemispheric and global time series and trend curves for CO (from Novelli et al. 1998)

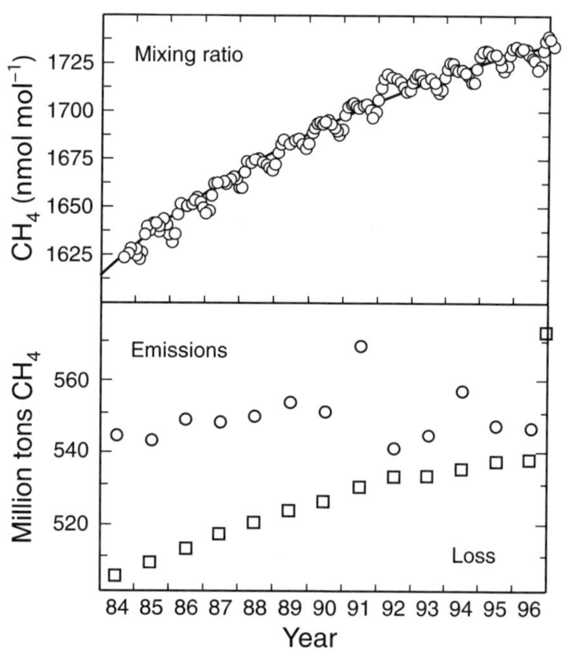

Fig. 3.11. Time variation of global average atmospheric methane (*top*) and calculated emissions and loss assuming a methane lifetime of 8.9 years (*bottom*) (from *http://www.cmdl.noaa.gov/ccgg/news.html*)

Figure 3.12a and b respectively show C_2H_6 at 2–4 km and 6–8 km. The highest values are seen over the continents at 2–4 km, decreasing over the oceans and with altitude. The relatively long lifetime of ethane (weeks in summer) allows it to be transported some distance from its sources, such as over the western Pacific during March–May (PEM-West-B). Mixing ratios above 400 pmol mol^{-1} are seen over the southern Pacific and Atlantic Oceans during September–November, indicating transport of biomass burning emissions from Africa and South America (TRACE-A and PEM-Tropics-A).

In the near future column CO measurements from the Measurement of Pollution in the Troposphere (MOPITT) satellite should become routinely available on a twice-weekly basis (the GOME NO_2 measurements provide monthly depictions, at best). In addition, GOME has demonstrated the capability to measure formaldehyde in regions of enhanced concentrations. Thus, in the early part of the first decade of the new millennium, the capability to observe key tropospheric constituents that contribute to understanding the oxidising capacity of the troposphere will become a reality (see Chap. 5 for more detail).

Table 3.3. Background continental NMHC mixing ratios in several rural and remote terrestrial ecosystems (surface and boundary layer)

Ecosystem	Ethane	Ethene	Propane	Benzene	Isoprene	α-Pinene
Sub-Alpine forest[a]	2.2	0.46	1.27	0.24	0.63	0.14
Tropical rainforest[b]	0.98	0.97	0.37	0.08	2.04	0.10
Tropical rainforest[c]	1.10	0.70	0.16	0.17	5.45	0.20
Tropical rainforest[d]	0.73	0.29	0.10	0.07	1.21	0.06
Wooded savanna[e]	0.65	0.33	0.11	0.16	0.04	<0.01
Wooded savanna[f]	0.26	0.09	0.05	0.04	<0.01	<0.01
Mixed temperate forest[g]	–	–	0.88	0.13	4.33	0.28

[a] Niwot Ridge, CO, USA, August/September, surface (Greenberg and Zimmerman 1984).
[b] Amazon Tropical Forest, Brazil, dry season, boundary layer (Zimmerman et al. 1988).
[c] Amazon Tropical Forest, Brazil, wet season, boundary layer (Zimmerman et al. 1988).
[d] Nigerian Tropical Forest, wet season, surface (Zimmerman et al. 1988).
[e] Kenya, dry season, surface (Greenberg et al. 1985).
[f] Kenya, wet season, surface (Greenberg et al. 1985).
[g] Jacquin, Alabama, July 1990, boundary layer (Guenther et al. 1995).

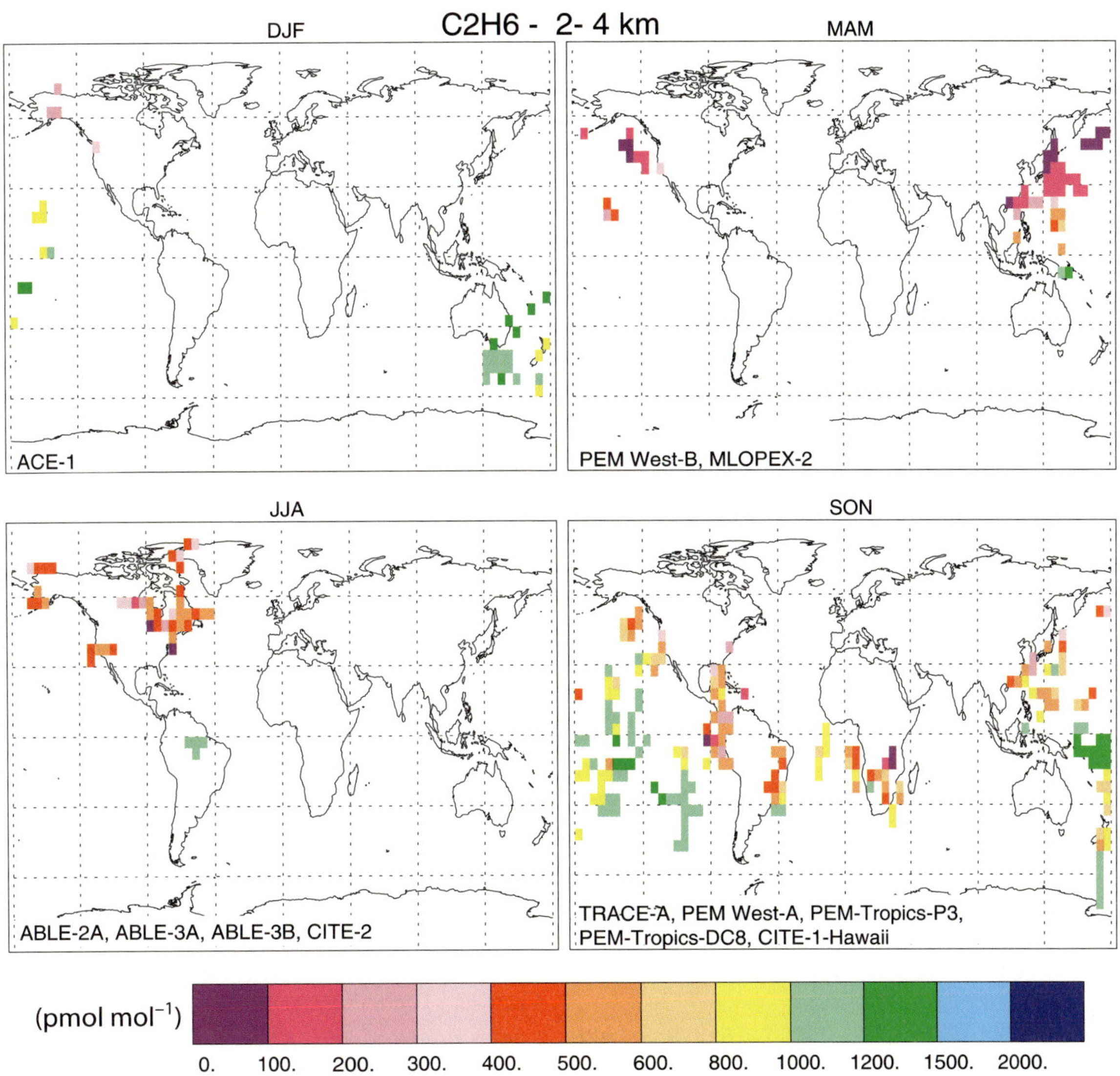

Fig. 3.12 a. Global distribution of ethane at 2–4 km (courtesy of L. Emmons, University of Michigan, and NCAR, on January 20, 1999) (from *http://aoss.engin.umich.edu/SASSarchive/Species.html*)

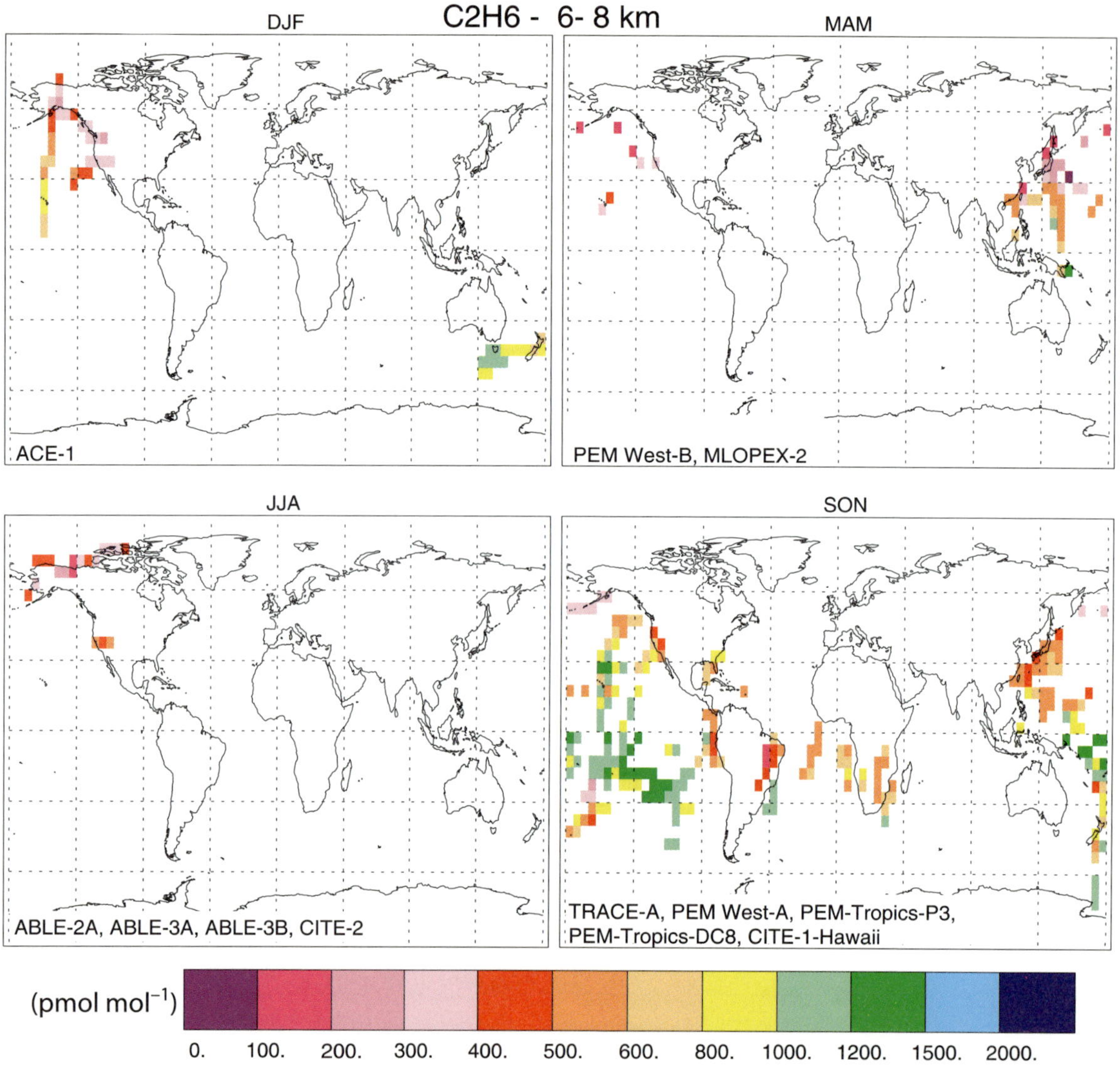

Fig. 3.12 b. Global distribution of ethane at 6–8 km (courtesy of L. Emmons, University of Michigan, and NCAR, on January 20, 1999) (from *http://aoss.engin.umich.edu/SASSarchive/Species.html*)

3.3 Photochemistry in the Troposphere

3.3.1 Background

Photochemical generation of ozone in the troposphere was first identified in the work carried on in California in the '50s (Haagen-Smit 1952) and until the 1970s was thought to be a local phenomenon associated with air pollution. Photons at wavelengths shorter than ~290 nm are not present in the troposphere so that O_2 cannot be photolysed and O_3 cannot be produced via the Chapman mechanism i.e.:

$$O_2 \xrightarrow{h\nu} O + O \quad (\lambda < 242\text{ nm}) \tag{3.0}$$

$$O + O_2 + M \longrightarrow O_3 + M \tag{3.1}$$

However NO_2 can be photolysed in the troposphere, yielding ground state O which recombines with O_2 to form O_3 i.e.:

$$(j_2) \quad NO_2 \xrightarrow{h\nu} NO + O \quad (\lambda < 400\text{ nm}) \tag{3.2}$$

NO also reacts with ozone reforming NO_2:

$$(k_3) \quad NO + O_3 \longrightarrow NO_2 + O_2 \tag{3.3}$$

The coupled reaction cycle 3.1–3.3 leads to the establishment of a photostationary state between NO, NO_2, and O_3 in the sunlit atmosphere on a timescale of ~100 s. Since Reaction 3.1 is effectively instantaneous we then obtain:

$$O_3 = \frac{j_2[NO_2]}{k_3[NO]} \tag{A}$$

Although O atoms are reactive chemically, their concentration is suppressed by reaction with molecular oxygen (Reaction 3.1), and they do not play a significant role in trace gas oxidation. This research pointed to the possible importance of free radicals of the HO_x family (H, OH, HO_2) and related radicals derived from organic species (e.g. CH_3O_2, CH_3O) in atmospheric chemistry. In 1972 Weinstock and Niki proposed that OH radicals led to the atmospheric oxidation of CO:

$$OH + CO \longrightarrow H + CO_2 \quad (3.4)$$

and in 1971 Levy outlined a new theory which predicted significant OH concentrations in the normal sunlit troposphere and pointed out its significance for the chemical removal of many minor constituents. This theory involves production of OH from the small amount of highly reactive excited atomic oxygen, $O(^1D)$, produced by photolysis of O_3 at wavelengths less than $\lambda \sim 320$ nm, which reacts with water vapour, present at high concentrations in the lower troposphere, in competition with quenching to the ground state:

$$(j_5) \quad O_3 \xrightarrow{h\nu} O(^1D) + O_2 \quad (\lambda < 320 \text{ nm}) \ (j5) \quad (3.5)$$

$$(k_6) \quad O(^1D) + M \longrightarrow O(^3P) + M \ (k6) \quad (3.6)$$

$$(k_7) \quad O(^1D) + H_2O \longrightarrow OH + OH \ (k7) \quad (3.7)$$

OH rapidly converts to HO_2, primarily (~70% of the time in unpolluted air) by reaction with CO:

$$OH + CO \longrightarrow H + CO_2 \quad (3.4)$$

$$H + O_2 + M \longrightarrow HO_2 + M \quad (3.8)$$

and at low NO_x concentrations, HO_2 is converted back to OH by reaction with O_3:

$$HO_2 + O_3 \longrightarrow 2O_2 + OH \quad (3.9)$$

About 30% of the OH radicals convert to HO_2 via a more complex chain of reactions initiated by reaction with methane:

$$OH + CH_4 \longrightarrow CH_3 + H_2O \quad (3.10)$$

followed by oxidation of the CH_3 radical to HCHO and HO_2. A steady state between HO_2 and OH is rapidly established in sunlight, allowing us to define the odd hydrogen family $HO_x = [OH] + [HO_2]$. HO_x is lost through formation of H_2O_2 and HNO_3:

$$HO_2 + HO_2 \longrightarrow H_2O_2 + O_2 \quad (3.11)$$

$$OH + NO_2 + M \longrightarrow HNO_3 + M \quad (3.12)$$

These products are highly soluble, and are removed from the atmosphere fairly rapidly by absorption into cloud water and rainout, along with gaseous deposition in the boundary layer. Both H_2O_2 and HNO_3 can also be photolysed, releasing HO_x (and NO_x). Using steady state approximation, a knowledge of the photolysis frequency of ozone, and rate coefficients for the radical loss reaction, Levy (1972) estimated the [OH] present in the sunlit surface atmosphere to be 3×10^6 molecules cm^{-3}.

Crutzen (1973) and Chameides and Walker (1973) first suggested that photochemical generation of ozone might be important in the global troposphere. They pointed out that the oxidation of CO and CH_4 initiated by OH radicals would lead to ozone production, whenever NO_x was present, due to the reactions of peroxy radicals with nitric oxide. NO competes with O_3 for the available HO_2, in an extended reaction cycle:

$$OH + CO(+O_2) \longrightarrow HO_2 + CO_2 \quad (3.8)$$

$$HO_2 + NO \longrightarrow OH + NO_2 \quad (3.13)$$

NO_2 can be photolysed, yielding ozone i.e.:

$$NO_2 \xrightarrow{h\nu} NO + O \quad (3.2)$$

$$O + O_2 + M \longrightarrow O_3 + M \quad (3.1)$$

Thus the net effect of incorporating Reactions 3.8–3.13 is:

$$\text{net:} \quad CO + 2O_2 \longrightarrow CO_2 + O_3$$

so that ozone is produced by the overall oxidation cycle. Note that both OH and NO are regenerated during the cycle.

The oxidation of methane is initiated by:

$$OH + CH_4(+O_2) \longrightarrow H_2O + CH_3O_2 \quad (3.10)$$

The principal fate of CH_3O_2 in a low NO_x environment is:

$$HO_2 + CH_3O_2 \longrightarrow CH_3OOH + O_2 \quad (3.14)$$

CH_3OOH is removed from the atmosphere in a similar way to hydrogen peroxide, leading to a net loss of HO_x. In this situation the oxidation of CO and CH_4 leads to a loss of ozone at a rate equal to the rate of the $O[^1D] + H_2O$ reaction. The reaction of methyl-peroxy with NO leads to the formation of NO_2 (3.15), which leads to further ozone formation via Reaction 3.2 and 3.1 and to formaldehyde which can also be photolysed in the troposphere to liberate HO_x.

$$CH_3O_2 + NO \longrightarrow CH_3O + NO_2 \quad (3.15)$$

$$CH_3O_2 + O_2 \longrightarrow HCHO + HO_2 \quad (3.16)$$

$$HCHO + h\nu \longrightarrow HCO + H \quad (3.17)$$

$$HCO + O_2 \longrightarrow CO + HO_2 \quad (3.18)$$

Thus, the presence of NO completely shifts the balance of the oxidation process in favour of production of both ozone and additional HO_x. There is also a critical concentration of NO which leads to net ozone production and it can be shown to be approximately equal to the ratio of the rate constants for Reaction 3.9 and 3.13 multiplied by the ambient ozone concentration. For a background ozone concentration of 35 nmol mol^{-1} this approximates to an NO concentration close to 10 nmol mol^{-1} (Fishman et al. 1979).

These developments showed the potential importance of photochemistry in determining the budget of ozone in the troposphere and the critical importance of NO_x. Thus there is a sink for ozone (the $O[^1D]$ reaction with H_2O) and a source for ozone (from HO_x/NO_x coupling). The net balance between these processes modifies the ozone fields, which would otherwise be determined only by transport and surface deposition (see Sect. 3.5 for quantitative estimates of the magnitudes of the various chemical and transport-related terms in the ozone budget). The mechanism also explains the anthropogenic perturbation of the ozone amounts by man-made emissions of NO_x and volatile organic compounds (VOC). Quantitative understanding of tropospheric ozone production has been largely dictated by the extent of knowledge of the amounts and distribution of tropospheric NO_x and VOCs which has slowly been improving over the period of IGAC.

Following the introduction of tropospheric free radical theory, much interest was directed to determine the lifetime of pollutant gases due to reaction with OH. Of particular interest were the halocarbons because it was realised that their potential for depleting stratospheric ozone depended on their tropospheric lifetime, which in turn depended on the global mean OH concentration. In the mid 1970s Singh and others used a simple box model of the troposphere together with observational data for methyl chloroform, and an inventory of its industrial production and release, to establish its tropospheric lifetime. This information, combined with a knowledge of the rate coefficient for its reaction with OH, allowed an estimate of the global mean concentration of OH.

Subsequently more sophisticated models of tropospheric chemistry were developed, in which two-dimensional time dependent OH fields were computed using chemical schemes with ozone photochemistry, hydrocarbon and NO_x gas phase chemistry, and physical scavenging by aerosols and rainout. Some of the first examples of these models were reported in 1977 by Crutzen and Fishman, and by Derwent and Curtis (1977). They are now widely used tools for tropospheric research.

At the same time the first attempts to measure OH directly in the atmosphere were being made. Field instruments based on two techniques were developed utilising spectroscopic properties of OH in the ultraviolet region. The first used laser-induced fluorescence (LIF) where the emission was detected from OH following excitation by pulsed laser radiation near 280 nm. This method presented many difficulties for tropospheric conditions, most notably the unavoidable production of OH due to photolysis of O_3 in the incoming air by the probe laser. The second technique utilised long path absorption spectroscopy (LPAS), where the characteristic absorption features of OH in its UV spectrum were detected in a broadband laser beam propagating over a long atmospheric path, up to 10 km. This method lacks intrinsic sensitivity and is subject to interference due to absorptions by other atmospheric gases present at much higher abundance. However calibration of this measurement is straightforward since the absorption cross-sections for OH are well known. Both these methods have subsequently benefitted from technical improvements and during the period of IGAC have provided a substantial amount of data which have demonstrated the basic validity of the tropospheric photochemical theory summarised in Fig. 3.2.

Thus by 1988 at the start of IGAC there existed a well developed theory of tropospheric OH which enabled estimates to be made of local time-varying and global mean concentrations. This provided a working basis for determining the tropospheric lifetime of halocarbons, which were being considered as substitutes for CFCs and of climate gases, e.g. CH_4 and DMS, and the budget of tropospheric ozone. It also provided a mechanism for the efficient production of acidic substances (H_2SO_4 and HNO_3) in the atmosphere and for the production of sulphate aerosols, both by gas-phase and multi-phase methods (see later). The theory was supported by the limited amount of observational data on relevant atmospheric composition available at that time, such as measurements of CH_3CCl_3 and ^{14}CO, and embodied the most up-to-date knowledge of atmospheric chemistry. The first reliable measurements demonstrating the presence of OH in the sunlit atmosphere using the LPAS method had been made (Perner et al. 1976), and indirect detection of peroxy radicals using a chemical amplification technique had been reported (Cantrell and Stedman 1982; Cantrell et al. 1984). However, these observations were neither extensive nor reliable enough to validate the photochemistry schemes being used in models. The nitrate radical had been detected in the night-time boundary layer, but its reactivity was considered to be low. Oxidation by halogen radicals was not considered an issue in tropospheric chemistry.

More importantly there was no consensus as to the extent to which photochemical processes were active in determining tropospheric ozone amounts, and hence tropospheric oxidising efficiency. A major role perceived for IGAC at the outset was to extend knowledge of the tropospheric free radical chemistry needed to formulate models with better diagnostic and predictive power for global change issues involving tropospheric ozone and chemically reactive greenhouse gases.

3.3.2 IGAC Activities Related to Testing the Theory of HO_x Photochemistry

One of the main aims of IGAC was the creation and coordination of major field campaigns to provide observational data on atmospheric chemical species to test theory. Photochemical oxidant production in the troposphere was the major focus of a number of large field projects worldwide. These initiatives also stimulated some major improvements in the instrumental capability for measurement of trace species involved in the photochemical ozone balance, and many important observational campaigns have been carried out in remote areas (Ridley et al. 1987; Cantrell et al. 1996c,d; Jaffe et al. 1996; Carsey et al. 1997). In the first years of IGAC the emphasis was on precursor species (hydrocarbons, nitrogen oxides) and the reservoir species for HO_x and NO_x (peroxides, nitrates, HNO_3, and PAN). In recent years, measurements of HO_x free radical species, oxygenated VOCs and other radicals such as NO_3 and halogen oxides have also been achieved. Measurements have been made at meteorologically and climatologically well-characterised ground observation stations as well as from ships and aircraft. Airborne measurements have expanded recently, allowing concentrations of HO_x and related species to be determined throughout the free troposphere. These have provided new opportunities to test the photochemical theories of tropospheric free radical chemistry and its relation to the net production and loss of ozone throughout the atmosphere. These activities have made a significant and prominent contribution to the overall success of IGAC.

3.3.2.1 *HO_2/OH Measurement Campaigns*

The following summary traces the progress in atmospheric HO_x radical measurements. Successful direct measurement of HO_x (OH and HO_2) at a remote site at high altitude on the Island of Hawaii was accomplished in 1991 and 1992 during the MLOPEX 2 experiments. Maximum concentrations of OH of 4 to 6×10^6 molecules cm^{-3} were measured using SICIMS (Eisele et al. 1996), which was a factor of two or three lower than modelled values. Other aspects of the MLOPEX experiment, which was one of the first to characterise local ozone photochemistry in the remote troposphere, are published in special sections of the Journal of Geophysical Research (e.g. Atlas and Ridley 1996; Hauglustaine et al. 1999).

Two important intercomparison campaigns took place in the US in 1993 and in Germany in 1994. The TOHPE intercomparison at Idaho Hill in the US involved measurement of OH by SICIMS, LIF (OH / HO_2), and LPAS. Only two systems worked satisfactorily; a fourth OH technique was used but the data were not published. This was the wet chemical salicylic acid method. Problems were encountered with the interpretation of long path OH absorption data. The calculated [OH] was typically a factor of one and a half greater than observed and the calculated [HO_2] was up to a factor of ten greater than observed. The HO_2/ OH ratio was dependent on local NO_x; ratio values of 15–18 were observed for $NO_x > 100$ pmol mol^{-1}, which agreed well with model calculations, while for clean air the ratios (three to four) were lower than predicted (Eisele et al. 1997). The POPCORN campaign involved intercomparison between a FAGE-LIF and LPAS in a multi-reflection probe at a rural site in North Germany. Good agreement between the different techniques was observed (see Chap. 5). The recorded maximum [OH] was 1×10^7 molecules cm^{-3} and [OH] correlated well with $j_{(O^1D)}$ (see Fig. 3.13) (Dorn et al. 1996; Hofzumahaus et al. 1996, 1998; Plass-Dülmer et al. 1998).

As further measurements of OH and HO_2 at ground level have been conducted in recent years, the quality of OH measurements has improved dramatically, giving much lower detection limits and freedom of interferences. Full diurnal profiles are now measured routinely. Several intercomparisons have now been made between physically distinct techniques, but there have been no blind intercomparisons of HO_x measurements.

The first reported aircraft measurements of OH in the lower atmosphere were conducted in 1995 as part of the ACE 1 campaign in the Southern Ocean in the vicinity of Tasmania and New Zealand (Mauldin et al. 1998a). A SICIMS instrument mounted in an aircraft was successfully deployed. An important observation was the much higher [OH] (8 to 15×10^6 molecules cm^{-3}) above clouds compared to clear skies (3 to 5×10^6 molecules cm^{-3}). Comparison of results with model calculation showed greatest errors in clouds and in the marine boundary layer (Mauldin et al. 1997, 1998b).

The first aircraft measurements of OH and HO_2 in the upper troposphere were made using the LIF technique in the STRAT campaigns in 1995–96 (Jaeglé et al. 1997). The concentrations of OH and HO_2 were substantially higher than predicted by models, in which the $O(^1D) + H_2O$ reaction was the dominant source of radicals, leading to the suggestion that acetone photolysis was a major OH source in the dry upper troposphere-lower stratosphere. Acetone has been found to be a wide-

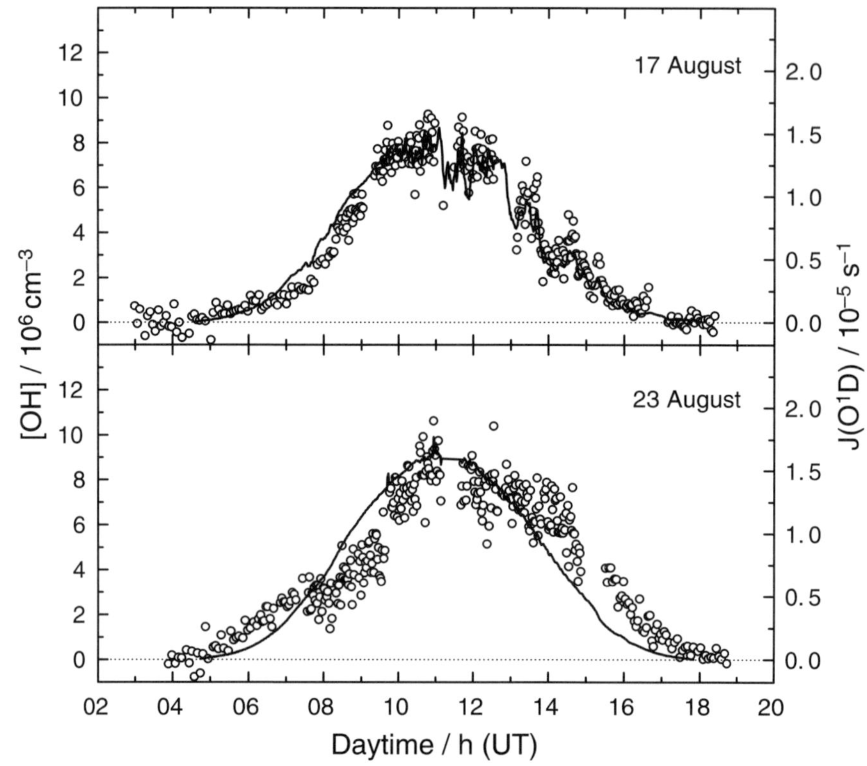

Fig. 3.13. Relationship between diurnal variation of [OH] and $j[O^1D]$ (Hofzumahaus et al. 1996)

spread constituent of the troposphere and is believed to result from degradation of various volatile organic compounds, both natural and man-made. For example concentrations of oxygenated organics (including methanol, acetone, acetaldehyde, and formaldehyde, CH_3OOH) in PEM-Tropics B were found to be considerably higher than those of non-methane hydrocarbons (NMHCs) and to play an important photochemical role. They showed relatively little spatial gradients and were only weakly correlated with pollution. Better understanding of the factors controlling the abundances of these oxygenated organics emerges as a major issue in the wake of PEM-Tropics B.

Sources of OH other than ozone photolysis in the mid-upper troposphere were also indicated by airborne measurements of OH and HO_2 by LIF in the SUCCESS campaign conducted in 1996 (Brune et al. 1998). Observed midday [OH] was 2 to 10×10^6 molecules cm^{-3}, and $[HO_2]$ was 6 to 30×10^7 molecules cm^{-3}. [OH] was sensitive to cloud and the HO_2/OH ratio, measured in the upper troposphere in clean air and in aircraft contrails, was observed to be a function of NO_x, and varied from >100 at low NO_x (<20 pmol mol^{-1}) to ~2 at very high NO_x (~1 nmol mol^{-1}). The agreement of the ratio with a model was good. Further airborne measurements with a SICIMS instrument, conducted in 1996 on the PEM-TROPICS A campaign, showed midday marine boundary layer [OH] of 2 to 10×10^6 cm^{-3} with higher levels (× 2–3) seen above cloud (Mauldin et al. 1999).

In 1997 the Subsonic Assessment: Ozone and Nitrogen Oxide Experiment (SONEX) assembled the most complete measurement complement to date for studying HO_x (OH and HO_2) chemistry in the free troposphere. This mission included flying in contrails in the North Atlantic flight corridor. The measured HO_x, and also the measured HO_2, depended upon NO_x. Observed and modelled HO and HO_2 agree on average to within experimental uncertainties (see Fig. 3.14). However, significant discrepancies occur as a function of NO and at solar zenith angles, some of which appear to be removed by model adjustments to HO-NO chemistry (Brune et al. 1999; Jaeglé et al. 1999).

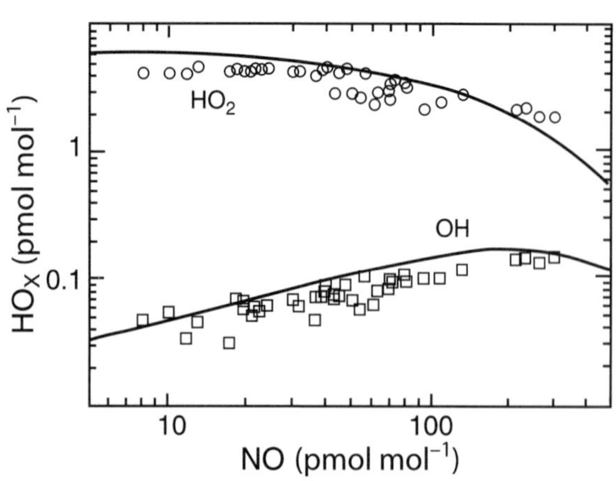

Fig. 3.14. Calculated and observed OH and HO_2 as a function of [NO] (data from SONEX experiment; Brune et al. 1999)

There is now a considerable database for OH at many surface locations with differing characteristics and also in the free troposphere from aircraft. Recently the campaigns have provided comprehensive supporting data (e.g. CO, CH_4, CH_2O, $j_{(O^1D)}$, $j_{(NO_2)}$), other photolysis rates from spectroradiometric measurements, and aerosol size and surface area, enabling case study modelling to be performed in order to validate theory.

3.3.2.2 *Comparison Between Observations and Models*

3.3.2.2.1 *Ground*

In general the models overpredict observed (OH) and (HO_2) measured at the surface (e.g. TOPHE, ACSOE campaigns). A number of possible reasons have been suggested, including heterogeneous losses, missing gas phase sinks, poor co-location of instruments, etc. In SOAPEX-2, preliminary modelling shows that under baseline conditions, model and experiment agree to within 20%, but agreement deteriorates (overprediction of (OH) and (HO_2)) as air masses influenced by continental sources move over the site at Cape Grim, Tasmania (Cox et al. 1999). This suggests that the primary production of HO_x in clean air is accurate, but the secondary production of radicals, e.g. via $HO_2 + NO$, is not correctly modelled. The opposite behaviour was found in AEROBIC, a forested site, where the model underpredicts measured OH and HO_2, which is thought to be due to additional HO_x sources from O_3-terpene reactions, which may be underestimated in models. However the mechanism is likely to be incomplete, as all terpenes are converted to equivalent α-pinene. The modelling of OH and HO_2 observed in the polluted air of the Los Angeles basin seems remarkably good considering the simple parameterised treatment of VOC (George et al. 1999).

3.3.2.2.2 *Airborne*

Initial model-experiment comparisons with airborne measurements showed deficiencies in OH and HO_2. When acetone-peroxide sources of HO_x are included in the models following PEM Tropics-B agreement is reasonably good in the remote upper troposphere-lower stratosphere, but in and around clouds there are discrepancies. This is attributed to poorly measured actinic flux above and missing in-cloud heterogeneous processes such as HO_2 loss. In SONEX, measurements were made in contrails of aircraft in the North Atlantic flight corridor. A huge variation in NO_x was observed, and OH and HO_2 were observed to be a strong function of NO_x. Clearly models need to calculate accurate NO_x fields in order to represent HO_x.

3.3.3 Measurements of Total Peroxy Radicals ($HO_2 + RO_2$)

Measurement of the atmospheric steady state concentrations of peroxy radicals provides a direct insight into the local production or loss rate of ozone, as well as a diagnosis for HO_x photochemistry. Improved field measurements of RO_2 have been an objective of several projects in IGAC. Two techniques are used: the PERCA (Peroxy Radical Chemical Amplifier) and MIESR (Matrix Isolation Electron Spin Resonance), and much effort has been made in obtaining reliable field intercomparison measurement using these techniques. PERCA makes no distinction between RO_2 (organic peroxy radicals) and HO_2, whereas the MIESR method can give information on specific peroxy radicals, although with low resolution. Stand-alone deployments suffer from uncertainty in absolute calibration. There are still a very limited number of ground level HO_2 measurements mainly using FAGE. The measurement is not as straightforward as for OH.

3.3.3.1 *Field Measurements of Peroxy Radicals*

To date only terrestrial surface measurements have been reported for RO_2 and these have been mainly conducted using the PERCA technique and consequently do not distinguish between HO_2 and organic peroxy radicals. In the US measurements have been made at a forested location in Alabama (ROSE I Campaign, Cantrell et al. 1992), at Mauna Loa (Cantrell et al. 1996a,b), and in Colorado (TOPHE Campaign, Cantrell et al. 1997a). In the latter campaign HO_2 was also measured by LIF and the ratio $[RO_2]/[HO_2]$ was typically a factor of ten larger than predicted by case study models.

European groups have conducted an intercomparison of five different PERCA and MIESR instruments measuring RO_2 radicals at Schauinsland in Germany (Heitlinger et al. 1996). The results show no systematic difference between PERCA instruments although MIESR measured systematically higher.

PERCA instruments have been deployed at a coastal location in Tasmania (SOAPEX-1 campaign, Monks et al. 1996; Penkett et al. 1997) and in Ireland (ATAPEX, Carpenter et al. 1999, and ACSOE-EASE campaigns, Salisbury et al. 2001). These results show clear confirmation of simple photochemical theory for HO_x production in clean and polluted air situations, in terms of the dependence of RO_2 on the actinic UV radiation for $O(^1D)$ production from O_3 photolysis.

Measurements of RO_2 have also been conducted at the Jungfraujoch, Switzerland (Zanis et al. 1999), allowing seasonal variation of HO_x chemistry in the free troposphere to be established, after account has been

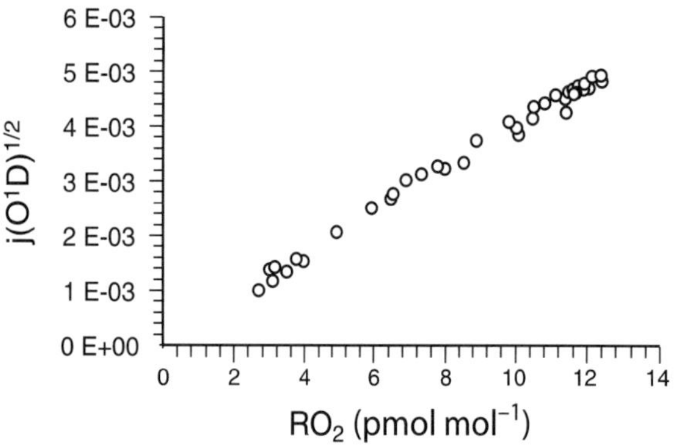

Fig. 3.15. Plot showing the linear dependence between the total peroxy radical concentration [RO_2] and $j[O^1D]^{1/2}$ for baseline conditions at Cape Grim, Tasmania (Penkett et al. 1997)

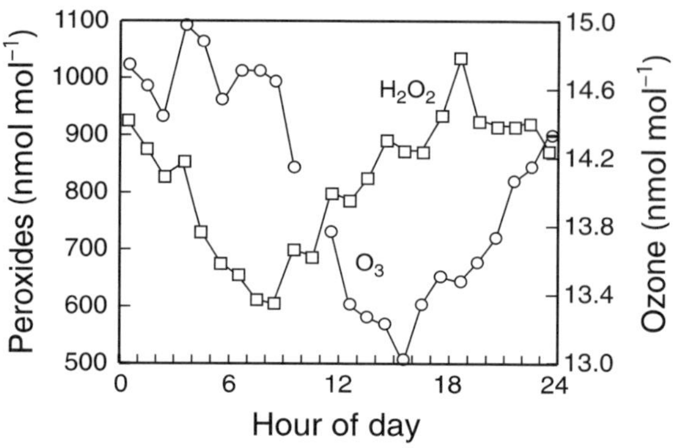

Fig. 3.16. Diurnal variation in ozone and total hydroperoxide (ROOH = $CH_3OOH + H_2O_2$) for baseline conditions at Cape Grim, Tasmania (Ayers et al. 1992)

taken of upslope-downslope effects, first observed in MLOPEX (Cantrell et al. 1997b).

Overall the results from the field measurements of RO_2 are broadly consistent with photochemical theory of HO_x radicals in the troposphere. However, difficulty arises when the detailed amounts of HO_2 and RO_2 (total peroxy radicals) are examined. Some calibration and sampling issues remain with the PERCA technique. In comparison with case study models, observed RO_2 concentrations in the surface atmosphere are mostly lower than calculated in the models. Other points which emerge from the measurements are: (*1*) RO_2 radicals are present at night-time; (*2*) Total peroxy radical concentration shows a square root dependence on $j_{(O^1D)}$ in clean air, as expected from theory (Fig. 3.15); this relationship changes towards linear dependence as NO_x increases; and, (*3*) Measurements are often complicated by effects of local emissions and diurnal flows.

3.3.4 Measurements of Ozone and Peroxide Climatologies in Clean Air

3.3.4.1 *Evidence for Ozone Photochemistry*

Long-term measurements and intensive measurement campaigns at meteorologically and climatologically well-characterised ground observation stations have provided new opportunities to test the photochemical theories of tropospheric ozone. Sites in the Southern Hemisphere (Cape Grim, Tasmania, 40.7° S, 144.7° E) and in the Northern Hemisphere (the Oki Islands in the Sea of Japan, 36° N, 133° E, and Mace Head, Ireland, 53.3° N, 9.9° W) have been exploited in this way (Ayers et al. 1992, 1996b; Penkett et al. 1997; Carpenter et al. 1997; Jaffe et al. 1996; Kanaya et al. 1999). These sites are exposed to unpolluted maritime air for long periods, interspersed with periods when more or less polluted continental air advects into the region. The homogeneity of the marine air masses in terms of chemical composition provides optimum conditions for testing atmospheric photochemistry in the surface atmosphere; also the contrast between the behaviour of maritime and continental air provides information on the perturbation of the unpolluted atmosphere by terrestrial emissions, especially man-made pollutants.

This approach has been particularly successful at Cape Grim where Ayers et al. (1992, 1996b) have observed diurnal modulation of ozone and hydrogen peroxide concentrations which are consistent with a simple model in which O_3 photolysis depletes ozone in the marine boundary layer (MBL) during daytime with associated production of peroxides (H_2O_2 and organic peroxides, mainly CH_3OOH) from the free radical chemistry (Fig. 3.16). Changes in patterns when polluted air is advected over the Cape Grim site and also at Mace Head are quantitatively consistent with photochemical theory (Carpenter et al. 1997). This includes experimental determination of the critical NO concentration required to switch from ozone destruction to ozone production in marine air (see also Sect. 3.7).

3.3.4.2 *Indications of the Importance of Water Vapour in Tropospheric Free Radical Chemistry*

Hydrogen peroxide is one of the major products of HO_x recombination reactions and as such it is a good indicator of the extent of tropospheric free radical chemistry. The controlling role of water vapour on the chemistry in the troposphere is shown rather well in some measurements of H_2O_2 and water vapour over the Pacific and Atlantic Oceans between the surface and 8 km (Penkett et al. 1995, 1998). The left-hand panel of Fig. 3.17 shows that these two species are highly correlated in the vertical profiles. The reason is revealed in Expression B below for the equilibrium concentration of hydrogen peroxide produced by HO_x chemistry in the absence of NO_x assuming that all the peroxy radicals are produced from the oxidation of CO by OH (Penkett et al. 1998).

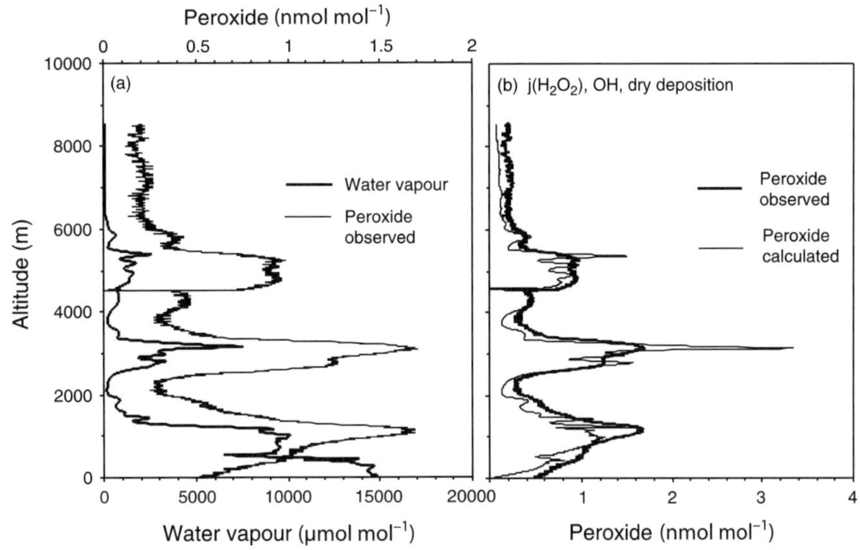

Fig. 3.17. Vertical profiles of peroxide and water vapour and a comparison of measured and calculated peroxide between the surface and 7.5 km. The measurements were collected in the OCTA campaign in September 1994, which was a component of IGAC/NARE (Penkett et al. 1998)

$$[H_2O_2] = \frac{k_7 j_5 [O_3][H_2O]}{k_6[M](j_{20} + k_{19}[OH] + k_{het})} \qquad \text{(B)}$$

where hydrogen peroxide is removed by the following reactions:

$$(k_{20}) \quad H_2O_2 + OH \longrightarrow H_2O + HO_2 \qquad (3.19)$$

$$(j_{20}) \quad H_2O_2 + h\nu \longrightarrow OH + OH \qquad (3.20)$$

$$(k_{het}) \quad H_2O_2 + \text{surface} \longrightarrow H_2O + ½O_2 \qquad (3.21)$$

There is often good agreement between measured peroxide concentrations and those calculated using Expression B (right-hand panel of Fig. 3.17). This only occurs in low NO_x conditions when ozone is being destroyed by the HO_x chemistry rather than being created, and there is an anticorrelation between ozone and peroxide in the vertical profile. In conditions where a positive correlation is observed between ozone and peroxide in the profiles, due to both being produced simultaneously by the free radical chemistry outlined earlier, the close agreement between calculated and measured peroxide shown in Fig. 3.17 disappears (Penkett et al. 1998).

3.3.5 Reactive Nitrogen Chemistry

Nitrogen oxides are emitted into the atmosphere in the form of NO and NO_2 (see Sect. 3.2). These are chemically modified by HO_x chemistry and photolysis with the formation of ozone, through Reactions 3.2 and 3.3 and, the simultaneous formation of many other forms of oxidised nitrogen compounds. These include nitric acid formed by Reaction 3.12 and many organic forms including peroxyacetyl nitrate (PAN) and organic nitrates which are formed in the course of oxidation of hydrocarbons and carbonyls. The sum total of NO and NO_2 is referred to as NO_x, and the sum total of all oxidised nitrogen compounds (except N_2O) is referred to as NO_y. NO_z is the difference between NO_y and NO_x (see Fig. 3.2), and is often highly correlated with ozone in the troposphere (Dentener and Crutzen 1993; Jaeglé et al. 1999).

PAN results from the oxidation of various carbonyls including acetaldehyde and acetone.

$$CH_3CHO + OH(+O_2) \longrightarrow CH_3C(O)OO + H_2O \qquad (3.22)$$

$$CH_3COCH_3(+O_2) \xrightarrow{h\nu} CH_3C(O)OO + CH_3 \qquad (3.23)$$

$$CH_3C(O)OO + NO_2 \longleftrightarrow CH_3C(O)OONO_2 \text{ (PAN)} \qquad (3.24)$$

with, for Reaction 3.24, a ratio between the forward and reverse reaction rate of

$$k_f / k_r = 1.11 \times 10^{28} \exp(-14\,000 / T) \text{ molecule cm}^{-3}$$

Other commonly observed molecules in this series are peroxypropionyl nitrate (PPN, $C_2H_5C(O)OONO_2$), peroxybenzoyl nitrate (PBzN; $C_6H_5C(O)OONO_2$), and peroxymethacryloyl nitrate (MPAN; $CH_2 = C(CH_3)C(O)OONO_2$) (Williams et al. 2000). A unique property of PANs is that they are not directly emitted from any known source. They are all products of atmospheric photochemical reactions involving hydrocarbons and nitrogen oxides (NO_x). This makes them excellent indicators of photochemical activity. As is shown above, PAN is in thermal equilibrium with NO_2. However, the equilibrium rate constant is an extremely strong function of temperature. At warm temperatures, as may be encountered in the boundary layer, PAN has a lifetime of only a few hours and the equilibrium is shifted towards NO_2. At colder temperatures, as may be encountered in the upper troposphere, PAN is essentially thermally stable.

These properties allow PAN-like materials to be effective reservoirs and carriers of atmospheric NO_x (Singh 1987; Roberts et al. 1996). Other reservoir species such as HO_2NO_2 are also formed but they are stable only in the cold temperatures of the upper troposphere (>7 km).

$$HO_2 + NO_2 \longleftrightarrow HO_2NO_2 \quad (3.25)$$

More stable organic nitrates are also formed via the reaction:

$$RO_2 + NO \longrightarrow RONO_2 \quad (3.26)$$

This process is in competition with the reaction that converts NO to NO_2 ($HO_2 + NO \longrightarrow OH + NO_2$) and a higher percentage of $RONO_2$ is made as the carbon number of the RO_2 radical increases. Alkyl nitrates are relatively less abundant in the atmosphere and photolyse extremely slowly. While they are not significant reservoirs of NO_x, they do provide useful information as tracers of photochemical reactions.

An important role for the HO_x radicals is in the removal of reactive nitrogen from the atmosphere. These radicals can quickly oxidise NO_2 to HNO_3 that can be removed from the atmosphere via processes of dry and wet deposition. In the upper troposphere where the lifetime can be of the order of a week, HNO_3 can be converted back to NO_2 prior to its removal by deposition.

$$HNO_3 \xrightarrow{h\nu} OH + NO_2 \quad (3.27)$$

Thus while NO may be emitted as a simple molecule, it is quickly converted to a complex set of species such as PAN, HNO_3, HO_2NO_2, N_2O_5, $RONO_2$, RO_2NO_2, and aerosol nitrate (Singh 1987; Roberts et al. 1996). A full understanding of the reactive nitrogen chemistry is made difficult by the fact that many of the products are hard to measure, have varying lifetimes, and under certain conditions can act as reservoirs and provide secondary sources of NO_x. In spite of this a rather complete set of data on the NO_y composition of the upper troposphere was obtained by the SONEX experiment (Talbot et al. 1999) where nearly 90% of total NO_y as measured by the instrument was accounted for by individual measurements of the species making up NO_y (Fig. 3.18). These included particularly NO_x, PAN, organic nitrates, and HNO_3.

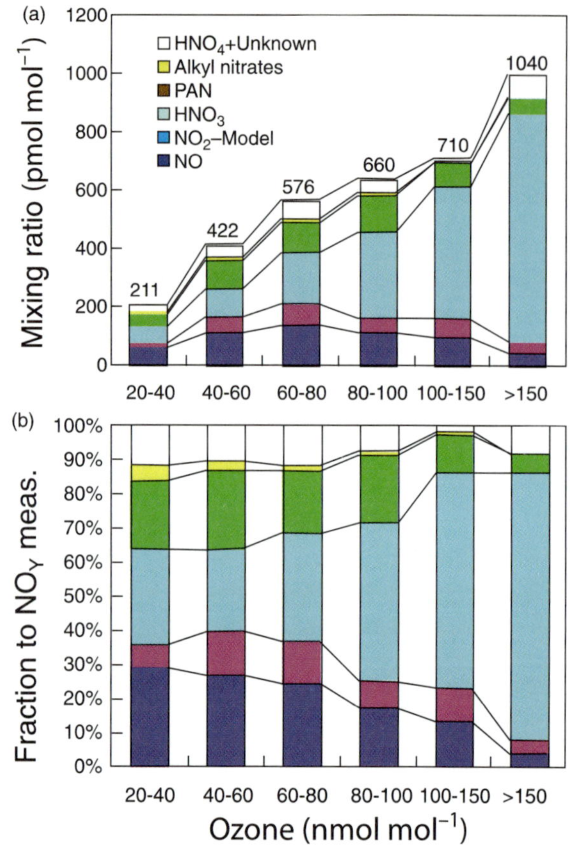

Fig. 3.18. Reactive nitrogen budget during the NASA SONEX mission (Talbot et al. 1999)

3.3.6 The Role of Tropospheric Chemistry in the Production of Particulate Matter in the Troposphere

Particles of different sorts (see Chap. 4) are formed in the atmosphere by chemical transformations to low volatility products. Examples include nitrates formed from the reaction of N_2O_5 with water, organic aerosols formed from the oxidation of terpenes and other naturally emitted hydrocarbons, and sulphates formed from the oxidation of SO_2. The latter reaction has been extensively studied in the course of increasing understanding of the problem of acid rain, which occurred before the setting up of IGAC (National Acad. of Sciences, Report on Acid Rain Formation 1985).

The main reaction responsible for gas phase oxidation of SO_2 is:

$$OH + SO_2 \longrightarrow HSO_3 \quad (3.28)$$

$$HSO_3 + O_2 \longrightarrow SO_3 + HO_2 \quad (3.29)$$

$$SO_3 + H_2O \longrightarrow H_2SO_4 \quad (3.30)$$

All other gas phase processes are of little significance. Of major significance are the reactions of SO_2 dissolved in droplets. Here the hydrolysis products HSO_3^- and SO_3^{2-} are oxidised efficiently by peroxides and by ozone to produce sulphuric acid and neutralised sulphate in cloud droplets, so much so that up to 80% of all atmospheric sulphate has been estimated to be produced this way (Langner and Rodhe 1991).

$$HSO_3^- + H_2O_2 \longleftrightarrow H_2SO_3O^- + H_2O \quad (3.31)$$

$$HSO_3O^- + H^+ \longrightarrow H_2SO_4 \quad (3.32)$$

also

$$HSO_3^- + O_3 \longrightarrow HSO_4^- + O_2 \quad (3.33)$$

and

$$SO_3^{2-} + O_3 \longrightarrow SO_4^{2-} + O_2 \quad (3.34)$$

The process of formation of sulphuric acid by reaction with peroxide is acid catalysed, which overcomes the tendency for SO_2 to become less soluble due to suppression of hydrolysis as the cloud droplets become more acid. In the case of the ozone reaction the rate of Reaction 3.35 (see Sect. 3.1.3.9) is several orders of magnitude faster than Reaction 3.34 meaning that ozone oxidation is most efficient in neutral or basic solutions. The switch-over point for the peroxide or the ozone reaction to predominate occurs at a pH slightly above 5, which is higher than the pH of most cloud droplets (Penkett et al. 1979). For a fuller discussion the reader is referred to Calvert et al. (1985) and references therein. The significance of this multiphase oxidation of SO_2 is also described in more detail in Chap. 4 concerned with atmospheric aerosols.

3.3.7 Modelling of Radical Chemistry and Ozone Production and Loss

Because the time constants for radical photochemistry are short compared with transport times, photochemical theory can be readily tested by comparison to OH and RO_2 measurements using simple zero-dimensional models, providing the boundary condition imposed by the concentrations of relevant stable constituents can be defined. Numerous examples of modelling studies associated with the OH and HO_2 measurements have been reported. The more recent HO_x measurements, with fast time resolution, compare well with the general features of the photochemical theory (see for example Hofzumahaus et al. 1998; Carslaw et al. 1999a, b). Ayers et al. (1997) were able to describe the peroxy radicals, ozone, and hydrogen peroxide observed at Cape Grim with a simple one-dimensional model of the marine boundary layer.

Box modelling has also been used to investigate the critical nitric oxide concentration NO_{crit}, the so-called "compensation point," where ozone loss and production due to reactions:

$$HO_2 + O_3 \longrightarrow 2O_2 + OH \quad (3.9)$$

$$HO_2 + NO \longrightarrow OH + NO_2 \quad (3.13)$$

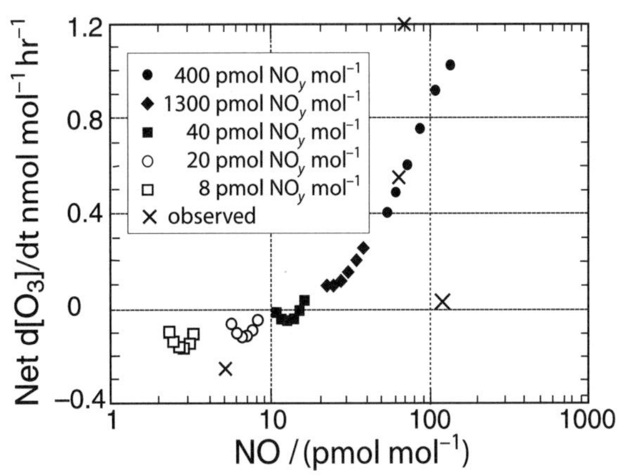

Fig. 3.19. Net ozone production as a function of [NO], in clean oceanic air at Cape Grim, Tasmania (Cox 1999)

are balanced. Cox (1999) investigated the influence of NO_x in the clean marine boundary layer in a box model with simple chemistry. Figure 3.19 shows the mean hourly rate of change of ozone concentration during the period 0900–1400 hours as a function of local NO concentration calculated for summertime conditions at Cape Grim. Net loss of ozone occurred at [NO] = 15 pmol mol^{-1} changing over to net production above this level. On the basis of their observations of O_3 and NO at Cape Grim, Galbally et al. (1996) estimate a compensation point of 20 ±5 pmol mol^{-1} NO. Similar model calculations and observations at Mace Head give a higher value of [NO] at the compensation point of 30–55 pmol mol^{-1} (Cox 1999; Carpenter et al. 1997), which is consistent with the control being primarily due to a competition between the above reactions, taking account of the higher climatological ozone concentrations at the Northern Hemisphere site.

Davis et al. (1996) have conducted a comprehensive box modelling study of ozone photochemistry in the northwest Pacific region, making use of aircraft measurements from the PEM-West A mission to constrain the model. They calculated the ozone photochemical tendency term, $P(O_3)$ (i.e. the difference between the total gas phase production and loss terms for ozone) using measurements as input to the model. $P(O_3)$ values were generally negative below ~6 km altitude and positive above 8 km throughout the study region, although significant positive values were also found in fresh boundary layer air of continental origin. Davis et al. (1996) also estimated the critical nitric oxide concentration NO_{crit} for which $P(O_3)$ was zero (equivalent to the compensation point), based on median values of all model input parameters in 13 altitude-latitude bins. In the lowest 2 km, NO_{crit} was between 9 and 17 pmol mol^{-1}, while O_3 was in the range 10–30 nmol mol^{-1}. Above 1 km, NO_{crit} tended to decrease with increasing altitude. These results indicate somewhat lower compensation points than in the boundary layer, where physical loss proc-

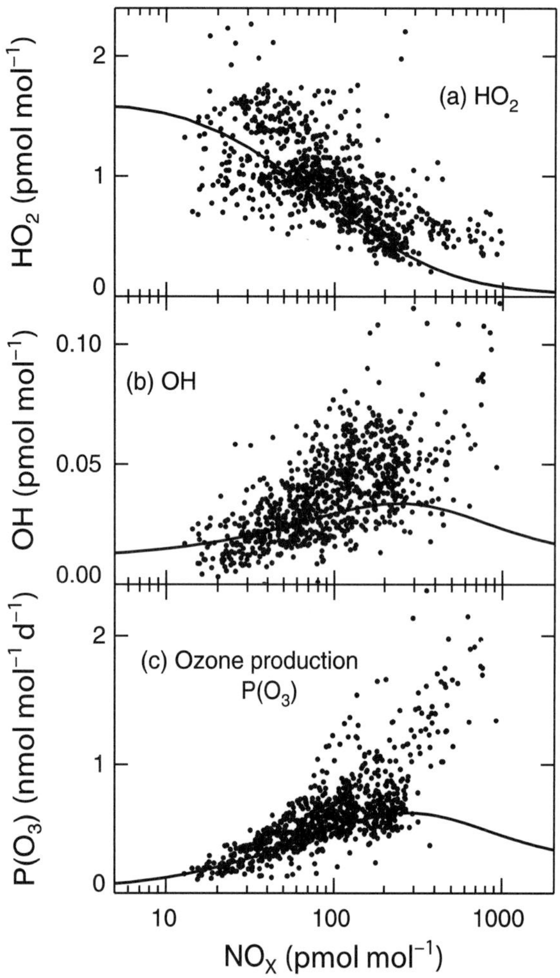

Fig. 3.20. Mixing ratios of HO_2 (*upper panel*), and OH (*middle panel*), and net ozone production rate (*lower panel*) as a function of NO_X mixing ratio (from Jaeglé et al. 1999)

esses have an influence. Klonecki and Levy (1997) have also discussed the effect of water vapour on the compensation point. Overall these studies confirm the basic theory for NO-mediated ozone production in clean tropospheric air.

Observations of free radicals have also been used to calculate ozone budgets in the troposphere. Ozone production in the upper troposphere has been studied by Jaeglé et al. (1999). Simultaneous observations of NO, HO, and other species were obtained as part of the SONEX campaign in the upper troposphere over the North Atlantic (40–60° N). These were used to derive ozone production rates, $P(O_3)$, and to examine the relationship between $P(O_3)$ and the concentrations of NO_x (= NO + NO_2) and HO_x (= OH + peroxy) radicals. A positive correlation is found between $P(O_3)$ and NO but after filtering out transport effects, $P(O_3)$ is nearly independent of NO for NO > 70 pmol mol^{-1}, showing the approach of NO-saturated conditions (Fig. 3.20).

3.3.8 Progress in Modelling the Global Budget of OH

The OH concentration responds almost instantaneously to variations in sunlight and local trace gas composition changes and the OH field varies by orders of magnitude in space and time. Observations can be used to test photochemical theory in specific circumstances but are not capable of providing globally averaged values of [OH]. We must therefore rely on numerical models and surrogates to provide the global distributions of OH that are required to calculate, for example, trace gas lifetimes and global ozone production and loss rates. The development of such models has advanced significantly in the past decade. An example of the output of such models for calculated global concentration fields of OH in July and January for the whole earth surface layers, is given in Fig. 3.21. Broad features of seasonal and latitudinal variation are superimposed by smaller scale variations caused by a multitude of factors such as UV penetration, and trace gas variations, which affect HO_x radicals. Discussion of these model developments is beyond the scope of this synthesis and we concentrate here on recent progress in determination of globally averaged [OH] based on new observations of methyl chloroform and ^{14}CO, obtained in IGAC, and using more sophisticated models than those used earlier. For a general discussion of atmospheric chemistry models, see Chap. 6.

3.3.8.1 *Global OH Estimates from CH_3CCl_3 Measurements*

Methyl chloroform (CH_3CCl_3) is an industrial gas for which analyses of atmospheric concentrations and trends have been used to derive concentrations of OH on broad scales. This technique provides estimates of "global" OH concentrations that are weighted heavily towards the Tropics and subtropics of the lower atmosphere since this is where [OH] is highest and most of the CH_3CCl_3 is oxidised. Estimates made in the presence of substantial methyl chloroform emissions suggest a mean global lifetime of 4.8 ±0.3 yr over the period 1978–1994 (Prinn et al. 1995). This represents a significant decrease in the estimated methyl chloroform lifetime, and the lifetimes of other gases such as CH_4 and CFC-substitutes, compared to those made ten years ago. Such estimates, however, depend heavily upon the accuracy of measurement calibration and emission figures. As a result these estimates have changed continuously over the past 15 to 20 years owing to improvements in calibration and in estimated emissions. More recently emissions have declined dramatically owing to production restrictions on substances that deplete the ozone layer outlined in the Montreal Protocol. The resulting decline of methyl chloroform observed in the atmo-

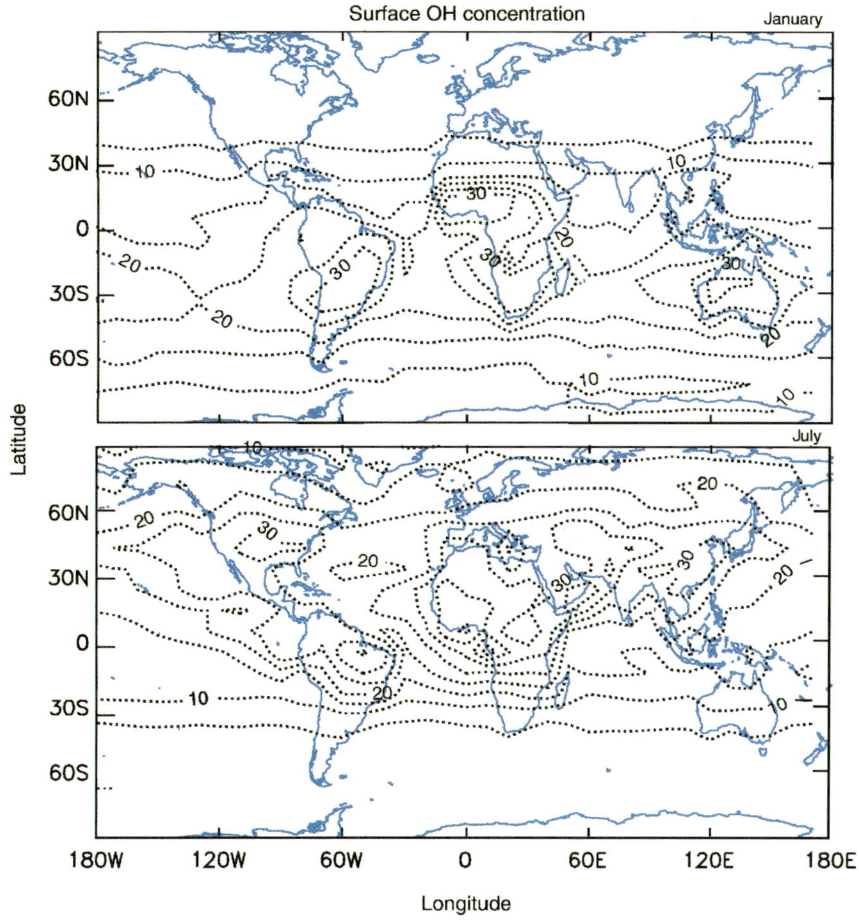

Fig. 3.21. Global concentration fields of OH (10^5 cm^{-3}), in July and January for the whole Earth surface layers (Spivakovsky et al. 2000)

sphere allows estimates of global and hemispheric lifetimes of this gas that are much less sensitive to uncertainties in measurement calibration and emission figures. This approach suggests a global mean lifetime for methyl chloroform of 5.2 [+0.2/−0.3] yr (Montzka et al. 2000). Using the latest information on deriving mean [OH] from methyl chloroform lifetime (Spivakovsky et al. 2000) gives a global mean value [OH] of 1.1×10^6 molecules cm^{-3}.

3.3.8.2 *Global OH Estimates from ^{14}CO Measurements*

Atmospheric carbon monoxide exists in several isotopic forms, one of which, ^{14}CO, has been useful in understanding tropospheric OH, which is the main sink for CO. ^{14}CO is produced in the atmosphere from cosmic rays, mainly in the stratosphere and upper troposphere, with a reasonably well known source strength. A substantial amount of new data for the global distribution of ^{14}CO has been obtained in recent years using a new technique based on accelerator mass spectrometry (Mak et al. 1992; Brenninkmeijer et al. 1992). Comparison of the observed ^{14}CO distributions with those obtained using photochemical-transport model calculations (Mak et al. 1994; Derwent et al. 1994) shows a good general level of agreement, with an implied global [OH] of ~9×10^5 molecules cm^3.

3.3.8.3 *Interhemispheric Differences in OH*

An independent method for the evaluation of the OH distribution and seasonality on the global scale is based on observations of ^{14}CO. ^{14}CO is primarily produced within the atmosphere by cosmic ray interaction. The smaller secondary source comprises "biogenic" ^{14}CO, containing ^{14}C which was recycled through the biomass. It was found that the cosmogenic ^{14}CO could be regarded as independent tracer, with a small perturbation from the biogenic contribution (e.g. Jöckel et al. 2000). The primary global average source strength of ^{14}CO is modulated by the changing solar activity. The atmospheric response to this varying global source strength has been calculated (Jöckel et al. 2000). Further, the relative solar modulation of the source has been quantified by using observations of atmospheric neutron count rates, sunspots, and calculations of the heliospheric potential (Jöckel and Brenninkmeijer 2001). These results can be used if ^{14}CO observations of different epochs are to be compared, and have been used for compiling a ^{14}CO climatology, i.e. a zonally averaged seasonal cycle at the surface comprising 1088 ^{14}CO observations from four institutes (Jöckel and Brenninkmeijer 2001). This climatology has been used for the evaluation of two three-dimensional atmospheric models (MATCH and TM3) in terms of the simulated OH distribution and seasonality and the simulated stratosphere-troposphere exchange (Jöckel et al. 2001). Especially, the intensively discussed issue of the interhemispheric asymmetry of OH has been revisited. Evidence for a higher OH abundance in the Southern Hemisphere is no longer supported, since the observed interhemispheric ^{14}CO asymmetry can consistently be explained by interhemispheric differences of the exchange rate between the stratosphere and the troposphere (Jöckel et al. 2001).

Interhemispheric differences in OH derived from model analysis of methyl chloroform data have also been inconclusive, although the contemporary distribution of this gas arising from the virtual cessation of emissions offers prospects for answering this question. The most recent analysis of the methyl chloroform data indicates mean [OH] is 15% higher in the Southern Hemisphere (south of the ITCZ) (Montzka et al. 2000).

3.3.8.4 *OH Temporal Trends*

The question of temporal trends of OH has been discussed by several workers. This would be an important diagnostic for model performance, particularly of interest for prediction of the oxidising efficiency of future atmospheres. Prinn et al. (1995) derived a trend of 0 ±0.2% yr^{-1} for OH based on model analysis of CH_3CCl_3 data for the period 1978–1994. Krol et al. (1998) used a three-dimensional chemical transport model scaled to the same CH_3CCl_3 data to derive a trend of 0.46 ±0.54% yr^{-1}. Both methods have uncertainties due to calibration of the halocarbon data and the significance of the derived trends must be accepted with caution. Nevertheless, the present evidence points to a zero or slightly positive trend which implies that OH production may have increased over the period. Considering the known upward trends (~11% change) in CH_4 and CO, the main sinks for OH, increased penetration of UV accompanying the known reduction in stratospheric ozone over this period could have contributed to increased OH production.

3.3.9 Night Time Free Radical Chemistry – NO_3

Field measurements of HO_x radicals have consistently indicated the presence of free radicals during nighttime, implying a non-photochemical source. Two important non-photochemical sources of radicals have been discovered. Firstly, the reaction of ozone with alkenes leads to production of OH. This source is particularly efficient for terpenes and other biogenic alkenes (Atkinson et al. 1992), and there are indications that it can be significant in terrestrial forested regions (Cantrell et al. 1997b). The second non photochemical source is the nitrate radical, NO_3, which is formed throughout the atmosphere by the reaction:

$$NO_2 + O_3 \longrightarrow NO_3 + O_2 \qquad (3.35)$$

In daylight it is removed instantly by photolysis and by reaction with NO, but at nighttime it is removed from the atmosphere either directly (e.g. by reaction with DMS and unsaturated organic molecules such as terpenes), or indirectly through its equilibrium with di-nitrogen pentoxide (N_2O_5), which reacts with H_2O on aqueous aerosol surfaces to form HNO_3.

$$N_2O_5 + H_2O(\text{surface}) \longrightarrow 2HNO_3 \qquad (3.36)$$

The reactions of NO_3 with organic molecules generate peroxy radicals (RO_2) as secondary products, and a correlation between NO_3 and RO_2 has been observed at night (Carslaw et al. 1997a). Box models of NO_3 chemistry in the marine boundary layer, when constrained by a comprehensive suite of measurements of all these loss processes, are able to reproduce the observed NO_3 levels satisfactorily (Allan et al. 1999). Such models show that nighttime NO_3 chemistry provides a route for converting to NO_x to HNO_3 which could be as high as 50% of the daytime route via OH + NO_2. This also constitutes a major loss for atmospheric NO_x with significant consequences for the tropospheric ozone budget (Dentener et al. 1996).

Atmospheric measurements of the nitrate (NO_3) radical have mostly been made by the technique of differential optical absorption spectroscopy (DOAS), using the strong absorption band at 662 nm. The exception to this is a data set obtained using Matrix Isolation – Electron Spin Resonance (MIESR) (Mihelic et al. 1993). DOAS instruments operating with an artificial light source in the lower boundary layer over optical path lengths up to 10 km have now achieved detection limits of 1 pmol mol^{-1}. DOAS measurements using either the moon or scattered pre-dawn sunlight as a light source have also been employed to measure NO_3 columns in the free troposphere (Aliwell and Jones 1998). There has also been one direct measurement of free tropospheric NO_3 at a high altitude site, Izaña de Tenerife, Canary Islands (Carslaw et al. 1997b).

Although there have been a few published reports on measurements of continental NO_3 during the past ten years (Platt and Heintz 1994; Smith et al. 1995), most field campaigns have focused on the behaviour of the radical in the marine boundary layer. These have been driven by the close coupling of the NO_x and sulphur cycles through the rapid reaction between NO_3 and dimethylsulphide (DMS) (Yvon et al. 1996a; Carslaw et al. 1997b; Heintz et al. 1996; Allan et al. 1999). All these studies, as well as more recent measurements in the remote marine boundary layer (Tenerife and Mace Head), have shown that NO_3 is present at up to approximately 5 pmol mol^{-1} even under low NO_x conditions ($[NO_x] < 150$ pmol mol^{-1}), when the turnover lifetime of the radical is well over an hour. By contrast, the lifetime can be less than a minute in semi-polluted air where reactive organics and aerosol loadings are higher. It should be noted that when the NO_3 concentration is larger than about 1 pmol mol^{-1}, DMS is oxidised by OH more rapidly at night than during the day. This seems to be the case even in the quite remote marine boundary layer (Allan et al. 1999).

3.3.10 Halogen Chemistry in the Troposphere

Observational data on tropospheric ozone over the past 40 years show that there are a number of anomalies in its distribution which are not easily explained by current theory. One example is the episodic and rapid depletion of ozone in the surface atmosphere in polar regions, particularly in the boreal and austral spring (Bottenheim et al. 1990). Another example is the observation of almost complete absence of ozone in the troposphere in the tropical oceanic regions of the eastern Pacific Ocean (Kley et al. 1996). Recent observations from Mace Head indicate a seasonal correlation of low ozone with phytoplankton activity (Carpenter et al. 1999).

Of the known species which could destroy ozone catalytically in the troposphere, the halogens, particularly bromine, are potentially the most efficient. It is now fairly clear that rapid loss of ozone in the Arctic boundary layer is due to catalytic cycles involving BrO radicals (Barrie et al. 1988).

$$BrO + BrO \longrightarrow 2Br + O_2 \qquad (3.37)$$

$$Br + O_3 \longrightarrow BrO + O_2 \qquad (3.38)$$

$$\text{net} \quad 2O_3 \longrightarrow 3O_2 \qquad (3.39)$$

Catalytic destruction of ozone can also occur in the troposphere by reaction of XO radicals (X = Br or I) with HO_2 in the following reaction sequence:

$$XO + HO_2 \longrightarrow HOX + O_2 \qquad (3.40)$$

$$HOX + h\nu \longrightarrow X + OH \qquad (3.41)$$

$$OH + O_3 \longrightarrow HO_2 + O_2 \qquad (3.42)$$

$$X + O_3 \longrightarrow XO + O_2 \qquad (3.43)$$

$$\text{net} \quad 2O_3 \longrightarrow 3O_2 \qquad (3.27)$$

This mechanism with X = I has been postulated to account for low ozone in the marine boundary layer (Davis et al. 1997; Stutz et al. 1999). Although there is no direct evidence, a recent model study (Vogt et al. 1999) based on observed IO concentrations indicates that iodine could have removed up to 12% of the ozone per day, making it more important than dry deposition or odd hydrogen photochemistry. Recent work shows that OIO may be important in iodine chemistry (Cox et al. 1999) and this has been confirmed by observations at coastal sites in the Northern and Southern Hemispheres (Stutz et al. 1999; Allan et al. 2000).

3.3.10.1 *Release of Halogens into the Troposphere*

Halogens are released into the troposphere from a variety of sources which are summarised in Table 3.4. The photochemical breakdown of organic halides such as CH_2I_2, CH_2ICl, CH_3I, and $CHBr_3$ (Schall and Heumann 1993) is probably the major source of reactive iodine species in the marine boundary layer. Modulation of tropospheric ozone and oxidising capacity can thus be achieved by the release of biogenic source gases.

ClO_x and BrO_x can also be produced from the oxidation of Cl^- and Br^- in sea salt (aerosol, sea ice, salt pans). The processes leading to loss of halide from the condensed phase have only recently been diagnosed, but it is clear that this source of reactive Br and Cl is likely to be more significant than thought hitherto. Currently there are two theories: (*1*) auto catalytic release of Br_2

Table 3.4. Sources of reactive halogen species found in various parts of the troposphere

Species, site	Likely source mechanism
ClO_x in the polar boundary layer	By-product of the "bromine explosion"
BrO_x in the polar boundary layer	Autocatalytic release from sea salt on ice ("bromine explosion" mechanism)
BrO_x in salt pans (e.g. the Dead Sea Basin)	Bromine explosion on salt pans
BrO_x in the free troposphere	Photodegradation of halogen-containing organic halogen species (e.g. CH_3Br); "Spill-out" from the boundary layer; Transport from stratosphere
IO_x in the marine boundary layer	Photodegradation of iodocarbons (e.g. CH_3I, CH_2I_2)

(and probably BrCl and IBr) from sea salt bromide (Vogt et al. 1994) the 'bromine explosion' mechanism; and (2) formation of $BrNO_2$ and $ClNO_2$ by the reaction of gas-phase N_2O_5 with sea salt halides (Vogt and Finlayson-Pitts 1994), which could be important in polluted regions.

However, the evidence for these mechanisms in the field is rather circumstantial and there appear to be (yet unknown) sources for ClO_x in oceanic regions. The role of iodine in the release of Cl (and Br) requires further study.

3.3.10.2 *Reactive Halogens in the Troposphere*

Free tropospheric BrO has been observed both in the Arctic as well as at midlatitudes by the technique of Differential Optical Absorption Spectroscopy (DOAS) (Hebestreit et al. 1999). Amounts of BrO observed are consistent with BrO-catalysed decomposition being responsible for the ozone loss observed in the local boundary layer. Combined balloon, ground-based, and satellite observations, using the same technique, indicate a residual component in the total column BrO of up to 1–2 pmol mol^{-1} BrO distributed in the free troposphere at midlatitudes and in the polar regions. This implies catalytic ozone loss via the BrO + HO_2 reaction cycle described above which is likely to be significant in the tropospheric ozone budget.

IO is the only halogen oxide so far observed in the low to midlatitude marine boundary layer. IO was measured for the first time in 1997 at Mace Head by the DOAS technique (Alicke et al. 1999; Allan et al. 1999). Figure 3.22 illustrates measurements off the north coast of Tenerife, where IO displays a clear diurnal cycle. It has now been observed in both hemispheres (Allan et al. 1999), implying a significant contribution to ozone loss in the local marine boundary layer.

3.3.10.3 *Cl Atoms in the Troposphere*

Halogen atoms are efficient oxidising species in their own right and can lead to the degradation of volatile organics. Cl atoms are of interest because they are up to 10^3 times more reactive than OH and may significantly enhance trace gas removal from the atmosphere. High concentrations of Cl atoms may destroy O_3 in certain regions of the atmosphere, influence the oxidation of DMS in the marine boundary layer, and initiate processes that could lead to the production of stratospheric ozone destroying chemicals (e.g. $C_2Cl_4 + Cl \longrightarrow \longrightarrow CCl_4$).

Prior to 1988 it was believed that the abundance of Cl atoms in the troposphere was extremely small (10^2–10^3 molecules cm^{-3}) and the main source was oxidation of HCl (Singh and Kasting 1988). Experimental studies in the laboratory (Zetzsch and Behnke 1992) and in the field (Keene et al. 1990; Pszenny et al. 1993) provided tantalising evidence that inorganic chlorine in the form of Cl_2 or HOCl may be present in an environment that contained wet sea salt, ozone, and sunlight. Indirect evidence for the presence of Cl atoms in the arctic BL (Jobson et al. 1994), tropical-Pacific-boundary layer (Singh et al. 1996a), and subtropical-Atlantic-BL (Wingenter et al. 1996) was deduced from the loss rate of C_2-C_4 NMHC. These studies suggested that Cl atoms in the vicinity of 10^3–10^5 molecules cm^{-3} may be present in the MBL. Singh et al. (1996b) also concluded that Cl atom formation was associated with mechanisms that did not involve acid displacement of sea salt chlorine. The mechanisms for the release of Cl from sea salt are only now being elucidated (Finlayson-Pitts and Hemminger 2000; Vogt et al. 1996). In none of these studies Cl atoms were thought to be present at concentrations high enough to cause O_3 destruction.

Complementing these MBL Cl-atom studies were inferences of global Cl from the budgets of C_2Cl_4 and

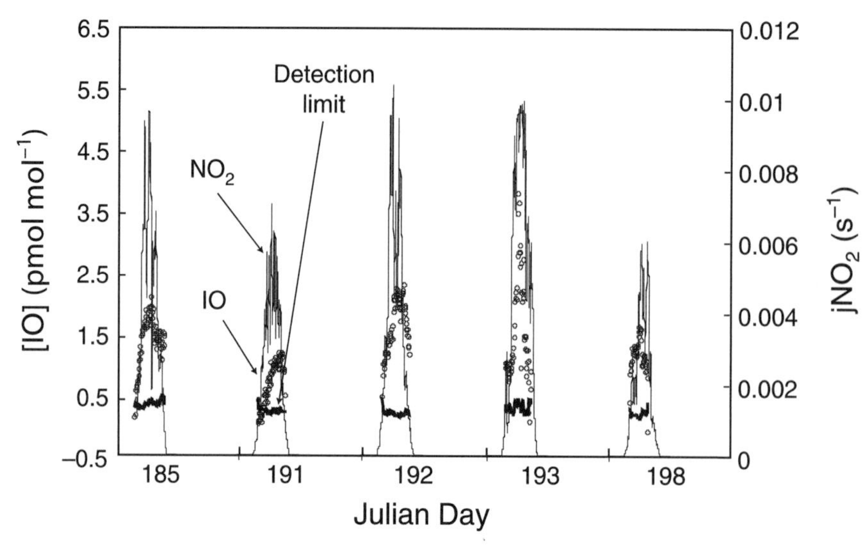

Fig. 3.22. Observed IO concentration profiles obtained at Tenerife during the period of 04–17 July 1997. The IO concentrations (*broken symbols*, courtesy of B. J. Allan, UEA) are plotted with the measured jNO_2 values (*solid dark lines*, courtesy of G. McFadyen, ITE). The average error for the IO concentrations is 0.2 pmol mol^{-1} (2σ) for this period with an average detection limit of 0.2 pmol mol^{-1} (2σ)

C_2H_6, both highly sensitive indicators of Cl atoms (Singh et al. 1996c; Rudolph et al. 1996). These studies concluded that the mean global abundance of Cl atoms was low (10^2–10^3 molecules cm^{-3}), but did not rule out the presence of substantially high Cl concentrations in the MBL. This implied that Cl atoms may enhance CH_4 oxidation by <2% and DMS oxidation by >20%.

There are no direct measurements of Cl or ClO although in one study substantial concentrations of Cl_2 were detected during the night (Spicer et al. 1998). No mechanism for nighttime release of Cl_2 is currently proposed. At the present time, the role of Cl atoms in tropospheric chemistry must be considered speculative although there is much progress. There is a great need for instrumentation that can directly detect Cl / ClO along with a need for laboratory studies to elucidate the processes that are involved in the release of Cl / Br atoms from sea salt.

3.4 Transport and Mixing Processes

In the previous sections, the emissions of O_3 precursors and photochemical processes important for the production and destruction of O_3 were discussed. Another important component of the tropospheric system which influences trace gas distributions and budgets is the transport of trace gases within the troposphere. The distributions of trace gases emitted into the troposphere are affected by both dynamical and chemical processes. Traditionally, the extent to which a particular trace gas is affected by transport or chemistry is given by its lifetime with respect to photochemical loss such as the reaction with OH. While this classification provides a useful term of reference, the distribution of a particular trace gas will depend on chemical lifetime relative to characteristic transport times and physical loss processes such as wet and dry deposition. For example, the distribution of CO, which has a chemical lifetime on the order of months in the free troposphere, will be governed much more by transport and mixing processes than by chemical processes, compared to, for example, isoprene which has a chemical lifetime of hours or less. Gases such as O_3, which are photochemically produced and destroyed in substantial amounts (see Sect. 3.3), are also affected by transport processes because their photochemical lifetime is also of the order of months in the free troposphere. Therefore, in order to quantify the budget of a gas such as O_3 it is necessary to understand the dynamical as well as chemical processes which govern its distribution.

Dynamical processes occur on a range of different scales starting with mixing and exchange between the planetary boundary layer (PBL) and the free troposphere. This includes exchange associated with day-to-day variations in the height of the PBL and rapid transport of air vertically by deep convection often into the mid and upper troposphere. Transport of air by synoptic-scale frontal systems is also an important mechanism for moving air out of the PBL into the free troposphere as well as for transporting air downwards from the upper troposphere and lower stratosphere. The latter is also important for O_3 which has a large source in the stratosphere. It should also be noted that deposition at the surface is an important sink for many trace gases including ozone. This process is intimately linked to PBL exchange and the rate at which air comes into contact with Ear rface. Dry deposition is discussed in more detail in Chap. 2. The treatment of dry deposition and transport processes in chemistry transport models is described in Chap. 6.

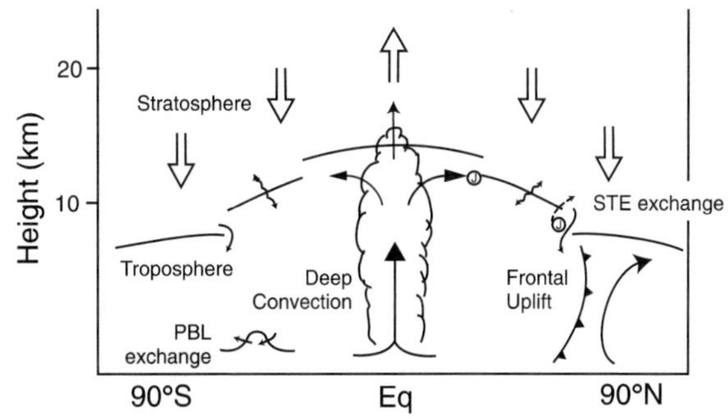

Fig. 3.23. A schematic representation of various transport processes in the troposphere (courtesy of K. S. Law, University of Cambridge, England)

The processes discussed in the following sections all play a role in transporting emissions and photooxidants from source regions into the free troposphere. They also form an integral part of the general circulation of the troposphere as shown in Fig. 3.23. For example, large regions of deep convection associated with the ITCZ are very efficient at moving trace gases from the surface into the upper troposphere. This forms the upward part of the Hadley circulation which then subsides in the subtropics. In midlatitudes, synoptic scale transport and coincident STE is intimately linked to the large-scale winds in the upper troposphere which move material from west to east around the globe. This large-scale flow is also important for transporting long lived trace gases northwards and southwards.

The recognition that O_3 and its precursors can be transported very large distances throughout the free troposphere is a major finding of several IGAC field experiments which are discussed in Sect. 3.6. In summary, the emerging picture is a complex one with air masses of different origin interleaved with each other in a multi-layered troposphere. Quantification of the role of transport and mixing on the budget of O_3 and its precursors remains an on-going topic of research.

3.4.1 Boundary Layer Mixing and Exchange

Pollutants are emitted and deposited in the boundary layer and transported away from the surface into the free troposphere making boundary layer mixing and exchange important processes to be considered in the study of photooxidant formation and removal. Here we describe the evolution of the boundary layer, how air can be exchanged with the free troposphere, and the importance of these processes for redistributing emissions and photooxidants.

The boundary layer can be defined as the lowest part of the troposphere which is directly influenced by the presence of Earth's surface. The evolution and depth of the PBL responds to changes in surface processes such as evapotranspiration, sensible heat transfer, orographic forcing, and frictional drag on time scales of an hour or less (Stull 1988). The depth of the boundary layer is also affected by synoptic scale weather systems; it will be thinner in regions of high pressure where air is subsiding. In low pressure regions, frontal uplift complicates the picture often resulting in the rapid transfer of boundary layer air to the free troposphere (see Sect. 3.4.3). Convection is also an efficient mechanism for moving air out of the boundary layer (see Sect. 3.4.2). Local circulations, such as land-sea breezes, can also affect the nature of boundary layer evolution.

In the absence of these processes there are diurnal variations in the boundary layer height which depend largely on the surface below. Over the sea there is only a small diurnal cycle because the underlying sea surface temperatures vary only slightly. The boundary layer heights during the middle of the day when heating is at its maximum are only slightly higher than at night. This cycle is much more pronounced over land where the day to night contrasts in surface heating are much larger. Figure 3.24 shows the typical evolution of the boundary layer under a high-pressure system (Stull 1988). During the day a well-mixed convective layer builds up and the depth of the boundary layer deepens to a maximum around noon. Air from aloft can be entrained into the boundary layer during this time and clouds can form at its top. Towards dusk, the boundary layer collapses to form a rather stable nocturnal boundary layer leaving formerly mixed air in a residual layer just above.

At the top of the boundary layer there is a stable entrainment zone where air can mix between the boundary layer and the free troposphere. Sometimes this layer becomes very stable and acts as a capping inversion preventing any movement of air between these two regions. In this case, which often occurs under high-pressure systems, pollutant concentrations can buildup to high levels over a period of several days. For example, Fig. 3.25 shows the buildup of O_3 at Mace Head during a period of several days in July 1996 (Julian day 198 to 203) when there was an anticyclone positioned over Europe and high concentrations of pollutants were advected westward over the UK and Ireland. Also shown are results from a photochemical trajectory model which, even though it included a rather simple treatment of boundary layer mixing, reproduces the observed O_3 levels reasonably well (Evans et al. 2000).

Theoretical treatment of boundary layer mixing and exchange of trace gases is dealt with in different ways depending on the scale of the phenomena being studied. Gaussian plume models have been used to study the evolution of point sources whereas Large Eddy Simulation (LES) models have been used to study PBL processes on the scales of 50 m or more. The latter resolve

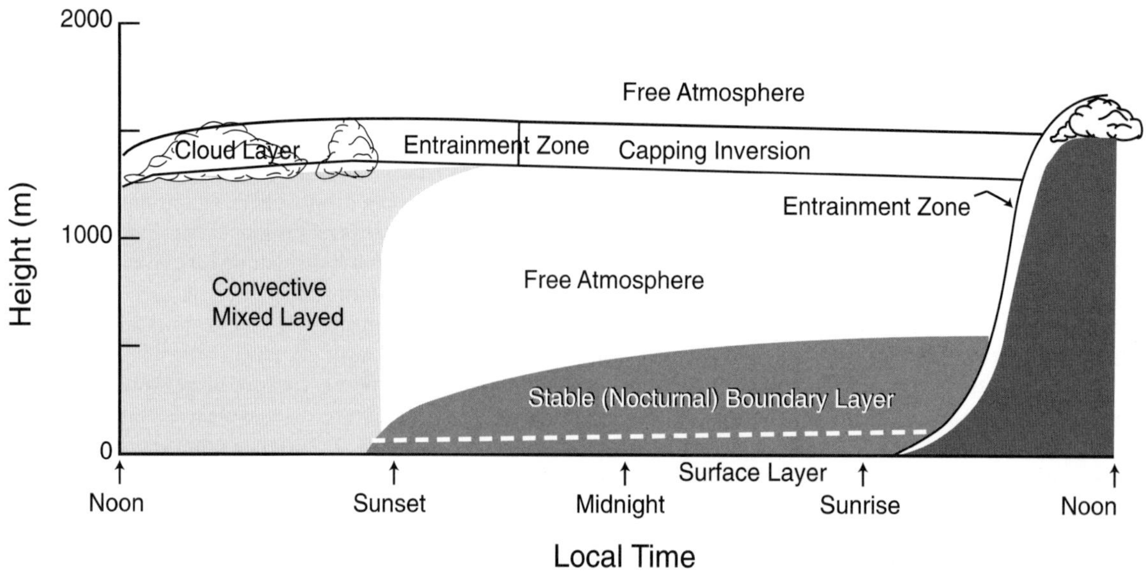

Fig. 3.24. Typical evolution of the boundary layer under a high-pressure system (Stull 1988)

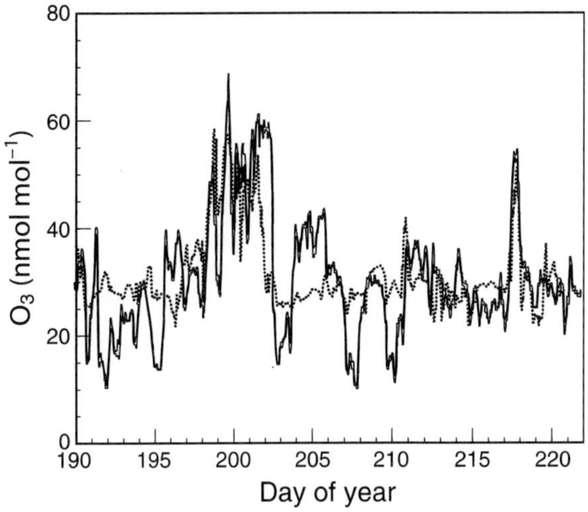

Fig. 3.25. A comparison of calculated and measured buildup of ozone during the ACSOE experiment at Mace Head in Ireland in July 1996 (Evans et al. 2000)

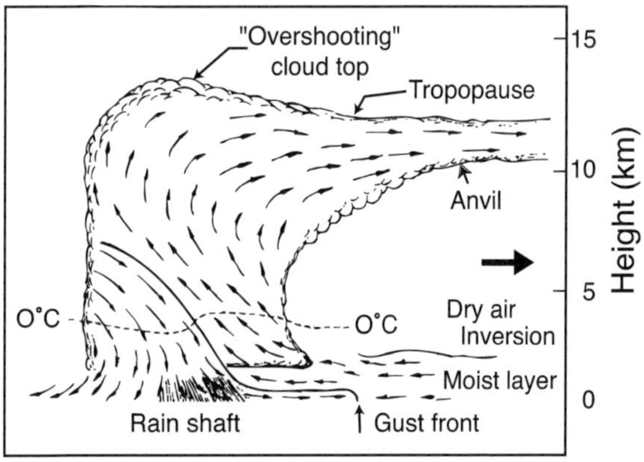

Fig. 3.26. An illustration of the role of a cumulonimbus cloud in transporting air from the surface to the tropopause (from Wallace and Hobbs 1977)

eddies down to smaller scales than most other techniques which parameterise these processes, often by assuming similarity between vertical fluxes of quantities such as heat, momentum, and moisture or by relating the vertical flux to an eddy diffusivity and the vertical gradient of a particular quantity. This method, known as K-theory, was also first used in chemical transport models but to a large extent has been superseded by more complex parameterisations developed for numerical weather prediction models which also take into account counter-gradient flow. This can be important in unstable and convective conditions when the largest transporting eddies may have a similar size to the boundary layer itself and the vertical flux can then be counter to the local gradient (e.g. Holtslag and Boville 1993). Not surprisingly, it was found during a World Climate Research Programme (WCRP) comparison exercise when models were compared against Radon-222 data (Jacob et al. 1997) that models with a more realistic treatment of boundary layer processes were able to simulate the data more accurately. Realistic treatment of convection was also an important factor (see Chap. 6).

3.4.2 Convection

Convection associated with thunderstorms, and sometimes with organised ascent within frontal systems, plays an important role in vertical exchange between the lower and upper troposphere. Large, organised ascents occur in warm, humid regions of the globe such as around the ITCZ in the Tropics. Summertime continental regions in midlatitudes are also prime areas for deep convection. It is the process by which air, heated at the surface, becomes buoyant and rises rapidly through the depth of the troposphere forming a cumulonimbus cloud as it develops. Updraft velocities are large and there is considerable entrainment and detrainment at the cloud boundary. There is outflow at the top of the cloud, with a characteristic anvil cloud defining this region. These processes are shown schematically in Fig. 3.26 (Wallace and Hobbs 1977).

Observations have shown that short lived trace gases are efficiently redistributed in deep convective clouds (e.g. Dickerson et al. 1987; Pickering et al. 1988) giving rise to C-shaped altitude profiles with high concentrations in the boundary layer and in the upper troposphere. Early modelling studies indicated that the vertical redistribution of trace species in rapid convective updrafts is important for their chemical processing (Chatfield and Crutzen 1984; Pickering et al. 1992). For example, convective venting is important for transporting NO_x out of the continental boundary layer. Since the lifetime of NO_x in the upper troposphere is up to a factor ten longer than in the boundary layer, its potential to contribute to O_3 formation is enhanced (Ehhalt et al. 1992). Convective uplift is also important for transporting sources of HO_x such as H_2O_2, CH_3OOH etc. into the tropical upper troposphere (e.g. Mari et al. 2000). This recent study also indicated that downward transport of air in and around convective clouds could also be important.

Since deep convection is often associated with lightning and consequent NO_x production, fresh NO_x of natural origin can also be added to boundary layer pollutants in the anvil outflow, yet another mechanism to contribute to upper tropospheric O_3 formation (see also Sect. 3.2). Recent field campaigns, such as the European Lightning Oxides Project (EULINOX) and STERAO-A in the United States have led to estimates that a significant fraction of upper tropospheric NO_x can be traced back to convective regions with high lightning activity and enhanced O_3 production downwind (Höller and Schumann 2000; Huntreisser et al. 2000; Dye et al. 2000). Moreover, convective mixing with background air can lead to NO_x dilution, which also increases the O_3 formation efficiency. For example, a recent modelling study, based on observations from the STERAO-A campaign over the central United States, estimated that lightning NO_x could have increased O_3 production rates by a maximum of 7 nmol mol^{-1} per day downwind of a convective storm in the upper troposphere (Skamarock et al. 2000).

It is very difficult to distinguish between the various upper tropospheric NO_x sources, including downward transport from the stratosphere and *in situ* emissions by aircraft. From both measurements and modelling studies it is nevertheless evident that several NO_x sources play a major role in the upper tropospheric O_3 budget. This has been clearly shown in NO_x data from the NOXAR programme collected in the North Atlantic flight corridor during summer and autumn 1995 and 1997 (Brunner et al. 1998; Jeker et al. 2000). As well as spikes in NO_x that were attributable to aircraft emissions, more widespread peaks exceeding 500 pmol mol^{-1} NO_x were measured which were attributed to uplift of boundary layer pollution or lightning emissions of NO_x over the western North Atlantic in the upper troposphere. These enhancements can be seen in Fig. 3.27 (from Jeker et al. 2000). They were also found during the SONEX campaign (Thompson et al. 1999). Model calculations suggest that while, in certain cases, 30–50% of NO_x derives from aircraft emissions other sources certainly contributed a similar amount.

Whether the convective updrafts of boundary layer NO_x and other O_3 precursors or mesoscale descent of upper tropospheric O_3 and NO_x dominate the overall effect of convection depends, to a large extent, on the geographical location, e.g. marine or continental (Lelieveld and Crutzen 1994). Also, considering the relatively small scales of thunderstorm convection (≤10 km) and the grid size of chemistry-transport models (≥100 km), it has proven difficult to account for these processes in global model simulations. This not only applies to deep convection, but as well to shallow and mid-level convection that also contributes to boundary layer venting. It remains a challenge to represent these processes accurately in large-scale models (see Chap. 6).

An important additional feature associated with convection is that radiative and evaporative cooling in the cirrus anvil can cause mesoscale descent and therefore downward transport of upper tropospheric trace gases towards the surface where their lifetime is generally shorter. This involves, for example, O_3 and NO_x of stratospheric origin. Gravitational settling of ice crystals below the cirrus anvil could also be an important sink for species such as HNO_3 (Lawrence and Crutzen 1998).

3.4.3 Synoptic Scale Transport

Synoptic scale processes affect the transport of trace species largely at midlatitudes in both hemispheres. Upper level winds drive the lower level pattern of high and low pressure weather systems which track from west to east around the globe. In high pressure systems the flow is clockwise (in the Northern Hemisphere) and the air is slowly subsiding. In the summertime, large regions of high pressure form over the sub-tropical oceans and can act as effective sinks for O_3 due to the higher concentrations of water vapour. The flows associated with low pressure systems or cyclones are more complex. When air motion is considered relative to the movement of the front or cyclone, several airstreams or conveyor belts can be identified moving along vertically sloping isentropic surfaces. This is shown schematically in Fig. 3.28 (Cooper et al. 2001). These airstreams or conveyor belts tend to be well defined and are called the warm conveyor belt (WCB), and the cold conveyor belt (CCB) along with the dry airstream or intrusion (Carlson 1991; Browning 1990).

Dry intrusions, which occur behind a cold frontal surface, can lead to the transfer of lower stratospheric

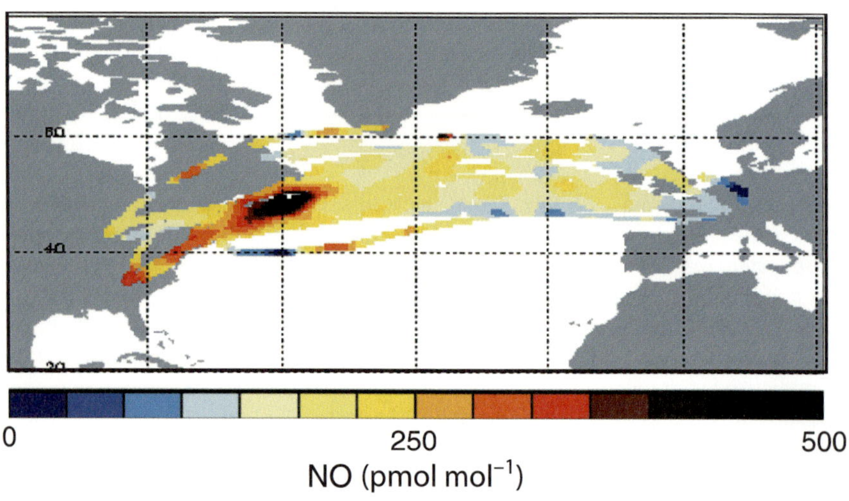

Fig. 3.27. Composite of all 2-min averaged NO measurements performed at cruising altitude (300–190 hPa) in the period 13 August to 23 November 1997 (Jeker et al. 2000)

air or upper tropospheric air to lower altitudes in the troposphere. Occasionally, the dry intrusion will overrun the cold frontal zone. The importance of stratosphere-troposphere exchange is discussed in more detail in the next section. The WCB originates in the boundary layer on the equatorward side of the cyclone and moves rapidly upward and ahead of a cold front in the upper troposphere. As it ascends, it can turn anticyclonically above the warm front. The CCB originates ahead of the warm front and travels westward (in the Northern Hemisphere) ahead of this front ascending into the mid-troposphere. This airflow often moves underneath the WCB flow aloft.

The role of frontal systems as a mechanism for transport of pollutants away from continental regions has been an area of study that IGAC has contributed to through its NARE and the East Asia-North Pacific Regional Study (APARE) activities (see Sect. 3.6). Studies following the first IGAC NARE experiment in 1993 noted the potential importance of frontal systems for the transport of ozone and its precursors into the upper troposphere using back trajectory analysis (e.g. Berkowitz et al. 1996; Wild et al. 1996). Aircraft measurements of chemicals in the vicinity of mid-tropospheric fronts reveal sharp air mass differences and fine interleaving of air from different source regions (Bethan et al. 1998). Further field campaigns have highlighted the widespread existence of multiple layers with different O_3 and pollutant concentrations. Often these layers display characteristics that lie between polluted boundary layer air and stratospheric air. The amount of mixing which occurs between different air masses and its impact on photochemistry still has to be resolved but it is possible that mixing between dry O_3-rich stratospheric air and humid boundary layer air could enhance OH concentrations (e.g. Esler et al. 2001).

Frontal systems also contribute to the long-range transport of O_3 and precursors from one continent to another because they develop over a period of several days while moving from west to east. An example is shown in Fig. 3.29 when O_3 lidar data collected over Garmisch Partenkirchen in southern Germany showed a region of elevated O_3 concentrations in the upper troposphere; these were attributed to long-range transport by a WCB of air from the continental North America boundary layer into the upper troposphere and then across the North Atlantic to Europe (Stohl and Trickl 1999). The data also show a filament of lower stratospheric air being transported downwards in an intrusion behind a cold front several hours earlier.

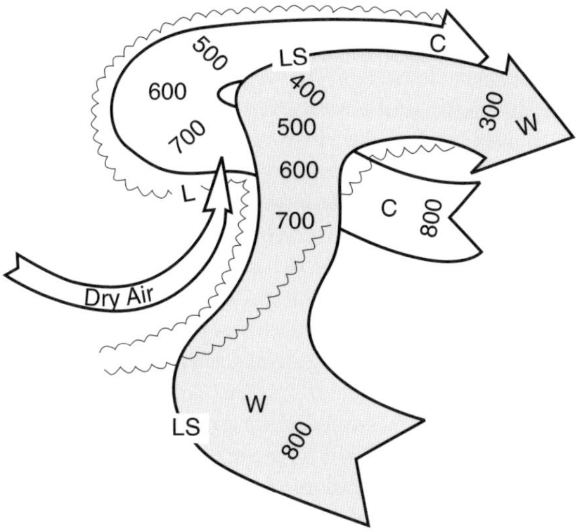

Fig. 3.28. A representation of mixing of air streams and conveyor belts associated with frontal systems (from Cooper et al. 2001)

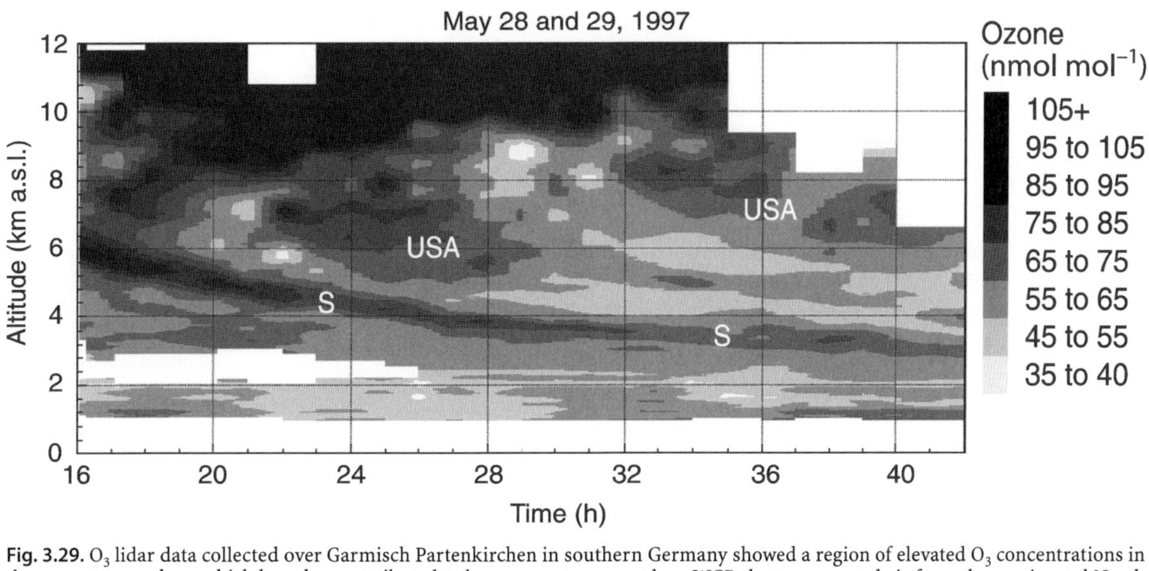

Fig. 3.29. O_3 lidar data collected over Garmisch Partenkirchen in southern Germany showed a region of elevated O_3 concentrations in the upper troposphere which have been attributed to long-range transport by a WCB that transported air from the continental North America boundary layer into the upper troposphere and then across the North Atlantic to Europe (Stohl and Trickl 1999)

3.4.4 Stratosphere-Troposphere Exchange

Previous to the discovery that photochemistry is of global scale importance, it was widely assumed that stratosphere-troposphere exchange (STE) controlled the distribution and abundance of tropospheric ozone (Junge 1962; Danielsen 1968). This assumption was based on the observed increase of O_3 with altitude and a negative correlation between O_3 and H_2O at extra-tropical latitudes, being suggestive of downward O_3 transport across the tropopause, and removal by dry deposition on the surface. Field campaigns and global modelling studies within the framework of IGAC have underscored the significance of photochemical processes. Nevertheless, STE provides an O_3 "background", required to initiate OH formation. While several field campaigns have been undertaken to examine processes related to STE (e.g. WMO 1999), the overall impact of STE on the tropospheric O_3 budget has yet to be fully quantified.

Stratospheric ozone is transferred to the troposphere in the Brewer-Dobson circulation that is driven by wave disturbances that propagate upward from the troposphere (Haynes et al. 1991). This "wave forcing" causes a mass flux that carries O_3-rich air from the tropical upper stratosphere poleward and downward, and ultimately through the extra-tropical tropopause. The mass balance of the stratosphere is maintained by suction of tropospheric air across the tropical tropopause (Holton et al. 1995). The wave forcing is most efficient in winter and early spring, since meridional temperature gradients are strongest then, and thus wave disturbances are most frequent. In addition, entrainment of relatively O_3-rich lower stratospheric air into the troposphere is strongest in spring, associated with increasing tropopause altitudes in the winter-summer transition (Appenzeller et al. 1996).

The two major STE processes are tropopause folds which occur along the edges of upper-level troughs (Danielsen 1968; Appenzeller et al. 1996) and cut-off lows (Price and Vaughan 1993). These events occur year-round in the midlatitudes (Austin and Follows 1991; Beekmann et al. 1997). Additionally, transport from the stratosphere along the subtropical jetstream may be important (Langford 1999). There is also evidence from theoretical calculations and from field measurements for transfer of trace gases from the upper troposphere into the lower stratosphere (e.g. Bregman et al. 1997; Fischer et al. 2000). For example, during the Stratosphere-Troposphere Experiment by Aircraft Measurements (STREAM) extensive CO plumes from wild fires over Canada were observed over Ireland. Trajectory analyses showed that this air had become entrained into the lower stratosphere, transported large distances, and then transported back into the troposphere (Waibel et al. 1999).

Satellite water vapour imagery can be used as an effective dynamical tracer to identify regions of stratospheric-tropospheric exchange. Derived products, such as specific humidity and isentropic potential vorticity, have been used to identify dry intrusions and cut-off lows in the mid to upper troposphere (Appenzeller et al. 1996; Cooper et al. 1998b). These features are often associated with O_3 enhancements as already shown in Fig. 3.29 (Stohl and Trickl 1999). On occasions, layers with elevated O_3 which originate from the stratosphere have also been observed in the mid and lower troposphere, for example, in ozonesonde data collected over the North Atlantic in post cold-frontal air masses (Cooper et al. 1998b).

Recently, fully coupled dynamical-chemical models that simulate downward transport of ozone on a global scale have become available (e.g. Roelofs and Lelieveld 1997). While models perform well in certain regions when comparing against observations near the tropopause where our understanding of the meteorology is good (e.g. over the North Atlantic) they perform rather less well in other regions (e.g. subtropics) (Law et al. 2000). State of the art chemistry transport models have also been used to diagnose the stratospheric contribution to the tropospheric O_3 budget (see Table 3.5). Estimates

Table 3.5. Ozone budget (Tg-O_3 yr^{-1}) below 300 hPa, except mentioned differently, as computed by the thirteen global three-dimensional models in the IGAC/3-D CTM O_3 intercomparison exercise

MODEL O_3 budget	IMAU-3	IMAGES	HARVARD[a]	UKMETO	ECHAM	MATCH	MOZART	TM3	MOGUNTIA	GFDL[b]	Average
Chemical production	2495	4763	4100	3890	3060	3580	3018	3979[c]	4061	4528	3447
Chemical destruction	2821	4898	3680	3276	3191	3550	2511	4065[c]	3923	4379	3629
Deposition	724	1253	820	1199	559	641	898	681	1017	898	869
Cross 300 hPa flux	1074	1388	400	401	582	594	391	768[c]	878	748	722
Burden	179	238	310	207	220	280	193	245	240	336	245

[a] Below 150 hPa.
[b] Poleward of 30° S and 30° N budgets are calculated below 241 hPa, between 30° N and 30° S budgets are calculated below 150 hPa.
[c] In 1E10 molecules cm^{-2} s^{-1}.

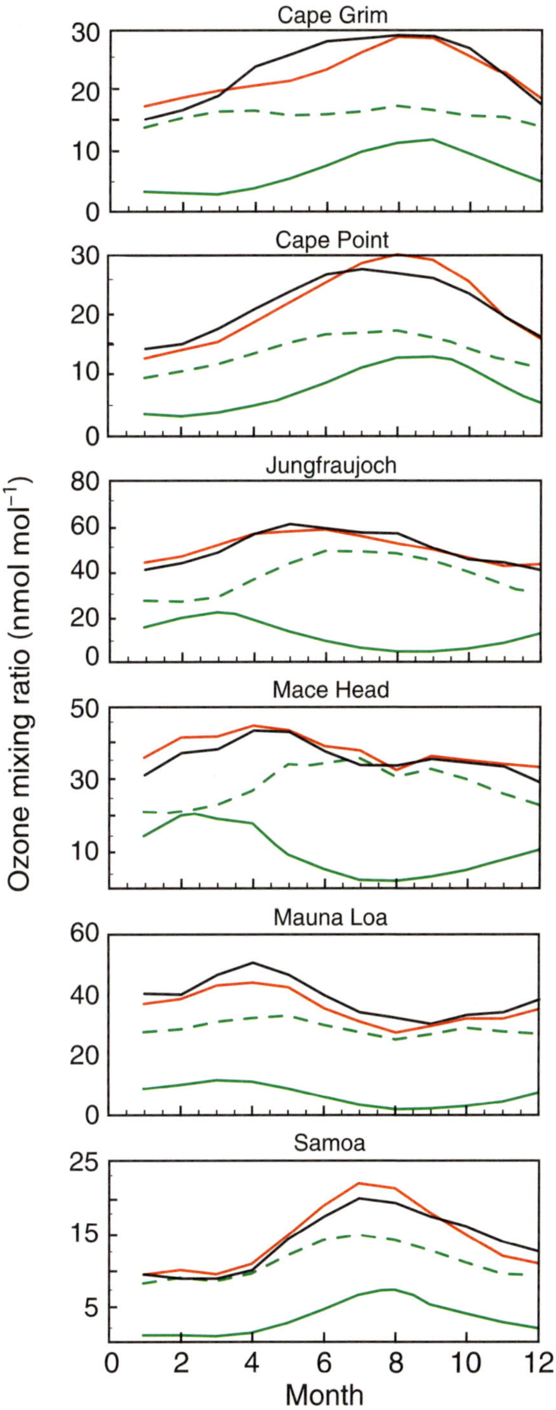

Fig. 3.30. Observed (*black*) and simulated (*red*) surface O_3 at several background monitoring sites. *Cape Point* is located 42° S, 18° E, *Jungfraujoch* (Switzerland) is located at 47° N, 8° E (3.6 km altitude), *Mace Head* (Ireland) is located at 53° N, 10° W, and *Mauna Loa* (Hawaii) is located at 19° N, 155° W (3.4 km altitude). Average seasonal cycles. Model-calculated ozone of stratospheric origin (O_3s) is indicated with the *green solid line*, and O_3 from *in situ* tropospheric formation is indicated with the *green dashed line*. Measurements are by courtesy of the National Oceanic and Atmospheric Administration (NOAA), S. Oltmans, E. Brunke, H. Scheel, and J. Stähelin (*http://www.badc.rl.ac.uk/data/toms/references.html*) (taken from *J. Geophys. Res.*, *105*(3), p3540, 2000)

range from 400–1 400 Tg yr^{-1} for this flux illustrating that significant uncertainties still exist in these estimates. Models have also been used to investigate the causes of the spring O_3 maximum at remote sites (see Sect. 3.3.5.1). For example, Lelieveld and Dentener (2000) indicate that while STE influences the seasonal cycle by contributing to the spring maximum, photochemical production of O_3 clearly dominates the seasonal cycle at the surface, especially in the Northern Hemisphere. Figure 3.30 shows that for three locations in the Northern Hemisphere, Jungfraujoch (JFJ), Mace Head (MH), and Mauna Loa (ML), the dominant source in spring and summer is tropospheric photochemistry, with the stratosphere making a roughly equal contribution in the winter at the two more northerly sites (MH and JFJ). In the Southern Hemisphere at Cape Point (CP) tropospheric production again predominates but the influence of the stratosphere is highly significant in the Southern Hemisphere spring.

STE can be important on regional and relatively short time scales, associated with tropopause folding events etc. However, results emerging from recent field campaigns and modelling studies suggest that both concentrations and distributions of ozone in the troposphere are, to a large extent, controlled by *in situ* photochemistry (see Sect. 3.5).

3.5 A Climatology of Tropospheric Ozone

3.5.1 Factors Controlling the Global Distribution of Ozone

The distribution of ozone in the troposphere is controlled by the processes that have been discussed in previous sections, that is, emission of ozone precursors, *in situ* photochemical processing, and transport within the troposphere. Ozone is directly transported into the troposphere from the stratosphere but as has been emphasised previously this is not the main source. Similarly the main sink is not deposition to Earth's surface. Dry deposition is not discussed in detail here since it is dealt with in Chap. 2. It is a very important process for many tropospheric trace gases and in the case of ozone is much more effective over land than over ocean, such that approximately two-thirds of the deposition occurs in the Northern Hemisphere.

As discussed in Sect. 3.3, the main sources and sinks of ozone in the troposphere are *in situ* photochemical production from photolysis of NO_2 assisted by the HO_2-OH chain reaction, and photochemical destruction from ozone photolysis and the same chain reaction. The switch from net destruction to net production is determined mostly by the levels of NO_x, which because of the very limited database on these species (see Sect. 3.2), makes a quantitative prediction of the distribution of ozone in the troposphere by theoretical models very difficult. Water also plays a role in the switch-over of the chemistry due to its large influence on the rate of ozone destruction (Klonecki and Levy 1997).

In spite of these difficulties many models are now producing realistic representations of the global distribution of ozone. Table 3.5 scales up the various terms involved in the global ozone budget as calculated by nine models in a model intercomparison exercise organised recently by the European IGAC Project Office (EIPO) as part of the Global Integration and Modelling (GIM) subproject (see also Chap. 6) (Kanakidou et al. 1999). The individual terms for production and destruction are both very large with average values of 3 747 Tg O_3 yr^{-1} and 3 629 Tg O_3 yr^{-1} respectively. Both these terms are significantly larger than either the average cross tropopause flux (722 Tg O_3 yr^{-1}) or the average deposition of ozone to the surface (869 Tg O_3 yr^{-1}). A graphical representation of the chemical and physical processes along with quantification of the various terms has been shown earlier in Fig. 3.2.

The process of injection of ozone from the stratosphere is clearly visible in the troposphere and there is strong experimental evidence that the cross tropopause flux of ozone lies somewhere between 300 and 800 Tg O_3 yr^{-1} (Murphy and Fahey 1987). This is considerably larger than the average net production in the troposphere calculated by models (~120 Tg O_3 yr^{-1}). The *in situ* chemical processes are also visible and their nature and large scale are shown very effectively from composite LIDAR pictures obtained by the NASA DC8. Figure 3.31a shows the distribution of ozone over the South Atlantic, where ozone is produced in biomass plumes transported from the continents of Africa and South America. Figure 3.31b shows the ozone distribution over the western Pacific where ozone destruction is occurring in uplifted ma-

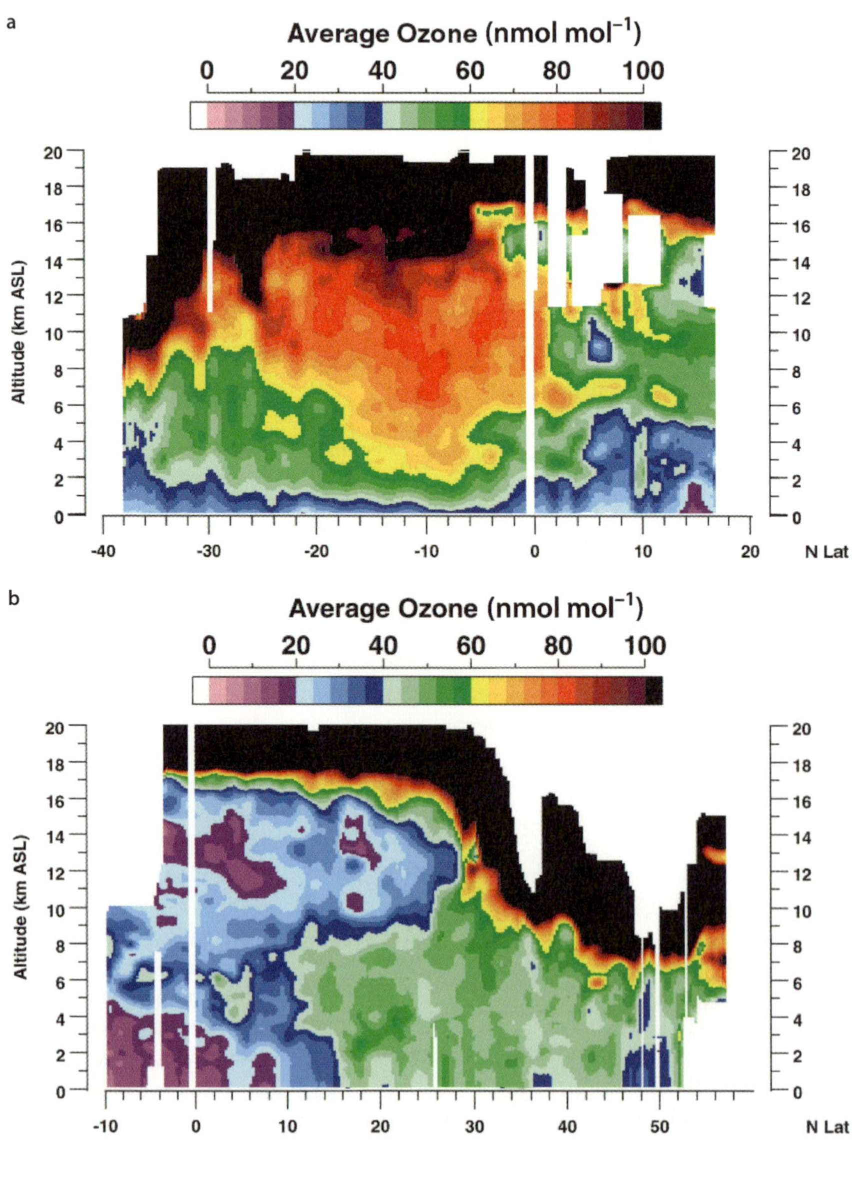

Fig. 3.31. **a** Average ozone distribution over the South Atlantic basin obtained from ozone lidar measurements during the NASA Global Tropospheric Experiment (GTE)/Transport and Atmospheric Chemistry Near the Equator-A (TRACE-A) field experiment conducted in September–October 1992. *Enhanced orange* associated with biomass burning in Africa and South America can be readily seen across the entire troposphere between 0° and 25° S (Browell et al. 1996a); **b** Average ozone distribution over the Western Pacific obtained from airborne lidar measurements during the NASA GTE/Pacific Exploratory Mission West (PEM-West II) conducted during February-March 1994. Ozone destruction near the surface in the Tropics produces the observed low ozone levels in the lower troposphere, and deep cloud convection in the Tropics transports low ozone from near the surface to the upper troposphere (courtesy of E. V. Browell, NASA Langley Research Center, Hampton, Virginia)

rine boundary layer air, with low NO_x and high water vapour (Browell et al. 1996).

It is clear, however, that large uncertainties remain about the relative magnitudes of ozone sources and sinks and it is very instructive therefore to examine the database on the distribution of ozone in the troposphere against this background. This database historically has been made up of ground-based data and sonde records that have perhaps been misinterpreted in the past by considering only physical processes. Increasingly very valuable records are being collected both by research and commercial aircraft and by satellites which emphasise the role of *in situ* processing and transport. It is hoped that much more data will be available from these latter sources over the next decade.

3.5.2 Global Measurements of Ozone and Sonde Data

3.5.2.1 *Ground-Based and Sonde Observations*

Surface ozone measurements in a global perspective have been surveyed by Oltmans and Levy (1994) and are summarised in some detail along with data from the ozone sonde network in a recent paper by Logan (1999). These publications have performed a very valuable service while at the same time indicating the inadequacy of the extent of the data record, some aspects of which are discussed below.

At remote sites in mid and high latitudes in the Northern Hemisphere, surface-ozone displays a distinct maximum in spring (March to May) (Fig. 3.32a,b), which could result from a variety of causes (Penkett and Brice 1986; Monks 2000). Downward transport from the stratosphere carrying O_3-rich air into the troposphere maximises in this season (Danielsen 1968), but photochemical formation in the troposphere has also been proposed due either to increasing insolation and the presence of ozone precursors which have accumulated during the winter (Penkett et al. 1993), or buildup of photochemically-formed ozone during the winter when its atmospheric lifetime is long enough to allow accumulation (Liu et al. 1987).

Global chemical transport model results suggest that the spring maximum of ozone is the result of the superimposition of a stratospheric component, which maximises in late winter to early spring at the surface, and a photochemical component in the troposphere which maximises in summer (Wang et al. 1998c; Lelieveld and Dentener 2000). Wang et al. (1998c) also argue that tropospherically-produced ozone is transported more effectively in spring than in summer because of a longer lifetime, again emphasising the role of transport. An exception to the normal spring maximum in Fig. 3.32a is Barrow, Alaska, where a minimum of ozone occurs in April caused by photochemical destruction by bromine species over the ice-covered Arctic region (Barrie et al. 1988; Oltmans et al. 1989). While this phenomenon is now observed generally in the boundary layer of the northern polar region (Oltmans and Levy 1992), the precise mechanism, particularly the source of the active bromine, has not been well established (Oum et al. 1998; Hirokawa et al. 1998).

In rural areas subject to regional pollution on the European and North American continents the highest values of ozone are found in summer, undoubtedly due to *in situ* production in the boundary layer (Fig. 3.32b). A gradient toward higher concentrations from west to east in the United States, and from northwest to southeast in Europe has been noted (Logan 1999). At coastal remote sites in Asia the summer maximum is emphasised due to large-scale air mass exchange by the Asian monsoon showing the critical importance of transport in determining the distribution of ozone in the troposphere (Logan 1985; Ogawa and Miyata 1985). Remote marine sites at midlatitudes experience a summer minimum due to ozone destruction by strong insolation associated with high water vapour concentrations (see Fig. 3.32c), although at high altitude sites, such as Mauna Loa (Hawaii Islands), Izana (Tenerife) and Niwot Ridge (Colorado), where the influence of high water vapour of marine air is less pronounced, a summer minimum is not observed.

At Barbados and Venezuela in the Northern Hemisphere Tropics, surface ozone maximises in winter and minimises in summer (Fig. 3.32d). At Samoa, a clean tropical site in the Southern Hemisphere, a similar seasonality is observed but shifted six calendar months compared to the Northern Hemisphere. Measurements made from ships often observe very low concentrations of ozone (less than 10 nmol mol^{-1}) over the equatorial Pacific (Piotrowicz et al. 1991; Johnson et al. 1991). While OH radical chemistry is generally thought to be mainly responsible for photochemical loss of ozone in the marine boundary layer (e.g. Ayers et al. 1997), there is also evidence that bromine chemistry may play a role (Sander and Crutzen 1996; Nagao et al. 1999). Losses due to iodine chemistry have also been proposed to account for the extremely low observations of ozone in the tropical marine boundary layer (Evans et al. 2000)

Surface ozone concentrations display a distinct maximum in the austral spring which is particularly pronounced at continental sites in the Southern Hemisphere subtropics such as Brazzaville in the Congo and Cuiaba and Natal in Brazil (Fig. 3.32e). Spring in the Southern Hemisphere is the dry season, when large-scale biomass burning occurs in tropical and subtropical areas in South America, Africa, and Indonesia, releasing ozone precursors which cause extensive ozone production. At Brazzaville surface ozone is about 30 nmol mol^{-1} in the wet season between December and July, whereas it is only ~10 nmol mol^{-1}

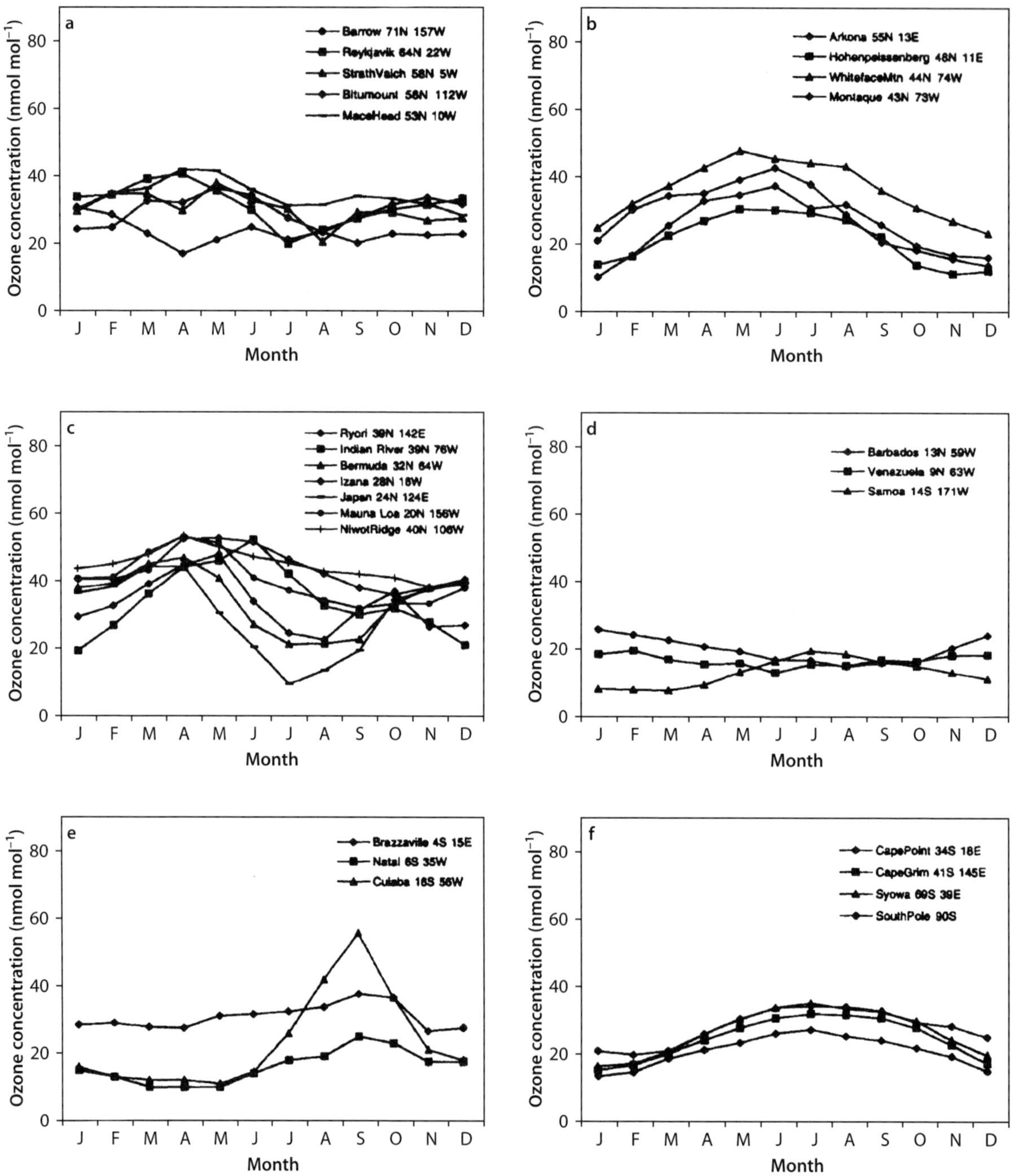

Fig. 3.32. Average annual ozone cycles at the surface for background global measurement sites (Logan 1999)

at Natal and Cuiaba which are both strongly influenced by marine air masses transported from the Atlantic Ocean (Kirchoff and Rasmussen 1990). A similar seasonal variation, but shifted by six months, is observed in the Northern Hemisphere Asian subtopics, with Thailand, for instance, giving a distinct maximum in March and April due to regional-scale biomass burning (Pochanart et al. 2001).

Finally, the ozone seasonality at locations in the Southern Hemisphere middle and high latitudes displays a winter maximum and a summer minimum (Fig. 3.32f) driven by photochemical destruction of ozone, which is most efficient during the summer months. In addition, stratospheric-tropospheric exchange will bring large amounts of ozone to the lower troposphere in late winter when photochemical destruction is at a minimum.

In the sonde data at 500 hPA in the mid troposphere most stations in the Northern Hemisphere north of the Tropics show a broad summer maximum between April

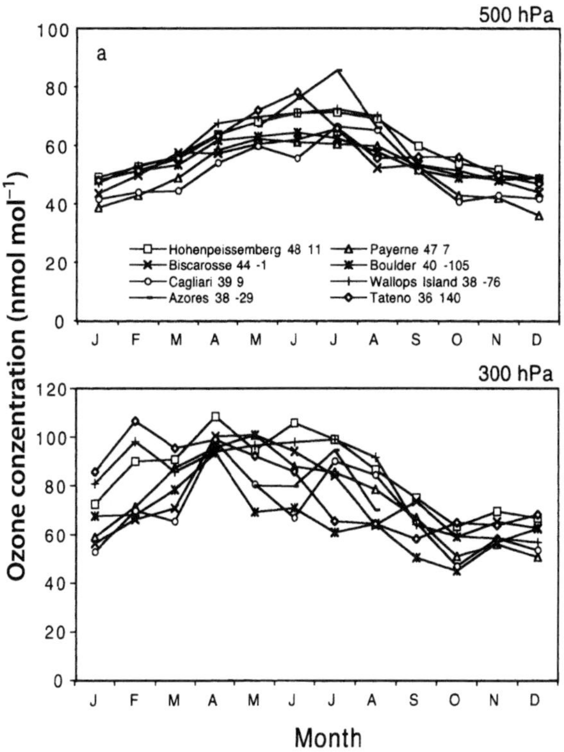

Fig. 3.33 a. Ozone seasonal cycles at 500 hPa and 300 hPa for Northern Hemisphere mid latitude sonde launch sites (Logan 1999)

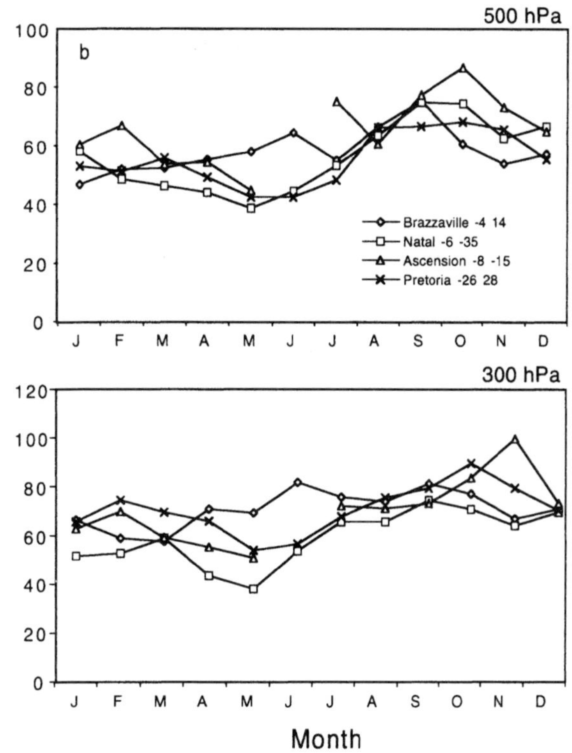

Fig. 3.33 b. Ozone seasonal cycles at 500 hPa and 300 hPa for Southern Hemisphere sonde launch sites strongly influenced by biomass burning (Logan 1999)

and September (Fig. 3.33a); this is almost certainly caused by efficient photochemical ozone formation in summer aided by industrial emissions of NO_x. Also NO_x emissions from lightning, which are most pronounced over continents in summer, may contribute significantly. The peak in ozone occurs somewhat earlier at higher latitudes than shown in Fig. 3.33a probably due to different contributions of stratospheric-tropospheric exchange (STE), maximising in spring, and photochemical formation, maximising in summer. The Northern Hemisphere ozone concentrations at 500 hPa all lie between 35–50 nmol mol^{-1} in winter and 50–75 nmol mol^{-1} in summer consistent with recent model calculations (e.g. Lelieveld and Dentener 2000; Wang et al. 1998c). At 300 hPA the ozone concentration starts to increase in February at many stations in the Northern Hemisphere (Fig. 3.33a), no doubt due to a significant contribution from the stratosphere. The tropospheric influence is still present though extending the broad maximum in the summer months for stations south of 60° N.

At Southern Hemisphere tropical latitudes, such as for Natal and Ascension Island, high ozone concentrations occur from August to November in both the lower and the upper troposphere (Fig. 3.33b). The same pattern is also noticeable at Pretoria (South Africa). Ozone minimises in the wet season between April and June when the *in situ* photochemical activity decreases. The high values of ozone between August and November are

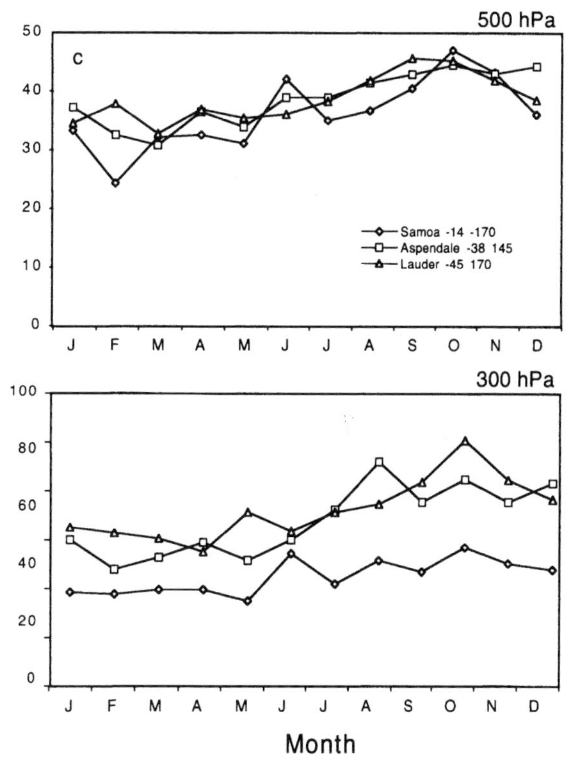

Fig. 3.33 c. Ozone seasonal cycles at 500 hPa and 300 hPa for Southern Hemisphere tropical and mid latitude sonde launch sites (Logan 1999)

caused by emissions of precursors from biomass burning in Brazil and South Africa (Logan and Kirchoff 1986; Cros et al. 1992). The July maximum in the lower troposphere at Brazzaville reflects the peak burning season in May to August in eastern Africa south of the Equator, while the secondary peak in January reflects the burning season north of the Equator (Nganga et al. 1996). The biomass burning influence is also noticeable at 500 and 300 hPa for Aspendale (near Melbourne, Australia) and Lauder (New Zealand). At 500 hPa in the free troposphere over Samoa in the tropical Pacific (Fig. 3.33c), ozone concentrations are low between January and April and maximise between August and November, i.e. at the same time as ozone concentrations maximise over Africa and South America. This hemispheric spread of ozone from biomass burning on the continents is now recognised as a major feature of the ozone budget in large parts of the Southern Hemisphere (Blake et al. 1997).

3.5.2.2 *Aircraft Data*

Over the last decade several projects have been initiated to use commercial aircraft as platforms to make regular measurements of various trace gases including ozone. These represent a major advance in our measuring capability and have provided new information on trace gas distributions. The projects include: MOZAIC (Measurement of Ozone and Water Vapour by Airbus In-Service Aircraft) (Marenco et al. 1998); NOXAR (Nitrogen Oxide and Ozone along Air Routes) (Dias-Lalcaca et al. 1998); and, CARIBIC (Civil Aircraft for Remote Sensing and *In situ* Measurements in the Troposphere and Lower Stratosphere Based on the Instrumentation Container Concept) (Brenninkmeijer et al. 1999). The MOZAIC data are discussed below in more detail. The NOXAR project made measurements of O_3 and NO_x on board a commercial aircraft flying across the North Atlantic (Jeker et al. 2000) (see Sect. 3.4.2) and the CARIBIC project makes measurements of a wide range of trace gases from a container carried on an aircraft flying from Germany to the Maldives and Southern Africa. CARIBIC is particularly suitable for examining the contribution made by various processes (industrial emissions, biomass burning, etc.) to the ozone levels observed in the free troposphere at cruise altitudes. An indication of its capabilities is shown in Chap. 5.

MOZAIC is an EU project, which started in 1994, consisting of automatic instrumentation measuring ozone and water vapour, installed on five long-range Airbus A340 aircraft in normal airline operation (Marenco et al. 1998; Thouret et al. 1998a,b). In addition to vertical profiles collected over major airports, the MOZAIC data have provided a climatology of the O_3 distribution around the tropopause, in particular over the North Atlantic. For example, Fig. 3.34 shows seasonally averaged data obtained between September 1994 and Au-

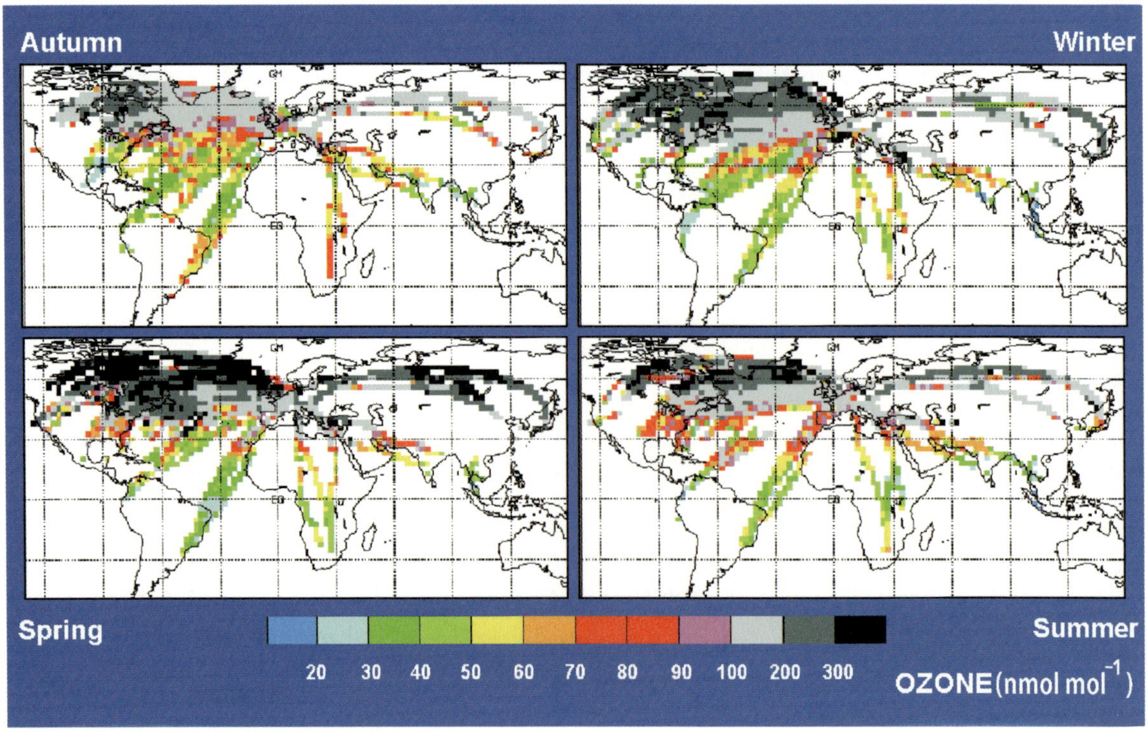

Fig. 3.34. Seasonally averaged ozone data between September 1994 and August 1996 at approximately 11 km (courtesy of A. Marenco, CNRS Laboratoire d'Aérologie, Toulouse, France)

gust 1996 at approximately 11 km. In the Northern Hemisphere midlatitudes, MOZAIC O_3 observations show a relatively well-defined seasonal variation with maximum concentrations in the late winter and spring in line with the sonde data at 300 hPA and with the magnitude of the maximum dependent on the position of the measurements relative to the height of the tropopause (Thouret et al. 1998a; Law et al. 2000). For example, peak concentrations over Europe are lower than over eastern North America and eastern Asia. In areas where little or no sonde data have been available previously, such as over northern India and neighbouring countries, a spring maximum is also found. This is probably produced by incursions of stratospheric air across the subtropical tropopause. Data collected at these locations also show a summer minimum with monthly mean concentrations as low as 40 nmol mol^{-1} in the upper troposphere. Further south (e.g. over Burma), O_3 levels are lower throughout the year and there is no apparent influence from the stratosphere. Occasionally, very high concentrations (greater than 400 nmol mol^{-1}) have been observed in the MOZAIC data. These were first noted by Suhre et al. (1997) and were found to be small-scale features existing in the upper troposphere. Further study by Cammas et al. (1998) showed that these events are most likely to be due to transport across the subtropical tropopause; they have been observed over the Atlantic and the Indian Ocean and sub-continent.

In the Tropics, MOZAIC data are sparser but nevertheless have given new insights into the distribution of ozone around the tropopause. Generally, ozone concentrations decrease towards the Equator and show less seasonal variation. Occasionally, ozone concentrations as low as a few nmol mol^{-1} have been observed in the upper troposphere over tropical oceans. Very low levels of ozone have also been observed in ozonesonde data collected on ship cruises over the Indian Ocean (Kley et al. 1996). Examination of MOZAIC data shows that this phenomenon is more widespread with concentrations of less than 10 nmol mol^{-1} also found above the Atlantic Ocean. Even so, the lowest concentrations (less than 1 nmol mol^{-1}) do appear to occur over the Indian Ocean. The mechanism leading to such low concentrations is still unclear.

The MOZAIC vertical profile data collected over major cities also show many interesting features with the major airline centres being the most well documented (e.g. Paris, Frankfurt, New York City). The large number of profiles collected over several years has enabled, for the first time in some cases, examination of day-to-day variations, seasonal variability, and interannual variability of the O_3 distribution. For example, Fig. 3.35 shows a time series of profiles collected over Frankfurt from 1994 to 1999. In some cases more than one profile was collected on any one day. The broad seasonal variations with a summer maximum are in agreement with those seen in sonde data. In addition, the high temporal resolution of the MOZAIC data clearly shows the considerable day-to-day variations in O_3 concentrations which are superimposed on the seasonal cycle. Studies of the data have shown that convective uplift and stratospheric intrusions affect O_3 levels on particular days (Law et al. 1998; Cammas et al. 1998). Another interesting feature is the buildup of O_3 during the summer months, especially in the upper troposphere, which appears to be largely due to photochemical production of O_3 (Plantevin et al. 2000). The flux from the stratosphere is rather continuous throughout the year with a larger contribution in the late winter and early spring. The summer maximum is seen over many northern midlatitude locations in the MOZAIC and sonde data, suggesting that photochemical production in the troposphere is affecting O_3 levels on a hemisphere-wide scale (Law et al. 2000; Logan et al. 1999).

Similar to the sonde and ground based data at coastal sites (Simmonds et al. 1997; Oltmans and Levy 1994), MOZAIC data collected at coastal locations in the Northern Hemisphere also show the influence of advection of clean maritime air producing a summer minimum in O_3 in the lower and mid troposphere (e.g. Miami and Toyko). This has also been observed in MOZAIC data collected in the Tropics (e.g. Madras, India, and Caracas, Venezuela). Over Madras air masses containing low levels of O_3 are advected from the south as part of the monsoon circulation. These data again emphasise the role of transport in controlling tropospheric ozone, in this case transport of air in which ozone has been photochemically destroyed.

Vertical profile data collected over continental regions in the Tropics and the Southern Hemisphere over South America and southern Africa also show the influence of biomass burning and industrial emissions on the O_3 distribution. In particular, the seasonal influence of biomass burning produces layers with elevated O_3 concentrations which can be transported many thousands of kilometres from source regions. The MOZAIC data show that this is a feature which exists from year to year, as illustrated in Fig. 3.36 for data collected over Johannesburg, South Africa, where O_3 concentrations peak in the austral spring in the upper troposphere. This phenomenon was referred to when considering the ozonesonde data and is attributed to convective uplift of ozone and its precursors from biomass burning regions into haze layers, which can exist for many days and be transported long distances away from the African continent (Garstang et al. 1996; Diab et al. 1996). The MOZAIC data in this region also confirm earlier results from regular sonde measurements and field campaigns such as TROPOZ I-II (Jonquières et al. 1998), TRACE-A, and SAFARI. Over South America (e.g. Sao-Paulo-Rio) an austral spring peak is observed throughout the entire depth of the troposphere. Deep convection combined with northwesterly flow out of the South Ameri-

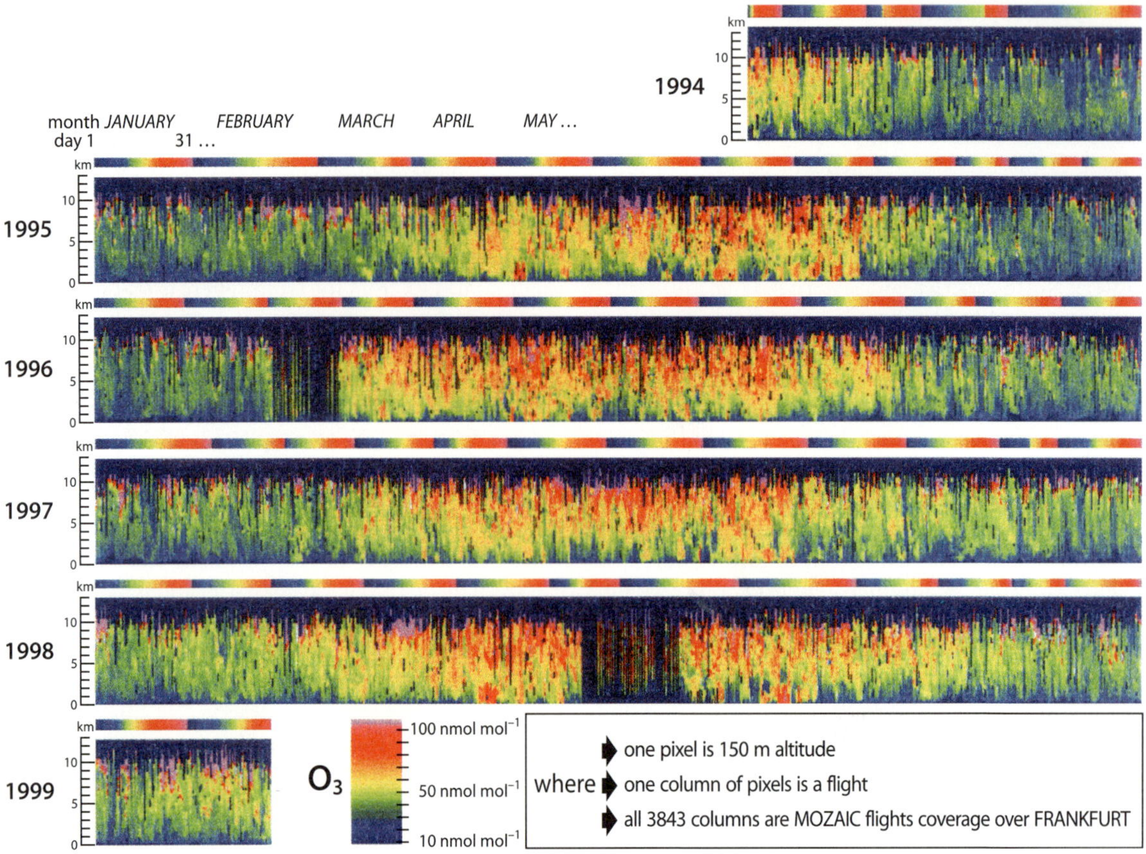

Fig. 3.35. The seasonal behaviour of ozone throughout the troposphere in a continuous series of vertical profiles collected by the MOZAIC programme over Frankfurt for the period July 1994 to February 1999 (courtesy of A. Marenco, CNRS Laboratoire d'Aérologie, Toulouse, France)

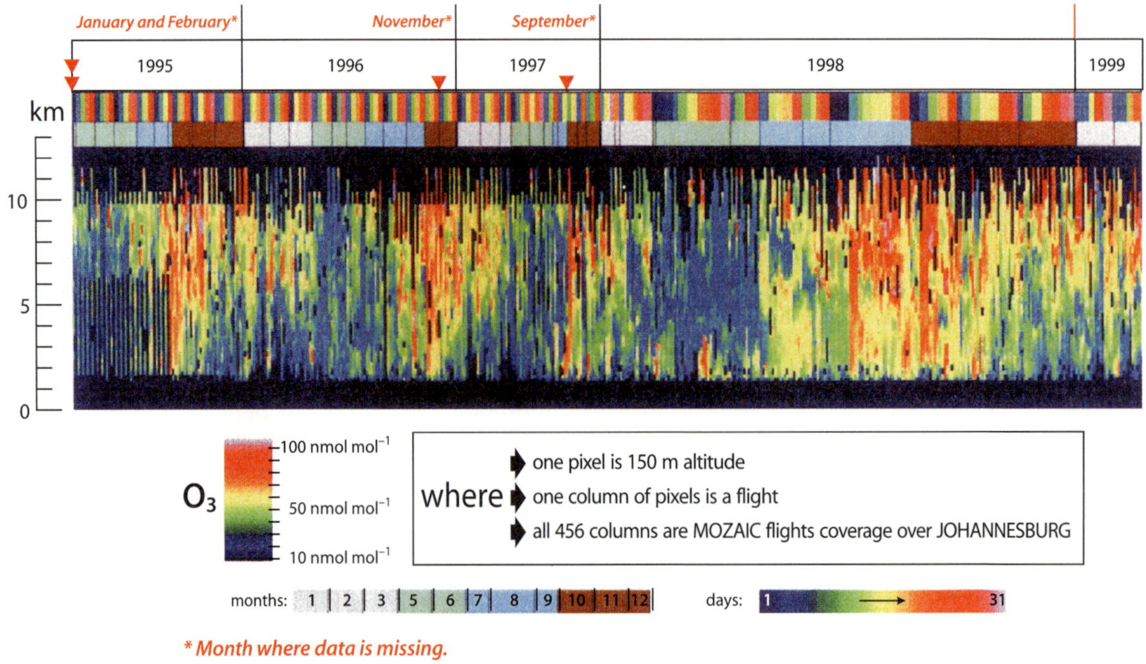

Fig. 3.36. The vertical distribution of ozone over Johannesburg between 1995 and 1999 as collected by the MOZAIC programme (courtesy of A. Marenco, CNRS Laboratoire d'Aérologie, Toulouse, France)

can continent transports O_3 and precursors away from areas where burning is taking place in the Amazon forest and savannah regions to locations further south (Thompson et al. 1996b; Singh et al. 1996d, 2000; Law et al. 2000). The influence of biomass burning on trace gas emissions and distributions is also discussed in Chap. 2; it clearly has an important influence on O_3 concentrations throughout the Tropics and the Southern Hemisphere. In summary, the MOZAIC data emphasise the fact that the O_3 distribution is affected by long-range transport of O_3 and/or its precursors in both hemispheres.

3.5.2.3 *Satellite Data*

In a series of papers over the last decade, Fishman and co-workers used concurrent measurements from SAGE (Stratospheric Aerosol and Gas Experiment) and TOMS (Total Ozone Mapping Spectrometer) to produce global maps of the amount of ozone in the troposphere. Vertical ozone profiles from SAGE were used to provide an independent measurement of the stratospheric ozone content which could then be subtracted from the collocated TOMS total ozone amount to derive the tropospheric ozone residual (TOR). This methodology was then applied to more than 40 000 concurrent measurements over more than a decade to derive the climatological seasonal depiction of tropospheric ozone (Fishman et al. 1990; Fishman and Brackett 1997). The seasonal distributions derived summarise many of the features observed by the *in situ* techniques. This includes large areas of low ozone over the tropical oceans, enhanced ozone throughout large parts of the Northern Hemisphere during the summer, and an enhancement during austral spring in the southern Tropics and subtropics resulting from widespread biomass burning (Fig. 3.37). More recently, the residual technique has been refined to produce daily maps of TOR using TOMS measurements in conjunction with stratospheric ozone data from the Solar Backscattered Ultraviolet (SBUV) instrument (Fishman and Balok 1999).

Another algorithm using TOMS data that provides information about tropospheric ozone is the convective-cloud differential (CCD) technique of Ziemke et al. (1998). In the CCD technique total column ozone is inferred from low cloud reflectivity measurements and the amount of ozone in the stratosphere is derived from column measurements taken above the tops of nearby, very high tropopause-level clouds under conditions of high reflectivity. The difference between these quantities yields the total tropospheric column ozone (TCO). A unique property of the CCD technique is that it is not affected by inter-instrument calibration errors in ozone. The CCD method works best in the Tropics where there are large quantities of strong tropopause-level convective systems throughout the year.

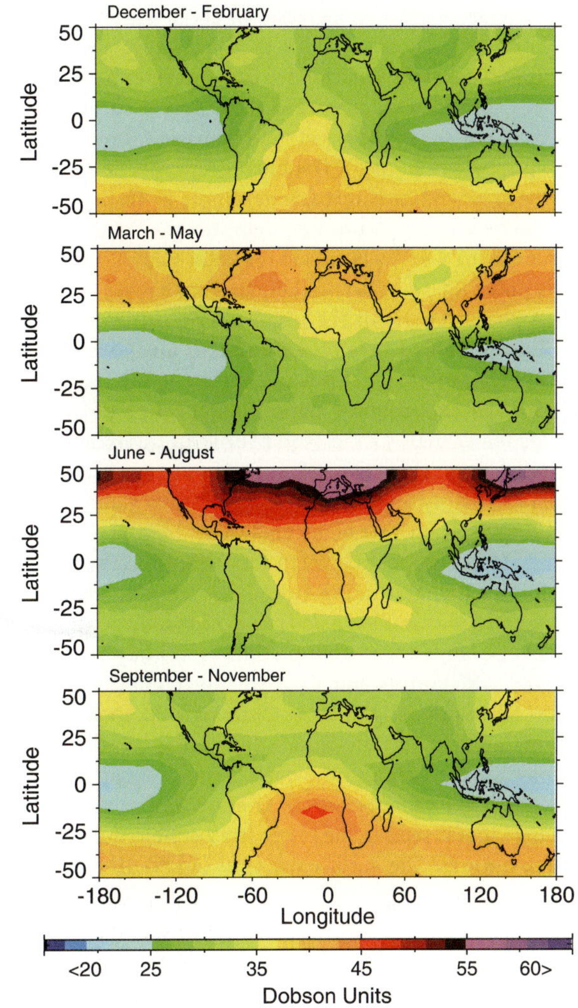

Fig. 3.37. Composite seasonal distribution of the tropospheric ozone column as determined from TOMS and SAGE (courtesy of J. Fishman, NASA Langley, US)

3.5.2.4 *Tropospheric Ozone Trends*

There is clear evidence that tropospheric ozone has increased at least by a factor of two to three from the end of 19th century as a result of increases in emissions of ozone precursors from anthropogenic sources (Volz and Kley 1988). Observational evidence is primarily based on several European records of either surface ozone measurements or ozonesonde data (Tiao et al. 1986; Staehelin et al. 1994; Marenco et al. 1994), which demonstrated that tropospheric ozone has increased mostly in the 1960s and 1970s. A similar increasing trend in the 1970s and 1980s has also been observed in the ozonesonde data from Japan (Akimoto et al. 1994).

A comprehensive analysis of trends in the ozone vertical profile and in surface ozone has been made by Logan (1994), Oltmans et al. (1998), and SPARC (2000), respectively. Typically, at European stations, such as Hohenpeissenberg, Germany, increases of about 2% yr^{-1} below 10 km have been recorded for the period 1971–1991. The temporal behaviour of ozone at Hohenpeissenberg since 1978 is similar to that recorded at the nearby surface stations at the Zugspitze (3 km) and Wank (1.8 km) (Logan 1994; Scheel et al. 1997). The bulk of increase took place in the 1970s and early 1980s. For example, at the Zugspitze the increasing trends of ozone in the first half of the 1970s is 3.06 ±1.01% yr^{-1} whereas in the latter half of the 1980s it is only 0.63 ±0.42% yr^{-1} (Oltmans et al. 1998).

A characteristic of the tropospheric ozone increase in Europe is that it is observed at all altitudes below 10 km (Logan 1994). This is in contrast to the increases of ozone at the Japanese stations that occur only at altitudes lower than 5.5 km (500 hPa) with the largest increase being below 2 km (Akimoto et al. 1994). It is important to note that the increasing trend at Naha is still continuing at the rate of 2.5 ±0.42% yr^{-1} in the 1990s, in contrast to Europe and North America; this may be ascribed to the continuing regional increasing trend of NO_x emissions in East Asia (Lee et al. 1999).

In the United States no distinct trend has been observed at Wallops Island, Virginia and Whiteface Mountain, New York (Logan 1994; Oltmans et al. 1998). Also three Canadian stations show a decrease in tropospheric ozone for the 1980s to the early 1990s. The long-term decline in the Canadian arctic troposphere parallels the change that is seen in the lower stratosphere during the same periods. The surface ozone at the South Pole has declined from the late 1970s until the 1990s, which may be associated with additional photochemical ozone loss in the troposphere due to the increased UV radiation (e.g. Ayers 1992). However, the exact reasons for the regional and temporal variations in ozone trends still remain to be determined.

3.6 Long-Range Transport of Pollution and Impact on the Ozone Budget

3.6.1 Introduction

The preceding section showing the global distribution of ozone as measured by various systems indicated that high ozone concentrations are widespread throughout the troposphere in spring and summer and that the cause is primarily tropospheric photochemistry. It is also very likely that large-scale transport of ozone is occurring which quite probably has its origin in the boundary layer over continents. This is a key finding from many of the IGAC aircraft campaigns which have been undertaken over the last decade. It is also clear that this process is not well quantified and a new IGAC subproject entitled Intercontinental Transport and Chemical Transformation (ITCT) has been formulated to investigate these processes in a systematic manner over the next decade.

In this section the intention is to highlight from various field campaigns, many with an IGAC label, which

Box 3.1. Looking at tropospheric chemistry from space

Making measurements of tropospheric composition from space is a great challenge to instrument designers. The presence of clouds and the proximity of the surface are two considerable impediments to proposed projects. Only recently – about 20 years after similar studies of the middle atmosphere became possible – has significant progress been made in this area.

The advantages of a satellite measurement are the total global coverage that may be obtained and the length of time that the measurements can be sustained, in many cases over many years. The disadvantages are the lack of measurement density in time at any point, and the bias of the measurements to clear-sky conditions, which is why satellite measurements can never totally replace other measurement techniques.

Measurements are made using either a limb or a nadir geometry. The limb path has excellent vertical resolution but poor horizontal resolution. Cloud, aerosol, and spectral interference effects also render measurements at altitudes below 8 km very problematic. Nadir geometry gives better horizontal resolution and is capable of sensing down to the surface, but the vertical resolution is poor; sometimes only a column measurement is possible.

Similarly, the gases that can be measured are restricted. Strong and unique signatures are necessary to sort out the gaseous signal from other effects. A middle atmosphere overburden can also be a problem. Measurements of ozone and CO in the troposphere have been made on a "regular" basis, but BrO, NO_2, HCHO, and SO_2 have been seen in long-term averages or when their concentrations have been very high.

The first major measurements of tropospheric composition from space were of CO made by the MAPS instrument from the shuttle in 1984 and 1994 (Reichle et al. 1986), but a technique developed by Fishman and Brackett (1977) using the differences between two satellite instruments and measurement techniques has also yielded measurements of tropospheric ozone.

More recently, instruments specifically designed for long-term measurements have been developed and flown. On the Terra spacecraft, there is the Measurements of Pollution in the Troposphere (MOPITT) instrument; on the ERS-2 satellite, we have the Global Ozone Monitoring Experiment (GOME). These will be joined by the Tropospheric Emission Spectrometer (TES) instrument on the Aura spacecraft in 2003 and the SCanning Imaging Absorption SpectroMeter for Atmospheric CHartographY (SCIAMACHY) instrument on Envisat in 2002. Each instrument has different target gases and operating scenarios, and therefore yields different views of the tropospheric system. These vary from the specific two-gas (CO and CH_4) monitoring role of MOPITT to the broad survey capabilities of the TES instrument.

The results of these early tropospheric measurements have been quite revolutionary. The early detection of an ozone feature in the troposphere off the east coast of equatorial Africa was sufficiently unusual that the TRACE-A (Transport and Atmospheric Chemistry near the Equator-Atlantic) experiment was designed to investigate the truth of the measurement. The measurements of CO from MAPS showed clear evidence of the outflow from biomass burning in the equatorial belt, and the recent MOPITT maps show not only the time evolution of those flows, but additional features due to industrial and other sources. As time goes on and the monitoring of tropospheric composition becomes more comprehensive, we can expect considerable evolution in our understanding of the global picture of the chemistry of this region.

show where ozone is being formed in the troposphere and the manner in which it is transported. This is probably the most important outcome of IGAC in the area of photooxidants. Thus there are brief discussions of the findings from studies in the continental boundary layer, and outflow from continental regions over the Atlantic and Pacific Oceans. Highlights of the experiments designed to study the impact of aircraft emissions on ozone production in the upper troposphere and lower stratosphere are also presented. These experiments taken as a whole consider a range of processes which have a strong influence on the ozone budget in the free troposphere, including chemistry, dynamics, wet and dry deposition, and free tropospheric emissions.

3.6.2 Continental Pollution

While this topic has not been a major focus of IGAC field activities, it is important to know how much and by what mechanisms O_3 is produced over continental regions in order to compare with the amount produced in the free troposphere away from source regions. An issue of particular importance, which has received much attention in the form of dedicated field campaigns, is whether the photochemical production of O_3 is controlled by the availability of NO_x or VOCs in the boundary layer or lower troposphere over continental regions (see also Box 3.2). In the United States, the impetus for many of these studies was the 1991 report of the US National Research Council Committee on Atmospheric Chemistry entitled, "Rethinking the ozone problem in regional and urban air pollution" (NRC 1992). This report drew attention to serious difficulties in understanding the nature of photochemical air pollution, particularly in the southern US, and in measures then being adopted to ameliorate the problem. This was stimulated by results that showed that much of the hydrocarbon material oxidised to produce photooxidants was emitted naturally in the form of isoprene (Trainer et al. 1987).

Box 3.2. Health effects of ozone and its precursors

Tropospheric ozone, along with multiple pollutants involved in ozone formation, has been associated with adverse health outcomes at current ambient concentrations. For ozone, evidence is principally drawn from time-series epidemiological studies (which evaluate correlations between daily concentration changes and daily changes in morbidity or mortality) and controlled human or animal experiments.

The primary mechanisms by which ozone influences respiratory health are fairly well defined. Ozone and other oxidants can react easily in the fluid lining the lung, an effect more pronounced in individuals with asthma or chronic obstructive pulmonary disease (Levy et al. 2001). Ozone has been shown to lead to increased airway resistance, increased reactivity to bronchoconstrictors, and airway inflammation (Mudway and Kelly 2000). Long-term exposures have also been shown to lead to sustained decrements in lung function, with the potential for persistent developmental changes given exposure at a young age (Paige and Plopper 1999).

These mechanisms support the observational evidence of respiratory-related health outcomes. In epidemiological studies, ozone has been linked with premature mortality, hospital admissions and emergency room visits for respiratory causes, minor restrictions of activity, and possible development of asthma (Thurston and Ito 1999; Levy et al. 2001). The approximate magnitudes of some of the health effects associated with ozone are summarised in Table 3.6.

Although it is difficult to separate the independent effects of ozone from other correlated pollutants or weather (since hot and humid days lead to both higher ozone formation and heat-related health problems), studies that evaluate the relationship between temperature and health outcomes more carefully have generally documented greater effects than studies with more naïve models (Thurston and Ito 1999). Interpretation of findings is also complicated by the range of averaging times considered in the literature (e.g. one-hour maximum, eight-hour average, 24-hour average). Because of these and other issues, the magnitudes of many ozone effects remain somewhat uncertain. As a result, the US EPA has thus far declined to incorporate ozone mortality into its benefit-cost analyses, to avoid potential double counting with particulate matter mortality (US EPA 1999).

Along with these effects from ozone exposures, precursors such as carbon monoxide (CO), nitrogen oxides (NO_x), and various non-methane hydrocarbons (NMHC) can have their own independent effects. It is well known that CO preferentially reacts with haemoglobin in the blood, reducing oxygen transport and leading to symptoms ranging from headaches to depressed heart and respiratory rates to death at extremely high concentrations. At lower ambient levels, there is weak but suggestive evidence of possible links with premature mortality, with stronger evidence of an influence on cardiovascular morbidity for susceptible sub-populations (e.g. heart disease hospital admissions for elderly individuals with pre-existing disease) (US EPA 2000). Understanding CO health effects at low ambient levels is complicated by the high correlation with other automotive-related pollutants, which may have independent effects on health.

NO_x has generally been linked with fewer health outcomes at ambient levels than ozone, although the pollutants share the same potential for oxidative activity in the lungs. At extremely high levels (often found in poorly ventilated indoor settings with combustion sources), NO_x has been linked with shortness of breath, cough, and hemoptysis (Hedberg et al. 1989). Increased rates of respiratory illness and symptoms have been found in children exposed to elevated indoor NO_x (Hasselblad et al. 1992), but evidence of significant health effects from outdoor NO_x has been somewhat inconsistent. As a result, the US EPA has only considered respiratory symptoms and hospitalisations for selected causes in its benefit-cost analysis of the Clean Air Act (US EPA 1999).

Finally, determining the health effects of NMHC is complicated by the fact that numerous compounds are involved, with the composition varying by location. Some compounds, such as benzene and 1,3-butadiene, are considered to be human carcinogens at ambient concentrations. Benzene has been classified as a known human carcinogen, based on evidence of increased rates of leukaemia in exposed workers (WHO 2001). 1,3-butadiene is considered a probable human carcinogen, with sufficient rodent studies but limited information in humans (WHO 2001). On the other hand, many NMHC associated with ozone formation (such as *d*-limonene) have no documented carcinogenic or non-carcinogenic risks at ambient concentrations. Volatile organics can also adsorb onto particles, forming carbonaceous particles that may be associated with numerous particulate-related health effects.

Table 3.6. Selected central estimates of the effects of short-term or long-term ozone exposure (adapted from Levy et al. 2001)

Outcome	% change in outcome per 10 µg m^{-3} increase in daily average O_3
Premature mortality (all ages)	0.5
Respiratory hospital admissions (all ages)	2
Minor restricted activity days (age 18+)	3
Chronic asthma (males age 27+)	8[a]

[a] Based on 20-year average concentrations; percentage refers to incident cases of adult asthma.

Many experimental campaigns have taken place over the last decade that have addressed these issues. Only a couple of examples are discussed here as this is not the main focus of this chapter. For example, in the United States, the Southern Oxidants Study (SOS), which carried out a series of large field experiments in the 1990s, assessed the relative importance of biogenic emissions, emissions from cities in the form of urban plumes, and emissions from power plants, which are rich in nitrogen oxides but lean in reactive hydrocarbons (Cowling et al. 1998, 2000).

Several campaigns have also taken place over Europe. For example, during the EU Testing Atmospheric Chemistry in Anticyclones (TACIA) project it was found that over the most polluted regions of northeast Europe (southern England, northern France, Belgium, and The Netherlands) O_3 production was VOC limited rather than NO_x limited which is more generally the case (Hov et al. 1998). This was due to the very high levels of NO_x that had built up under the anticyclonic conditions. It was also found that large emission rates of short lived hydrocarbons, including biogenics such as isoprene, have the potential to produce significant quantities of O_3 over Europe particularly in warm summer conditions, a finding that had already been confirmed by SOS for the United States during the summer months. A full review of the natural emissions over Europe can be found in an article by Simpson et al. (1999).

3.6.3 North Atlantic Ocean Field Campaigns

The North Atlantic region has been the focus of several large field campaigns (carried out principally under the auspices of the IGAC North Atlantic Regional Experiment, NARE) investigating the distribution and budget of O_3 and its precursors. NARE was a joint US and European experiment to investigate whether pollution plumes from North America could in fact be observed over the North Atlantic, and if the pollution levels are high enough to be significant for the regional O_3 budget. Two major international aircraft experiments took place; NARE1 in summer 1993 and NARE2 in 1997. In addition, other smaller campaigns took place at other times together with regular monitoring at surface sites. Some data from NARE, emphasising the role of frontal uplift of pollution, have already been described in Sect. 3.4 concerned with transport.

Surface measurements of CO and O_3 provided an extremely useful tool for initial investigations of the anthropogenic impacts on the tropospheric O_3 budget in the North Atlantic. Three sites were operated in the Maritime Provinces of Canada at distances progressively further downwind from the eastern US seaboard. North American pollution plumes were readily detectable at all three sites, the furthest being 1 500 km downwind from the sources (Parrish et al. 1993a) as shown in Fig. 3.38. Subsequently, measurements at a site in the Azores located 3 000 km downwind from the sources established that North American pollution enhances O_3 in the central North Atlantic, at least in the spring (Parrish et al. 1998).

These qualitative observations of O_3 and CO transport were important, but they also provided the basis for the quantitative assessment of the amount of anthropogenic O_3 transported from North America. Based upon the observed ratio of O_3 to CO and emission inventories of CO, Parrish et al. (1993c) concluded that in the summer, O_3 transported from North America exceeded natural O_3 from the stratosphere in the lower troposphere over the North Atlantic. From these measurements, further refinements of the quantitative calculations indicate that the total flux of ozone from North America to the North Atlantic in the summer is on the order of 1.0–1.6 Gmol day^{-1} (Chin et al. 1994; Berkowitz et al. 1996). These studies have conclusively shown that during the summer the O_3 budget in the lower troposphere above the area near the western seaboard of the North Atlantic is dominated by photochemically produced O_3 derived from North American sources.

The form in which continental outflow of oxidants takes place is shown rather well in some model simulations used in support of the European NARE studies in 1997. Figure 3.39 shows high levels of ozone at ground level over the eastern part of the US in mid August 1997, extending into the Atlantic in patterns similar to those observed on the ground by the US studies in the Maritime Provinces of Canada (Parrish et al. 1993b). Three days later the same air had been lifted up to an altitude of 7 km and had carried substantial amounts of ozone over to the European side of the ocean by similar processes to those discussed in Sect. 3.4.3 (Flatøy et al. 2000). Layers of air containing high concentrations of ozone

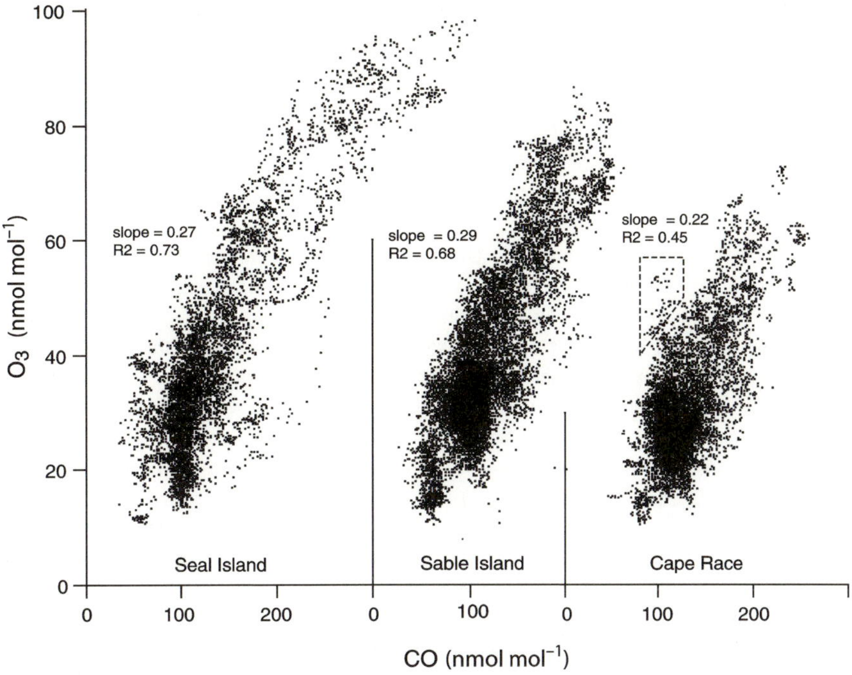

Fig. 3.38.
Correlations between ozone and carbon monoxide in North American pollution plumes detected at three sites in the Canadian Maritime Provinces at increasing distances from the source (From Parrish et al. 1993c)

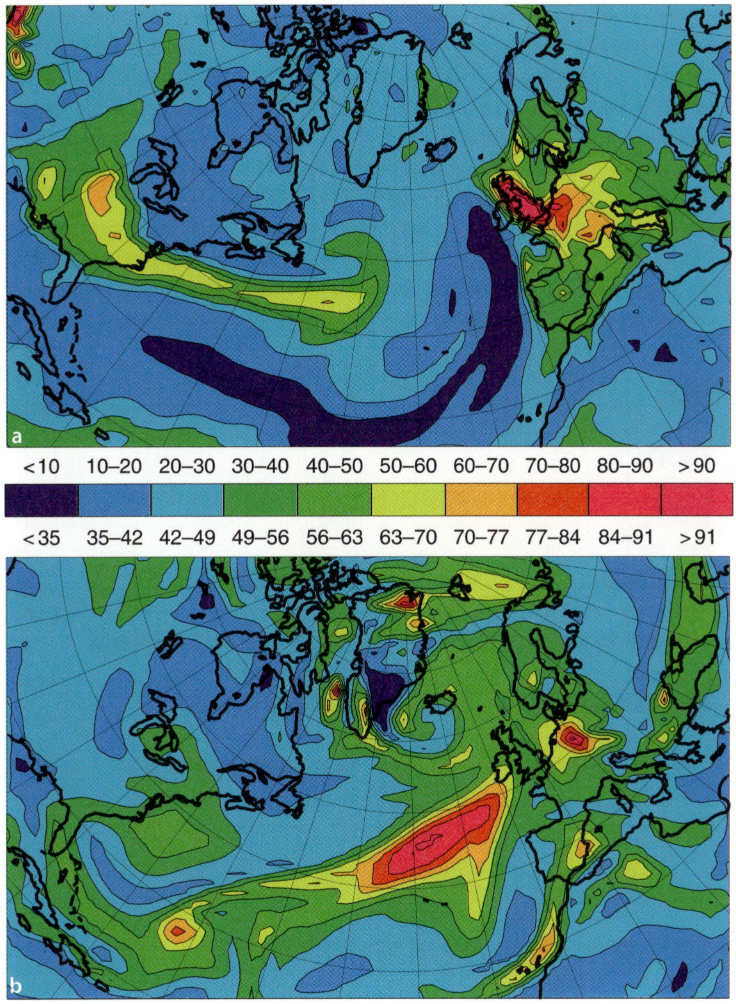

Fig. 3.39.
The development of a plume of ozone pollution (nmol mol^{-1}) over the Atlantic in August 1997 as depicted by a three-dimensional chemistry and transport model. **a** 19 August 1997, O_3 (nmol mol^{-1}) at ground level; **b** 21 August 1997, O_3 (nmol mol^{-1}) at 7 000 m (Flatøy et al. 2000)

with an American origin are frequently observed in aircraft experiments over the North Atlantic and Europe (e.g. Wild et al. 1996; Hov et al. 2000). LIDAR measurements over southern Germany have also detected the presence of high ozone levels in air uplifted by a warm conveyor belt from the continental boundary layer over the US (Fig. 3.29) (Stohl and Trickl 1999). Interestingly, observations of layers containing low ozone concentrations have also been made in the upper troposphere over Aberystwyth in west Wales by Davies et al. (1998) who concluded that their origin was the tropical Pacific Ocean. The observations of both high and low ozone in layers with a distant origin emphasise the role of transport in determining the ozone budget over the Atlantic.

3.6.4 Pacific Ocean Field Campaigns

Several large field campaigns have also taken place over the Pacific region organised under the auspices of NASA GTE Pacific Exploratory Missions (PEM) and the IGAC activity, East Asia/North Pacific Regional Study (APARE).

3.6.4.1 *Northwest Pacific*

The PEM-West missions characterised in detail the chemical composition of air masses and the meteorological factors that controlled their distribution over the northwest Pacific (e.g. Gregory et al. 1997; Talbot et al. 1997; Blake et al. 1997). Transport in this region was influenced strongly by the position of the westerly jet, and the Pacific High, and by vertical mixing from convective processes over continental Asia, associated with typhoons (Merrill et al. 1997). Because O_3 production is closely linked to the distribution of NO_x (and other precursors), the observed distributions and transport patterns had a strong influence on the regional and altitudinal distribution of net ozone production (e.g. Davis et al. 1996; Crawford et al. 1997a,b). For example, Crawford et al. (1997a) examined the impact of regions characterised by high and low levels of NO_x observed during PEM-West B. The high NO_x regime was characterised by convective transport from continental regions, including a significant source of NO_x from lightning; the low-NO_x regime was associated with air masses of tropical marine origin. The differences in NO_x (and H_2O) resulted in a significant contrast in the photochemical ozone budgets of the two regions. Ozone production in the high NO_x region exceeded that in the low-NO_x region by factors of two to six. Ozone destruction rates, though, were as much as three times higher in the low NO_x regime. Extrapolating a model to include altitudes up to 17 km (tropopause height) showed that the high NO_x regime had a small net ozone production while the low NO_x regime strongly destroyed ozone.

Seasonal aspects of ozone photochemistry in the northwestern Pacific were also examined from the data in PEM-West A (Sept.–Oct. 1991) and B (Feb.–March 1994). Crawford et al. (1997b) showed that the ozone budget of the region was strongly impacted by continental emissions in the lower atmosphere during the early spring. Thus, while the net ozone production during PEM-West A switched from negative at low altitudes to positive at altitudes >6 km, net ozone production during PEM-West B was positive at all altitudes (Fig. 3.40). The authors note that the observed ozone production in the marine boundary layer may be ex-

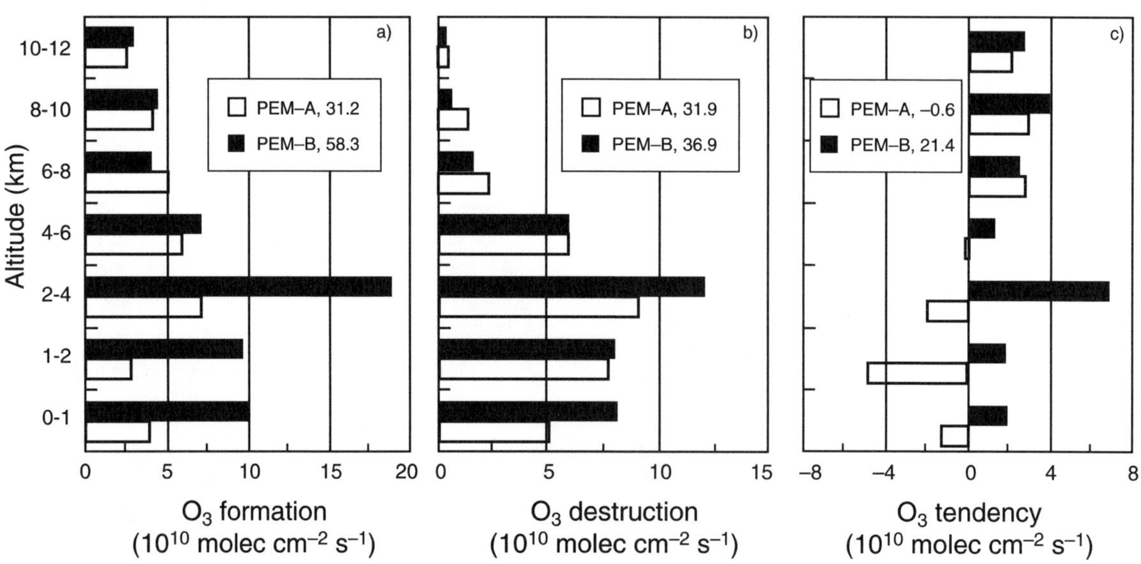

Fig. 3.40. Diurnal-average column-integrated values of **a** $F(O_3)$, **b** $D(O_3)$, and **c** $P(O_3)$. Values are calculated from median conditions for PEM-West A (18–42° N) and PEM-West B (20–30° N). Total tropospheric column amounts are annotated at the bottom of the figure (from Crawford et al. 1997b)

pected near coastal regions, but there are suggestions that such ozone production could extend far from the Asian Rim. Consequences of continental outflow on ozone budgets may be expected on large regional or global scales. Significantly, the measurements and the model results from the PEM-West studies indicated the predominance of *in situ* ozone production compared to stratospheric ozone influx on the tropospheric ozone budget. The column-integrated ozone production rate is nearly six to 12 times the estimated northern hemispheric stratospheric flux of ozone.

3.6.4.2 *Tropical Pacific*

PEM-Tropics A flew at the end of the dry season of the southern Tropics and observed extensive biomass burning emissions transported 10 000 km or more from fires in South America and southern Africa (Sept.–Oct. 1996). PEM-Tropics B, conducted in the wet season of the southern Tropics, observed exceedingly clean atmosphere over the South Pacific (March–April 1999). This contrast is illustrated in Fig. 3.41 with mean ozone concentrations measured remotely by DIAL as a function of altitude and latitude. The large ozone enhancements over the South Pacific during PEM-Tropics A are due to biomass burning and are absent during PEM-Tropics B. These results reinforced conclusions from the earlier PEM-West missions that the tropospheric ozone budget is determined mostly from a balance between *in situ* production and loss processes, and by transport from continental regions (Hoell et al. 1999; Schultz et al. 1999).

A significant finding from PEM-Tropics A was that biomass burning emissions from Africa and possibly South America had a large impact on the levels of various trace gases and on the oxidising capacity of the South Pacific. NO_x levels were sustained in the lowest 4 km by decomposition of PAN transported from biomass burning regions. Furthermore, it was suggested from model analyses that westerly transport of ozone from continental pollution is approximately equal in magnitude to the *in situ* chemical production of ozone. Considering *in situ* processes only, photochemical destruction of ozone was found to be approximately twice that of ozone production. Net photochemical loss of 18×10^{10} molecules $cm^{-2}\,s^{-1}$ was balanced by influx from continental sources (Schultz et al. 1999).

Observations during PEM-Tropics B indicated two major barriers for atmospheric transport, the ITCZ and the South Pacific Convergence Zone (SPCZ). Pollution layers were almost exclusively confined to the north of the ITCZ, while the SPCZ to the south separated clean equatorial and midlatitude air masses. Considerable vertical mixing by convection was observed in the equatorial wedge between the ITCZ and the SPCZ with rather low O_3 concentrations (4–20 nmol mol^{-1}) observed throughout the troposphere (Browell et al. 2001). These data from PEM-Tropics B confirm previous balloon sonde measurements made by Kley et al. (1996).

North of the ITCZ measurements over the tropical North Pacific had relatively high O_3 concentrations, revealing a complex mix of combustion influences. A major component was long-range transport from the Eurasian continent at midlatitudes in the prevailing westerly flow, followed by subsidence around the Pacific High and transport to the Tropics in the lower troposphere. This Eurasian influence extended over the full longitudinal extent of the Pacific and appeared to include major pollution contributions from both Europe and Asia (Staudt et al. 2001). Long-range transport from western North America at low altitudes around the Pacific High and into the easterly trade wind circulation was found to make a significant contribution to pollution levels in the tropical boundary layer over the north Pacific. Biomass burning from fires in southeast Asia (particularly active during the PEM-Tropics B period) made a major and perhaps dominant contribution to pollution in the free troposphere.

This combined transport of pollution from a multiplicity of sources over the tropical north Pacific, channelled by subsidence and transport around the Pacific

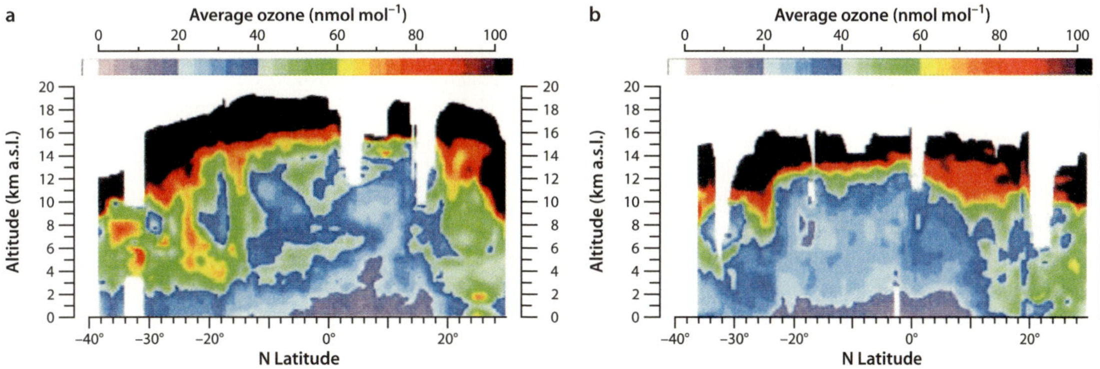

Fig. 3.41. Average latitudinal ozone distributions over the tropical Pacific Ocean as measured by LIDAR. **a** PEM Tropics A, August–October 1996, latitudinal ozone distribution (–170 to –120° E long.); **b** PEM Tropics B, March–April 1999, latitudinal ozone distribution (–170 to –120° E long.) (Browell et al. 2001)

High, resulted in a continuous and persistent air pollution "river" in the boundary layer. These air masses, with elevated levels of trace gases such as CO were carried southwestward by the trade winds (Bey et al. 2001a,b). Evidence for interhemispheric transport of trace gases, based on concentration gradients of hydrocarbons and halocarbons, was also found in the upper troposphere in the region of the ITCZ and, to a lesser extent, near to the SPCZ (Staudt et al. 2001).

Therefore, it is clear that emissions from tropical Asia, in particular the Indonesian maritime continent and northern Australia, can have a large impact on chemical composition over the tropical Pacific. Combustion, lightning, and biomass burning sources all contribute. For example, during the Biomass Burning and Lightning Experiment (BIBLE), conducted between 1998 and 1999, enhanced concentrations of O_3 precursors over Indonesia (Kita et al. 2001; Koike et al. 2001) were observed. Elevated concentrations of O_3 and its precursors were found in air masses originating from further away over southern Africa, South America, the Indian Ocean, and southeast Asia (Kondo et al. 2001). In 1997, during an El Niño, enhanced O_3 concentrations related to large-scale biomass burning over Indonesia were also observed (Kita et al. 2000).

3.6.5 Studies of Aircraft Emissions

Several aircraft campaigns have studied the impact of aircraft emissions on the levels of NO_x and ozone in the North Atlantic Flight Corridor (NAFC). Radiatively, the upper troposphere is a region which is particularly sensitive to perturbations in ozone concentration (Lacis et al. 1990; Forster et al. 1997). The ozone budget of this region is still very uncertain with significant contributions from photochemistry, and transport of ozone and precursors from the lower troposphere and the stratosphere. Production of ozone from natural emissions of NO_x from lightning (see Sect. 3.3.2 and 3.3.4) and associated convective transport of surface pollutants (see Sect. 3.3.4) are also important factors which have received considerable attention over the last decade. In particular, the POLINAT (Pollution from Aircraft Emissions in the North Atlantic Flight Corridor), and SONEX (SASS-Subsonic Assessment-Ozone and NO_x Experiment), and NOXAR campaigns (see Sect. 3.4.2) investigated the contribution of aircraft NO_x emissions to the budget of NO_y and O_3 in the upper troposphere (Schlager et al. 1999; Schumann et al. 2000; Jeker et al. 2000; Singh et al. 1999, and related publications in these special sections).

Measurements during both SONEX and POLINAT-2 in 1997 showed clear evidence of enhanced concentrations of NO_x, SO_2, and CN in the NAFC as a result of recent emissions from aircraft in the summer and autumn. This was also observed in previous campaigns over the North Atlantic as shown in Fig. 3.42 which is a composite figure of all the NO data collected during POLINAT (and related campaigns) from 1994 to 1997 (Schumann et al. 2000). In 1997 the flights also identified large-scale enhancements of NO mixing ratios inside the corridor of about 50–150 pmol mol^{-1} producing a local maxima in NO_x and NO_y between 10.5 and 11.0 km. Occasionally, enhancement in NO_x of up to 3 nmol mol^{-1} was observed. While no corresponding O_3 accumulations could be detected in the NAFC, direct measurements of HO_x and NO_x were used to show that *in situ* O_3 production occurs in this region and is rather insensitive to NO_x concentrations at high levels of NO_x (see Sect. 3.3). The mean net production rate of O_3 was calculated to be 0.5 nmol mol^{-1} per day which can be contrasted to rates of 10 nmol mol^{-1} per hour or more in the continental boundary layer.

Models have been used to diagnose the contribution from different sources to the budget of NO_y in the upper troposphere. An example is shown from Allen et al. (2000) in Fig. 3.43. Here, a global model with increased resolution over the NAFC was run for the SONEX period. Results show that aircraft, lightning, and surface sources of NO_x all make significant contributions to NO_y in the upper troposphere. Many model calculations have been carried out to assess the impact of aircraft NO_x emissions on O_3 concentrations in the troposphere. In general changes are predicted to be rather small when comparing simulations with and without aircraft ranging from 2–10% increase in O_3 in northern midlatitudes at cruise altitudes (IPCC 1999; Brasseur et al. 1998a). It is clear that these calculations will have to be revisited as models improve and our understanding about other important processes, such as uplift of pollutants from the surface, increases.

3.7 Summary of Principal Achievements and Remaining Uncertainties

The IGAC Project of IGBP has been directly responsible for stimulating a large amount of research into virtually all aspects of the chemistry of the troposphere. In the area of photochemistry, which has been the subject of this chapter, the research has had many successes that have considerably improved our understanding of the nature of atmospheric chemical processes, and of their overall impact on the composition of the troposphere.

3.7.1 Achievements

A list of the main achievements includes the following:

- Showing that the distribution and extent of ozone in the troposphere is mostly controlled by *in situ* chemistry and transport, rather than simply by injection from the stratosphere and deposition at the surface.

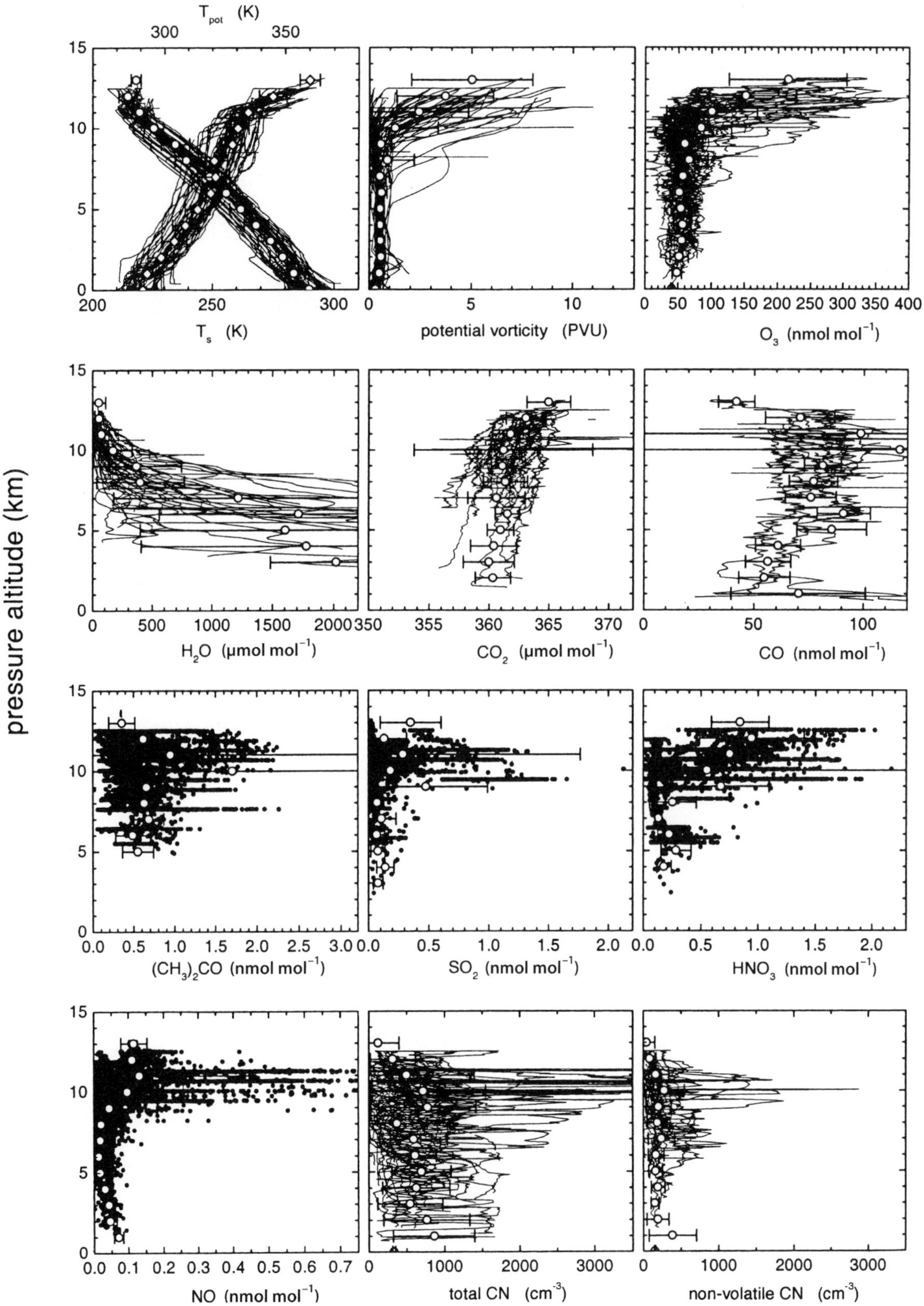

Fig. 3.42. Composite vertical profiles of temperature, potential vorticity, and concentrations of chemical compounds vs. pressure altitude from all POLINAT flights in the NAFC domain in 1994, 1995, and 1997. The *open circles* with *error bars* represent mean values over 1 km altitude intervals and the standard deviation (Schumann et al. 2000)

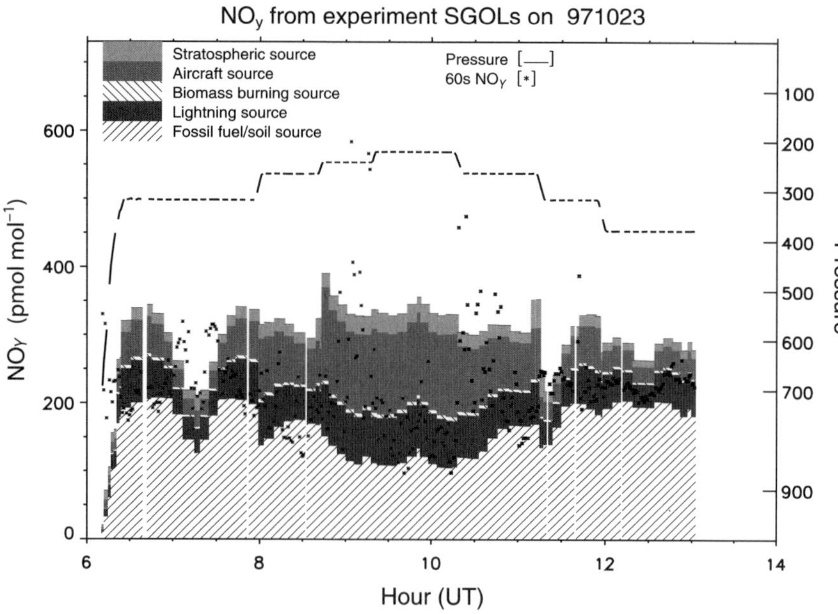

Fig. 3.43. Model-calculated vs. measured NO$_y$ (pmol mol^{-1}) on October 23, 1997. NO$_y$ (pressure) data after averaging over 60 s are shown by *asterisks* (*a dashed line*). Model-calculate NO$_y$ is shown for the grid volume containing the measurement. The value for each 60-s period is obtained by linearly interpolating model output that is available at 0000, 0600, 1200, and 1800 UT. *Shading* and *patterns* are used to show the model-calculated contribution from each term as well as total NO$_y$. Time periods where data are missing are *unshaded*. Model output from experiment SGOLs (Allen et al. 2000)

- Validating the basic free radical chemistry theory of tropospheric chemistry involving HO$_x$ under conditions of low NO$_x$.
- Demonstrating that transport of pollution from the continents has a large impact on the composition of most of the troposphere. This includes ozone and its precursors, and associated photochemically-produced products, particularly nitrogen compounds.
- Proving the existence of other types of free radical chemistry including halogen and nitrate radical chemistry.
- Carrying out coordinated field campaigns, both ground-based and from aircraft, which have led to the establishment of large databases on tropospheric composition.
- Developing and applying 3-D global models of tropospheric chemistry capable of successfully replicating observations, and providing estimates of the global budgets for ozone and other major trace gases.
- Developing important new instrumentation for measurements of stable gases and free radicals which have been used extensively in IGAC field campaigns.
- Obtaining the first global scale maps for trace gases from satellites and from commercial aircraft.

3.7.2 Uncertainties

In spite of these achievements, which have revealed the true nature of chemistry in the troposphere on a global scale, many uncertainties remain that require further research. These include:

- The links between chemistry and climate on the one hand and climate and chemistry on the other.
- Determining the impact of large-scale pollution of the troposphere by ozone and by reactive nitrogen compounds on climate and regional air pollution, and on fertilisation of the oceans and land surfaces.
- Quantifying by experiment the various processes known to affect global tropospheric ozone, in particular how and where ozone is made in the troposphere.
- Determining trends for reactive molecules such as ozone and free radicals, particularly OH.
- Detecting the presence of heterogeneous chemistry and quantifying its influence on free radical chemistry and ozone destruction.
- Quantifying the global influence of halogen radical chemistry, particularly in ozone destruction, and nitrate radical chemistry, particularly in the marine boundary layer.
- Studying the impact of increased emission of pollution from developing regions in Asia, South America, and Africa on global tropospheric ozone.
- Improving the understanding of the transport processes that affect the ozone budget such as STE, frontal systems, and convection.
- Removing the large uncertainties in the magnitude of many natural processes, including emissions from vegetation, lightning, and the ocean and the potential impact of climate change.
- Identifying the full range of compounds which are emitted into the atmosphere from both natural and anthropogenic processes.
- Determining detailed global distributions of the most important trace gases in tropospheric chemistry including H_2O, NO$_x$, NO$_y$, CO, NMHCs, peroxides, and carbonyl compounds.
- Obtaining global-scale measurements of many free radicals including OH, HO_2, RO_2, BrO, IO, ClO, and NO_3.

Chapter 4

Tropospheric Aerosols

Lead authors: Jost Heintzenberg · Frank Raes · Stephen E. Schwartz
Co-authors: Ingmar Ackermann · Paulo Artaxo · Timothy S. Bates · Carmen Benkovitz · Keith Bigg · Tami Bond · Jean-Louis Brenguier Fred L. Eisele · Johann Feichter · Andrea I. Flossman · Sandro Fuzzi · Hans-F. Graf · Jeremy M. Hales · Hartmut Herrmann Thorsten Hoffmann · Barry Huebert · Rudolf B. Husar · Ruprecht Jaenicke · Bernd Kärcher · Yoram Kaufman Geoffrey S. Kent · Markku Kulmala · Caroline Leck · Catherine Liousse · Ulrike Lohmann · Beatrice Marticorena Peter McMurry · Kevin Noone · Colin O'Dowd · Joyce E. Penner · Alex Pszenny · Jean-Philipe Putaud · Patricia K. Quinn Ulrich Schurath · John H. Seinfeld · Herman Sievering · Jeffrey Snider · Irina Sokolik · Frank Stratmann · Rita van Dingenen Douglas Westphal · Anthony S. Wexler · Alfred Wiedensohler · David M. Winker · Julian Wilson

4.1 Introduction

Between 1970 and 1990 the major advances in atmospheric chemistry were made in gas-phase photochemistry, except perhaps for a brief intermezzo of "nuclear winter" studies. This focus is now shifting, as it is recognised that natural and anthropogenic aerosols play a substantial role in the radiative properties of the atmosphere and Earth's climate. In addition, studies on the causes of the Antarctic ozone hole have demonstrated the large role of reactions that take place on ice and particulate surfaces. If such reactions occur in the stratosphere, they must take place also in the troposphere, with its abundance of various types of aerosol. Considering these factors, and especially because of various breakthroughs in experimental techniques, it is likely that aerosol research will be prominent in atmospheric chemistry in the coming decades. This research will involve process studies both in the atmosphere and in laboratories, studies on the sources and sinks of aerosols, chemical analyses of the particulate matter (PM), modelling, and especially regional (campaigns) and global (satellites) observations on the distribution of the atmospheric aerosol. This is all the more important because climate models, which in most cases currently consider only sulphur chemistry, cannot be tested sufficiently for want of data, despite the potentially great climate effects of aerosols. Aerosol particles may already be significantly counteracting the radiative forcing by the greenhouse gases (Ramaswamy et al. 2001).

One of the most important advances in aerosol research in the past decade has been the realisation that it will not be possible to understand (and even less so to be able to predict) the chemical state of the atmosphere without taking into account its multi-component and multi-phase nature. Figure 4.1 shows in a generic way that atmospheric aerosol particles exhibit a wide range of sizes, from nanometers to micrometers, and a range of shapes, and that chemical composition usually differs among the size ranges and even among individual particles within a given size range. These properties distinguish aerosols from trace gases, for which knowledge of the mass concentration or mixing ratio is sufficient to characterise their abundance and to permit specification of their properties, reactions, and effects. Particle size is central to the description of the radiative effects of aerosols and to their influences on clouds. Likewise, composition affects the hygroscopic growth of aerosol particles, again affecting their radiative influences and ability to form cloud droplets. For these reasons the ability to characterise particle size and size-dependent composition and to represent these properties in large-scale chemical transport models is essential.

As aerosol particles are strongly coupled with gas-phase chemistry and clouds, an understanding of their properties and effects, and of the atmosphere in general, requires that gases, aerosol particles, and clouds be treated as a single system. The scales to be considered in this system range from molecules and nanometer-sized aerosol particles to frontal cloud systems spanning hundreds of kilometres.

An example for the gas-particle-cloud coupling is that aerosol particles and cloud droplets can influence gas-phase chemistry by acting as sinks of reactive species and by decreasing or increasing actinic flux. Clouds also serve as reaction media that can release reactive species to the gas-phase. As such, aerosol particles and cloud droplets affect the so-called oxidising capacity of the atmosphere, which determines the chemical lifetimes of atmospheric trace substances including condensed phases (see Chap. 3).

An example highlighting the range of scales that must be considered is that clouds cover roughly half of the earth's surface; yet any individual cloud is composed of billions of individual droplets or crystals, each of which started its life as an aerosol particle. Thus, understanding and predicting the characteristics and behaviour of clouds on a global scale will not be possible without an understanding of the microscale processes that create and control clouds.

Aerosol particles have a multitude of sources. They derive from primary sources, involving direct emissions of particles, and from secondary processes, i.e. reactions of gaseous precursors in the atmosphere, to form particles. Some of the natural processes leading to particle production and the anthropogenic processes leading to

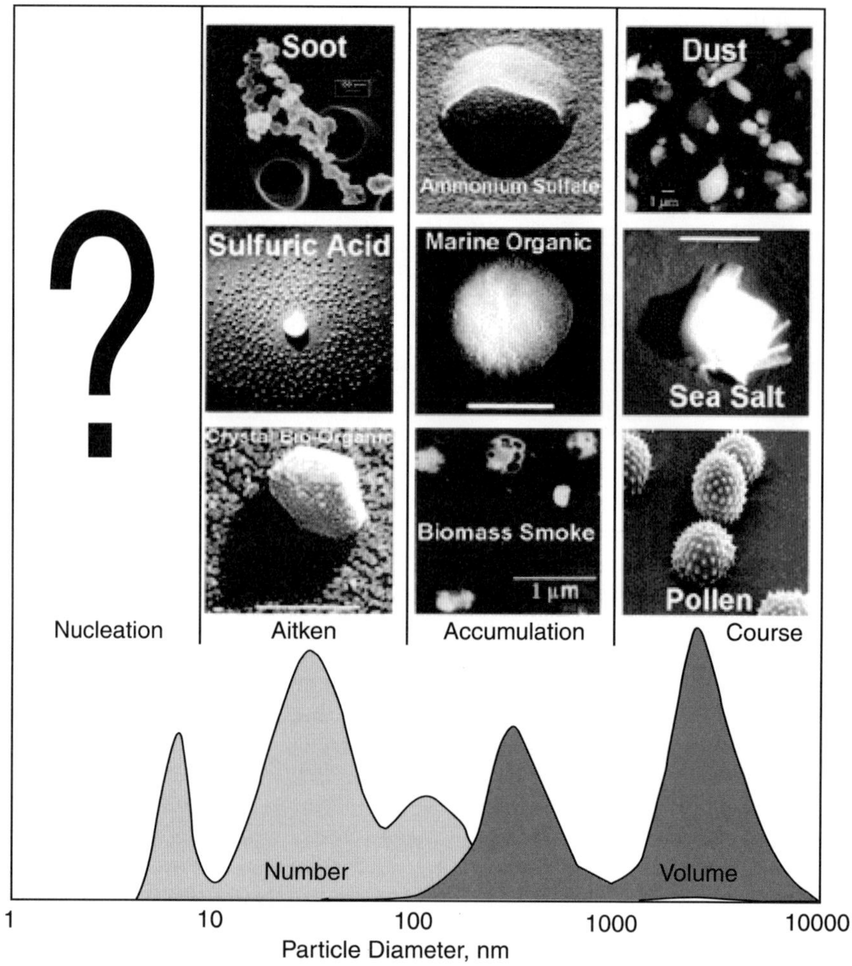

Fig. 4.1.
Generic presentation of the range of particle shapes and chemical composition of the atmospheric aerosol

Table 4.1. Natural processes leading to particle production and anthropogenic processes leading to changes of the natural aerosol

Species	Natural processes	Anthropogenic processes	Present day particle burden compared to pre-industrial time	Elements of climate change potentially affecting emissions of species in column 1
Primary particle				
Mineral dust	Wind erosion	Land use change; Industrial dust emissions	Increased dust	Changing winds and precipitation
Sea salt	Wind			Changing winds
Biological particles	Wind, biochemical processes	Agriculture	?	Changing winds
Carbonaceous particles	Vegetation fires	Fossil fuel; Biomass burning	Increased carbonaceous particles	Changing precipitation (droughts…)
Precursors of secondary particles				
Dimethylsulphide	Phytoplankton degradation	Increased oxidising capacity	Increased sulphate	Changing winds hence air-sea exchange
SO_2	Volcanic emissions	Fossil fuel combustion	Increased sulphate	
NH_3	Microbial activity	Agriculture	Increased ammonium nitrate	
NO_x	Lightning	Fossil fuel combustion	Increased nitrate	Change in convective activity hence lightning
Volatile organic compounds	Emission from vegetation	Increased oxidation capacity; Industrial processes	Increased organic aerosol	

changes in natural aerosol are summarised in Table 4.1. Anthropogenic effects on the atmospheric particle burden can be direct, through changing and adding emissions of particles or their precursors. The effects can also be indirect, as consequences of changing the transformation processes leading to secondary aerosol particles (e.g. altered oxidising capacity of the atmosphere), or of climate change. The effects of climate change relevant for particle production are exemplified in the last column of Table 4.1.

The discussion of recent developments presented in this chapter points to a highly exciting future of this subject as an integrative component of atmospheric chemistry.

Box 4.1. Definitions

Optical thickness: A measure of extinction of direct solar beam by transmittance through the atmosphere. The fractional transmittance of the direct beam for solar zenith angle, θ, is expressed in terms of the optical thickness, τ, at a given wavelength as $\exp(-\tau / \cos\theta)$; that is, τ is the optical thickness for overhead sun. Optical thickness consists of additive contributions from Rayleigh scattering, gaseous absorption, and aerosol scattering and absorption.

Ångström exponent: A measure of the dependence of aerosol optical thickness τ_a on wavelength λ, $\alpha = -d \ln \tau_a / d \ln \lambda$. A greater value of Ångström exponent corresponds to smaller size aerosol particles.

S(IV): The Roman numeral IV in parentheses indicates the oxidation number of +4 of the sulphur atom S in SO_2 and its family of aqueous dissolution products, i.e., bisulphite (HSO_3^-) and sulphite (SO_3^{2-}). The S atom in sulphate (SO_4^{2-}) has a higher oxidation number of +6 and is a member of the S(VI) family. Oxidation numbers are discussed in most general chemistry texts.

4.2 Integrated View of the Present State of Knowledge of Atmospheric Aerosols

During the past decade, model calculations, *in situ* observations, and measurements from satellites have looked in different ways at the atmospheric aerosol. The first space-borne radiometers that probed the troposphere revealed the transport of aerosols on regional and global scales. Measuring campaigns focussing on areas of interest revealed by those satellite images documented a large variety of aerosol physical and chemical properties, which guided the way to a better understanding of the processes controlling the evolution of the aerosol particles. Model calculations helped to identify links between aerosol sources, atmospheric chemistry, transport, and some of the characteristics of the observed fields.

What emerged from these different perspectives is that there is no such thing as a global aerosol. Instead there exists a superposition of largely independent regional aerosol plumes and layers, each having unique sources and correspondingly differing spatial and seasonal patterns as well as specific microphysical and chemical characteristics.

4.2.1 Space-Borne Measurements

4.2.1.1 *Maps of Aerosol Optical Depth from Space-Borne Radiometers*

Global horizontal patterns of aerosol optical thickness and Ångström exponent (see Box 4.1 for definitions) over the oceans derived from the POLDER instrument for December 1996 and June 1997 are shown in Fig. 4.2a,b. The oceanic aerosol optical depth maps reveal two distinctly different spatial patterns: aerosol plumes originating from continents and oceanic aerosol patches that are detached from the continents. Near coastal areas the continental aerosol plumes are characterised by high concentrations, which decrease with distance from the coast. This pattern arises from aerosol emissions from the continents, followed by atmospheric dispersion, transformation, and decay in the downwind direction. In large-scale flow fields, such as the trade winds, these continental plumes persist over several thousand kilometres.

With some *a priori* knowledge of source distributions, the plumes illustrated in Fig. 4.2 can be attributed to mineral dust, biomass burning, and industrial emissions. The dominant plumes are in the Tropics and subtropics and are due to mineral dust emissions and biomass burning; the aerosol levels generally peak in the dry seasons in the respective hemispheres.

The Ångström exponent shown in Fig. 4.2a exhibits a very different pattern from that of the aerosol optical thickness; specifically, it exhibits high values in or near industrialised regions and regions of biomass burning, indicative of small particles arising from direct emissions from combustion sources and/or gas-to-particle conversion, and low values associated with large particles in plumes of soil dust from deserts and in sea salt aerosols.

Sand dunes are shown on Fig. 4.2b as yellow patches over the Sahara and northwestern China. However, it should be noted that sand dunes are not necessarily strong or exclusive sources of wind-blown dust. The emission of wind-blown dust is seasonal and it is driven by highly episodic windy conditions. Frequently the sources of such aerosols are pans of dry silty soil that have resulted from prior rain events. The characteristic dust particle is in the coarse particle mode (see Fig. 4.4i). Its composition does not always reflect the local sand composition, but might be enriched by anthropogenic sulphates and nitrates.

Biomass fires are shown as red dots. Note that the fire locations in December (Fig. 4.2a) are distinctly dif-

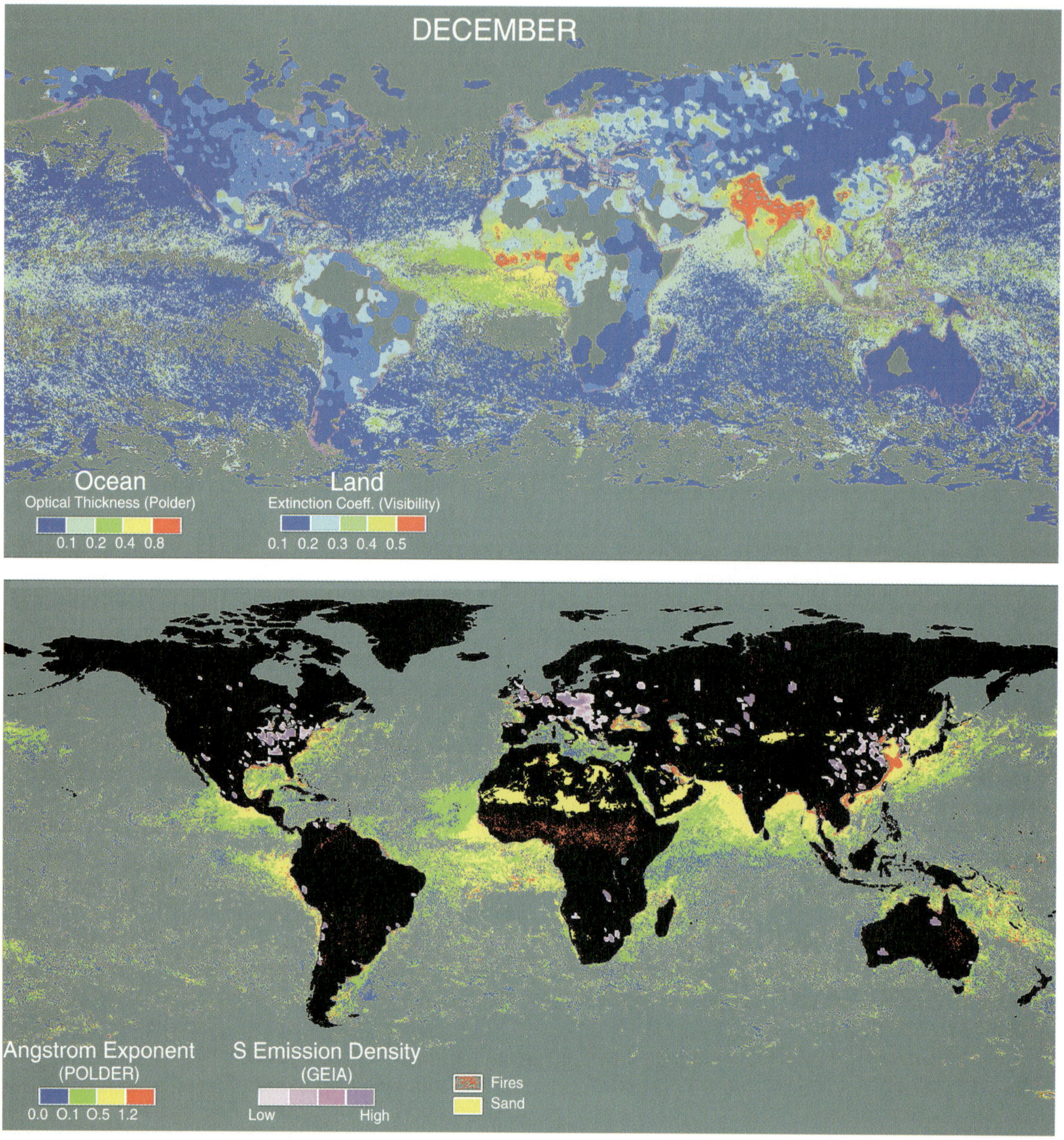

Fig. 4.2a. Global horizontal patterns for December 1996 of aerosol optical thickness and Ångström exponents over the oceans from the POLDER instrument aboard the ADEOS satellite and extinction coefficients over land areas from visibility data (reproduced with permission of Laboratoire d'Optique Atmospherique (LOA), Lille; Laboratoire des Sciences du Climat et de l'Environnement (LSCE), Gif-sur Yvette; Centre National d'Etudes Spatiales (CNES), Toulouse, France; and the National Space Development Agency (NASDA), Japan)

ferent from those recorded in June (Fig. 4.2b). The regular yearly biomass burning regions of the world are distributed over central and sub-Saharan Africa, and Central and South America, as well as Indonesia and Indochina. More sporadic fires are recorded over the boreal forests of Canada and Siberia. Biomass smoke particles are composed of organic species, elemental carbon, ionic species (sulphate, ammonium, and potassium), and other combustion residues, and they are generally submicrometer in size (see Fig. 4.4d).

Anthropogenic sulphur sources obtained from the IGAC GEIA inventory (see Sect. 4.3.2.3), shown as purple areas, are broadly representative of other types of industrial aerosol emissions and their precursors (e.g. organic species, nitrates, black carbon, etc.). Anthropogenic emissions are located mainly in the Northern Hemisphere and mainly in midlatitudes. The three major anthropogenic sulphur-emitting regions are eastern North America, Europe, and eastern Asia. Anthropogenic sulphate particles are sub-micrometer in

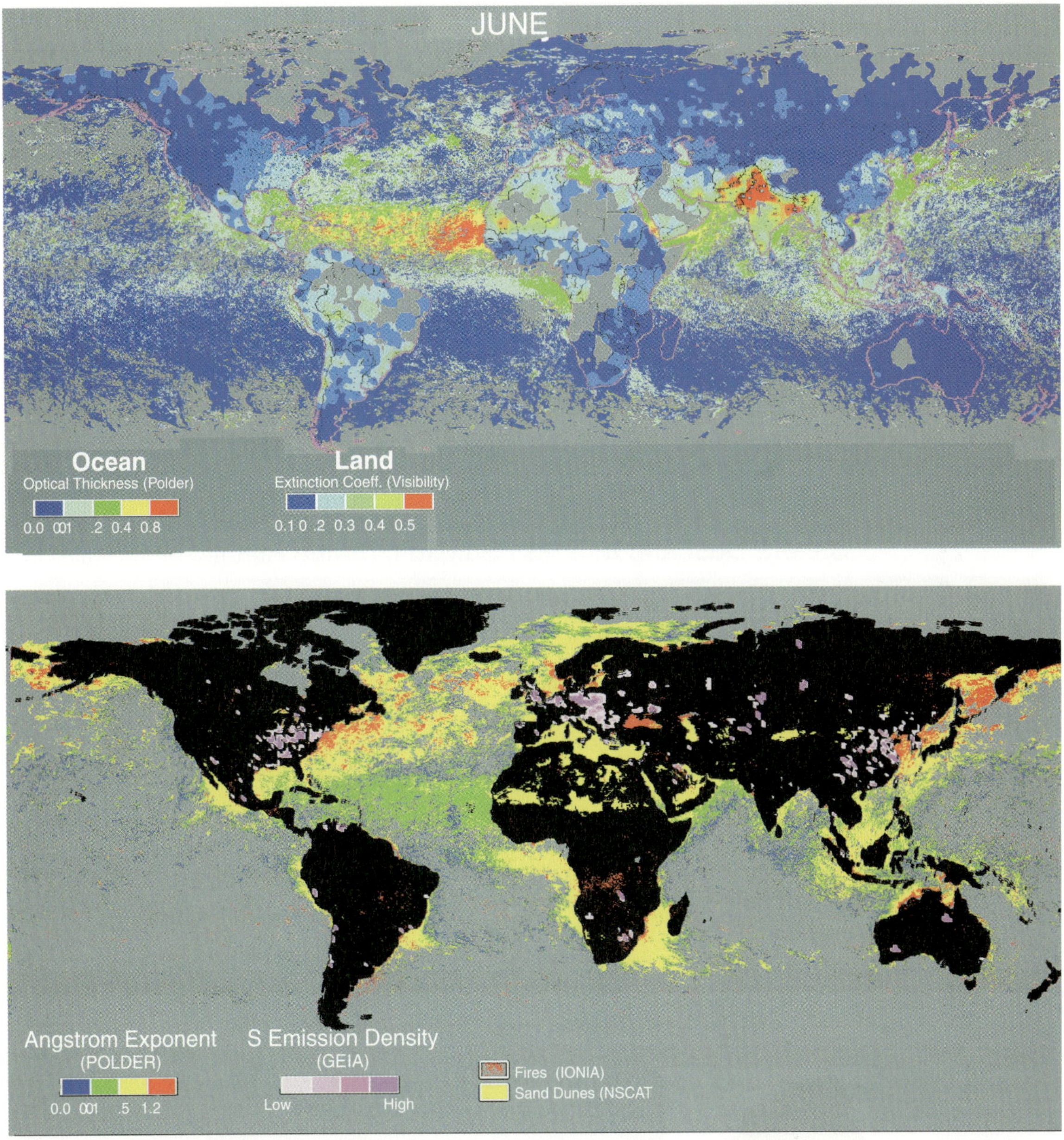

Fig. 4.2b. Global horizontal patterns for June 1997 of aerosol optical thickness and Ångström exponents over the oceans from the POLDER instrument aboard the ADEOS satellite and extinction coefficients over land areas from visibility data (reproduced with permission of Laboratoire d'Optique Atmospherique (LOA), Lille; Laboratoire des Sciences du Climat et de l'Environnement (LSCE), Gif-sur Yvette; Centre National d'Etudes Spatiales (CNES), Toulouse, France; and the National Space Development Agency (NASDA), Japan)

size and often co-exist with condensable organic species (see Fig. 4.4c).

The oceans are also sources of aerosols and precursor gases, especially dimethylsulphide (DMS). DMS emissions are seasonally and regionally dependent with the highest emissions in the middle and high latitudes during the summer months (Kettle et al. 1999). Unfortunately, there are no satisfactory proxy indicators for oceanic sulphur emission patterns. For example, oceanic emissions of DMS do not appear to be related to chlorophyll distributions in the ocean (e.g. Leck 1990), which can be measured by satellite. The sea is also a source of primary particles to the atmosphere through bubble bursting and shearing of wave crests. Although the total number concentration of mechanically produced super-micrometer sea salt particles is relatively small even in the remote marine boundary layer (MBL, cf. Fig. 4.4f), they can dominate light scattering by aerosol particles (Quinn et al. 1996c).

4.2.1.2 *Vertical Aerosol Distributions from Space-Borne Lidar*

The vertical distribution of aerosol plumes (and clouds) over the Atlantic Ocean is shown in Fig. 4.3a. By a favourable coincidence, the LITE (Lidar in Space Technology Experiment, September 1994) lidar recorded the vertical distribution of both the Sahara dust plume and the biomass smoke plume emanating from southern Africa.

The picture provides clear illustrations of aerosol-cloud interactions. For instance at 12° N the marine boundary layer (MBL) extends to about two km and there is evidence of dust entrainment into the MBL and the clouds. The cloudiness associated with the Inter-Tropical Convergence Zone is also clearly visible at about 5° N. The entire convergence zone is relatively aerosol-free, possibly because of the intense cloud processing and removal of particles by precipitation. The picture, however, also shows that significant amounts of aerosol are transported above clouds and become well separated from the boundary layer. The slow removal within the free troposphere allows the long-range transport of these plumes over many thousand kilometres until they are entrained into precipitating cloud systems.

The biomass burning haze layer over the Amazon basin is depicted in Fig. 4.3b. The layer extends from the basin floor to about 3 000 m altitude. It is remarkable that the aerosol layer is spatially homogeneous over the entire 1 000 km cross section along the track of the spacecraft. During the LITE passage over this part of South America, the free troposphere above 5 km was remarkably aerosol-free. The Andes constitute a strong barrier to the dispersion westward of the smoke layers.

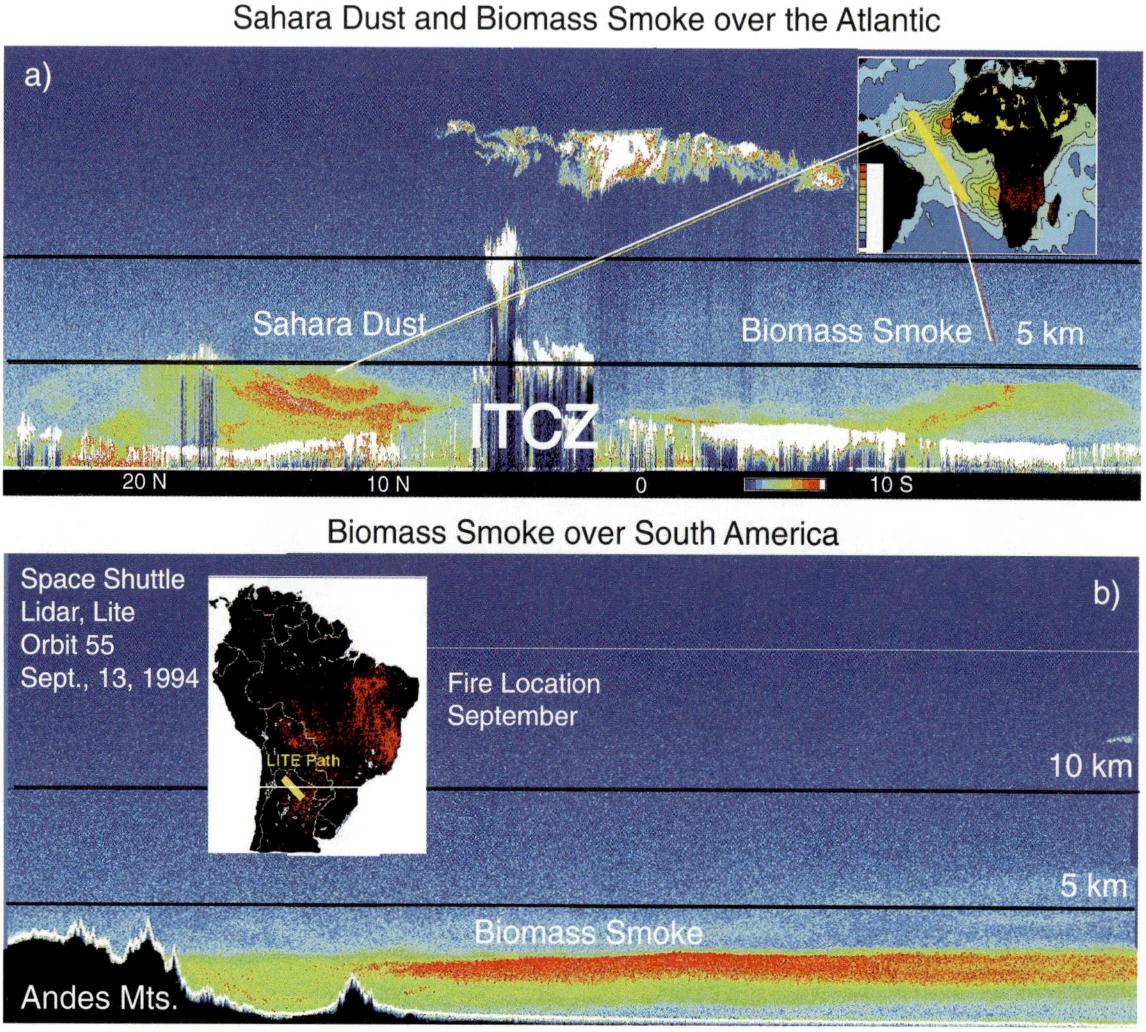

Fig. 4.3. Vertical distribution of aerosol plumes (and clouds) from the Lidar in Space Technology Experiment (LITE) instrument, over **a** the Atlantic Ocean, and **b** South America

4.2.2 *In situ* Measurements

There has been extensive development in instrumentation for the physical and chemical characterisation of aerosol particles. Implementation of these instruments in well-designed field studies in IGAC and elsewhere (see Sect. 4.4) has provided considerable new information on aerosol microphysical and chemical characteristics, at least in those areas visited by these experiments. A compilation of particle size distributions and their chemical composition is shown in Fig. 4.4, and discussed below. The degree to which they lead to a three dimensional picture of the global aerosol is discussed subsequently.

4.2.2.1 *Particle Size Distributions*

In the global atmosphere, the size distribution of aerosol particles, while highly variable, generally can be represented as a superposition of more or less distinguishable modes. These modes are often approximated as log-normal distributions. Following the original suggestions of Whitby (1978), these modes are generally referred to as the nucleation mode (particles with diameter $D_p < 0.01$ µm), the Aitken mode ($0.01 < D_p < 0.1$ µm), the accumulation mode ($0.1 < D_p < 1$ µm), and the coarse mode ($D_p > 1$ µm). Each mode results from specific emissions and atmospheric processes. The nucleation mode is the result of recent nucleation of new particles from gases. The Aitken mode results from condensation on and coagulation of nucleation-mode particles, as well as from primary emissions during combustion. The accumulation mode typically results from coagulation and from the formation of particle mass by chemical reactions in cloud droplets. Primary emissions from the mechanical break-up of bulk material also contribute to accumulation mode particles, but they dominate the mass of the coarse mode.

Particle number concentrations typically decrease for particles with diameters below 10 nm, because the smallest particles and clusters have a high mobility and therefore rapidly diffuse to surfaces or coagulate with other particles. The presence of high numbers of nucleation-mode particles, for example, in urban and suburban environments (Fig. 4.4) and (on occasion) in the clean MBL of the Southern Ocean (Fig. 4.4f), indicates that nucleation has taken place recently. Over oceans, this is thought to happen aloft in the neighbourhood of clouds (e.g. Hegg et al. 1990). The nucleating particles are 1 nm and less in diameter and they are not detectable with the present instrumentation, which has a lower limit of 2–3 nm.

Number concentrations also decrease for particles with diameters above about 500 nm, because of impaction or gravitational settling. Significant number concentrations of coarse particles are found in the presence of primary emissions, e.g. of sea salt (Fig. 4.4c–f) or crustal dust (Fig. 4.4b). The Saharan dust layer in the free troposphere (Fig. 4.4i) contains relatively high numbers of coarse particles even several hundreds of kilometres from the emission area. In the diameter range 10 to 500 nm, the number size distributions exhibit one or more maxima. When particles resulting from industrial activity (Fig. 4.4c) or biomass burning (Fig. 4.4e) age during transport from their sources, primarily because of coagulation and cloud processing, a single accumulation mode develops with diameter typically centred near 100 nm. These aged aerosols are also present over the continents where they can mix with freshly produced aerosols (e.g. Fig. 4.4a,b).

In the clean MBL, far from anthropogenic influences, the sub-micrometer size distribution is typically bimodal (Fig. 4.4d,f,g). The Aitken mode is thought to arise from entrainment of free tropospheric aerosol (Fig. 4.4h) into the boundary layer, whereas the accumulation mode results from the growth of smaller particles, mainly by chemical reactions in non-precipitating clouds. In the summer Arctic three sub-micrometer modes have consistently been found (Covert et al. 1996b).

4.2.2.2 *Chemical Composition*

Within a given environment the chemical composition of sub-micrometer particles ($D_p < 1$ µm) is generally different from that of coarse mode particles ($D_p > 1$ µm), reflecting the different formation mechanisms (Fig. 4.3). In the following discussion we will focus on the sub-micrometer fraction, which is most relevant for many of the aerosol effects.

Urban aerosols are often dominated by sulphate and by organic species that derive from the incomplete combustion of fossil fuels. Other products from fossil fuel burning, such as nitrate and black carbon, are also present in this aerosol. The acid fraction of the particles is partially neutralised by ammonia. Figure 4.4a and b show data for the city of Milan (Italy) during summertime. Aerosols in cities in northern Europe contain a smaller contribution of organic particulate matter and exhibit significant seasonal variations. Outside of the urban environment, the regional continental sub-micrometer aerosol (Fig. 4.4c) is generally dominated by sulphuric acid and its partial neutralisation products with ammonia $(NH_4)_xH_ySO_4$, where x and y range from zero to two and two to zero, respectively. The dominant source of the sulphur in this aerosol is industrial fossil fuel combustion.

Although the measurement of the organic constituents is fraught with large uncertainties, there are indications that organics also form a large fraction of the

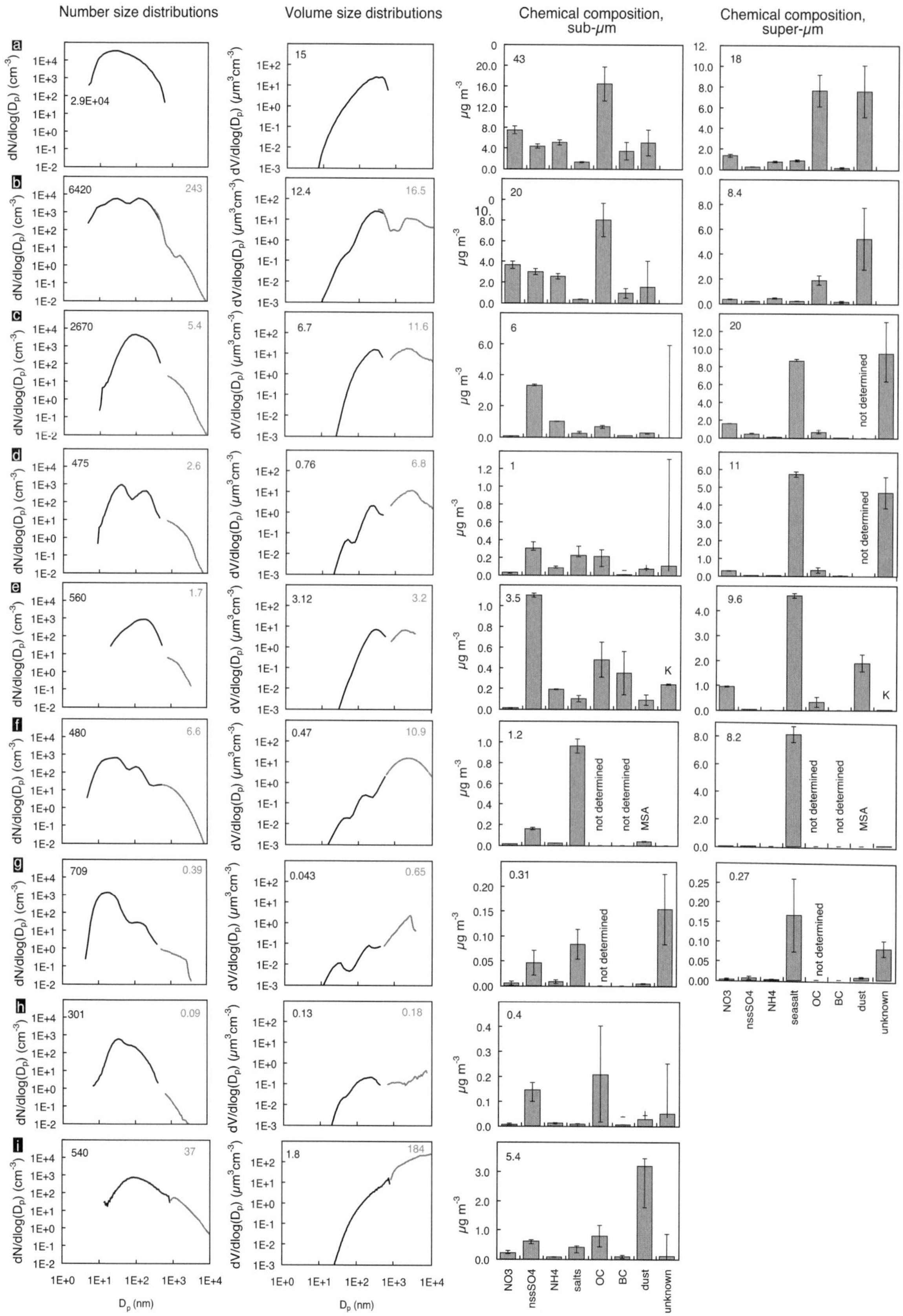

Number size distributions
Volume size distributions
Chemical composition, sub-μm
Chemical composition, super-μm
a
2.9E+04
15
43
18
b
6420
243
12.4
16.5
20
8.4
c
2670
5.4
6.7
11.6
6
20
not determined
d
475
2.6
0.76
6.8
1
11
not determined
e
560
1.7
3.12
3.2
3.5
9.6
K
K
f
480
6.6
0.47
10.9
1.2
not determined
not determined
MSA
8.2
not determined
not determined
MSA
g
709
0.39
0.043
0.65
0.31
not determined
0.27
not determined
h
301
0.09
0.13
0.18
0.4
i
540
37
1.8
184
5.4
dN/dlog(Dp) (cm-3)
dV/dlog(Dp) (μm3cm-3)
μg m-3
Dp (nm)
1E+0
1E+1
1E+2
1E+3
1E+4
NO3
nssSO4
NH4
salts
OC
BC
dust
unknown
NO3
nssSO4
NH4
seasalt
OC
BC
dust
unknown

sub-micrometer aerosol in the free troposphere (see Fig. 4.4h) (Novakov et al. 1997; Putaud et al. 2000).

The particulate products of biomass burning (see Fig. 4.4e) are enriched in black carbon, potassium, ammonium, and chlorine. Some chlorine can be displaced by sulphate over short distances from the fires (Liu et al. 2000).

Sea salt generally dominates the mass of super-micrometer particles over the ocean (Fig. 4.4c,f) and can dominate the mass of the sub-micrometer particles in the remote marine environment (Fig. 4.4f). Single particle analysis during ACE-1 revealed that over 90% of the aerosol particles with diameters >130 nm (Murphy 1998) and up to 70% of the particles with diameters >80 nm (Kreidenweis et al. 1998) contained sea salt. In the subtropical North Atlantic (ACE-2) the contribution of sea salt to the sub-micrometer aerosol was less dominant (compare Fig. 4.4e and f), even in air masses of marine origin.

Over the oceans, in plumes originating over the polluted continent, the sea salt becomes depleted in chloride (55% on average during ACE-2, see Fig. 4.4c), indicating the interaction of sea salt with anthropogenic acids (Quinn et al. 2000) or perhaps other mechanisms (Keene et al. 1990). In such plumes, the contribution of nitrate to the sub-micrometer aerosols is also very small. This can be explained by the displacement of the equilibrium $NH_4NO_3 \longleftrightarrow NH_3 + HNO_3$ through the reaction of HNO_3 with : $NaCl + HNO_3 \longrightarrow NaNO_3 + HCl$.

During transport events out of North Africa, mineral dust is the main component of sub-micrometer aerosol particles in the free troposphere (see Fig. 4.4i). However, the presence of nitrate and sulphate and the high SO_4^{2-} / Ca^{2+} ratio suggest that dust plumes over the Atlantic tend to be mixed with aerosol resulting from industrial activity.

◀ **Fig. 4.4.** A compilation of particle size distributions and their chemical composition representative of different atmospheric settings. Size distributions below 1000 nm are obtained with differential mobility analyser (*black line*), above 1000 nm with optical particle counter or aerodynamic particle sizer (APS) (*gray line*). Aerodynamic diameters have been converted to geometric diameters for consistent representation. *Numbers* inside the size distribution panels indicate integrated number (resp. volume) concentration for the sub-µm (*black*) and the super-µm (*gray*) size range in the same unit as indicated in the Y-axis. *Numbers* inside the chemical composition panels indicate gravimetric mass (µg m^{-3}) for the respective size ranges. Labels for the composition bars are as in the lower panels, unless indicated otherwise (*black bars*). The unknown fraction is the difference between gravimetric mass and the sum of quantified compounds. **a** Urban background during rush hour; **b** Suburban at local noon; **c** N. Atlantic MBL, continentally influenced; **d** N. Atlantic MBL, background; **e** N. Atlantic MBL, biomass burning influenced; **f** Southern Ocean MBL, background; **g** Arctic MBL, background during clear sky; **h** N. Atlantic free troposphere, background; **i** N. Atlantic free troposphere, Saharan dust plume; (courtesy of R. Van Dingenen, J. P. Putaud, EC Joint Research Centre, Ispra, Italy; T. S. Bates, P. K. Quinn, NOAA PMEL, Seattle, WA; C. Leck, MISU, Stockholm, Sweden, H. Maring, RSMAS, University of Miami, FL)

4.2.2.3 *The 3-Dimensional Distribution of the Atmospheric Aerosol Based on in situ Measurements*

Measurements such as those shown in Fig. 4.4 must be compiled to construct global aerosol climatologies. Such climatologies are necessary for testing models, for verifying satellite observations and, eventually, for quantitatively assessing any regional or global aerosol effects in the Earth system. This goal is far from being achieved.

A compilation of surface-based aerosol measurements has recently been published for marine areas (Heintzenberg et al. 2000). Well-calibrated and parameterised submicrometer particle size distributions are available for only about a quarter of the oceanic surfaces. For bulk chemical composition the corresponding coverage is somewhat better (≈60%), but still far from satisfactory. Significant progress in marine aerosol characterisation came through the large IGAC field experiments of the 1990s: ASTEX (Huebert et al. 1996); ACE-1 (Bates et al. 1998; Bates 1999); TARFOX (Russell et al. 1999); and ACE-2 (Raes et al. 2000a).

For continental aerosols, no comprehensive parameterisation of the size distribution of continental aerosols has been accomplished since the pioneering work of K. T. Whitby (see Whitby 1978), despite the availability of commercial and specialised methodology that can reveal many more details than was possible 25 years ago (e.g. Heintzenberg et al. 1998). Regional air pollutant networks such as EMEP in Europe (*http://www.nilu.no/projects/ccc/emepdata.html*) and IMPROVE in the US (*http://vista.cira.colostate.edu/IMPROVE*) have been collecting data on aerosol chemical composition (typically sulphates, nitrates, and black carbon), which are of some use, and the Global Atmospheric Watch programme (GAW) of WMO has started archiving various aerosol parameters at its World Data Centre for Aerosols (*http://www.ei.jrc.it/wdca/*). The latter have been used for a validation exercise of global chemistry and transport models of the sulphur cycle (Barrie et al. 2000).

Knowledge of the vertical aerosol distribution is even more limited than that of the surface distribution. Over the continents there has been only one long-term effort in vertical aerosol profiling by *in situ* measurements (Hofmann 1993); valuable data were obtained on the effects on the upper troposphere and lower stratosphere of volcanoes, of the increase of air traffic, and of decreasing surface emissions. A series of research flights complemented the continental data to some extent with measurements over the Pacific and North Atlantic region (e.g. Clarke 1993; Heintzenberg et al. 1991; Hobbs and Yates 1985). As an example, all Pacific measurements from major aircraft experiments made over a ten-year period are summarised in Fig. 4.5. The measurements

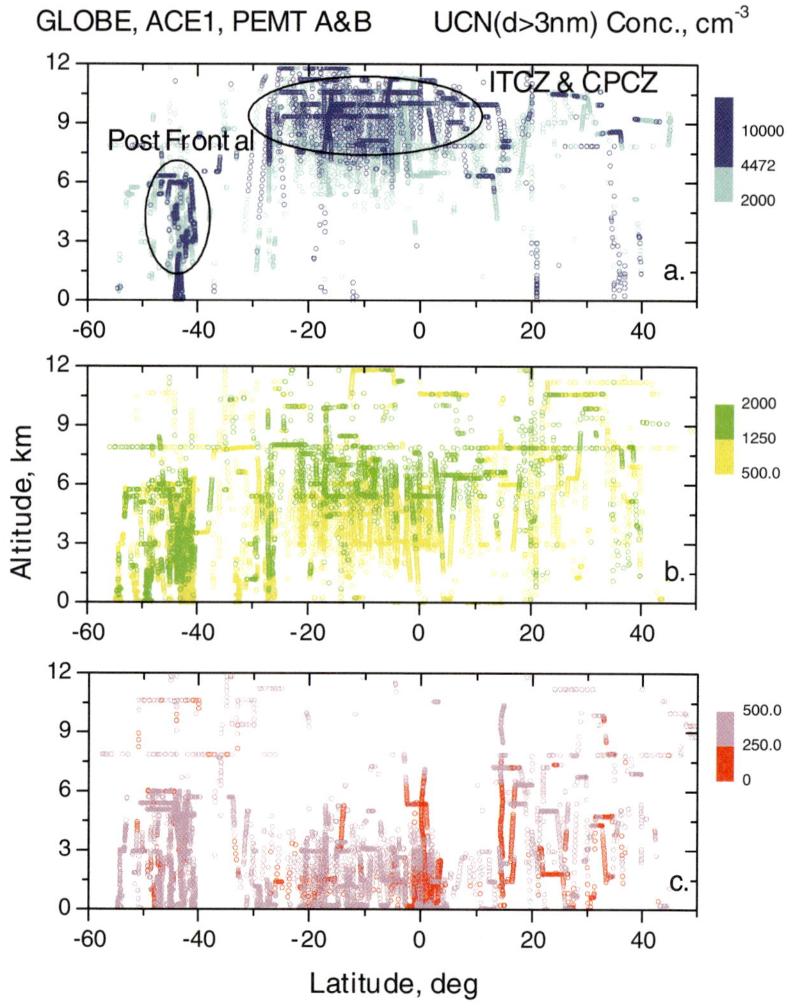

Fig. 4.5.
Ultrafine particles (diameter $D_P > 3$ nm) in particles cm^{-3}, STP as a function of latitude and altitude from five major aircraft experiments: GLOBE (1989, 1990); ACE-1 (1995); and PEM-Tropics A and B (1996, 1999, respectively) above the Pacific Ocean. Because of data point overlap, the data have been separated into three panels with indicated concentration ranges of 0–500 cm^{-3} (*red-pink*); 500–2 000 cm^{-3} (*yellow-green*) and 2 000–10 000 cm^{-3} (*light-dark blue*) (data courtesy of A. Clarke, University of Hawaii, and B. Anderson, NASA, Langley Research Center)

reveal the occurrence of large-scale natural aerosol production over the Pacific. For these extensive regions the highest concentrations are generally most common aloft in the Tropics and near deep convective regions near the south polar front and elsewhere. These are regions of well scavenged air with very low particle surface area and new particle production that appears to be linked to cloud pumping of precursors aloft where low surface area, low temperatures, and high humidity favour nucleation (Clarke et al. 1998, 1999a,b; Weber et al. 1999). Lower concentrations are more common at higher latitudes in both hemispheres. In spite of fewer flights in the higher altitude and latitude regions, it appears likely that these differences reflect air masses de-coupled from the tropical latitudes, as would be expected for typical characteristics of the Hadley circulation.

Lower number concentrations but larger particles were frequently observed in the North Pacific midlatitude free troposphere (FT). Here aged "rivers" of continental aerosol are present frequently, often internally mixed with a non-volatile residual indicative of soot and/or externally mixed with dust. The higher concentrations of small volatile nuclei often observed near combustion source regions tend to be depleted in the aged plumes, perhaps because of coagulation with the high available surface areas. The FT in the subtropics tends to show frequent and marked transitions and mixing between these clean and continental aerosol types.

Vertical aerosol distributions have also been monitored over several years by a few lidars over the European continent (Ansmann et al. 1997; Jäger and Carnuth 1994) and more recently by Raman lidar at a site in central North America (Peppler et al. 2000).

After about five years of technical development, an aerosol payload is now flying frequently on a commercial aircraft (Brenninkmeijer et al. 1999). An aerosol climatology for the tropopause region has been accumulated for several years during scheduled flights between Europe and the Indian Ocean (Hermann 1999). As over the Pacific (cf., Fig. 4.5) there is a clear influence of surface aerosol sources in the Tropics and in the midlatitudes over Europe (Fig. 4.6). In the latter region aircraft emissions contribute to aerosol sources in the tropopause region.

Box 4.2. Aviation-produced particles

Aircraft affect atmospheric aerosols and cirrus (ice) clouds by releasing particles directly into the upper troposphere and lower stratosphere and by producing contrails (condensation trails) (Penner et al. 1999). At cruise altitudes (9–12 km), concentrations of atmospheric aerosol particles are much lower, and their associated residence times are much longer, than for regions closer to Earth's surface. Two major particle types can be distinguished in aircraft exhaust (Kärcher 1999): (*1*) ultrafine liquid particles (diameter range 5 to 10 nm) formed by nucleation during cooling and dilution of the exhaust, which are composed of water, sulphuric acid, and organic matter; and (*2*) solid soot particles (20 to 60 nm) composed of carbonaceous agglomerates containing sulphur and organic compounds formed during combustion.

Charged molecular clusters (chemi-ions) produced via high temperature chemical reactions in engine combustion chambers are believed to play a central role in the formation and evolution of the liquid particles (Yu and Turco 2000). In the dispersing aircraft plumes, the particles grow by coagulation and condensation to form internally mixed soot-sulphate particles.

Perturbations in particle number concentrations in aircraft corridors have been observed at regional scales. The aircraft fleet may increase cirrus cloudiness (Jensen and Toon 1997). Satellite and *in situ* observations show that persistent contrails may develop into cirrus clouds. Aircraft-produced particles may also trigger cirrus indirectly, after the disappearance of short lived contrails. Exhaust soot particles coated with hygroscopic sulphates could be more efficient freezing nuclei than ambient aerosol particles. Concerns have been raised about the role of aircraft soot in modifying existing or stimulating formation of new cirrus (Jensen and Toon 1997) because there is the potential for heterogeneous nuclei to cause ice formation at relative humidities that are lower than those needed to form ice homogeneously on sulphate particles.

Observational evidence exists for aircraft soot influencing cirrus (Ström and Ohlsson 1998), but the physical mechanisms are not understood. Whereas direct radiative forcing associated with aircraft-produced particles is estimated to be small compared with that originating from greenhouse gases now, the effect of these particles on cloud formation and modification could become substantial in the future. A better understanding of physico-chemical and ice-forming particle properties is required before the impact of aviation on cloud formation and heterogeneous chemistry can be addressed fully.

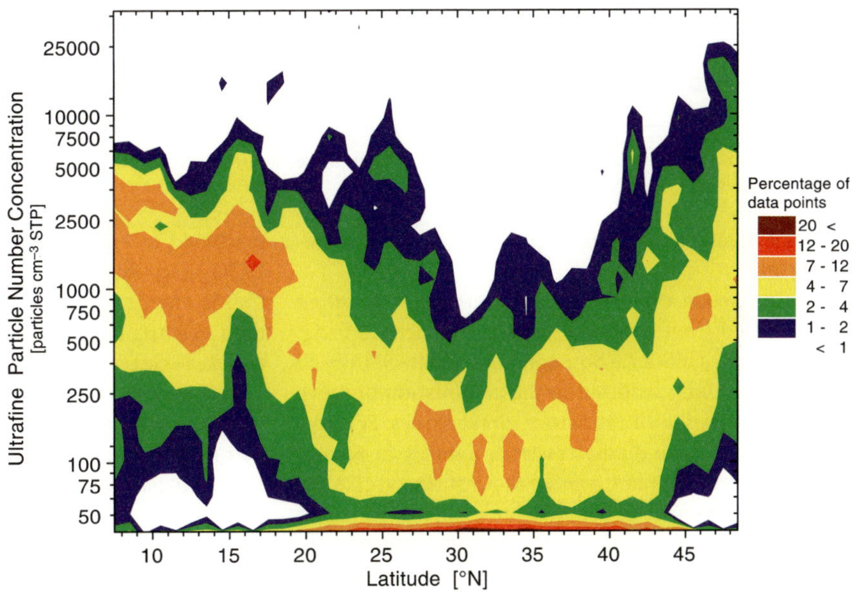

Fig. 4.6.
Climatology of particles ($4\ \text{nm} \le D_p \le 12\ \text{nm}$) in ($\#\ \text{cm}^{-3}$, STP) in the middle and upper troposphere as a function of latitude from scheduled flights during the period 1997–2000 with a commercial aircraft between Europe and the Indian Ocean (Herman et al. 2002)

4.2.3 Process Understanding

An increased understanding of the processes underlying the observed variability in Fig. 4.2, 4.3, and 4.4 has evolved over the past 10 to 20 years. This includes both the microphysical and chemical processes involved in aerosol particle formation, evolution, and removal, as well as the role of large-scale meteorological processes.

4.2.3.1 *Microphysical and Chemical Processes*

The main microphysical and chemical processes that influence the size distribution and chemical composition of atmospheric aerosol particles are depicted schematically in Fig. 4.7. The figure illustrates the existence of primary and secondary particles and also the wide range of sizes involved in their formation and evolution. How aerosol particles participate in atmospheric chemical processes through homogeneous, heterogeneous, and in-cloud reactions is also highlighted in Fig. 4.7.

Primary particles that derive from the comminution of bulk material and subsequent suspension by the wind, such as sea salt, soil dust, and biological material, have most of their mass in the coarse particle mode. However, their highest numbers occur in the sub-micrometer range. Because of their small numbers and large sizes, these primary particles generally do not coagulate. They can, however, become more heterogeneous

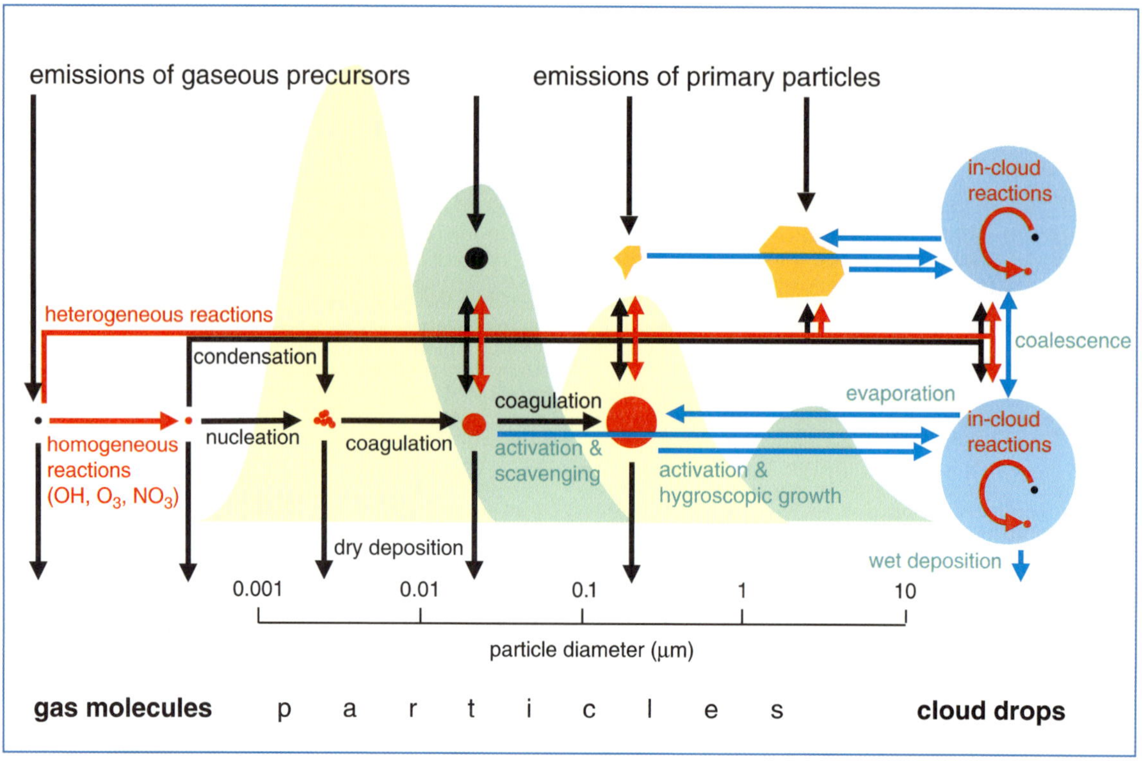

Fig. 4.7. Major microphysical and chemical processes that influence the size distribution and chemical composition of atmospheric aerosol particles

through uptake and chemical processing of condensable gases.

An important type of primary particle is so-called soot emitted from combustion of carbonaceous fuels. Soot consists of black carbon-containing material, which has not been fully oxidised in the combustion process; it is often mixed with refractory metal oxides. Primary soot particles have diameters from about 5 to 20 nm. Such particles coagulate rapidly to form fractal-like aggregates, which in turn collapse under the influence of capillary forces of condensing vapours to form more compact structures having diameters of up to micrometer size.

Gas-to-particle conversion (i.e. nucleation of new particles or condensation on existing particles) occurs when a volatile species reaches a concentration that exceeds its equilibrium vapour pressure, resulting in a thermodynamic driving force for condensation. In the atmosphere, this situation can be driven by chemical reactions leading to products with very low equilibrium vapour pressures (e.g. $SO_2 \longrightarrow H_2SO_4$, α-pinene ⟶ pinonic acid). Alternatively, this situation also can be reached by a reduction in temperature, which reduces equilibrium vapour pressures. Another suggested mechanism is the co-condensation of several substances for which the equilibrium vapour pressure is lowered by the presence of other compounds in the particle. For example, the co-condensation of small amounts of ammonia has been predicted to enhance greatly the binary nucleation of H_2SO_4 and water vapour (Kulmala et al. 2000). Nucleation is energetically less favourable than condensation onto existing surfaces because of the free energy penalty associated with formation of new surface area. Therefore, nucleation is facilitated by the absence of pre-existing particulate surface. After nucleation occurs, the new particles grow further by condensation and coagulation. As particles reach a diameter of the order of the mean free path of the condensing molecule, typically ca. 0.1 μm, condensation becomes diffusion limited and slows down. Also, coagulation eventually slows down as the number concentrations of particles decrease. Under background tropospheric conditions particles formed initially by nucleation require from days to weeks to grow larger than about 0.1 μm solely by condensation and coagulation (Walter 1973). Under polluted conditions this growth can occur within a day because of the strong supply of condensable material (Raes et al. 1995).

Another growth process is by chemical processing in non-precipitating clouds (e.g. Bower et al. 1997; Laj et al. 1997). This process begins with the uptake of water vapour by the particle with increasing relative humidity (RH). According to traditional Köhler theory, a critical supersaturation exists slightly above 100% RH. For RH exceeding this critical value, water vapour condensation is energetically favoured, leading to rapid growth of sub-micrometer aerosol particles to supermicrometer cloud drops. This critical supersaturation depends on size and chemical composition of the particle as well as the concentrations of soluble gases in the surrounding air. Some soluble gases, most prominently SO_2, are oxidised in the aqueous phase. As most clouds evaporate before the onset of precipitation, larger particles result from the additional oxidised material, e.g. sulphate (Birmili et al. 1999b; Yuskiewicz et al. 1999).

Reactions that occur in cloud water also occur in wet particles under sub-saturated (non-cloudy) conditions. However, reaction rates are generally much lower because of lower liquid water content; reaction rate expressions also differ because of the higher ionic strengths in the wet particles. Additionally, adsorbed gases react on the particle surfaces yielding products that might either remain on the particle or return to the gas phase.

Particles are removed from the atmosphere by dry and wet processes. For particles whose diameter is less than 0.1 μm, the dominant dry removal mechanism involves turbulent diffusion followed by Brownian diffusion to the surface through a viscous sub-layer, typically about 1 mm thick; this process becomes less efficient as particle size increases. Coarse particles ($D_p > 1$ μm) settle gravitationally, a process that becomes less efficient as particle size decreases. Coarse particle dry deposition is enhanced by impaction, which is wind speed dependent (Slinn and Slinn 1980). In the range $0.1 < D_p < 1$ μm, dry removal is very slow, resulting in long atmospheric residence times (up to weeks). Provided their atmospheric residence time is not limited by any of the above sink mechanisms, these particles are removed mainly by growth to cloud drops during cloud formation and subsequent removal from the atmosphere in precipitation.

4.2.3.2 *Horizontal and Vertical Transport*

The aerosol plumes and layers depicted in Fig. 4.2 and 4.3 result from meteorological transport, including exchange between the lower and upper troposphere by dry or wet convection. The characteristic times of many of the microphysical aerosol processes depicted in Fig. 4.7 are days up to several weeks. This is typically longer than the time between emission/formation of aerosols in the boundary layer and their transport into the free troposphere. Hence, to understand aerosol properties one cannot confine the discussion to such compartments as the continental or marine boundary layers or the free troposphere. Instead, one needs to view aerosol microphysical processes in the context of the atmospheric transport processes that connect those compartments (Raes et al. 2000b). Meteorological processes also affect microphysical processes and influence the size distribution and chemical composition of the particles. For example, during vertical transport in precipitating clouds, a separation occurs between soluble compounds that are rained out and insoluble compounds that are pumped aloft (Rodhe 1983). In the outflow regions of such clouds, optimal conditions exist for nucleation, the consequences of which have been both observed (e.g. Perry and Hobbs 1994; Clarke et al. 1999b) and modelled (e.g. Hegg et al. 1990; Wiedensohler et al. 1997). Aerosol layers observed in the free troposphere are possibly the result of such cloud outflows, the structures of which are maintained by meteorological stratification. On the other hand, entrainment of free tropospheric aerosol into the boundary layer may be a source of Aitken-mode particles, which may quench nucleation of new particles. Because strong gradients of particle concentration will form above the inversion layer, turbulent mixing will cause very large changes in particle concentrations and size distributions observed at the surface (Bigg et al. 1996).

4.2.4 Large-Scale Aerosol Models

The advances in understanding of aerosol processes are making their way into numerical models describing the emissions, transport, transformation, and deposition of aerosol particles and their precursors on a variety of scales from local to global (see Chap. 6). Major progress has been achieved over the last decade in simulating the global distribution of tropospheric particulate mass using chemical transport models (CTMs). Simulation of the global distribution of biogenic and anthropogenic sulphur (Langner 1991) led to the recognition that anthropogenic sulphate aerosols may have a significant impact on the global radiation balance (Charlson et al. 1991). This spurred a large interest in simulating global mass distributions of the aerosol types listed in Table 4.1. Despite the simplification of considering each aerosol type independently, these studies were important in several respects. They related emissions to global aerosol distributions, constructed global and regional budgets, estimated the contributions of anthropogenic sources to the aerosol burden, and drew attention to elements of the general circulation that are important for aerosol transport, particularly deep convection (Feichter and Crutzen 1990). As an example, global fields of mineral dust, sulphate, and "smoke", calculated with a CTM driven by observation-derived meteorological data, are shown in Fig. 4.8. Many of the calculated features can be compared with the observations from space, shown in Fig. 4.2a and b. A sub-hemispheric sulphate model driven by observation-derived meteorological data has been evaluated by thousands of daily comparisons with surface measurements (Benkovitz et al. 1994; Benkovitz and Schwartz 1997).

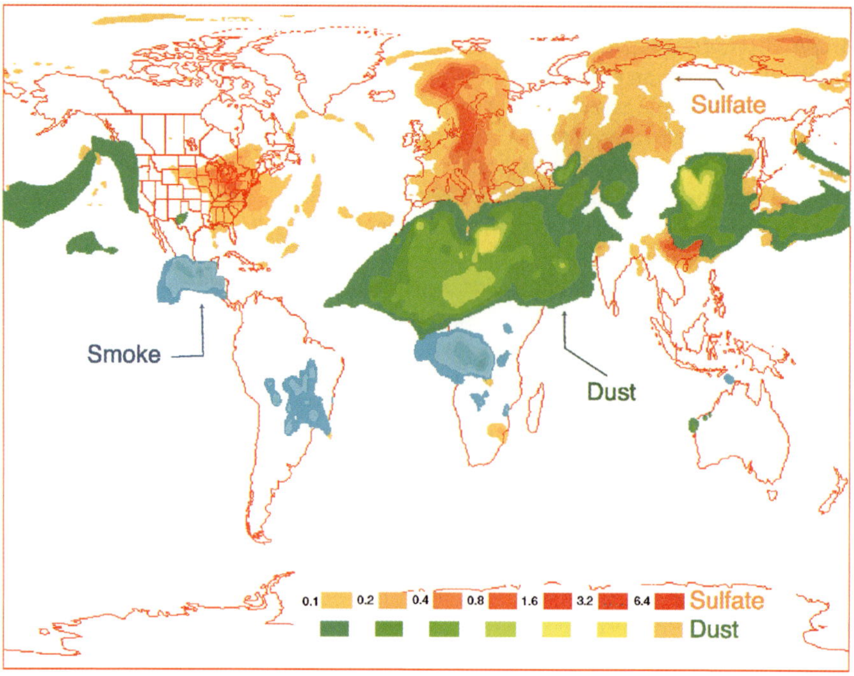

Fig. 4.8.
Global fields of mineral dust, sulphate, and "smoke", calculated with a CTM using actual meteorology (*http://www.nrlmry.navy.mil/aerosol/*)

4.3 Selected Recent Developments

4.3.1 Primary Emissions

4.3.1.1 *Soil-Derived and Industrial Dust*

The majority of soil-derived dust particles is lofted into the atmosphere by wind in arid and semi-arid regions, which comprise approximately one third of Earth's land area. However, any type of land surface is a potential source of dust particles. Besides production mechanisms, surface properties (e.g. texture, roughness, composition, moisture, and vegetation) are key factors that determine the emission of dust for specific meteorological events (Gillette et al. 1980).

Estimates of global annual mean dust production range from 1 000 to 5 000 Tg yr^{-1} (Duce 1995). The large uncertainties are due mainly to the complexity of dust production mechanisms, which exhibit a large spatio-temporal variability. Despite progress in developing dust emission schemes for Saharan desert dust (Marticorena and Bergametti 1995), these schemes cannot be extended directly to other source regions due to a lack of data on surface properties. Emission data and models need to include explicit information on the particle size distribution and mineralogical composition to allow for better physically-based treatments of key processes (Sokolik et al. 1998; Claquin et al. 1999).

Anthropogenic dust is of special interest in climate change studies because human activities such as land use practice and construction, etc., can modify the geographical area of dust sources and increase the dust loading of the atmosphere. Recent estimates show that the anthropogenic fraction of dust could be as much as 30 to 50% of total dust production, but these percentages are very uncertain (Sokolik and Toon 1996; Tegen and Fung 1995). The dependence of dust emissions on meteorological parameters, such as wind speed and rainfall, suggests that atmospheric dust concentrations could be affected significantly by climate change.

Another type of anthropogenic aerosol is industrial (fly ash) produced by most combustion processes, cement manufacturing, metallurgy, and so on (Flagan and Friedlander 1978). Current estimates give a global emission for industrial dust of about 130 Tg yr^{-1} (Andreae 1995). The rapid economic expansion and industrialisation in developing countries may result in additional industrial aerosol emissions.

Both natural and anthropogenic fractions of dust must be better quantified to predict the overall effects of dust on atmospheric chemistry and on the climate system.

4.3.1.2 *Primary Particles from the Oceans*

The ocean is a source of primary sea salt particles to the atmosphere through the bursting of bubbles (yielding film and jet droplets), mechanical tearing (spume droplets), and spillover (splash droplets) of wave crests (reviewed by Andreas et al. 1995). Number concentration is dominated by the smallest, sub-micrometer salt particles, whereas surface area is dominated by jet drops, and volume by spume drops when present. Although the total number concentration of mechanically produced super-micrometer sea salt particles is relatively small even in the remote marine boundary layer (MBL, *cf.*

Fig. 4.4f), these particles can dominate the mass size distribution and thus have significant effects on chemical reactions occurring in the MBL (Sievering et al. 1992), the nucleation of new particles (Covert et al. 1996a), and the formation of clouds (O'Dowd et al. 1997b). The magnitude of the ocean-atmosphere flux of sea salt particles depends on the wind-dependent sea state (Woodcock 1953). The instantaneous atmospheric sea salt size distribution depends on the present and prior ocean-atmosphere flux, mixing between the MBL and free troposphere (FT), size-dependent removal processes, and advection. Typically, number and mass concentrations increase with increasing wind speed. Estimates of the source strength of aerosol are quite uncertain. Andreae (1995) gives a best estimate of the annual global mass emission rate of 1 300 Tg yr^{-1} with an uncertainty range of 1 000 to 10 000 Tg yr^{-1}; the estimate of Tegen et al. (1997) is 5 900 Tg yr^{-1}. Estimates of local production flux and its dependence on controlling variables, such as wind speed, are also quite uncertain as are estimates of the size-dependent production flux.

Sea salt aerosol particles also contain biological material. Blanchard (1963) reported surface-active components of the marine aerosol, which he demonstrated were transported into the atmosphere by bubble bursting. Subsequently, Gershey (1983) found that the production of particles by bubbling in seawater discriminated against the more soluble low molecular weight compounds in favour of the more surface-active high molecular weight compounds. These latter compounds are most likely concentrated in the walls of the bubble and are liberated as film drops when the bubble bursts, leading to a considerable enrichment of these organic components in the aerosol relative to the parent water. As a result, primary biological particles in the marine environment will usually contain a wide range of biogenic compounds such as long chain fatty acids, alcohols, esters, and soluble proteins. Some of these compounds are oxidised in the atmosphere to form other products (e.g. fatty acids to oxalic acid and proteins to amino acids).

4.3.2 Emissions of Particle Precursors

4.3.2.1 *Dimethylsulphide*

The emission of dimethylsulphide (DMS) from the oceans is a major sulphur source to the atmosphere (Barnard et al. 1982; Kettle et al. 1999) and contributes to the sulphur burden in both the marine boundary layer and free troposphere (over areas of active cloud convection) (Chin et al. 1996). Recent global estimates of DMS flux from the oceans range from 8 to 51 Tg S yr^{-1} (Kettle et al. 1999; Spiro et al. 1992). This accounts for at least 50% of natural sulphur emissions from ocean, plants, and soils taken together (Bates et al. 1992). Altogether the natural sulphur flux is estimated to be of the same order of magnitude as the anthropogenic sulphur emissions, mainly from fossil fuel combustion, which are currently estimated at 76 Tg S yr^{-1} (Benkovitz et al. 1996) (see also Sect. 4.3.2.3). The relatively large uncertainty in the reported estimates of the oceanic DMS flux is due partly to differences in the transfer velocities used in the sea-to-air calculations, but mainly to the different assumptions involved in seasonal and latitudinal extrapolations of the DMS seawater measurements. In particular, there is a paucity of data for seawater DMS concentrations in the winter months and at high latitudes. This very important natural precursor of sulphate particles and its biogeochemical cycling in the water column is discussed in more detail in Chap. 2.

The chemical and physical pathways that lead from atmospheric DMS to sulphur particles are complex and still poorly understood. Over the past ten years, DMS chemistry has been studied extensively by comparing model simulations with laboratory experiments and atmospheric observations. One of the main findings of these investigations has been the important role of heterogeneous reactions on pre-existing secondary and primary (sea salt) particles and cloud drops in producing particulate non-sea salt sulphate (nss-SO_4^{2-}) and methanesulphonate (MSA). The latter two compounds are the most important end products of DMS oxidation via the formation of gaseous sulphur dioxide, sulphuric acid, and methanesulphonic acid (Campolongo et al. 1999; Capaldo and Pandis 1997; Davis et al. 1999; Yin et al. 1990a,b). Additional oxidation products are dimethylsulphoxide (DMSO) and dimethylsulphone ($DMSO_2$).

Initial reaction products of gas-phase reaction of DMS include CH_3SO_2 and CH_3SO_3. The unimolecular decomposition pathways of both compounds are considered important in affecting the yields of SO_2, H_2SO_4, and MSA. The branching between these pathways determines whether DMS oxidation leads to particle nucleation from H_2SO_4, or to condensation of H_2SO_4 and MSA on existing particles. Unfortunately, the decomposition rates and their temperature dependencies are poorly known and subject to an intense debate. For example, reported CH_3SO_3 decomposition rates range over five orders of magnitude (Campolongo, et al. 1999). Temperature is thought to strongly affect the decomposition rates. Future research on the atmospheric life cycle of DMS should concentrate on reducing these uncertainties and on their implications for a potential DMS-climate feedback.

4.3.2.2 *SO_2 Emissions from Volcanoes*

Volcanoes are a major natural source of atmospheric sulphur species. Most active volcanoes are in the Northern Hemisphere (≈80%). The strongest source region is Indonesia. The most important products of volcanic degassing are H_2O, CO_2, and sulphur (as SO_2, H_2S, and SO_4^{2-}) in varying fractions depending on the magma type. The measured concentration of SO_2 in volcanic plumes can be combined with wind speeds to estimate the SO_2 fluxes. However, emissions have been measured from only a few of the approximately 560 potential volcanic sources, and only a handful of these have been observed more than episodically. For some important regions (e.g. Kamchatka, Russia) there exist no data at all. Based on published observations, (and excluding important regions) the total amount of volcanic tropospheric sulphur emissions has been estimated at 14 ±6 Tg yr^{-1} (Graf et al. 1997) in accordance with other recent estimates (e.g. Andres and Kasgnoc 1998). This is much less than the 25 Tg yr^{-1} estimated by Lambert et al. (1988) from the $SO_2/^{210}Pb$ ratio.

As active volcanoes generally reach considerable elevations, most of their emissions are injected into the free troposphere. There, removal processes are slower and, consequently, the volcanic sulphur species have longer boundary layer residence times than anthropogenic and biogenic species, leading to more efficient large-scale transport than for low-elevation emissions.

Mean volcanic sulphur emissions are of comparable importance for the atmospheric sulphate burden as anthropogenic sources because they affect the sulphate concentrations in the middle and upper troposphere whereas anthropogenic emissions control sulphate in the boundary layer. Volcanic emissions are highly variable in space and time, so there is need to quantify these sulphur sources better. There is also need to estimate transformation rates of SO_2 to SO_4^{2-}, in warm, moist volcanic plumes, and the contribution of H_2S, SO_4^{2-}, and SO_2 associated with plume particles, because these compounds are not measured by conventional methods. Finally, there is need to determine three-dimensional SO_2 and sulphate fields.

4.3.2.3 *Industrial SO_2 Emissions*

A key advance that has enabled the modelling of spatially and temporally varying aerosol concentrations has been the development of spatially disaggregated inventories. The Global Emissions Inventory Activity (GEIA) has been particularly successful in engaging researchers from different countries in preparing separate country-level inventories (Graedel et al. 1993). For the industrial SO_2 inventory, in regions where local inventories were not available, GEIA specified an inventory based on a single emission factor for fossil fuels and distributed emissions according to population. Where available, the individual country-level emissions were used (Benkovitz et al. 1996). This method of inventory preparation suffers from the lack of uniform methods for different countries but is expected to be more accurate than previous inventories which used a single emission factor for fossil fuel emissions from different locations. In addition, emissions from industrial activities (e.g. smelting) can be more accurately estimated. A remaining uncertainty with the inventories is the estimate of sulphur released as primary sulphate. Estimates range from less than one percent (Dietz and Wieser 1983; Saeger et al. 1989) to five percent (Eliassen and Saltbones 1983).

The global inventory developed for calendar year 1985 (see Fig. 4.9) indicates that about 81% of anthropogenic sulphur emissions are from fossil fuel combustion, 16% from industrial processes, 3% from large-scale biomass burning or waste treatment activities, and 1% from the combustion of biofuels. The current inventory is in need of substantial revision for other years. Total emissions of SO_2 from fossil fuel burning and industrial activities for 1985 have been estimated as 76 Tg S yr^{-1}, accurate to 20–30% (Benkovitz et al. 1996; Penner et al. 2000).

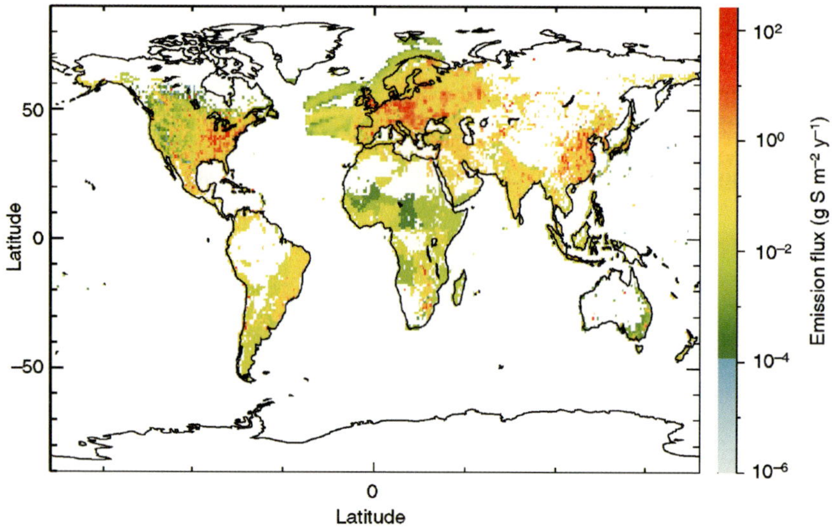

Fig. 4.9. Global distribution of anthropogenic SO_2 emissions for 1985 as estimated by the IGAC/GEIA Activity (Benkovitz et al. 1996)

4.3.2.4 *Nitrogen Oxides and Ammonia*

In addition to their importance in atmospheric oxidant chemistry (Chap. 2 and 3) nitrogen oxides are important as precursors for nitric acid, which, despite its high vapour pressure, can be bound into the particle phase by the equilibrium reaction with NH_3:

$$HNO_3(g) + NH_3(g) \longleftrightarrow NH_4NO_3(s)$$

Ammonia is important as well because of its role in neutralization of aerosol sulphuric acid and possible role in new particle formation. Rather complex dynamic equilibria will usually exist in the atmosphere between aqueous aerosols and the gas phase $(H_2O, HNO_3, H_2SO_4, NH_3)_{gas} \longleftrightarrow$ (water, sulphate, nitrate…) particle. The emissions of ammonia (NH_3) and nitrogen oxides (NO_X) are discussed in Chap. 2 and 3, respectively.

4.3.2.5 *Organic Carbon and Black Carbon*

There are two main sources of combustion aerosols: fossil fuel burning and biomass burning. The main components are organic and black carbon and, for fossil fuel combustion, sulphate. The formation of particulate matter depends strongly on the conditions of combustion, especially temperature, combustion efficiency, and the nature of the fuel. Higher temperature favours the production of larger amounts of black carbon (BC) and a lesser organic fraction. Inefficient combustion increases total particulate matter production.

Biomass burning sources include savannah and forest burning associated with both land clearing and deforestation, burning of agricultural wastes, and domestic burning of biofuels. Development of an understanding of the temporally and spatially varying emissions requires an estimation of total fuel burned and emission factors.

A number of IGAC-sponsored field studies have helped to quantify emissions factors from biomass burning, especially in the major savannah and forest burning regions in the Tropics (e.g. Andreae et al. 1998; see also Chap. 2). These emission factors have been used to develop inventories of both total particulate matter and BC from biomass burning (Liousse et al. 1996). The amount of fuel burned has been based on statistics from the Food and Agricultural Organisation (FAO) of the United Nations for wood and agricultural grain production, amount of deforestation, and estimates of savannah burning frequency. Data from satellites, together with vegetation maps, promise to provide more quantitative estimates of area burned (Barbosa et al. 1998). Development of emissions estimates from other biomass activities may require data on individual country practices.

Current estimates for total biomass burning, expressed as dry mass, range from 4 000 to 10 400 Tg yr^{-1} (see Table 2.4). The total soot production associated with these estimates ranges from 60 to 100 Tg yr^{-1}, with approximately 10% of this total associated with BC.

Fossil fuel emissions of combustion particles depend on burning practices in individual regions and the type of fuels burned. Rough estimates for BC emissions are available from, e.g., Cooke and Wilson (1996). Bond et al. (1998) suggested that new emission factors should be produced, based on the absorption properties of the emitted compounds, but insufficient data of this nature are available to develop such an inventory.

A simple estimate of organic carbon (OC) emissions from fossil fuel combustion was based on observed ratios of BC to OC in source regions (Liousse et al. 1996). Cooke et al. (1999) attempted a more detailed estimate based on emission factors and fuel use. However, their total inventory was about half that estimated in Liousse et al. (1996) and underestimated observed OC when used in a model study. The currently estimated range of emissions of OC is 10 to 30 Tg yr^{-1}. Using the measured ratios of BC to OC summarised by Liousse et al. (1996) yields a source strength for fossil fuel BC ranging from 2.3 to 7 Tg C yr^{-1}.

4.3.2.6 *Volatile Organic Compounds*

Volatile organic carbon compounds (VOCs) are emitted into the troposphere from anthropogenic and biogenic sources. Anthropogenic compounds include organics such as alkanes, alkenes, aromatics, and carbonyls, whereas biogenic gases include isoprene, mono- and sesquiterpenes, and a variety of oxygen-containing compounds. Because the volatility of the oxidation products is one of the most important parameters that determine the significance of precursor gases in terms of their particle-forming potential, only hydrocarbons with more than six carbon atoms are considered to contribute to secondary organic particles under atmospheric conditions (Seinfeld and Pandis 1998). Thus, natural secondary organic aerosol (SOA) formation is believed to result mainly from the oxidation of C_{10}- and C_{15}-terpenoids. Based on emission inventories and laboratory data of biogenic VOC particle yields, Andreae and Crutzen (1997) estimated 30 to 270 Tg yr^{-1} for the production of secondary organic particulate matter from natural VOCs. Because the estimated global emissions of anthropogenic VOCs are significantly lower, their contribution to organic particulate matter on a global scale would appear to be small in comparison. However, in the urban atmosphere during severe smog episodes, anthropogenic SOA formation can make a major contribution to fine particulate mass (see Fig. 4.2b). Among the anthropogenic precursor gases, aromatics have been identified to dominate the process of SOA formation; this suggests that anthropogenic SOA formation in an urban air shed can be modelled based on the aromatic content of the complex organic mixture (Odum et al. 1997).

Box 4.3. The CLAW hypothesis

In 1989, Charlson, Lovelock, Andreae and Warren (Charlson et al. 1987) wrote an influential paper that hypothesised about the role that sulphate particles derived from DMS emissions play in climate control. (The hypothesis is called "Claw", after the first initial of each man's last name.) In order to explain the relatively stable temperature on Earth over geological times, they proposed a negative feedback cycle with the following elements: DMS emission from oceanic phytoplankton are supposed to increase with increasing temperature. The increase in DMS then leads to an increase in the number of non-sea-salt (nss)-sulphate aerosol particles, which were thought to be the main nuclei on which cloud droplets form in the marine boundary layer. At a given liquid water content, more cloud droplets result in an increased reflectivity of the clouds (Twomey 1974). The latter leads to a cooling, counteracting the initial warming.

The CLAW hypothesis linked oceanic biology, atmospheric chemistry, cloud physics, and climate in an elegant way. As such it spurred extensive new atmospheric and biological research and improved understanding of aspects of the biogeochemical cycling of sulphur and its relationship to climate. In particular it promoted the atmospheric aerosol as a possible key player in the climate system through its indirect radiative effect on clouds. In fact, it was the basis of most of the aerosol work initiated and performed within IGAC.

The CLAW hypothesis was criticised early on the grounds that its proposed sensitivity of cloud albedo to sulphur emissions together with anthropogenic sulphur emissions largely in the Northern Hemisphere would result in a much greater interhemispheric contrast in cloud albedo and temperature change than has been observed (Schwartz 1988). The hypothesis was criticised as well on the grounds that the energy requirements for DMS production by marine phytoplankton were inconsistent with the benefits that would be derived by the individual organisms (Caldeira 1989). Nonetheless after 15 years of research the CLAW hypothesis still stands. However, a number of related mechanisms have been proposed that might dampen the effectiveness of the controlling feedback loop, or that at least might help to explain why a relationship between DMS emissions and cloud-forming particles remains elusive. Figure 4.10 summarises the present thinking, indicating some of these mechanisms:

1. Whereas the CLAW hypothesis assumed that the nss-sulphate particles derived from DMS oxidation are the main contributors to cloud condensation nuclei, it is now accepted that some of the nss-sulphate mass in the MBL is internally mixed with sea salt (e.g. Chameides and Stelson 1992; Sievering et al. 1992). In combination with the analysis of individual sea salt particles during ACE-1, this suggests that a significant fraction of the conversion of DMS-derived SO_2 to sulphate occurs on or in sea-salt particles over the oceans and therefore does not lead to new particle production. Several pH-dependent sulphur oxidation pathways in sea-salt particles have been hypothesised that involve either ozone, hydrogen peroxide, halogens, and/or trace metals (see review by Keene et al. 1998). Sulphur oxidation also occurs in the marine stratiform clouds (e.g. Ayers and Larson 1990; Hegg and Larson 1990; O'Dowd et al. 2000), where it is also pH-dependent. Coarse-mode sea-salt particles dry-deposit to the ocean surface relatively quickly (e.g. Slinn and Slinn 1980) and, thus, remove nss-sulphate from the MBL efficiently. This multiphase processing and removal of biogenic sulphur could be an effective shunt that prevents the formation of new sulphate particles by nucleation, for which gas-phase oxidation of SO_2 is prerequisite.
2. In the original CLAW hypothesis all processes that transform DMS into new or better cloud-forming particles were supposed to take place within the marine boundary layer. A key link in this chain of processes was the nucleation of sulphuric acid into new particles and eventually new cloud-forming particles. Box models of the MBL that described these processes depicted in Fig. 4.7 dealing with sea-salt and sulphate particle formation were in better agreement with the observed size distributions only when entrainment of aerosols from the free troposphere was considered as a source of particle number in the MBL (Raes 1995; Katoshevski et al. 1999). Some measurements have been reported of the frequent occurrence in the summer Arctic (Covert et al. 1996b) of large numbers of ultra-fine particles following rain events or subsidence of free tropospheric air (Covert et al. 1992). Other observations indicate that *in situ* nucleation is absent and that entrainment from the free troposphere is the source of MBL particle number concentration (Clarke et al. 1996).
3. An alternative feedback mechanism has been proposed that involves convection in the MBL rather than the active participation of marine phytoplankton (Shaw et al. 1998a). Based on the accepted view that the frequency of occurrence and intensity of convection increases with sea surface temperature, they suggest that such an increase in convection would increase the formation of free tropospheric sulphate particles, which after subsidence and entrainment would interact with the MBL clouds and increase their reflectivity, leading to cooling of the sea surface. It has been estimated that in the upper troposphere, in particular in the Tropics and the midlatitudes of the Southern Hemisphere, up to 100% of the SO_2 might be derived from DMS (Raes et al. 2000b). Hence a mechanism by which changes in the mass and number of free tropospheric sulphate particles result from changes in convective transport of boundary layer DMS into the free troposphere is a possibility.

4.3.3 Formation, Evolution, and Removal of Condensed Material

4.3.3.1 *Nucleation*

4.3.3.1.1 *Laboratory Studies*

Homogeneous nucleation has been investigated in the laboratory for several inorganic and organic chemical systems pertinent to the atmosphere.

The inorganic system that has received the most attention is the binary sulphuric acid-water vapour system. Recent advances include the direct measurements of molecular clustering in the H_2SO_4-H_2O nucleating system (Eisele and Hanson 2000; Hanson and Eisele 2000). Despite reasonable agreement between prediction of binary H_2SO_4 nucleation theory and laboratory experiments (Viisanen and Kulmala 1997), large discrepancies have been reported between this theory and field data. The addition of ammonia (NH_3) was shown to increase nucleation rates dramatically and to reduce the dependence on H_2SO_4 (Ball et al. 1999). These laboratory studies, however, have all been done at sulphuric acid vapour concentrations that are much higher than those measured in the atmosphere.

Several other laboratory studies of multi-component nucleation significant to the atmosphere have been reported. Binary nucleation in the MSA-water system oc-

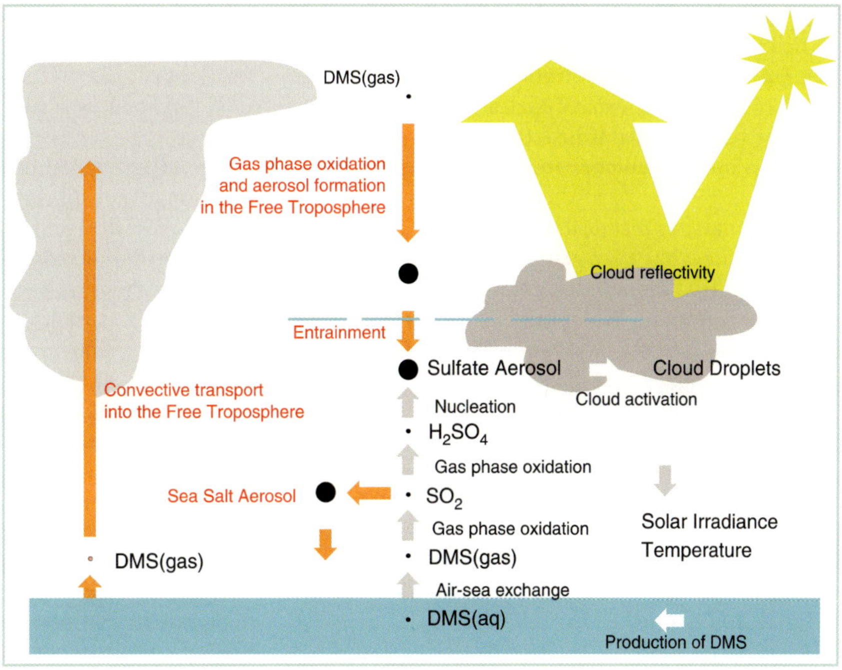

Fig. 4.10.
The original CLAW hypothesis (*in black*) and major additional mechanisms proposed such as: (*i*) the role of sea-salt; (*ii*) the quenching of nucleation by aerosols entrained from the free troposphere; and (*iii*) the role of convective transport into the free troposphere

curs at partial pressures below those required for either pure species (Kreidenweis et al. 1989). It is unlikely, however, that atmospheric concentrations of MSA are high enough for binary MSA-H_2O nucleation to play an important role. Likewise, it is unlikely that ternary nucleation involving H_2SO_4-MSA-H_2O is important (Van Dingenen and Raes 1993). Recent experiments have shown that multi-component nucleation rates are enhanced even for immiscible substances (Strey and Viisanen 1995).

Recent studies of particle production by ionising radiation (He and Hopke 1995; Kim et al. 1998; Mäkelä 1992) have shown that small ions enhance nucleation rates in systems that are chemically similar to the atmosphere and that NH_3 enhances nucleation rates when SO_2 is irradiated in air containing it and H_2O. Understanding of clustering on small ions is needed to permit quantitative estimates of the importance of ion-induced nucleation in the atmosphere; such studies are proceeding (Yu and Turco 2000).

Data from smog chamber studies of the formation of organic aerosol particles have been analysed to make inferences about homogeneous nucleation (Pandis et al. 1991; Stern et al. 1987; Wang et al. 1992). However, organic precursors typically produce multiple particulate products, and the molecular identity and physical and chemical properties of these products are often unknown. Nevertheless, summary yields and product information can be gained through this approach (Odum et al. 1996).

4.3.3.1.2 *Field Experiments*

Nucleation is often observed during daylight hours in the vicinity of convective clouds in continental (Radke and Hobbs 1991) and marine environments (e.g. Clarke et al. 1998; Hegg et al. 1990). A definitive explanation for this phenomenon is not yet established, but H_2SO_4-H_2O nucleation has most often been invoked (Perry and Hobbs 1994). Atmospheric observations (Weber et al. 1999) suggest that nucleation in cloud outflows often occurs at H_2SO_4 relative acidity (ratio of H_2SO_4 partial pressure to equilibrium vapour pressure) consistent with that predicted by the classical binary theory for H_2SO_4-H_2O nucleation. Hoppel et al. (1994) also observed nucleation near clouds but found difficulties in matching the nucleation to expected production of H_2SO_4. De Felice and Cheng (1998) proposed that numerous smaller particles were produced from residues of rapidly evaporated cloud drops during the process of crystallisation. Leck and Bigg (1999) suggested that nucleation would also result if the cloud residues contained the amino acid L-methionine, derived from primary particles of marine biological origin. In the upper tropical troposphere observations suggest that nucleation is a significant global source of atmospheric particles (Brock et al. 1995; Clarke 1993). Evidence has also been reported for nucleation in the midlatitude upper free troposphere (Schröder and Ström 1997).

Nucleation has been observed occasionally in the cloud-free marine boundary layer (Covert et al. 1992; Clarke et al. 1998), and more frequently during on-shore flow in coastal areas that are exposed at low tide (e.g. O'Dowd et al. 1999).

Nucleation events also have been observed at many different ground-level continental sites (Birmili 1998b; Hörrak et al. 1998; Koutsenogii and Jaenicke 1994; Kulmala et al. 1998). These events typically follow regular diurnal patterns, with concentrations of nucleation-mode particles beginning to increase several hours after sunrise, reaching peak concentrations several hours later. The chemical mechanisms of nucleation in these regions have not yet been established, though one possible explanation involves sulphonate clusters (Kulmala et al. 2000).

In summary, a great deal of new insight into the nucleation process has been gained in the past decade. Binary H_2SO_4-H_2O theory may explain a limited number of field observations. When there is a discrepancy, nucleation rates inferred from atmospheric observations are typically higher than those predicted from theory. These findings suggest the presence of stabilising compounds such as ammonia, hydrocarbons, or electrically charged precursors that reduce the dissociation rate of the pre-nucleation molecular clusters. Laboratory studies have provided some support for these speculations. Clearly far more atmospherically relevant studies are needed in this area.

4.3.3.2 *Aerosol Growth through Concentrated Liquid-Phase Chemical Reactions*

Tropospheric aerosol particles are exposed to ubiquitous water vapour. The physical matrix of highly concentrated, or even supersaturated, aqueous electrolyte droplets in air contains a variety of compounds that may be involved in solution-phase aqueous reactions (see, e.g. Zellner and Herrmann 1995). A chemical reaction in a given electrolyte solution could be described correctly if either all activity coefficients involved or the rate constant for the reaction under the given set of conditions (pH, temperature, etc.) were known.

Kinetic salt effects in ion-ion reactions (type 1) can be treated by the Debye-Hückel theory and its extensions (for an overview, see Robinson and Stokes 1959). In reactions involving neutrals only, or one ion and a neutral, a primary kinetic salt effect is also clearly identified (Debye and McAulay 1925; Herrmann and Zellner 1998). This effect (type 2), cannot currently be treated other than by experimental examination of each reaction in each matrix separately.

Equilibrium constants are affected by ionic strength (secondary kinetic salt effect) as they may be viewed as the ratio of two rate constants, both of which are subject to the above primary kinetic salt effects (Harned and Owen 1958; Robinson and Stokes 1959). These effects have been considered in cloud models in the same manner as the type 1 effect for ion-ion reactions (Jacob 1986).

Ionic strength effects may be treated kinetically by the ion pairing approach. Stability constants for ion pairs are available (Högfeldt 1982; Perrin 1979) or may be calculated according to the Fuoss-Eigen equation (Davies 1962). The observed rate constant as a function of ion concentration then results from two parallel elementary reactions coupled by the ion pair formation equilibrium. Such treatment was suggested by Olson and Simonson (1949).

The experimental data set for a more complete description of aerosol chemistry is very sparse. In the case of non-radical reactions, various data sets relating to aqueous sulphur oxidation now exist (Lagrange et al. 1993, 1995, 1999a,b), and modelling the impact of ionic strength effects under tropospheric aerosol conditions should be possible in the near future. Concerning radical reactions, it is striking that even for OH no ionic strength dependency has been measured for a system of relevance in tropospheric aerosol chemistry. This is probably due to the widespread view that ionic strength effects involving one neutral species generally can be neglected. While this is true for the original Debye-Hückel theory in the concentration range for which that theory was derived, electrolyte contents of several mol l^{-1} may change a rate constant for a reaction involving a neutral by about one order of magnitude, as was shown for nitrogen trioxide (NO_3) reactions (Herrmann and Zellner 1998). Recently, ionic strength dependencies have been studied in H-abstraction reactions of chlorine radical anion (Cl_2^-) (Jacobi et al. 1999).

At present, there is no comprehensive tropospheric aerosol model that takes into account the available kinetic data described here. Further systematic laboratory studies are required. Apart from experimental data for single reactions, such studies may also yield correlations which, in the future, may allow the estimation of rate constants at elevated ionic strength in different cases for different electrolyte compositions (e.g. for a marine or continental tropospheric aerosol).

4.3.3.3 *Cloud Processes*

The life cycle of a cloud – its formation, development (both microphysical and chemical), and dissipation – is linked in part to the nature of the aerosol on which the cloud forms. Recent developments have increased knowledge of the underlying processes. The processing of compounds by liquid water droplets influences the chemical composition of the troposphere (e.g. Lelieveld and Crutzen 1991). A proper description of liquid phase chemistry is therefore necessary to assess the role of clouds in a changing atmosphere.

Aerosol scavenging. In most clouds, the fraction of aerosol that is taken up (scavenged) by cloud droplets is less than unity, both in terms of particle number and mass (Leaitch et al. 1992; Noone et al. 1992b). The ability of particles to act as cloud condensation nuclei is dependent on their chemical composition and size. Different compounds are scavenged with different efficiencies, either because the particle composition is size dependent, or because of the chemical nature of the substance itself (Facchini et al. 1999; Hallberg et al. 1992). This initial step determines the starting chemical composition of the cloud droplets.

Recent insights into the classical treatment of growth of aerosol particles into cloud droplets have expanded on the classical theory that chemical effects, and not simply particle size, are important factors in droplet formation. Slightly soluble substances (in particular certain organic compounds) may influence droplet surface tension and equilibrium vapour pressure, thus influencing the aerosol scavenging process (Laaksonen et al. 1998; Shulman et al. 1996). Also, this effect has important implications for indirect aerosol radiative forcing (see Sect. 4.3.4.2) by affecting the drop size distribution. Another potential factor that controls the growth of cloud droplets is co-condensation of soluble gases other than water onto aerosol particles. For example, model calculations have indicated that co-condensation of nitric acid (HNO_3) can increase the scavenging efficiency of particles in clouds (Laaksonen et al. 1997). The importance of these effects in the atmosphere has not been demonstrated experimentally.

Aqueous-phase processes. Clouds constitute an efficient reaction medium for chemical transformations. After cloud formation, the composition of the droplets can be modified by the dissolution of soluble gases. Two families of chemical compounds are key participants in cloud liquid-phase chemical reactions: sulphur and organic compounds. The chemistry of oxidised (e.g. NO_x) and reduced (e.g. NH_3) nitrogen compounds is of lesser importance, although their dissolution alters the pH of cloud droplets and thus affects aqueous chemistry.

Sulphur oxidation reactions occur in clouds at a much faster rate than in the clear air. Model calculations (Langner and Rodhe 1991) have shown that, on a global scale, tropospheric in-cloud SO_2 oxidation is from two to five times more important than out-of-cloud oxidation. Laboratory studies have identified the most likely reactions responsible for sulphur oxidation in the atmospheric liquid phase and in determining the associated rate coefficients (Warneck 1991).

Very little is known about atmospheric aqueous phase chemistry of organic compounds. In current box models, the description of organic chemistry in cloud water is limited to dissolved C_1 and C_2 compounds (Herrmann et al. 1999). Recent work has shown that cloud-forming particles contain a high percentage of soluble organic compounds (Facchini et al. 1999; Noone et al. 2000; Saxena and Hildemann 1996; Zappoli et al. 1999).

A further complication concerning cloud chemical processes derives from the size-dependence of cloud droplet chemical composition (Ogren and Charlson 1992). Because the concentration and composition of cloud droplets are not uniform, the transfer of gases into the droplets will depend on droplet size (Twohy et al. 1989). Model results (Hegg and Larson 1990; Pandis et al. 1990) have also shown the importance of size-dependent cloud droplet chemistry, demonstrating that use of bulk cloud water parameters can lead to substantial errors in description of processes taking place within individual cloud droplets.

Box 4.4. The utility of the cloud condensation nucleus (CCN) concept

The concept of CCN has been accepted by the cloud physics community, and by most of the aerosol research community, for many decades. Its attractiveness stems partly from the pioneering work of Twomey (1959), who introduced measured supersaturation spectra of droplets into cloud modelling. Visualising a distinct subset of the very complex atmospheric aerosol as a well-defined and measurable quantity, sufficient for a description of the interaction of particles and clouds, is another factor contributing to the attractiveness and the wide application of the CCN concept.

However, there are over-simplifications inherent in this concept that become increasingly difficult to justify with the rapidly advancing experimental and modelling capabilities of atmospheric aerosol research. The concept or definition of CCN is at best operational because any quantification of CCN with existing instrumentation is strongly dependent on the thermodynamic evolution forced upon an aerosol sample in any particular CCN counter (Nenes et al. 2001). An intrinsic problem is that the thermodynamic evolution occurring in any experimental CCN device bears rather little resemblance to the condensational growth that happens in any real cloud. Most CCN counters have not been characterised sufficiently to justify meaningful connection of their results to droplet growth in real clouds (Nenes et al. 2001). Many physical and chemical factors contribute to the subset of the total particle population that evolves into a cloud droplet population. Even certain soluble gases can strongly affect the growth of cloud elements (Kulmala et al. 1993).

Additionally, the concentration of CCN at a particular supersaturation is not necessarily applicable to the prediction of cloud droplet concentration. Even the generally accepted hypothesis that an increase in concentration of CCN leads to an increase in cloud droplet concentration is not necessarily correct, particularly in a mixed population of CCN (Feingold and Kreidenweis 2000). For example, an enhancement in CCN concentration measured at 0.2% supersaturation, typical of stratocumulus clouds, suggests a similar increase in cloud droplet concentration within this cloud type. However, if the additional CCN are activated at supersaturations considerably lower than 0.2%, the resulting depletion of water vapour and suppression of the peak supersaturation in a real cloud can inhibit activation of a significant fraction of the CCN population. This results in a reduction in the number of cloud drops (O'Dowd et al. 1999).

Avenues of research that may resolve the CCN issue include more detailed experimental studies of the condensational growth properties of size-resolved atmospheric particles, providing parameterisations for cloud models and cloud simulation experiments with more realistic flow-through cloud simulators.

Cloud dissipation. Upon evaporation, gases and particles contained in the droplets are released back to the cloud-free atmosphere. The residual particles are likely to be quite different (physically and chemically) from those that entered the cloud because of in-cloud processes. Field observations and modelling results have shown that sulphate size distributions before and after passage through a cloud differ significantly, with much larger concentrations in the outflows of the cloud systems (Bower et al. 1997; Hoppel et al. 1986; Laj et al. 1997; Wiedensohler et al. 1997). The effects of multiple cloud passages on cloud-processed particles have important implications for direct aerosol radiative forcing (see Sect. 4.3.4.1) by increasing the efficiency of light scattering due to in-cloud particle growth (Yuskiewicz et al. 1999).

4.3.3.4 *Aerosol Deposition Processes*

Wet deposition and dry deposition are important removal pathways for suspended particulate matter. As such they act as key determinants of the residence times and concentrations of tropospheric aerosol particles.

Aerosol deposition processes often involve multistage pathways and can include interactions with gases. This is especially true for wet deposition, which often occurs through a complex interaction involving various combinations of sorption, dissolution, particle capture, aqueous reaction, precipitation formation, and precipitation delivery. Dry deposition processes, while not so multi-faceted in nature, nevertheless can be complicated by condensational particle growth in humidity gradients.

The level of scientific effort in deposition research during the past ten years, while appreciable, was significantly less than corresponding efforts during the 1970s and 1980s. As a consequence, some of the older reviews of dry deposition (e.g. Hicks et al. 1990) and wet deposition (e.g. Hales 1990) still provide useful and relatively up-to-date synopses. Some of the more significant advances during the past decade are summarised below.

4.3.3.4.1 *Dry Deposition*

Wesely and Hicks (2000) present a review of recent dry deposition research which, combined with Hicks et al. (1990), provides an up-to-date summary of scientific understanding in this field. Brook et al. (1999a) review past measurements of deposition velocities and model results. Recent scientific progress in this area has been limited by complexities in four main areas, described below:

Measurement. Air-surface exchange fluxes to and from natural surfaces are exceedingly difficult to measure. This has resulted in a profusion of (relatively) direct as well as indirect surrogate and inferential measurement methods (Wesely and Hicks 2000). Techniques such as eddy correlation, eddy accumulation, and profile methods directed toward measuring Reynolds fluxes (i.e. those associated with turbulent transport) are generally favoured for gas flux measurement when practicable, but require monitoring equipment that has fast time response, high precision, or both. Such methods are applicable for measuring fine particle fluxes as well, but they are inappropriate for larger particles, whose gravitational settling rates contribute significantly to the total deposition flux. Much of the advance in air-surface gas exchange measurement during the past decade derives from the advent of newer, fast-response instrumentation, which allows Reynolds-flux measurement for a number of additional gases and particles (Gallagher et al. 1997; Shaw et al. 1998).

Conceptual microscale model development. Most recent development involved an extension of the traditional resistance analogy (see Chap. 2) through more detailed description of the micrometeorological and deposition processes within complex canopies, such as forests and croplands. Although founded on simple concepts that sometimes do not conform with actual two-way exchange processes, such conceptual models often involve considerable complexity.

Large-scale parameterisations. Significant advances have occurred in dry deposition parameterisation for large-scale CTMs during the past ten years through extension of traditional procedures (Brook et al. 1999b) as well as the employment of satellite land-use mapping (Gao and Wesely 1995a,b).

Advanced monitoring and field studies. Resulting in part from the advent of the newer instrumentation noted above (and discussed in Chap. 5), a number of advanced field studies have been conducted, which have contributed significantly to the general understanding of dry deposition rates and processes. The comprehensive investigation of dry deposition in the Speulder forest in the central Netherlands is a prime example of such efforts (Erisman et al. 1997 and companion articles in that issue).

4.3.3.4.2 *Wet Deposition*

Some of the more recent scientific advances in understanding wet deposition and precipitation scavenging have been associated with large, multi-participant efforts. Examples include the European Ground-Based Cloud Experiments (Fuzzi et al. 1992; Wobrock et al. 1994) and the High Alpine Aerosol and Snow Chemistry Study (Nickus et al. 1997).

Major features of the precipitation scavenging sequence were understood scientifically by the close of the 1980s. Because of this, advances since that time can be considered either as mechanistic extensions to existing theory or to composite modelling of the scavenging phenomenon. Particularly important advances have been made in the areas of aqueous-phase chemistry, aerosol cloud interactions, and composite modelling. Some of the advances in key categories during the past decade are summarised below.

Aqueous-phase chemistry. Jacob (2000) presents a review of progress in aqueous-phase chemical processes of importance to scavenging. Much of the more recent advance in this area pertains to the behaviour of in-cloud odd hydrogen (HO_x) chemistry. Aqueous reactions of a number of trace substances (e.g. sulphur and nitrogen oxides, soluble organics) are believed to be linked closely with this oxidant chemistry.

Aerosols and cloud formation. During the past decade a number of field, laboratory, and theoretical studies has evaluated cloud formation from mixed aerosol particles in competitive environments as functions of particle size, composition, and saturation conditions. These studies are particularly important because they can elucidate relative efficiencies for the scavenging of specific aerosol particle classes as functions of storm type and intensity. Several examples of field measurements of related phenomena appear in references associated with the multi-participant studies cited above.

Composite modelling. Although several global and regional models of the 1990s have incorporated increasingly complex representations of wet chemistry and precipitation scavenging (e.g. Rasch et al. 2000), the results of these efforts shed comparatively little light on physical and chemical mechanisms of the scavenging process. On the other hand, a number of mesoscale modelling efforts have been undertaken during the past decade that may provide valuable mechanistic insights, especially when linked with modern field studies. The study of Voisin et al. (2000) provides a recent example of such an effort.

4.3.4 Effects of the Aerosol on Radiation in the Atmosphere

Aerosols influence Earth's radiative balance both directly and indirectly, through their influence on cloud properties. The magnitude and sign of their influences are variable, and the controlling properties and processes behind the radiative effects are presently not well understood. Understanding these controlling properties and processes, and quantifying the influence of aerosols on the radiative balance of Earth, is expected to be a major component of aerosol research in the next decade. Aerosols also affect the actinic flux and, as such, photochemistry.

4.3.4.1 *Direct Radiative Effects*

Aerosol particles both scatter and absorb incoming solar radiation, effects that are visible to the naked eye. Figure 4.11a–c shows examples of visibility reductions over the Arabian Sea and Indian Ocean caused by transport of pollution from the Indian sub-continent. Figure 4.11d shows an almost pristine atmosphere south of the Inter-Tropical Convergence Zone.

If incoming radiation is scattered back to space, the energy contained in the scattered photons is lost from the system. In this case, aerosols would exert a net cooling effect on the Earth system. If the incoming radiation is absorbed in an aerosol layer above Earth's surface, the energy contained in the absorbed photons is transformed to heat, and can result in a warming of the absorbing layer. This backscattering and absorption in the atmosphere can reduce the amount of solar energy available at the surface for biological processes.

The mass scattering efficiency (scattering coefficient per mass concentration of aerosol) is strongly dependent on wavelength and particle size, reaching a maximum for particle diameter near the wavelength of the light. For radiation near the centre of the visible spectrum (0.53 µm) this maximum occurs at $D_p \approx 0.5$ µm for ammonium sulphate aerosol with value ca. 9 m^2 $(g\ SO_4^{2-})^{-1}$, and with the most efficient size range for scattering between particle diameters of about 0.1 and 1.0 µm (Ouimette and Flagan 1982; Nemesure et al. 1995; Schwartz 1996). Scattering can be reasonably accurately apportioned among the various chemical constituents that make up the atmospheric aerosol (White 1986; Sloane 1983; Zhang et al. 1994b; McInnes et al. 1998). The particular component that dominates scattering depends on location, time, and other factors.

Sulphate particles have received the most attention in assessing the direct radiative effects of aerosols (Charlson et al. 1990, 1992), but other chemical components are also important. Over the oceans, sea salt can contribute substantially to the total aerosol optical depth. Using sea salt distributions from a chemical transport model and similar mass scattering efficiencies, Tegen (1999) estimated an 87% relative contribution from sea salt in the 20° S to 20° N latitude band at 140° W, a 57% contribution for the world's oceans, and a 44% contribution to the global aerosol optical depth. Similarly, based on a comparison of clear-sky, top-of-the-atmosphere solar irradiances observed by the Earth Radiation Budget Experiment with those calculated from a general circulation model, Haywood et al. (1999) reported that sea salt is the leading particulate contributor to the global mean clear sky radiation balance over oceans. Organic aerosols can also be a significant contributor to particulate scattering. For example, from aircraft flights off the east coast of the US, Hegg et al. (1997)

March 21, 1999: Arabian Sea; Thick haze (9.2°N, 73.5°E)

March 25, 1999: Clouds under thick haze (3.0°N, 74.5°E)

February 24, 1999: Just north of ITCZ
Haze extends up to top of CU (0.5°N, 73.3°E)

March 24, 1999: Just south of ITCZ
Almost pristine clouds (7.5°N, 73.5°E)

Fig. 4.11. Aerosols and clouds as observed during the INDOEX experiment. **a** Arabian Sea, thick haze; **b** Indian Ocean, clouds under thick haze; **c** Indian Ocean just north of the ITCZ, haze extends up to the top of cumulus clouds; and **d** Indian Ocean south of ITCZ, almost pristine clouds (photographs from Center for Clouds, Chemistry and Climate, Scripps Institution of Oceanography, US; Sathees and Ramanathan 2000)

showed that organic particulate matter accounted for a substantial fraction of aerosol light scattering. Mineral dust also exerts a large influence on the radiative balance (e.g. Sokolik and Toon 1996).

Radiative forcing, defined as the change in the radiative fluxes at the top of the troposphere or top of the model domain due to a given aerosol component, has been calculated for a variety of particle distributions and emission scenarios. The most recent summary of forcing results is given by the Intergovernmental Panel on Climate Change (IPCC 2001); see Fig. 1.7. The wide range of values for the calculated global direct radiative forcing of aerosols clearly shows that more information is needed to constrain this forcing. Key information needs are: (*1*) the mass loading and geographical distribution of the aerosol; (*2*) the chemical components of the aerosol, their state of mixing in the atmosphere, their relative contributions to light scattering and absorption, and their response to varying relative humidities; and (*3*) a better understanding of the cycles of the various substances that make up the aerosol. The large uncertainty in direct aerosol forcing limits quantitative estimates of climate response to forcing over the industrial period.

4.3.4.2 *Indirect Effects*

The indirect aerosol effect refers to potential changes of cloud properties at the global scale due to anthropogenic perturbations of the concentrations and physical and chemical properties of the particles that form cloud drops or ice crystals. This indirect aerosol effect is presently the most uncertain of the known forcing mechanisms in the prediction of climate change over the industrial period (see Fig. 1.7). Bridging the scales involved is one of the major challenges in understanding and predicting the indirect radiative effect of aerosols. While the effect itself is global, the processes causing it occur on spatial scales as small as micrometers and temporal scales as short as seconds. Thus, the global-scale phenomenon cannot be understood and predicted quantitatively without adequate understanding of the microscale processes.

The basis of the indirect effect is a link that was established by Twomey (1974) between aerosol properties, cloud droplet concentrations, and cloud albedo. For a given cloud liquid water content and cloud depth, Twomey showed that an increase in aerosol particle number concentration could lead to an increase in cloud droplet number concentration, which would cause an increase in cloud albedo. The increase in cloud albedo would lead to cooling and partially counteract the warming due to greenhouse gases (Slingo 1990). The main contribution to the indirect effect is thought to come from marine stratocumulus clouds (Randall et al. 1984). However, cirrus clouds may also be influenced by anthropogenic aerosols, but for the cirrus case not even the sign of the global radiative effect is known. Additional feedback mechanisms potentially responsible for the uncertainty in the prediction of the indirect effect include increases in cloud cover with increasing global temperature (Arking 1991), reduction of the precipitation efficiency of clouds causing an increase in their lifetime or extent (Albrecht 1989), coupling between diabatic processes and cloud dynamics (e.g. Pincus and Baker 1994), and the radiative effect of in-cloud absorption on short wave radiation (Boers and Mitchell 1994).

A number of experiments have been carried to understand better the aerosol-cloud interactions responsible for the indirect radiative effect. Ship tracks in marine stratocumulus clouds have been used as a natural laboratory to understand better how anthropogenic aerosol emissions cause observable increases in cloud albedo (Durkee et al. 2000b; Ferek et al. 1998; King et al. 1993; Radke et al. 1989). In another approach, Brenguier et al. (2000) observed marine stratocumulus clouds during the second Aerosol Characterisation Experiment (ACE-2). In these studies, simultaneous *in situ* measurements of cloud microphysical properties and remote sensing of the cloud radiances in the visible and near-infrared provided direct evidence of changes in cloud radiative properties related to changes in the input aerosol. Drizzle suppression has also been observed in clouds influenced by anthropogenic aerosols, through both *in situ* measurements (Ferek et al. 2000) and satellite observations (Rosenfeld 2000).

Anthropogenic effects on cirrus cloud radiative properties have also been investigated in recent years, particularly in terms of the effects of aircraft emissions on cirrus cloud properties. Wyser and Ström (1998) showed that the balance between short-wave cooling and long-wave heating effects of cirrus clouds could be very sensitive to changes in crystal size caused by aerosol emissions from aircraft. Several recent research programmes have been aimed specifically at assessing the effects of aircraft emissions on cirrus cloud properties (Brasseur et al. 1998a; Schumann et al. 2000; Toon and Miake-Lye 1998).

Detecting large-scale changes in cloud properties from satellites has proven to be challenging. For example, Han et al. (1994) showed that effective radius in warm ($T > 273$ K) continental clouds was on average two to three µm smaller than that in marine clouds. They also showed that droplet radii are about one µm smaller in marine clouds of the Northern Hemisphere than in those of the Southern Hemisphere. Likewise, fairly strong spatial correlations have been reported of monthly mean aerosol optical depth and number concentration with effective radius (negative correlation) and optical depth (positive correlation) of low clouds, all of which would be consistent with the Twomey effect (Wetzel and Stowe 1999; Nakajima et al. 2001). However, the expected enhancement of cloud albedo due to the Twomey effect has not been identified in interhemispheric comparisons (Schwartz 1988; Han et al. 1998). Likewise, examination for aerosol enhancement of monthly mean cloud albedo as a function of distance from continents has been negative (Falkowski et al. 1992) or has shown only marginal indication of enhancement (Kim and Cess 1993). Evidence has been presented of enhanced reflectivity of marine stratocumulus clouds in the vicinity of copper smelters in Peru under conditions of offshore flow (Kuang and Yung 2000). Indications of the indirect effect were also presented by Rosenfeld and Lensky (1998), who analysed AVHRR satellite images of cumulus and derived the cloud top temperature and effective radius. Variations of the relationship between those two parameters were interpreted as a sign of changes in cloud radiative properties due to continental pollution.

Several global-scale model studies have been undertaken to explore the consequences of anthropogenic aerosol emissions on cloud radiative properties. While the parameterisations used in the models are limited by an incomplete understanding of the processes involved and by a lack of data on nearly all scales, the results as summarised by Penner et al. (2002) and Ramaswamy et al. (2001) indicate that the indirect effect may be very important (see Fig. 1.7).

The relationship between radiative forcing and climate response was examined for a wide range of artificial perturbations by Hansen et al. (1997b). The direct and indirect climate effects of anthropogenic sulphate were examined for equilibrium and transient responses to anthropogenic sulphate, and with prescribed or interactively calculated forcing. Tegen and Miller (1998) calculated the climate effect of mineral dust interactively in an equilibrium simulation. Climate models exhibit wide scatter in the climate sensitivity (defined as the ratio of the change in the global mean surface air temperature to the global mean radiative forcing). Factors causing this scatter include uncertainties in the distribution and the optical properties of aerosol particles and differences in approaches to treating cloud physical processes and in the assumed optical properties of clouds (e.g. Cess et al. 1996). In view of the many uncertainties associated with understanding and describing aerosol sources and processes and with their parameterisation, such forcing and the ensuing climate responses simulated by models must be considered tentative.

4.3.4.3 *Aerosol Effects on Actinic Flux*

Solar radiation, especially in the ultraviolet (UV) range, drives tropospheric photochemistry. The presence of tropospheric aerosols results in additional scattering and absorption of UV radiation and, hence, affects photolysis rates. Several recent studies have pointed out that aerosol particles might strongly perturb the formation of ozone and other trace gases in the lower atmosphere (Dickerson et al. 1997). For example, for given meteorological conditions and levels of photochemically active gases, ozone production is sensitive to the aerosol burden and optical properties. Strong UV-absorbing particles (such as soot and mineral dust) cause a decrease in photolysis rates, which, in turn, inhibits ozone production. The presence of purely scattering particles (such as sulphate) or clouds in the boundary layer can increase photolysis rates there and above cloud tops, which enhances ozone production aloft (He and Carmichael 1999). However, near-ground photolysis rates can decrease as particle concentrations increase. Because the distributions of tropospheric aerosols and clouds are highly variable in space and time, the quantification of the effects of the aerosol on photochemical processes remains a challenging problem.

4.3.5 Effects of Aerosols on Atmospheric Photochemistry

Aerosol particles can affect atmospheric photochemistry by changing the actinic UV flux (see Sect. 4.3.4.3), or more directly by acting as sinks or sources of reactive trace gases (Ravishankara and Longfellow 1999). Aerosol growth due to reactions in deliquescent particles at very high relative humidities has already been discussed (Sect. 4.3.3.2). Here we focus on heterogeneous reactions at lower relative humidities and/or on insoluble particulate matter. A textbook example is the hydrolysis of N_2O_5 on particles, which affects the oxidising capacity of the troposphere via the atmospheric residence time of NO_x. The process was shown to be important on a global scale by Dentener and Crutzen (1993) who assumed in their model calculations that the hydrolysis rate is first order in $[N_2O_5]$ and proportional to the surface area of the particulate matter, times a constant reactive uptake coefficient, γ. A value of $\gamma = 0.1$ was used in the base case calculations. More recent aerosol chamber studies yielded significantly lower uptake coefficients, which also depended on the nitrate activity of deliquescent aerosol particles (Mentel et al. 1999).

Considerable attention has been paid to reactions of oxidising species with soot particles from fossil fuel combustion. Several studies yielded relatively large values of γ for the decomposition of ozone on soot (Rogaski et al. 1997) which indicated that ozone destruction on aircraft soot may be significant (Bekki 1997). However, chamber studies yielded much lower time- and concentration-dependent values of γ, implying a complex reaction mechanism which eventually leads to surface passivation (Kamm et al. 1999).

The reduction of HNO_3 and/or NO_2 on soot was also studied in several laboratories, and surprisingly large values of γ were obtained (Rogaski et al. 1997). The latter reaction was found to produce HONO as a by-product (Kalberer et al. 1999), which is a photochemical source of OH in polluted atmospheres. However, simple thermodynamic considerations imply that neither NO_2 nor HNO_3 can be reduced on soot at a constant rate, because some reducing functionalities must be consumed in the process. This was confirmed by a recent aerosol chamber study which yielded much lower upper limits for the time-averaged reactive uptake coefficients of NO_2, HNO_3, and several other oxidising species on soot particles (Saathoff et al. 2001). Multi-day model calculations, assuming a large mass concentration of soot particles, resulted in up to 10% lower ozone maxima on the second day when the whole set of heterogeneous reactions was included. The reduction was mainly due to the loss of HO_2 on soot, for which $\gamma \leq 10^{-2}$ had been determined. More accurate measurements of $\gamma(HO_2)$ are needed to prove or disprove this result.

Trace gas-soot interactions are comparable to reactions with large polycyclic aromatic hydrocarbons. The oxidation by ozone of benzo-(a)-pyrene adsorbed on soot aerosol has recently been studied in an aerosol flow tube (Pöschl et al. 2001). The surface reaction with the adsorbate is preceded by the reversible adsorption of ozone on soot. The reaction rate decreases in humid air because of competition between water vapour and ozone for adsorption sites. Kirchner et al. (2000) reported that reactions of NO_2 and HNO_3 with soot induce permanent changes in the IR spectra of the particles, confirming the analogy with PAH reactions. They measured concentration- and time-dependent values of γ, which were limited by surface saturation, in good agreement with Kalberer et al. (1999), who observed saturation-limited NO_2-to-HONO conversion on soot particles. Clearly, the assumption that "initial" uptake coefficients, determined under idealised laboratory conditions, can be used to calculate reaction rates on aerosol particles under realistic atmospheric conditions is not justified in general.

The role of mineral particles as a reactive surface in the global troposphere was pointed out by Dentener et al. (1996). Recent laboratory investigations (Hanisch and Crowley 2001) yielded very large slightly humidity-dependent uptake coefficients for HNO_3 on $CaCO_3$ and on authentic samples of desert dust. The release of CO_2 confirms that carbonates are converted to nitrates, thereby changing the hygroscopic properties of the dust.

Sea salt particles are a potential source of photochemically active halogen compounds. It is now quite clear that bromine atoms are involved in polar tropospheric ozone destruction during springtime. Large tropospheric BrO-clouds have been observed by satellite in polar regions of both hemispheres while ozone destruction occurred (Richter et al. 2001). Details of the complex chemistry of sea salt activation have recently been reviewed (De Haan et al. 1999), but further work is needed to understand better the halogen source related to the sea salt aerosol.

In summary, there is now overwhelming evidence that aerosol particles can act as sources and sinks of reactive species in the gas phase. It is, however, important to note that reactions on particles not only impact on gas-phase photochemistry, but also change the bulk and/or surface chemical composition of particulate matter, with important consequences for their hygroscopic and optical properties.

4.3.6 Aerosols and Health

Although aerosol studies within IGAC primarily looked at aerosol effects on radiation and chemistry, it should not be forgotten that a major driver of aerosol research has been their effect on human health. It has been shown that cardio-pulmonary diseases and mortality are related to the presence of particulate matter (e.g. Dockery et al. 1993). Significant correlations have been established between mortality and the levels of PM10 (i.e. the mass concentration of particles with aerodynamic diameters less than 10 μm) and PM2.5 (mass concentration below 2.5 μm); see Table 1.3. As yet, no mechanism is known that relates specific chemical or physical characteristics of the particles to their effects on health. The fact that PM2.5 has a larger effect than PM10 may be explained by the fact that the PM2.5 fraction contains most of the combustion-derived and most soluble particles, or alternatively may be due to the greater depth of penetration of small particles into the lungs. Clearly, the advanced methodologies developed for IGAC experiments and the aerosol characteristics derived in these experiments will be useful for the understanding of aerosol effects on health.

4.4 Research Approaches

4.4.1 *In situ* Observations with Intensive Campaigns

To a great extent the progress in atmospheric aerosol research during the past decade can be attributed to a change in the way aerosol measurements and chemical atmospheric measurements in general are being conducted. Investigators and funding agencies recognised that meaningful progress in understanding atmospheric aerosol properties and processes requires a large number of measurement capabilities to be brought to bear simultaneously. This research strategy results in over-determined data sets that can be used in closure studies (e.g. Ogren 1995; Quinn et al. 1996a), which can summarise critically and quantitatively the state of knowledge about characteristics, processes, and effects of the atmospheric aerosol. This approach has been used successfully during several large, international campaigns (e.g. ACE-1, Bates et al. 1998; Bates 1999; TARFOX, Russell et al. 1999; and ACE-2, Raes et al. 2000a).

A second and related advance in the past decade has been the conduct of aerosol investigations over large areas (thousands of square kilometres or more) in order to capture significant aerosol processes that take place over such geographical scales. These studies require major resources – ships, aircraft, and surface stations, as well as large numbers of investigators and specialised instruments – that often exceed the capabilities of any single agency or even a whole country. With such resources it has been possible to conduct Lagrangian experiments, wherein an air parcel is identified and followed as it is transported. This strategy has been employed in several large-scale experiments (e.g. ASTEX-MAGE, Huebert et al. 1996; ACE-1, Bates et al. 1998; Bates 1999; ACE-2, Raes et al. 2000a; and INDOEX, Satheesh and Ramanathan 2000) and has yielded process information less confounded by effects of advection.

Large-scale projects also require the active participation of meteorologists and modellers who can provide descriptions of the transport fields that are necessary for the interpretation of the observations. Mathematical models are the integrator of understanding of atmospheric processes, and field data analyses are now being carried out hand-in-hand with model implementation and evaluation. Moreover, some models are being used prior to the field experiment in the design of the best possible measurement strategy. Recently, modelling in forecast mode has begun to aid the day-to-day planning of aircraft missions during large campaigns. The increasing interaction of experimentalists, modellers, and meteorologists is a third major advance in the past decade that is helping to advance understanding of aerosol processes.

A fourth related advance, and a major research strategy in large international experiments, has been the open sharing of data among research participants and with the broader research community, which greatly enhances the utilisation and value of the data. Similarly, much of the ability of meteorologists to contribute to these studies relies on the availability of synoptic-scale meteorological data from the operational weather forecasting community.

Quality control of the host of particle instruments remains a crucial problem that limits the gains of these large experiments as compared to focussed small scale process studies. In particular, there are no generally accepted sampling specifications for atmospheric aerosols, neither concerning the range of sampled sizes nor concerning the thermodynamic state of the aerosol. Also, for crucial particle properties such as carbonaceous matter or light absorption there are no internationally established reference methods and standards for calibration.

4.4.2 *In situ* Observations Using Long-Term Monitoring Networks

Although large, intensive projects such as ACE-1 and ACE-2 can contribute much to the understanding of aerosol characteristics and the underlying processes, they are not sufficient to produce needed aerosol climatologies. A major factor limiting the extent and quality of present aerosol climatologies is the rarity of well-defined and calibrated aerosol monitoring programmes. The successful networks operated by the University of Miami (Florida, US) until the mid 1990s (e.g. Prospero 1996; Prospero et al. 1983; Saltzman et al. 1986) provided a unique record of aerosol chemistry properties in the marine environment. NOAA-CMDL has records of aerosol physical properties dating from the 1970s at its baseline stations, but only limited aerosol chemistry data (Bodhaine 1989; Bodhaine and DeLuisi 1985; Bodhaine et al. 1986). Other noteworthy data sets exist for a few individual stations. Examples include Cape Grim, Tasmania (Ayers et al. 1991, 1995; Gras 1995); Jungfraujoch, Switzerland (Baltensperger et al. 1997; Weingartner et al. 1999); Hohenpeissenberg, Germany (Fricke et al. 1997); Spitsbergen, Norway (Heintzenberg and Leck 1994); and Alert (Ellesemere Island, Canada) (Sirois and Barrie 1999). In addition, regional acid deposition networks such as EMEP in Europe (*http://www.nilu.no/projects/ccc/emepdata.html*), CAPMon (Canadian Air and Precipitation Monitoring Network), NAPAP (US National Acid Precipitation Assessment Program) and CaSTNET (Clean Air Status & Trends Network) (*http://www.epa.gov/castnet/data.html*), and air pollution networks such as IMPROVE (*http://vista.cira.colostate.edu/IMPROVE*), make selective measurements of aerosol properties. Because the primary purpose of these networks, however, is not aerosol characterisation, the resulting data are not always appropriate for aerosol climatological use.

For lack of suitable methodology and local scientific capacity, the initial global aerosol monitoring programmes designed by WMO (1993) has been of somewhat limited value so far. These programmes had initially adopted a design philosophy restricted to monitoring stable long lived trace gases. However, there are renewed efforts to extend the Global Atmosphere Watch (GAW) programmes to detailed aerosol measurements (GAW 1994) and the World Data Centre for Aerosols (*http://www.ei.jrc.it/wdca/*) has made progress in constructing an archive of existing data. The latter has been used in the IGAC-WCRP intercomparison and validation study of global CTMs of the sulphur cycle (Barrie et al. 2000).

4.4.3 Remote Sensing of Aerosols

Although much aerosol research has traditionally focussed on *in situ* measurements, a major exception has been remote sensing by sun photometry. Networks of well-calibrated instruments capable of measuring aerosol optical thickness with high accuracy have been initiated over the past decade (e.g. AERONET; Holben et al. 1998). The availability of such data on the web greatly enhances their utility both in developing aerosol climatologies and in evaluating models.

Surface-based lidars are starting to be employed in networks, and limited experience thus far shows that they can probe aerosols and provide useful microphysical information with vertical resolutions of a few hundred meters.

An increasingly valuable source of data to the aerosol research community is satellite-borne instruments. Most data so far have come from passive instruments such as radiometers; often the aerosol signal is obtained as a residual after subtraction of other known or assumed contributions; and products, such as aerosol optical thickness, are somewhat dependent on assumptions (Kaufman et al. 1997a and papers in that issue; Wagener et al. 1997; King et al. 1999). Active devices such as satellite-borne lidars and the combination of radiometers and radars on the same satellite hold great promise for the future (see Chap. 5).

4.4.4 Aerosol Modelling

The need to address with models questions ranging from local and regional air pollution to the direct and indirect radiative influence of aerosols on a global scale is placing increasing demands on the aerosol modelling community (see also Chap. 6). Most work to date has focussed on particulate mass of a single substance, ignoring the dynamics governing its size distribution. This is a major shortcoming given the strong influence of particle size on aerosol radiative properties (Nemesure et al. 1995; Schwartz 1996; Boucher et al. 1998) and on cloud microphysical influences (Pruppacher and Klett 1980; Schwartz and Slingo 1996).

Likewise, from the perspective of developing strategies to control local and regional particulate air pollution, it is necessary to develop model-based representations of the processes responsible for aerosol burdens that can confidently relate these burdens to sources of particulate matter and to other controlling variables. Several global-scale models have represented mass concentrations of some key aerosol constituents, e.g. sulphate (Langner and Rodhe 1991; Pham et al. 1995; Chin and Jacob 1996; Chuang et al. 1997; Feichter et al. 1997; van Dorland et al. 1997; Graf et al. 1997; Lelieveld et al. 1997; Restad et al. 1998; Lohmann et al. 1999a; Barth et al. 2000; Rasch et al. 2000); nitrate (van Dorland et al. 1997); and organic and elemental carbon (Liousse et al. 1996; Cooke and Wilson 1996; Cooke et al. 1999; Penner et al. 1998b). Sectional models for a range of particle sizes have been used for sea salt (Gong et al. 1997a,b; Tegen et al. 1997) and soil dust (Tegen and Fung 1994); these models are essentially models for the mass of each of the several sections, as there is no dynamical evolution of these materials. Accuracies of these models are generally evaluated by comparison with monthly mean or annual measurements at surface stations. Often agreement is within a factor of 1.5 to 2 in such comparisons, but occasionally much worse. Comparison is occasionally made with aircraft flights, for which agreement is generally less. At the sub-hemispheric scale, for a model driven by observationally derived meteorological data, it has been possible to make comparisons for specific dates; measured and modelled daily-average mass concentrations of sulphate typically agree within a factor of two but with numerous outliers (Benkovitz and Schwartz 1997).

Considerable effort is now being directed toward developing models that explicitly represent aerosol microphysical processes to gain predictive capability of aerosol particle size and size-distributed composition. Approaches to this size-resolved modelling include explicit representation of the size distribution (Meng et al. 1998); modelling the evolution of the amount, peak radius, and standard deviation of each of several aerosol modes (Schulz, et al. 1998); and, modelling the evolution of the moments of the size distribution, from which aerosol physical properties can be calculated (McGraw 1997; Wright et al. 2000; Wright 2000). In view of the still limited understanding of many of the processes that must be represented in the models, it is clear that much effort must be directed to enhance understanding before these processes can be represented realistically.

A major objective of global-scale aerosol modelling is evaluation of aerosol influences on climate at present, over the industrial period, and for the future, for various assumed emission scenarios. Three types of evaluations may be distinguished. Until recently, assessments of aerosol influences on climate have used temporally-averaged aerosol fields calculated in transport-transformation models that are run separately from the radiation transfer model used to assess the climate influence; that is, the aerosol field is calculated off line. This approach may misrepresent key features of aerosol-climate interactions, for example, by not capturing correlations between particulate concentrations and relative humidity or cloudiness that might affect direct forcing. A second approach (e.g. Koch et al. 1999; Adams et al. 1999; Tegen et al. 1997) calculates the aerosol loading and distribution in a model for which the meteorological driver is the output of a GCM, thereby permitting such correlations to be captured. However, even these calculations fail to represent the influence of aerosols on the climate system, such as their influence on precipitation development and resultant changes in the hydrological cycle. In an early study Taylor and Penner (1994) found that inclusion of sulphate aerosol chemistry and transport in their GCM resulted in changes in precipitation patterns that in turn enhanced the sulphate loading. While it is premature to attribute such specific climatic effects and interactions to aerosols with confidence, the fact that such changes are exhibited in on-line aerosol-climate models underscores the need for aerosols ultimately to be represented on line in climate models and for the aerosol influences on the climate system being modelled to be taken into account. This will add considerable computational burdens, which poses a challenge to the aerosol modelling community to develop accurate and efficient methods of representing size- and composition-dependent aerosol processes suitable for incorporation in climate models.

4.5 Highlights and Remaining Challenges

The decade of the 1990s was one of major progress in the recognition of the role of aerosols in global atmospheric chemistry, in describing concentrations and properties of tropospheric aerosols and their geographical distribution, and in understanding the controlling processes. Much of this revolution has been fuelled by new measurement capabilities. Extension of the range of differential electrical mobility analysers and condensation particle counters down to particle diameters of a few nanometers has revealed the episodic nature of nucleation events, though the substances responsible for these events still remain elusive. Single-particle mass spectrometers have revealed that the internal and external state of mixing of particles is much richer than had previously been imagined. This richness is a challenge to explain but it provides a means of identifying the processes responsible for formation of particulate matter in the atmosphere. To a great extent the revolution in atmospheric aerosol research can be attributed also to a change in the way aerosol measurements and atmospheric chemistry measurements generally are being conducted. Investigators and funding agencies are recognising that meaningful progress in understanding atmospheric chemistry processes requires a large number of measurement capabilities to be brought to bear at the same time, as noted above.

The enhanced understanding is making its way into numerical models describing the emissions, transport, transformation, and deposition of aerosols and their precursors on a variety of scales, from urban air sheds to global.

During the past decade it has become clear that human activities have dramatically modified the regional distributions of atmospheric aerosols. It is also evident that anthropogenic aerosols play a major role in the determination of local and regional climatic regimes, through both direct and indirect radiative forcing of climate. Increases in anthropogenic aerosols can cause potentially disruptive changes to sociologically and economically important factors such as regional hydrological cycles.

While we have learned a lot about aerosol characteristics and the nature of many of the processes responsible for their impacts, in many cases our understanding is not yet sufficiently quantitative for accurate prediction of the severity of impacts. A few future challenges are outlined next.

4.5.1 Characterisation of the Atmospheric Aerosol

Characterisation of primary particulate matter from combustion should include divisions by mass into fractions that are relevant optically: ash, organic carbon, black carbon, and, possibly, sulphates. Particular attention is needed for the combustion practices thought to contribute the largest fractions to the global burden of primary particles, such as domestic coal and wood burning, and industrial combustion in less-developed countries. The hygroscopicity of these particles should be investigated as a first step in estimating their lifetimes and roles in cloud droplet formation.

A tabulation of combustion practices by region is needed for development of more accurate emission inventories. Information on practices in non-OECD countries is particularly sparse. There are no quantitative estimates of global fly ash emissions, in particular fly ash from biomass burning. Information is also lacking on primary mineral matter emissions from sources other than combustion for heat and power.

As volcanic sulphur emissions are comparable to anthropogenic emissions of sulphate, there is a need to estimate volcanic emissions better. Volcanic sources should first be grouped according to their respective magma provinces. Combination with their respective levels of activity will then yield better spatial and temporal emissions patterns. To obtain better data on the absolute emissions from volcanoes, more systematic measurements are needed, at least for those sources that contribute most. Exemplary long-term (a few years) measurements must be made at type-representative volcanoes.

Physical techniques based on single particle counting will continue to play a vital role in atmospheric aerosol science. The high counting sensitivity that can be achieved means that extremely small samples are sufficient. Bulk aerosol composition data provided by these techniques can be used for studies of environmental cycling of various elements, aerosol mass closure studies, and source apportionment based on receptor modelling.

Single-particle analysis can provide invaluable information on the origin and evolution of atmospheric particles, such as their degree of external and internal mixture and new particle formation. Single-particle compositional data still remain to be reconciled with measurements of other properties performed *in situ* on individual particles, such as hygroscopic and cloud-nucleating behaviour.

4.5.2 Formation and Growth of Particulate Matter

Laboratory studies are needed to test further binary and ternary nucleation theory and to investigate more quantitatively the influence of ambient levels of ammonia and various other potential biogenic and anthropogenic precursor compounds on particle nucleation rates. Also, a quantitative study of ion-induced nucleation needs to be undertaken. Models should be developed to help explain laboratory results and to help short-circuit the arduous path of investigating all possible combinations of nucleation precursors. Models should also be run to determine where climatic and health influences of this highly non-linear nucleation process are most sensitive to changes in precursor concentrations.

There is a need to assess the relevance of various oxidation pathways to form condensable species under atmospheric conditions, especially in the case of natural precursors, and to understand the chemical mechanisms leading to very low volatility oxidation products of reactive organic gases. The influence of humidity on product and particle yields requires further study. The speculation that aerosol particles (both organic and inorganic) might also affect the oxidising capacity of the troposphere, by serving as sources or sinks for gas-phase radicals, needs to be evaluated.

Further systematic laboratory studies are required to provide fundamental data (e.g. accommodation co-

Box 4.5. Effects of volcanoes

In addition to their considerable contribution to the tropospheric sulphur budget, episodic strong volcanic eruptions reaching the stratosphere cause measurable climatic effects over periods lasting a few years. Injections of volcanic material of more than about 3 Tg SO_2 to the stratosphere occur about once per decade. Such events cause an increase in stratospheric sulphate aerosol particles of one to two orders of magnitude and a reduction of solar radiation reaching the troposphere of ca. 2 $W m^{-2}$ in the global mean, thus compensating much of the anthropogenic greenhouse effect for one to two years. The radiative cooling resulting from the 1991 Mt. Pinatubo eruption was modelled to be −0.3 to −0.5 K (global annual mean) (Hansen et al. 1996); a concomitant global cooling of similar magnitude was observed. Discrepancies between observations and model results based on solar radiation effects alone have been overcome by the inclusion of terrestrial radiative heating of the stratospheric aerosol layer (Graf et al. 1993). The 2–4 K warming of the low-latitude lower stratosphere by this effect intensifies the winter polar vortex, with consequences for the structure of tropospheric planetary waves. This dynamic effect results in advective continental winter warming in the northern mid- and highlatitudes reaching maxima of +6 K over Siberia following the eruption of tropical volcanoes (Robock and Mao 1992).

The particle size distribution and optical properties of background and disturbed stratospheric aerosol are known better than for most tropospheric aerosols, and have been simulated in a CTM (Hamill et al. 1997) and in a climate model (Timmreck et al. 1999).

During volcanically disturbed episodes the gravitational settling of stratospheric sulphate particles contributes significantly to the upper tropospheric (UT) aerosol. Tropopause folds during other times bring stratospheric aerosol to the UT region and increase the frequency of occurrence of cirrus. After the eruptions of El Chichòn (1982) and Pinatubo (1991) subvisible cirrus increased by about 10%. By increasing heterogeneous reactions with anthropogenic chlorine the enhanced stratospheric sulphate aerosol also leads to a decrease of stratospheric ozone (Brasseur and Granier 1992), which reduces the net heating of the aerosol layer.

efficients, rate coefficients) needed for a more complete description of aerosol chemistry. In addition to experimental data for individual reactions, such studies should aim toward developing correlations that could allow the estimation of rate constants at high ionic strength in various electrolytes.

4.5.3 Aerosol-Cloud Interactions

The need to predict cloud droplet formation from given aerosol particle size distributions in models provides a strong motivation for identifying the source of the apparent disparities (e.g. Schwarzenböck et al. 2000) between aerosol physico-chemical measurements and cloud droplet number concentrations. Future investigations must examine more comprehensively the influence of soluble gases, partly soluble material, and surface active materials on water uptake by aerosol particles and drops. There is need for a coordinated effort to investigate the properties of particles on which cloud droplets form.

In the area of cloud chemistry, current theoretical research is centred on the interaction of radicals with organics that may influence the rate of S(IV) oxidation in droplets, alter the pattern of precipitation composition, and possibly lead to the production of harmful substances. The radical-driven oxidation of S(IV) to S(VI) is an extensive research topic of its own. Attention needs to be given to understand better the production and aqueous-phase chemistry of organic species via laboratory studies, focussed field campaigns, and modelling efforts. Current efforts to measure and model size-dependent cloud chemical composition should also be improved.

Potentially, dynamic and precipitation effects through ice nucleation and crystal multiplication processes have great global importance and must receive more attention in future aerosol research. There are large uncertainties in the understanding of basic ice formation processes and their connection with the atmospheric aerosol that need to be reduced before new experimental approaches for measuring ice nuclei can be developed. When improved methodologies are in place, global inventories need to be established that, in turn, must be connected to cloud microphysics and precipitation data.

A challenge for the prediction of indirect radiative forcing through aerosol-cloud interactions is to develop physically based GCM parameterisations from process parameterisations. The predictable parameters in a GCM are far fewer than the measured parameters in a closure experiment, and they characterise properties averaged over a scale much larger than the typical process scale. The basic information needed for simulation of aerosol-cloud interactions is the physical and chemical characteristics of aerosol particles and cloud droplets, interstitial gas composition, and the dynamic and thermodynamic properties of the cloud field. Numerous questions have to be answered collaboratively by experimentalists and modellers.

4.5.4 Modelling Challenges

Concerning aerosol process modelling, there is a strong need for improvements in the time efficiency and accuracy of the computational techniques used, especially for applications to atmospheric transport models. One of the most important questions is how to describe properly gas-to-particle mass transfer, especially for organic material (i.e. how to account for particle nucleation and growth processes when organic species are involved). Models to calculate heteromolecular nucleation rates for organic species in atmospheric systems are few and have large uncertainties associated with them.

Most global climate model studies so far have only considered the bulk particulate mass of few specific components, rather than the size distributions of an externally and internally mixed aerosol and size-dependent chemical composition. This is despite the recognition that the size distribution and the chemical composition of the particles control the activation of particles to cloud elements. Moreover, the cycles of aerosol particles are closely connected with hydrological processes. Models have to consider the complexity of a multiphase system. Certainly a weak point of the climate models is the parameterisation of cloud processes and properties as well as cloud-radiation feedbacks. Uncertainties in the calculated cloud properties are hard to quantify because satellite retrievals are not very accurate and *in situ* measurements cannot be extrapolated.

As always with large-scale models there is a continuing need for evaluation. Future work should take its cue from the recent successful large-scale, multi-investigator field projects that have brought together – in an arena covering thousands of kilometres and including a variety of platforms, instruments, and techniques, to provide detailed characterisation of aerosol sources, geographical distributions, and chemical and microphysical properties, and of the governing meteorological fields. Data sets resulting from these studies, together with satellite observations, provide the basis for a stringent test of the accuracy with which chemical transport models can represent aerosol loadings and properties is necessary if these models are to be used with known and reasonable accuracy to calculate aerosol influences on climate at the present, over the industrial period, and for future emissions scenarios.

Chapter 5

Advances in Laboratory and Field Measurements

Lead authors: Geoffrey S. Tyndall · David M. Winker · Theodore L. Anderson · Fred L. Eisele
Co-authors: Eric C. Apel · Carl A. M. Brenninkmeijer · Jack G. Calvert · David S. Covert · R. Anthony Cox
David R. Crosley · Clifford I. Davidson · James R. Drummond · William B. Grant · Alex B. Guenther
Barry Huebert · Geoffrey S. Kent · David C. Lowe · Wade R. McGillis · John M. Miller · Steve A. Montzka
John J. Orlando · Ulrich Platt · Eric Swietlicki · Neil B. A. Trivett · Anthony S. Wexler

5.1 Introduction

The global distribution and the regional and temporal variations of chemical compounds in the atmosphere, both gases and aerosols, are still not well known. A full understanding of the atmospheric system can only be achieved through an integrated use of field measurements, modelling, and laboratory measurements. The past ten years have seen an impressive increase in the number of field measuring systems in use, while laboratory studies of reaction rate coefficients and mechanisms have benefited from advances in technology. The advances are often synergistic, with field instruments frequently being based on laboratory systems, and new versatile detectors brought into the lab to increase the capabilities there.

A major focus of the atmospheric chemistry community has been on the measurement of atmospheric composition, from both *in situ* sampling devices and remote sensing instruments, which can be ground-, aircraft- or satellite-based. During the last decade it has been recognised that the deployment of a suite of coordinated measurement activities is required to provide an adequate understanding of the chemical composition of any region under study. Advances here include the development of well-calibrated instrument networks and the use of coordinated suites of instruments on multiple platforms deployed in field campaigns over large geographic regions. There is an increasing integration of observational and modelling activities, both for interpretation of observations and, recently, as an aid in developing data acquisition strategies and directing the conduct of field campaigns. These increasingly sophisticated strategies for using instruments have been as important to the advance of fundamental science as improvements in the instruments themselves.

This chapter examines the advances made in laboratory and field measurements over the last ten years or so, with an emphasis on the global troposphere. While significant steps forward have been made in stratospheric chemistry, particularly with regard to polar ozone, these aspects are not covered here. The chapter is not intended to be a comprehensive review of techniques, but more a synthesis of the ideas that are leading to the most notable steps forward in understanding the atmosphere. Several examples of successful long-term networking strategies are highlighted.

5.2 Laboratory Studies

5.2.1 Recent Advances

Our knowledge of the details of tropospheric chemistry has increased tremendously over the last ten years. This increase is in part related to the sheer volume of work done, but is also a consequence of new measurement techniques. The broadest source of data on oxidation mechanisms has been environmental chambers, which provide integrated pictures of the overall chemistry. However, the methodology often suffers from an inability to isolate the particular reaction of interest. Furthermore, wall reactions are often irreproducible, and differ from chamber to chamber. On the other hand, direct, time-resolved spectroscopic methods can be designed to study a single compound undergoing a single reaction. However, the conditions under which that compound can be studied are often very different from those encountered in the atmosphere, and care must be exercised when extrapolating the results to the real world. Clearly, the most progress has been made where direct and indirect methods have been integrated to test our understanding at both the microscopic and macroscopic levels. Some of the major advances in laboratory techniques are highlighted below, along with future trends.

5.2.1.1 *Ozone-HO_x Photochemistry*

The photolysis of ozone (O_3) to produce $O(^1D)$, followed by the reaction of $O(^1D)$ with H_2O, provides the largest source of hydroxyl radicals (OH) in the lower atmosphere.

$$O_3 + h\nu \longrightarrow O_2(^3\Sigma) + O(^3P)$$

$$O_3 + h\nu \longrightarrow O_2(^1\Delta) + O(^1D)$$

$$O(^1D) + H_2O \longrightarrow OH + OH$$

The study of the quantum yields for ozone photolysis (in particular in the "energy-deficient" region beyond 310 nm) has long fascinated physical chemists and has even been the subject of some controversy over the years as described by Ravishankara et al. (1998). However, in the last several years, major advances have been made in our understanding of the processes involved in ozone photolysis. The techniques used have included traditional methods such as resonance fluorescence or laser induced fluorescence, but also include more advanced techniques "borrowed" from the chemical physics community. For example, the excited $O(^1D)$ and $O_2(^1\Delta)$ photofragments have been detected using a resonance enhanced multi-photon ionisation (REMPI) technique (Ball et al. 1993) or by vacuum ultraviolet laser induced fluorescence (Takahashi et al. 1996). The direct detection of all the photofragments, $O(^3P)$, $O(^1D)$, $O_2(^1\Delta)$, and $O_2(^3\Sigma)$, has now been accomplished and the presence of a clear "tail" in the $O(^1D)$ quantum yields beyond 310 nm has been firmly established. Corroborative evidence for the existence of this long wavelength tail has been obtained from *in situ* measurements of the $O(^1D)$ production rate using chemical actinometry (Shetter et al. 1996). Calculated photolysis rates only show a good agreement with such actinometer measurements (particularly at high solar zenith angle) when the long-wavelength tail is incorporated into the calculation. Chapter 3 deals with advances in field measurements associated with atmospheric oxidants.

While the database for the kinetics of the reactions of OH with organic compounds (including hydrocarbons, halocarbons, and organosulphur compounds) has expanded (Atkinson 1997a), confidence in the accuracy of the measured rate coefficients can also be considered to have increased. Improvements in both laser technology and in the electronics used for detection systems have allowed OH to be measured in pressures up to 600 Torr O_2/N_2 in both flash photolysis (Hynes et al. 1995; Brown et al. 1999a) and flow tube systems (Abbatt et al. 1990), in spite of the rapid quenching of the fluorescence that occurs under these conditions. The versatility of flow tube experiments has been extended by the use of high pressure turbulent flow systems (Seeley et al. 1996; Donahue et al. 1997). The OH database has particularly been refined at low temperature; in the case of hydrocarbon reactions, this improvement has been achieved partly as a result of the care taken in sample purification (e.g., Vaghjiani and Ravishankara 1991).

While IGAC has not directly been involved in funding laboratory studies, it has supported evaluations of laboratory data. In conjunction with SPARC, it has facilitated two such evaluations. The first concerned the chemistry of peroxy radicals (Tyndall et al. 2001), while the second considered the quantum yields for ozone photolysis (Matsumi et al. 2002). Both brought together a group of the major researchers in each field to give specialist input on their topic and to provide an objective, hands-on evaluation giving a sound basis for future studies in these fields.

5.2.1.2 *Isoprene-Terpene Oxidation Mechanisms*

Much of the progress in isoprene chemistry has occurred through use of large photoreactors, which can be considered "descendents" of those developed at the University of California, Riverside, in the 1970s. While the major pathways from isoprene to methylvinyl ketone and methacrolein have been quantified using Fourier transform infrared (FTIR) spectrometry (Tuazon and Atkinson 1990) or gas chromatography (GC) (Paulson et al. 1992), there still exist many uncertainties in the yields of minor products, particularly under conditions of low NO_x (Miyoshi et al. 1994). The atmospheric pressure ionisation (API) mass spectrometric technique has been used by Atkinson's group to identify the presence of hydroxycarbonyls not detected by GC (Kwok et al. 1995). Such a technique offers the great advantage of direct sampling from the reaction chamber into the mass spectrometer, reducing the loss of compounds having low volatility or a tendency to stick to the walls of sampling tubes. Jeffries and coworkers (Yu et al. 1995) have implemented the technique of derivatisation using O-(2,3,4,5,6-pentafluorobenzyl) hydroxylamine hydrochloride (PFBHA). In derivatisation, carbonyl compounds are reacted with an amine (traditionally 2,4-dinitrophenyl hydrazine, DNPH) to form a new compound that can be detected by GC. A further technique allows the detection of alcohols and acids after reaction with N,O-bis(trimethylsilyl)-trifluoroacetamide. These new derivatisations enable the identification of many reactive carbonyls and acids which otherwise would not be amenable to detection by standard chromatographic methods (Yu et al. 1998, 1999). Both API mass spectrometry and PFBHA derivatisation need some development before they can be considered to be quantitative, but they offer the possibility of quantifying multi-functional products. The oxidation of terpenes has been less well studied, with each of the terpenes subject to only two or three investigations. The expected range of reaction products is wide, and the mass balances obtained are poor. The chemistry of the oxygenated reaction products also remains poorly quantified.

Much work in the last few years has been dedicated to understanding the chemistry of alkoxy radicals (Atkinson 1997b). Alkoxy radicals provide a branching point in the oxidation of most hydrocarbons, and the competition between the various reaction pathways determines the rate at which the carbon chain gets shortened, influences the degree of oxidation, and ultimately determines the solubility of the stable reaction products formed. The majority of studies have been performed in smog chambers by studying the relative rates of formation of the various products. However, recent chamber studies at lower temperatures have indicated that the oxidation chemistry of even simple organics such as ethene may be more complicated than inferred from room temperature studies (Orlando et al. 1998). A few studies using direct spectroscopic detection of alkoxy radicals have been performed, but these are limited to small radicals derived from alkanes (Devolder et al. 1999; Blitz et al. 1999). For larger radicals the spectra become congested, and alternative detection methods are desirable.

5.2.1.3 *Aerosol Formation-Heterogeneous Chemistry*

Several improvements in instrumentation have led to a better understanding of aerosol formation and growth at the microscopic level. Kulmala and coworkers have performed both experimental and theoretical studies on nucleation of sulphuric acid-water mixtures (Viisanen et al. 1997; Kulmala et al. 1998). Ball et al. (1999) made use of an ultrafine condensation nuclei counter to detect particles in the early stages of growth (diameter ≥ 3 nm), and thereby infer the nucleation rate. The use of chemical ion mass spectrometry (CIMS) made detection of very low concentrations of sulphuric acid possible. Both these groups concluded that the functional dependence of the nucleation rate on water and sulphuric acid was much stronger than given by older theory, but in reasonable accord with field experiments. Ball et al. (1999) also showed that nucleation was strongly enhanced in the presence of NH_3. The electrostatic balance (particle levitation) has been used to follow the deliquescence behaviour of NH_4HSO_4 and $(NH_4)_2SO_4$ particles (Imre et al. 1997; Xu et al. 1998). In this technique, a charged particle is trapped by an electric field, its mass is measured by the magnitude of the restoring force, and its physical state inferred from measurement of scattered laser light. Changes in the mass (and hence composition) of the particle can be related to the deliquescence behaviour of the solution.

Techniques for measuring particle size distribution have been applied to quantify the yield of aerosol in laboratory investigations of terpene oxidation. While the potential for the formation of aerosol particles from biogenic hydrocarbons has long been recognised, descriptions of particle growth are still largely empirical, and are limited to a description of the yield of particulate matter as a function of the extent of oxidation (Odum et al. 1996; Hoffmann et al. 1997). However, it is not clear whether this parameterisation can be simply transferred to atmospheric conditions. It is thought that particle nucleation, or initial growth on seed nuclei, involves the deposition of low-volatility products, often formed in small amounts, followed by ad- or absorption of more volatile products into the organic layer formed (Jang and Kamens 1999). Kinetic modelling has recently been used to predict the compounds likely to be deposited onto particles (Kamens and Jaoui 2001).

Tropospheric heterogeneous chemistry is a relatively new and untapped research area. Atmospheric pressure flow reactors have been used to measure the coefficients for reaction of gaseous N_2O_5 directly on monodisperse, sub-micrometer sulphuric acid aerosols (Mozurkewich and Calvert 1988; Fried et al. 1994; Lovejoy and Hanson 1995; Hallquist et al. 2000). Much progress has been made in understanding the uptake of organics onto aqueous or acidic aerosol using the falling droplet technique pioneered by Aerodyne/Boston College (Jayne et al. 1992). These experiments often yield values for both the uptake parameters and the solubility constants, and the suite of molecules investigated includes carbonyls, acids, alcohols, and sulphur gases. Less is known about the chemical transformations that occur in aqueous tropospheric aerosols. While the aerosol offers abundant surface area for uptake, the small volume may limit the extent to which gas-phase chemistry may be influenced. However, the very concentrated solutions encountered could lead to rapid reaction rates under the right circumstances. Recent advances in understanding the oxidation of SO_2 in solution are summarised by Georgii and Warneck (1999). The interaction of common atmospheric oxidants with organics, particularly the effects of ambient light, need to be investigated in detail before a full understanding of the impact of aerosol on biogenic carbon (and vice versa) can be determined. Developments in understanding the atmospheric aerosol are discussed in Chap. 4.

5.2.2 Future Needs

5.2.2.1 *Oxidant Chemistry*

The future will see an increased need for the detection of large atmospheric free radicals. The technique of cavity ring-down spectrometry has been shown to be a very sensitive way to detect both transient free radicals (Zhu and Johnston 1995) and weak transitions of stable molecules (Newman et al. 1999) with high resolution in the laboratory; however, it has not yet been tested in the field. Since it is an absorption technique it is not subject to the effects of quenching as encountered with fluorescence techniques, while the sensitivity may approach that of LIF for larger molecules. Other advances in laser technology such as near-IR diodes (Johnson et al. 1991) and difference-frequency or colour-center lasers in the mid-IR (e.g., Schiffman et al. 1991) should lead to improvements in techniques for detecting peroxy and alkoxy radicals. Part of the reason for the improvements in precision in the last ten years is related to the wider availability of lasers and signal acquisition hardware. It should also be noted that a better detection sensitivity allows lower radical concentrations to be used, which in turn minimises the effects of secondary reactions. As further developments in instrumentation occur they will undoubtedly be incorporated into laboratory systems too.

5.2.2.2 *Hydrocarbon Oxidation*

Much work still needs to be done on the chemistry of oxygenated reaction products, particularly multi-functional carbonyls. Derivatisation techniques for carbonyls are constantly being improved and incorporated into both laboratory and field studies. Mass spectrometric methods using ionisation by atmospheric pressure ionisation (Kwok et al. 1995), chemical ionisation (Huey et al. 1995), and proton transfer reaction (Lindinger et al. 1998a,b) have rapid time response and high sensitivity, and could be used in time-resolved laboratory studies if appropriate conditions can be found. The sensitivity and selectivity of CIMS has been demonstrated both for stable molecules (Lovejoy and Hanson 1995) and for free radicals (Villalta and Howard 1996).

The chemistry of alkoxy radicals remains central to our understanding of hydrocarbon chemistry. Most of the work to date has been done using competitive product studies. Some direct spectroscopic techniques have been used, but at present they tend to be labour intensive. The development of more general techniques, applicable to several different radicals, would be useful.

Finally, the new generation of environmental chambers should be mentioned. For example, the European community facility EUPHORE (Becker 1996; Klotz et al. 1998), which is near Valencia, Spain, includes a variety of optical *in situ* measurements in addition to the more conventional techniques such as gas chromatography. A similarly equipped large outdoor chamber, SAPHIR, has been constructed at the Forschungszentrum Jülich, Germany. The optical techniques can be used to measure transient free radical concentrations under well-controlled chemical conditions while utilising ambient solar irradiation, and should provide a means to validate photochemical theory with greatly improved accuracy.

5.2.2.3 *Dimethylsulphide (DMS) Oxidation*

There still exist many problems in identifying and quantifying the intermediate radicals in the oxidation of DMS. The intermediates studied directly include CH_3S by LIF (Tyndall and Ravishankara 1989; Turnipseed et al. 1992) and CH_3SO by mass spectrometry (Mellouki et al. 1988; Dominé et al. 1990). Both mass spectrometric and optical techniques should be developed to further investigate the reaction mechanisms. The key branching steps to the formation of SO_2, SO_3, and methanesulphonic acid (MSA), and their dependence on temperature and NO_x (Sørensen et al. 1996; Patroescu et al. 1999) need to be identified through direct laboratory studies in conjunction with chamber studies.

5.2.2.4 *Aerosol Chemistry*

Continuing laboratory studies should elucidate the details of nucleation of particles. Most studies to date have been on "model" mixtures, e.g., sulphuric acid-water. A better understanding of the conditions required for nucleation will enable the extent of new particle formation in the atmosphere to be assessed. Further studies of pre-nucleation clusters (Eisele and Hanson 2000) will aid this understanding.

The extent of formation of particles from the oxidation of organics is still a long way from being understood. It is particularly important to isolate the effects of the individual oxidants, e.g., OH, NO_3, or O_3, since these oxidants tend to be important in different regions of NO_x or solar intensity, and each leads to products of differing volatility. Better methods to speciate the organic components of aerosols need to be developed.

The question of chemistry inside, and particularly on the surface of, particles needs to be addressed more systematically. Currently, the identity and concentrations of oxidants within aqueous aerosol particles have not been thoroughly studied. The oxidation of partially oxygenated compounds removed from the gas phase could continue inside droplets and influence not only the measured concentrations but also the physical properties of the particles. Future studies should consider solubility, rate of uptake, and reaction inside the droplets. Finally, the occurrence of reactions on the surface of both organic aerosols and black carbon (soot) needs to be thoroughly investigated. This is a new field which has been the subject of speculation for many years, but which has not been quantified satisfactorily to date.

5.2.3 Summary

Tropospheric homogeneous and multi-phase chemistry have both seen impressive expansions in the last ten years. Tropospheric chemistry, particularly relating to biosphere-atmosphere interactions, tends to be complex and, to date, not enough studies have been carried out to fully test the reproducibility of many of the different systems studied in the laboratory. In this sense, the discipline is still immature, and much work remains to be done to validate and extend the current database. Our current understanding of heterogeneous and multi-phase systems is largely based on data from bulk systems. Further developments in measurement techniques will undoubtedly lead to a continued appreciation of the complexities and subtleties of the chemical reactivity in the atmosphere.

5.3 Field measurements: Gas Phase

5.3.1 Recent Advances

Major advances in instrumentation over the past decade, particularly in a few specific areas, have greatly enhanced understanding of gas phase chemical processing and are also providing a whole new view of the production, growth, and interaction of particles in the atmosphere.

5.3.1.1 *Measurements of OH and HO_2 Radicals*

The rate at which Earth's atmosphere cleanses itself is largely dependent on its ability to produce the hydroxyl radical (OH; see Chap. 3), yet only a decade ago very few if any credible *in situ* tropospheric measurements of this compound existed. By 1991, a sensitive new *in situ* ground-based OH instrument using selected ion chemical ionisation mass spectrometry (SICIMS) was successfully deployed. This technique employs a chemical conversion of OH radicals to isotopically labelled H_2SO_4, which is detected by CIMS (Tanner et al. 1997). The sensitivity and time response of the method exceed others, and even allow the detection of OH at night. The viability of the SICIMS technique was confirmed by comparing it to a long path UV absorption instrument in the Colorado Mountains (Eisele et al. 1994).

Large increases in the sensitivity of long-path absorption instruments have led to new folded-path instruments, which can make ground- or ship-based *in situ* measurements (Brandenburger et al. 1998). The technique of laser induced fluorescence (LIF), which has also seen substantial improvements, has been combined with the use of an expansion jet (Hard et al. 1995) in the method of fluorescence assay by gas expansion (FAGE). The expansion reduces the number of collisions suffered by the OH, extending the fluorescence lifetime, and allowing better discrimination against scattered laser light. Recent instruments use an excitation laser with low power and high pulse rate at a wavelength of 308 nm (rather than 280 nm), which reduces the extent of interference from secondary generation of OH caused by photolysis of O_3 by the detection laser (Heal et al. 1995; Hofzumahaus et al. 1996; Mather et al. 1997).

Instruments for the measurement of OH have also been adapted for the measurement of HO_2, by chemical conversion using the reaction $HO_2 + NO$ (e.g., Mather et al. 1997; Reiner et al. 1998). Since the HO_2 concentration is up to 100 times larger than that of OH the instruments do not need to be as sensitive as those dedicated to OH measurements. However, larger organic peroxy radicals can also be converted to OH in the presence of NO, so the technique suffers from uncertainties in calibration.

While still not commonplace, OH and HO_2 measurements have contributed significantly to several major aircraft and ground-based campaigns (e.g., Brune et al. 1998; Mauldin et al. 1998b). Simultaneous measurements of OH and HO_2 in the upper troposphere were used to provide an impressive constraint on uncertainties in kinetic rate parameters measured in the laboratory (Wennberg et al. 1998).

Alongside these developments for OH, the technique of peroxy radical chemical amplification (PERCA) has been further developed (Cantrell et al. 1996a; Carpenter et al. 1997). In this method, which provides a non-specific measurement of HO_2 and other peroxy radicals, CO and NO are added to an air flow, and the NO_2 produced in a chain reaction is measured by a chemiluminescence technique. The technique requires calibration, since the chain length is specific to the instrument used.

5.3.1.2 *Measurements of Organic Compounds*

Another area of major progress is the measurement of organic compounds. Over the past decade or so, the previously held view of vegetation as an emission source of a few fairly passive organic compounds has changed to one in which vegetation plays a central interactive role in ozone production in particular and atmospheric chemistry in general. The major problem is specificity because of the sheer number of compounds to be measured and the fact that many of the compounds are so similar chemically. In addition, many of the oxygenated product compounds are sticky and can be deposited onto surfaces. Chemical identification and the concentration measurement are important, as is flux of the compounds.

Much progress has been made using customised GC and GC-MS measurement schemes to inventory both primary biogenic emissions and their products (Guenther et al. 2000). While such techniques are highly specific they are typically quite slow, requiring 10^2–10^3 s. One of the most widely used techniques to measure gas-phase carbonyl compounds is the 2,4-dintrophenylhydrazine (DNPH)-coated cartridge technique (Fung and Grosjean 1981; Sirju and Shepson 1995; Apel et al. 1998). This technique, which has been around since the early 1980s, suffers from low sensitivity requiring large sample sizes (>50 l) and long sampling times (one to three hours). Detection limits are on the order of 0.5 nmol mol^{-1}. Direct gas chromatographic analysis of cryogenically collected samples was developed relatively early on (Jonsson et al. 1985; Pierotti et al. 1990) promising increased sensitivity (<50 pmol mol^{-1}) and reduced sample collection and analysis times (one-half hour). A lack of recognition of the importance of these compounds to atmospheric chemistry cycles and analytical difficulties associated with calibration and quantitative recovery led to slow general acceptance and usage of this technique. Recent advances in analytical standards preparation, inert coatings for tubing surfaces, faster chromatographic techniques, and improvements in mass spectrometer instrumentation have led to airborne GC/MS systems capable of measuring not only carbonyl compounds but also alcohols with a time response of five minutes and a limit of detection of less than 10 pmol mol^{-1}. Such instruments have been used to demonstrate the widespread distribution of oxygenated organic compounds in the upper troposphere (Singh et al. 2001). The role of such oxygenated compounds in tropospheric photochemistry and aerosol formation is still being elucidated.

Recently, a modified chemical ionisation mass spectrometry technique, which operates at low pressures and uses electric fields to enhance collision energies in the ion reactive region, was introduced into the atmospheric measurements arena. This instrument, the proton transfer reaction mass spectrometer (PTRMS) (Lindinger et al. 1998a,b), offers several advantages over existing hydrocarbon measurement techniques: fast measurements (in the one second time frame), a reasonable amount of specificity using mass analysis to identify compounds, sensitivity better than 1 nmol mol^{-1}, and the ability to be used on an aircraft (Warneke et al. 2001). The technique has been demonstrated to be sensitive for alkenes and carbonyls, as well as other compounds containing oxygen and nitrogen, which all have proton affinities in the appropriate range. To measure the large number of organic compounds present in the atmosphere in an unambiguous way may require more analytical power than is provided by the present PTRMS, but this can be greatly improved through the use of tandem mass spectrometers, ion traps, or time of flight mass spectrometry (TOF-MS).

Advances in diode laser technology have led to several new techniques for making faster measurements of compounds such as HCHO, NH_3, CO, NO_2, H_2O_2, N_2O, and SO_2 (Silver et al. 1991; Roths et al. 1996; Mackay et al. 1996; Fried et al. 2001). The use of Herriot cells allows a long absorption path length to be achieved in a small detection cell. The reduction in the volume of the cells leads to a concomitant decrease in sampling time. While these diode laser measurements are highly specific and have sensitivities for HCHO, for example, in the range of tens of pmol mol^{-1} (Fried et al. 2001), it is unlikely that they will be able to sort out and identify individu-

Table 5.1. Representative detection limits and sampling times for selected tropospheric molecules ca. 1990 and present day

Molecule	1990			Present		
	DL (time)[a]	Technique	Reference	DL (time)[a]	Technique	Reference
OH	0.1 (30 min)	Laser abs	Hofzumahaus et al. 1991	0.004 (30 s)	CIMS	Tanner et al. 1997
HO_2	5 (30 min)	MI/ESR	Mihelcic et al. 1990	0.04 (1 min)	LIF[b]	Mather et al. 1997
NO	3.5 (2 min)	LIF	Sandholm et al. 1990	0.6 (35 s) 0.5 (1 min)	LIF CL	Bradshaw et al. 1999 Ridley et al. 2000
NO_2	10 (6 min)	NO LIF[c]	Sandholm et al. 1990	2 (3 min) 1 (1 min)	NO LIF[c] CL[c]	Bradshaw et al. 1999 Ridley et al. 2000
HCHO	60 (5 min)	HPLC/enzyme	Heikes 1992	30 (1 min)	TDL	Fried et al. 2002
H_2O_2	30 (1 min)	HPLC/enzyme	Heikes 1992	10 (2.5 min)	HPLC/enzyme	Lee et al. 1998
PAN	4 (15 min)	GC-ECD	Gregory et al. 1990	5 (2 min)	GC-ECD	Roberts et al. 2001
C_5H_8	5 (30 min)	GC-FID	Zimmerman et al. 1988	100 (10 s)	PTRMS	Warneke et al. 2001
Carbonyls	200 (1 h)	GC-FID	Sirju and Shepson 1995	10 (5 min)	GC-MS	Singh et al. 2001
HNO_3	10 (30 min)	Filter/IC	Parrish et al. 1986	3 (1 s)	CIMS	Mauldin et al. 1998

[a] Detection limits (units of pmol mol^{-1}) are as stated in papers, corresponding to the time for sampling followed by analysis.
[b] After conversion to OH by reaction with NO.
[c] After conversion to NO by photolysis.
Techniques:
CIMS: Chemical ionisation mass spectrometry;
MI/ESR: Matrix isolation with electron spin resonance detection;
CL: NO-O_3 chemiluminescence;
HPLC: High performance liquid chromatography;
ECD: Electron capture detection;
IC: Ion chromatography;
GC: Gas chromatography; FID: Flame ionisation detection; MS: Mass spectrometric detection;
PTRMS: Proton transfer reaction mass spectrometry.

ally the bulk of major biogenic emissions and their products (particularly complex compounds with more than five atoms) over the next decade.

Table 5.1 shows some of the advances made in instrumentation capabilities over the last 10–12 years. While some of the progress has been incremental, other techniques are novel and did not exist at the beginning of IGAC. Such instruments have in general not been compared to others, so estimates of their accuracy can not be considered to be as good as those for more established techniques.

5.3.2 Techniques for the Measurement of Isotopes

Isotopic techniques are used in atmospheric chemistry to help constrain source strengths. Isotopic measurements are made in the atmosphere of both stable isotopes (e.g., D, ^{13}C, ^{18}O) and radioactive isotopes such as ^{14}C. Different biogenic and anthropogenic sources exhibit different ratios of isotopes, usually with a range of values typical for a given source. Furthermore, isotopic fractionation occurs in the atmosphere, since the different isotopic variants react at different rates. Thus, measurements of isotopic ratios can give information concerning both the source and the atmospheric processing of the particular molecule.

5.3.2.1 *Isotope Ratio Mass Spectrometry (IRMS)*

Most stable isotope measurements in atmospheric chemistry have been made using IRMS. The ion currents produced can be ratioed precisely to determine very small differences in isotopic ratios (relative to a reference) of the sample. Currently, IRMS is the only technique capable of resolving some of the very small fractionations of interest and importance in atmospheric chemistry, for example of the order of 1 part in $10^5 = 0.01‰$.

Conventional dual-inlet IRMS for trace gas analysis has the large disadvantage that, prior to analysis, the gas of interest must be quantitatively extracted from a large volume of air, the size of which increases as the gas mixing ratio decreases. Clearly in this approach there are potential problems as a result of incomplete recovery of the gas of interest or contamination. The use of GC-IRMS, in which the components are first separated using gas chromatography, offers huge advantages for atmospheric chemistry research when compared to conventional dual inlet IRMS. This is because of the very low sample volumes used and the potentially contaminant free on line preparation of sample gases.

Prototype instruments now provide simultaneous analyses of $\delta^{13}C$, $\delta^{18}O$, and mixing ratios in atmospheric CO_2 at ambient levels in 0.5 ml of air to precisions (1σ) of 0.02‰, 0.04‰, and 0.4 µmol mol^{-1}, respectively (Ferretti et al. 2000), or measure $\delta^{13}C$ in methane (CH_4) at ambient levels in 150 ml of air to a precision of 0.05‰. Future developments are likely to result in improved precision and an increase in the number of compounds studied using the technique.

5.3.2.2 *High Resolution Infrared Spectrometry*

In IRMS, high quantitative precision is obtained at the expense of high mass resolution so that isotopic molecules (isotopomers) of like mass, such as $O^{13}CO$ and ^{17}OCO, or $^{13}CH_4$ and CH_3D ($D = {}^2H$ = deuterium), cannot be distinguished. The infrared spectra of gases depend on both their mass and molecular structure, and IR spectrometry can sometimes be used to differentiate between isotopomers of the same molecule. Although the precision of the technique is poor compared to isotope ratio mass spectrometry, it has proved to be very useful in situations where the mixing ratios of atmospheric trace gases are elevated above ambient, for example at landfill sites or above a vigorously growing pasture (Esler et al. 2000). In addition Esler et al. (1999) have shown that FTIR may be used in a continuous baseline monitoring situation to monitor $\delta^{13}C$ in atmospheric CO_2 at the same time as the mixing ratios of CH_4, CO, and N_2O.

Infrared-based isotopic analysis of atmospheric gases has also been exploited using tuneable diode laser absorption spectrometry (TDLAS), for example to assay ^{13}C and D in CH_4 (Bergamaschi et al. 1994; Bergamaschi 1997). Although the ^{13}C data derived by this technique lack the precision needed for background studies of CH_4, the D data are the most precise (0.5–1.0‰) of any published method including IRMS (Bergamaschi et al. 1998a,b).

Optical absorption techniques for stable isotopic analysis offer many advantages such as semi-continuous measurement of several atmospheric compounds simultaneously, non destructive testing of samples, and lack of ambiguity for isotopomers of the same mass but different isotopic structure. It is likely that the precision of these techniques will improve in the future to the point where they offer a very attractive alternative to IRMS.

5.3.2.3 *Accelerator Mass Spectrometry (AMS)*

Accelerator mass spectrometry is a relatively new technique to derive the isotopic abundance of ^{14}C, which is produced by cosmic ray activity in the middle atmosphere. Since ^{14}C decays to ^{14}N with a half life of 5 730 years, measurements of its abundance provide a way to differentiate between carbon associated with fossil fuel (in which the ^{14}C has decayed away) and more recent sources that are derived from atmospheric CO_2. Tropospheric air contains only a few hundred ^{14}CO molecule cm^{-3}. However, using AMS (Litherland 1980) abundances of $^{14}C/^{12}C$ down to about 10^{-15} can easily be detected in samples containing from 100 to 1 000 µg of carbon. The first published AMS measurements of ^{14}C in atmospheric CH_4 were by Lowe et al. (1988) and Wahlen et al. (1989) and in atmospheric CO by Brenninkmeijer et al. (1992). With further reductions in sample size likely, AMS is certain to play a role in future developments in the use of ^{14}C in atmospheric chemistry. Measurements of ^{14}C in CO have been used to investigate downward transport from the stratosphere to the troposphere, and to estimate the abundance of tropospheric OH (Brenninkmeijer et al. 1992).

5.3.3 Use of Lidar on Airborne Chemistry Missions

Lidar (light detection and ranging) has been used on airborne atmospheric chemistry missions for the past 20 years (Browell et al. 1998) but it is only in the last decade that lidar has been fully integrated into the measurement strategies of these missions. A differential absorption lidar (DIAL) system operating in the ultraviolet to measure ozone profiles has come to be an integral part of the airborne Global Tropospheric Experiment (GTE) missions that study atmospheric chemistry in remote regions of the world. More recently, in the Laser Atmospheric Sounding Experiment (LASE), a water vapour DIAL system has been added to these and other missions (Browell et al. 1997).

DIAL systems make instantaneous profile measurements of aerosols, ozone, and/or water vapour, as well as the location and vertical extent of clouds. Profiles are important for determining features of atmospheric structure, such as the top of the mixing layer, and interpreting column measurements. Observation of plumes with elevated aerosol and ozone layers also provides insights into atmospheric dynamics. For example, during the Transport and Atmospheric Chemistry Near the Equator-Atlantic (TRACE-A) mission in 1992, the UV DIAL system clearly showed a fresh biomass burn plume with high aerosols and ozone coming off the coast of Africa below 6 km, with a layer of elevated ozone but not aerosols advected from Brazil at an altitude of 6–9 km (Browell et al. 1996a).

DIAL systems have also been used to classify air masses by type such as high ozone plumes, convective outflow, stratospherically influenced, near surface, etc. (Browell et al. 1996b; Fenn et al. 1999). Classifications are based on average aerosol and ozone profiles for the study region during the mission. Determination of the extent of the air masses from DIAL can be used to direct aircraft to regions of interest and optimise the placement and operation of *in situ* instruments. For example, during the Pacific Exploratory Mission-Tropics A (PEM-Tropics A), the UV DIAL system identified the extent of the largely unexpected biomass burning plumes that were found to be nearly ubiquitous over the South Pacific from late August to October 1996 (Fenn et al. 1999).

5.3.4 Flux Measurements

Advances in flux measurement techniques have enabled investigators to characterise the exchange of trace gases between the atmosphere and entire terrestrial landscapes using methods that do not influence the source. Previous measurements relied on the enclosure method that enabled measurement of only small components of an ecosystem and often considerably perturbed the environment surrounding the source. Eddy covariance is the most direct and accurate flux measurement method. The limited application of this method is due to the requirement for fast response chemical sensors. It has recently been used successfully to measure fluxes of several important trace gases including CH_4 (Fan et al. 1992), N_2O (Hargreaves et al. 1996), and isoprene (Guenther and Hills 1998). Future advances in fast response chemical sensors may provide eddy covariance measurements for most trace gases. Other techniques that have been developed and improved in the past decade, including relaxed eddy accumulation and periodic eddy accumulation (e.g., Hargreaves et al. 1996; Guenther et al. 1996; Rinne et al. 2000), have also significantly improved area-average flux measurement capabilities. Improvements have also been made in the application of indirect flux measurement methods including enclosures, surface-layer gradient, mixed-layer gradient, and mass balance techniques (e.g., Guenther et al. 1996).

Until recently, only a few attempts to measure air-sea gas fluxes directly have been made. Methods to measure air-sea gas flux, particularly CO_2, led to much controversy in defining the magnitude of the oceanic flux (Broecker et al. 1986). Because of the lack of direct flux measurements, parameterisations for air-sea gas exchange were based on indirect measurements. Models combine sea-surface measurements of concentration with parameterisations for the gas exchange rate. The number of parameterisations developed over the years to describe gas transfer across the sea surface is extensive. However, to date, all such relationships have sprung from observations made over large spatial or time scales, which smooth out even the synoptic-scale variability of atmospheric forcing. Algorithms relating gas exchange to wind speed are either developed from compilations of field data (Nightingale et al. 2000a), from controlled studies at a single field or laboratory site (Watson et al. 1991b), or from a combination of field and laboratory data (Liss and Merlivat 1986). Several recent gas exchange models are constructed to reconcile the budgets of radiocarbon and radon tracers (Wanninkhof 1992; Wanninkhof and McGillis 1999). However, lack of sufficiently decisive data from past field experiments has prevented confirmation of a single relationship for air-sea gas exchange (see Fig. 2.18).

Several advances in gas flux measurements address the concerns related to the application in oceanic conditions (Fairall et al. 2000). Advances in air-side gradient and covariance measurements have decreased the time scale to sub-hour, notably the ocean-atmosphere direct covariance method for CO_2 (McGillis et al. 2001) and the gradient method for DMS (Dacey et al. 1999). For air-sea fluxes for more soluble gases (e.g., DMS) the oceanic boundary layer is also considered (McGillis et al. 2000). Advances in air-sea gas flux measurements have relied on: (*1*) the ability to adequately remove the motion contamination; (*2*) the deployment of the sensor package at a location that minimises the effect of flow distortion around an ocean-going vessel; and, (*3*) having adequate signal levels and frequency response to compute the gas flux using currently available sensors. These concerns are now being addressed for air-sea flux systems and incorporated into analysis schemes for measuring air-sea fluxes.

5.3.5 Instrument Intercomparisons

Instrument calibration is vital, and is indeed as much a part of doing atmospheric research as is gathering data to test a geophysical hypothesis. A reliable estimate of the uncertainty in a measurement is as important as the measurement itself. While all investigators strive to calibrate their instruments, it is also vital to devise ways to compare instrument calibrations, either by sampling the same air mass or by the exchange of reliable standards. A great variety of intercomparisons have been carried out during the past ten years. Techniques and methods to measure many of the atmospheric gases have been evaluated quantitatively, for example NO_2 (Fehsenfeld et al. 1990); CH_2O (Gilpin et al. 1997); hydrocarbons (Apel et al. 1994, 1999); and SO_2 and other sulphur gases (Hoell et al. 1993; Gregory et al. 1993; Stecher et al. 1997). In the first airborne intercomparison of SO_2 instrumentation, CITE-3 held in 1989 (Hoell et al. 1993), no meaningful correlation was found among the five instruments below 200 pmol mol^{-1}. At a second blind ground-based intercomparison, GASIE (Stecher et al. 1997), much better agreement was found, with six of seven instruments being able to measure SO_2 levels of 40 pmol mol^{-1}.

The IGAC-sponsored NOMHICE Activity (Apel et al. 1994, 1999) has provided a world-wide standard for hydrocarbon measurements and has allowed scientists to compare their instruments, both in their own laboratories and in the field. Atmospheric scientists should continue to show an honest scepticism about the accuracy of measurements derived from any new measurement technique until such time that it has been proven through adequate testing and evaluation in blind intercomparisons and through other careful laboratory and field testing.

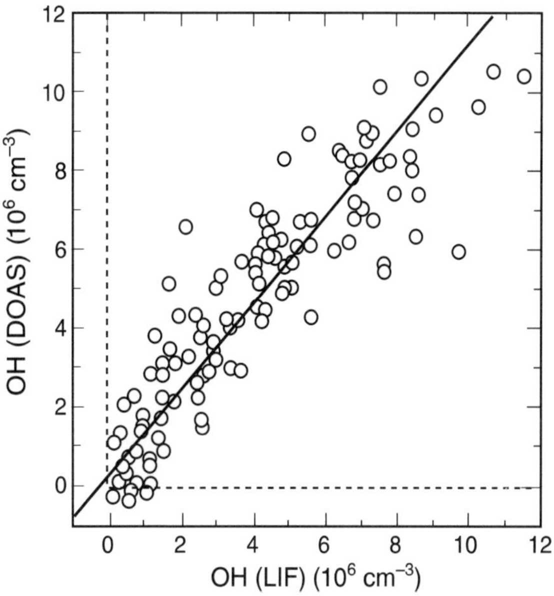

Fig. 5.1. Comparison of OH radical concentrations measured simultaneously using long-path differential absorption spectrometry and laser-induced fluorescence. The data show a good correspondence, and a 1:1 dependence using the calibrations available at that time (adapted from Brauers et al. 1996)

Several different techniques of measurement of the OH radical concentration in the troposphere have been developed (Sect. 5.3.1), such as the ion-assisted OH measurement (Tanner et al. 1997), laser induced fluorescence (Hofzumahaus et al. 1996; Mather et al. 1997), and long path absorption spectrometry (Dorn et al. 1996). Several informal intercomparisons which show promising agreement have taken place over the last few years, for example the 1993 Tropospheric OH Photochemistry Experiment, TOHPE (Crosley 1997), and Photo-Oxidant Formation by Plant Emitted Compounds and OH Radicals in North-Eastern Germany, POPCORN (Hofzumahaus et al. 1998). Figure 5.1, adapted from Brauers et al. (1996), shows a comparison of OH measurements made using laser absorption and fluorescence instruments. Techniques for OH have been improved greatly during the past ten years, and their use is now increasing. It is suggested that the atmospheric scientists plan an IGAC-sponsored intercomparison involving the OH radical. Such an intercomparison should use either blind laboratory tests in air mixtures prepared in a common manifold or a common calibration source, along with field studies at a common site under a variety of atmospheric conditions.

5.3.6 Future Needs

The better we come to understand atmospheric chemistry, the more complex we realise it is. This complexity takes on several forms, each of which poses a different set of measurement challenges. Part of this complexity occurs because observed chemical concentrations are typically a result of dynamics as much as chemistry. In order to understand the influences of each, more chemically speciated lidar and satellite measurements are needed. For example, DIAL profiling of ammonia seems feasible, at least in polluted conditions (Zhao 2000). The development of more rapid (preferably continuous) measurement techniques to determine fluxes would also contribute significantly, as would the measurement of a wider range of compounds, many of which could be used as tracers.

One of the more persistent problems in atmospheric chemistry has been the apparent discrepancy between measured and modelled NO_x and NO_y levels in the remote troposphere (e.g., Ridley et al. 1998; Schultz et al. 1999, 2000). It is often found that the measured ratio NO_x/HNO_3 is higher than predicted by models. However, it is not known to what extent this reflects problems associated with the measurement of NO_2 or HNO_3, or whether it is really an indication of poorly understood partitioning in the NO_y family. Nevertheless, this discrepancy limits our confidence in calculations of ozone production rates. More reliable measurements of HNO_3 and H_2O_2 would be particularly desirable in the study of ozone production and loss in the free troposphere, as would a more specific detection method for NO_2.

The use of multiple measurement platforms could also provide additional insight, especially near localised sources or convective systems. As measurements approach the surface, particularly in industrialised or urban areas, the complexity increases in part because the sheer number and types of sources increase. If such areas are to be studied in detail, far more compounds need to be measured, and if accomplished on an aircraft, the measurement time needs to be decreased to the order of a few seconds so that individual sources or plumes can be distinguished from each other. This will require new, more sensitive measurement techniques. Chemical ionisation mass spectrometry offers the required speed and sensitivity and can be applied to large, complex biogenic and anthropogenic molecules. More powerful analytical capabilities, such as tandem mass spectrometers and ion traps will, however, be required. Optical techniques can make fast sensitive measurements for small molecules although they have difficulty in speciating large organics, for example. Continued progress in this area is needed.

Another area that deserves far more attention in the next decade is related to aircraft inlets. As an ever increasing number of field studies is mounted from aircraft platforms, and as measurement capabilities are extended to include more reactive and/or sticky compounds, sample integrity becomes a major issue. An instrument that works perfectly on the ground or even in an aircraft will add little to an aircraft mission if the compound of interest cannot be transported in a well understood way between ambient outside conditions and the measurement instrument. This problem is particularly acute for aerosols (see next section), but affects many gas phase measurements as well. Some of the issues that need to be addressed are: (*1*) large air speed changes and the associated temperature changes, even if measurements are made outside the aircraft; (*2*) temperature changes associated with bringing air into the aircraft; (*3*) re-equilibration of compounds found both on particles and in the gas phase as temperature and relative humidity change; (*4*) loss to and evaporation from inlet surfaces as altitude, temperature and relative humidity change; (*5*) the influence of pressure changes, aircraft boundary layer thickness, inlet orientation and position relative to other inlets; and, (*6*) high speed droplets, droplet shattering, and icing.

5.4 Field measurements: Aerosols

In situ techniques are used to measure physical, optical, and chemical properties of aerosols. New techniques and improvements in older techniques have contributed over the last decade to improvements in understanding such areas as particle composition, the degree of internal and external mixing, and new particle formation. Further increases in understanding are expected to result as the single particle mass spectrometers that are under development come into regular service. The review article by McMurry (2000) contains a more detailed description of many of the instruments currently in use.

5.4.1 Chemical Analysis of Aerosol Samples

Numerous analytical techniques based on physical principles have been employed to study the composition of atmospheric aerosol particles. Electron and ion beam analytical techniques and mass spectrometric methods are powerful tools for analysing the composition of single aerosol particles, and can also offer isotopic information and surface analysis. The primary advances during the last decade are the lowering of the atomic mass which can be detected and increasing sensitivity so that smaller particles can be analysed.

5.4.1.1 *Ion and Electron Beam Analytical Techniques*

In PIXE (particle induced x-ray emission) analysis, characteristic x-ray quanta are emitted from a sample when it is irradiated with a beam of ions (normally protons) having energies of a few MeV / amu (Maenhaut 1992; Johansson et al. 1995). PIXE offers several advantages. It provides multi-elemental (Na-U) analysis, low detection limits, and quantitative results of good accuracy and precision. It also is non-destructive, fast, inexpensive, and its use requires minimal sample preparation.

For an aerosol sample collected in continental background air masses, 10–20 elements can normally be detected and quantified with PIXE within an error of 10–15%. Absolute and relative detection limits down to 10^{-12} g and 0.1 µg g^{-1} respectively can be achieved. The analysis itself can often be automated. These attributes make PIXE well-suited for long-term aerosol monitoring programmes (see Maenhaut 1990 for a review of PIXE and other competing trace element analytical techniques).

EPMA (electron probe micro-analysis) or SEM (scanning electron micrometry) if imaging is the main objective are the most widely used techniques for analysis of the composition of sub-micrometer single particles. EPMA is based on the emission of characteristic X-rays during electron bombardment; it offers multi-element analysis of elements above $Z = 13$ with reasonable accuracy and detection limits down to 0.1% in particles >50 nm. Self-absorption of the X-rays limits the size to 0.5–5 µm. Automated-EPMA works only on particles >150–200 nm. Important low-Z elements such as C, N, and O can be analysed in particles >50 nm if a windowless X-ray detector is used to avoid excessive attenuation of the soft X-rays. The aerosol sampling is then done on very thin aluminium or beryllium foils or on substrates on which a layer of $LiBH_4$ has been evaporated. The potential aerosol applications include analysis of organic compounds of both anthropogenic and biogenic origin (even purely organic particles) and nitrates (Ro et al. 1999). This combination of both high- and low-Z analysis capability in all but the very smallest particles would greatly facilitate studies of the degree of external or internal aerosol mixture.

Scanning transmission electron micrometry (STEM) can be combined with EELS (electron energy loss spectrometry) for analysis of sub-micrometer particles (in practice only for lighter elements, $Z < 30$). Reasonable data acquisition times and volatility losses can be obtained when the full electron energy loss spectrum is acquired simultaneously (Parallel-EELS, PEELS). The method has been demonstrated to work for 20 nm atmospheric aerosol particles (Maynard and Brown 1992) with detection limits of 4% for carbon, equalling 30 carbon atoms. EELS yields information regarding the chemical bonds and can, for instance, be used to distinguish between graphitic and amorphous carbon. Although very time-consuming, both STEM-EDX and PEELS might be worthwhile to pursue in special cases, e.g., to study particle nucleation events.

5.4.1.2 *Mass Spectrometric Methods*

Mass spectrometric techniques used for aerosol studies include ICP-MS (inductively coupled plasma mass spectrometry), SIMS (secondary ion mass spectrometry), and LAMMS (laser microprobe mass spectrometry). Of these, ICP-MS is rapidly growing in popularity. It is a fully quantitative multi-elemental technique capable of analysing nearly all elements in the periodic table with excellent detection limits (0.1 ng ml^{-1} or better). Isotopic information can also be obtained, since the ICP ionisation is almost complete. ICP-MS suffers from matrix effects that can be severe and difficult to predict or estimate, and the aerosol samples have to be dissolved in liquid (Sah 1995).

In TOF-SIMS, secondary ions are sputtered from the target particle during irradiation with an ion beam of low energy (keV) and analysed with TOF mass spectrometer. Secondary ions are only generated down to a depth of a few molecular layers, but repeated sputtering allows (destructive) depth profiling to be made with a lateral spatial resolution of a few micrometers.

In conventional LAMMS a high-power laser is used to ionise the sample collected on a backing substrate, and the ions produced are brought into a TOF-MS. The small laser focus (0.5–1 µm) allows analysis of single aerosol particles. The method is time-consuming and semi-quantitative at the very best, but can nevertheless provide useful information on single-particle inorganic and organic composition and state of mixture (e.g., Gieray et al. 1997).

5.4.1.3 *Single Particle Analysis*

A class of new instruments is becoming available for atmospheric sampling. These instruments provide chemical analysis of individual particles online and in real time (see Wexler and Prather 2000, and subsequent papers). The technique is referred to as single particle analysis (SPA). These SPA instruments are most commonly based on laser ablation-ionisation (Murphy and Thomson 1995; Noble and Prather 1996). A hot filament has also been used (Jayne et al. 2000). Most instruments use TOF-MS and some are able to analyse both positive and negative ions for each particle whereas others are only able to do one sign at a time. Other analysis techniques employed include ion trap mass spectrometry (Reilly et al. 1998), laser induced breakdown spectrometry (Hahn 1998), and quadrupole mass spectrometry (Jayne et al. 2000). All instruments provide some sizing information, but some are more accurate than others. Crude sizing is done by light scattering (Murphy and Thomson 1995) whereas more accurate sizing is accomplished by aerodynamic TOF (Noble and Prather 1996). A differential mobility analyser (DMA) has been employed to size ultrafine particles (Carson et al. 1997).

One primary advantage of SPA is the ability to elucidate the external mixing properties of the aerosol. That is, particles of a given size may have a wide range of compositions. SPA techniques are able to identify the different particle compositions that may occur in an aerosol. In addition, SPA provides composition on a number basis. Bulk analytical techniques are not readily able to assess composition of particles below a few hundred nanometers because the total mass of these particles is so small and a small amount of artefact from larger particles can bias the composition. Finally, SPA provides very high time resolution since sampling and analysis times are under one second. This is very important for labile particles, due to temperature or composition (Murphy et al. 1998b), or due to fast temporal transients as may occur at fronts. Figure 5.2 gives a flavour of the capabilities of measurements made in the upper troposphere and lower stratosphere by Murphy's instrument, showing a range of metallic elements, the sources of which are not well understood.

SPA still has several weaknesses, however. Laser or filament ablation techniques are not very quantitative. Although peak heights are related to concentration, they are also related to the matrix composition and so may be difficult to invert. Most organics break into small groups during the ablation-ionisation processes, so SPA is only able to distinguish elemental from organic carbon, and speciate aromatics. Super-micrometer aerosols are particularly difficult to analyse because of their relative sparseness and their greater losses in inlets.

In the near future, SPA instrumentation will become smaller and easier to use in the field. Using particle focussing techniques pioneered by Dahneke and Cheng (1979), particles can be size-selectively transmitted to the source region allowing a wider range of aerodynamic sizing and analysis, from 10 nm to 2 µm (Mallina et al. 2000). Using different wavelength lasers for ablation and ionisation gives better speciation of organic compounds, and may possibly have benefits for other compounds, at some expense in instrument complexity and cost (Cabalo et al. 2000; Tobias et al. 2000).

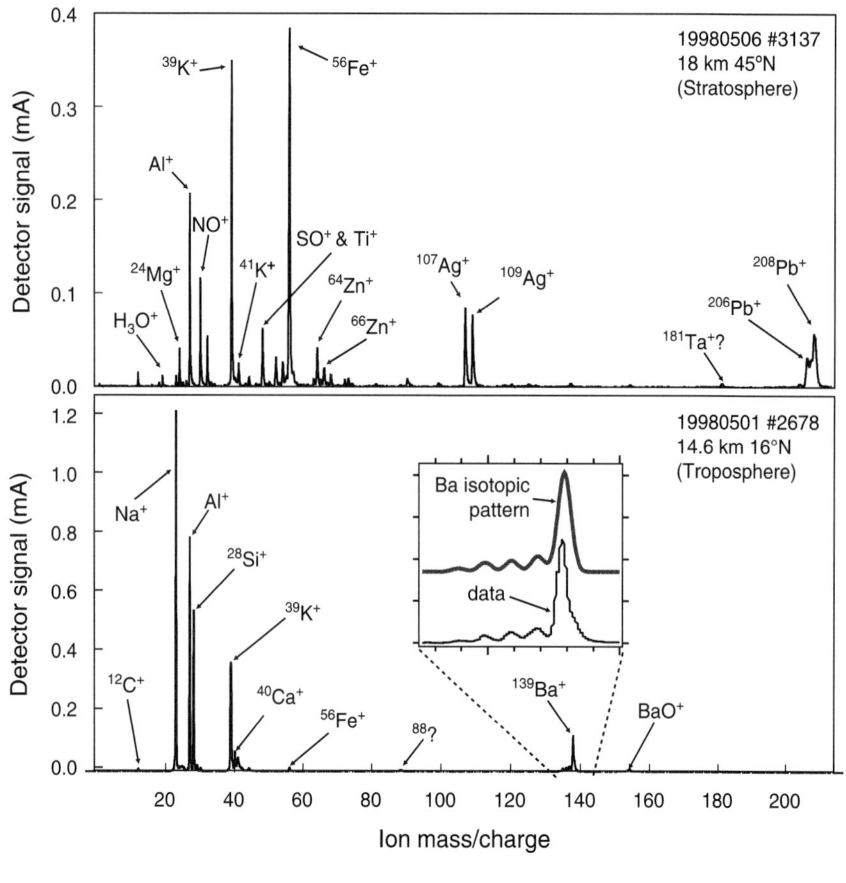

Fig. 5.2. Two examples of single-particle mass spectra using the PALMS laser ionisation instrument (Murphy et al. 1998b). The particles show the presence of unexpected metals in the free troposphere and lower stratosphere

5.4.2 Particle Number Concentration

The total number concentration of the aerosol particles as the integral of the number size distribution can be measured with a high time resolution. State of the art instruments to determine the particle number concentration are continuous flow condensation particle counters (CPC) (Agarwal and Sem 1980). This type of instrument counts each individual aerosol particle and therefore has an upper detection limit in particle concentration. Common CPCs have lower detection efficiency diameters of 10–15 nm, and an upper number concentration limit of about 10 000 cm^{-3} without significant coincidence. The development of the Ultrafine-CPC (Stolzenburg and McMurry 1991) enabled number concentration measurements down to 3 nm in diameter with an upper detection limit of 100 000 cm^{-3}. Measurements of particle number concentration and size distribution in the range below 8 nm became possible by the Pulse-Height-Analysis-UCPC (Weber et al. 1998).

5.4.3 Number Size Distribution

Optical particle counters (OPC) are instruments that size each individual aerosol particle according to its optical properties. The particles are aspirated into a scattering chamber and then directed into the centre of a laser or white light beam. The light is scattered by the particle and its intensity is related to the size of the particle via Mie theory for homogeneous spheres. With intra-cavity laser illumination, particles as small as 70 nm can be sized.

The development of DMA-based size spectrometers in recent years has enabled number size distribution measurements almost over the whole sub-micrometer size range. Two different sizing systems are presently used for atmospheric measurements. DMPS systems (differential mobility particle sizer) measure stepwise while SMPS (scanning mobility particle sizer) systems continuously scan through the size range. DMPS systems need a scan time of 10–15 min and provide relatively good counting statistics in each size bin. In contrast, SMPS systems can scan much faster (one to five minutes), possibly leading to higher statistical uncertainties and artificial distortions at the ends of the size distributions. Those artefacts may lead to large errors in calculations of higher moments of the size distribution such as surface area, scattering coefficient, volume, or mass.

However, DMPS systems, which have been available commercially since the 1980s, are limited to the size range above 10 nm. The development of DMAs for nanometer particles (as small as 2.5 nm) and the availability of the UCPC led to the design of "ultrafine"-DMA-based size spectrometers (Birmili et al. 1999a; Russell et al. 1996; Winklmayr et al. 1991). The availability of these ultrafine spectrometers has provided insights into new particle formation and for the first time provides a means of directly comparing nucleation theory to field observations (Weber et al. 1995). However, the statistical uncertainty in the range below 10 nm is still quite large for atmospheric measurements in each DMA-based sizing system due to decreasing DMA penetration efficiency, low charging probability, and the low aerosol flow rate of the UCPC (Wiedensohler et al. 1994).

5.4.4 Optical Properties

5.4.4.1 *Particle Light Scattering*

The particle volume scattering coefficient can be measured directly using nephelometers (Heintzenberg and Charlson 1996). In these instruments a white light source illuminates an air volume. The scattered light is collected within a certain scattering angle region (~7 to ~170 degrees) and certain spectral intervals. With some reasonable assumptions these data yield the volume scattering coefficient of the particles. They also allow a direct calculation of backscatter fraction, *b*, which is a proxy for the asymmetry parameter, *g*, used in many radiative transfer models (Marshall et al. 1995).

Nephelometer technology to measure aerosol light scattering properties has evolved steadily over the past 50 years (Heintzenberg and Charlson 1996), with demonstrated accuracy improving from roughly a factor of two in the 1960s to around 10% in the 1990s. A standardised commercial instrument, subjected to rigorous noise and calibration tests (Anderson et al. 1996), has been developed since the early 1990s. This instrument has been broadly embraced by the measurement community and is now routinely deployed in field experiments by a large number of research groups. Protocols for sampling, field calibration, and data reduction have been published in an attempt to unify standards for data collection (Anderson and Ogren 1998). Nephelometer techniques have recently been extended to the measurement of the 180° backscatter coefficient (Doherty et al. 1999), permitting *in situ* measurement of the extinction-to-backscatter ratio, S_a, which is the primary *a priori* aerosol parameter required for quantitative aerosol extinction retrievals from backscatter lidar data.

Because of its enclosed chamber, the nephelometer is suitable for controlled-humidity measurements (Covert et al. 1972), which allow a direct determination of the functional dependence of light scattering on relative humidity (RH). Knowledge of this functional dependence is of first order importance in estimating the radiative effects of atmospheric aerosols (Pilinis et al. 1995), most of which are hygroscopic. In addition, closed-cell nephelometers are frequently combined with size segregation at 1 μm diameter in order to determine, separately, the portion of scattering due to the coarse and fine modes of the aerosol. This information is strongly related to aerosol composition and source, since super-micrometer aerosol mass is generally dominated by mechanically-produced particles (e.g., dust, sea salt, fly-ash, or vegetative debris) while sub-micrometer mass is generally dominated by secondary or combustion-derived particles (e.g., sulphates, organics, or soot).

For experiments where *in situ* and remote measurements of the same air parcel are being compared, nephelometers have been operated outdoors at the ambient temperature such that the scattering measurement corresponds to the ambient RH. One approach is to use an open-cell nephelometer (e.g., Malm et al. 1996), while another is to eliminate all sources of sample heating in a closed-cell nephelometer (Heintzenberg and Erfurt 2000). The open-cell approach has the additional advantage of reducing artefacts caused by the sample inlet (see below), while the closed-cell approach retains the high sensitivity and accuracy of that technology.

Knowledge of the scattering phase function is important in retrievals of aerosol properties using remote sensing techniques. In spite of this, there is currently no instrument that can directly measure the scattering phase function of the atmospheric aerosol. The scattering phase function can be computed from fundamental microphysical properties such as particle size distribution, composition and mixing state, and particle shapes. Assumptions and simplifications are almost always made in performing these calculations, especially regarding particle shape, and direct measurements to confirm the validity of these assumptions are sorely lacking.

5.4.4.2 *Particle Light Absorption*

While the measurement of aerosol scattering is relatively mature and accurate, the measurement of aerosol absorption is not. Because of the low levels of absorption in the ambient aerosol, filter-based techniques are most commonly used and quasi-real time techniques have been developed since the early 1980s (Hansen et al. 1982; Reid et al. 1998). Recently, noise characterisation, calibration, and intercomparison tests (Bond et al. 1999; Reid et al. 1998) have been conducted. A detailed set of field use and data reduction protocols has been published for the PSAP instrument (Radiance Research, Inc.) by Anderson et al. (1999). Many research groups have adopted this instrument, bringing a measure of standardisation to the field.

Significant problems remain, however, regarding the accurate determination of aerosol absorption. Calculation of the optical properties of atmospheric particles from filter samples involves certain assumptions, and calibrations are uncertain due to variations in particle morphology and chemistry. Uncertainty of the PSAP absorption measurement has been estimated as ranging from ±20% (Bond et al. 1999) to ±40% (Reid et al. 1998). Finally, filter sampling techniques do not permit studies of the humidity dependence of light absorption. Indeed, there is no commonly available method for measuring this important parameter.

The photoacoustic method is an alternate technique, which measures the light absorption of suspended particles in an enclosed volume. Recent work has improved the sensitivity of this method by an order of magnitude (Arnott et al. 1995, 1999) allowing absorption measurements of the ambient aerosol in unpolluted as well as polluted conditions. Further refinements to the instrumental technique show promise for another order of magnitude improvement in sensitivity. The photoacoustic method also offers the potential for measuring the humidity dependence of light absorption.

5.4.5 Hygroscopicity

In the 1980s and 1990s, measurements of hygroscopic properties of aerosol particles were significantly improved in three ways. First, electrobalance techniques were used to make exact measurements of growth in diameter, density, and refractive index of pure compounds such as sulphates (Tang and Munkelwitz 1994) and sea salt (Tang et al. 1997). These data have proven to be very useful in modelling hygroscopic effects on aerosol physics and optics. Second, the development of the hygroscopic-tandem differential mobility analyser (Rader and McMurry 1986) allowed, for the first time, the size-resolved external mixture of aerosol particles in terms of hygroscopicity to be determined. However, these measurements are limited to the size range below 300 nm in diameter, and thus do not give the entire climate-relevant information set for cloud or fog processes, and scattering of solar radiation by aerosol particles under ambient conditions. The third development is the widespread use of the controlled humidity nephelometer (e.g., Kotchenruther et al. 1999), which allows hygroscopic effects on light scattering to be measured directly.

5.4.6 Aerosol Deposition Fluxes

Measurement of atmosphere-surface exchange of aerosol particles remains difficult. Nevertheless, the past decade has seen advances that have permitted somewhat better estimates of aerosol fluxes onto vegetative canopies, bodies of water, and surfaces in urban areas.

In previous decades, measurements of particle deposition onto vegetation had been conducted by numerous research groups but were of questionable validity. Recent advances in real-time particle sensors, more accurate chemical species measurement, and improved methods of characterising turbulence have permitted better estimates of particle fluxes using micrometeorological methods. For example, recent flux measurements by eddy correlation and by the gradient method over forests are reported by Beck et al. (2000) and Horvath et al. (2000), respectively. Improvements in measuring forest throughfall (water that falls to the ground beneath the trees during precipitation) have resulted in better estimates of dry deposition of aerosol chemical species on broader spatial scales (Beier et al. 1992; Butler and Likens 1995). Despite these advances, discrepancies remain between measurements and modelling of dry deposition to vegetation (Garland 2000). At least part of the discrepancy may be explained by the disproportional fraction of deposition from the largest airborne particles present (Zufall and Davidson 1998).

The recognition that many bodies of water are becoming contaminated by atmospheric deposition has resulted in considerable research efforts over the past ten years. Because of the difficulty of direct measurement of deposition onto water surfaces, many investigators have used surrogate surfaces positioned over the water despite their acknowledged inadequacies. These data, as well as deposition velocities found in the literature, have been used to estimate deposition fluxes to major bodies of water for several chemical species such as trace metals (e.g., Arimoto et al. 1997; Baker et al. 1997). In addition, Yi et al. (1997) have developed a surrogate surface containing water to assess the relative deposition rates of gaseous and particulate species and to explore the role of the humid boundary layer above the water surface.

Recent attention has also focussed on urban surfaces. Nemitz et al. (2000) have used airborne particle counters and ultrasonic anemometers to measure emissions and deposition of particles in the city centre of Edinburgh, Scotland, considering the city as a "canopy". Nicholson (2001) has explored deposition of particles to building materials such as brick and glass using wind tunnel techniques. The influence of particle deposition over long periods of time on limestone in an urban area has been explored by Davidson et al. (2000a). Overall, these studies have shown that particle deposition in urban areas can be significant and can degrade surfaces.

5.4.7 Issues in Aerosol Sampling

It is difficult to make accurate measurements of the concentrations and properties of ambient aerosols (Vincent 1989; Willeke and Baron 1993). Numerous potential artefacts can modify the ambient population as it is being sampled, so that the analytical system generates data that are a poor representation of reality. For example, due to changes in relative humidity, the aerosol is often measured under dry or undefined conditions which are not representative for the undisturbed ambient aerosol. These problems are subtle and poorly understood; accordingly users of analytical equipment are frequently unaware of them and/or unable to quantify the errors and uncertainties that are introduced. Thus it is likely that significant amounts of published aerosol data are misleading in one way or another. The potential for experiencing sampling artefacts depends strongly on the size and chemical properties of the aerosols being sampled.

One of the most thoroughly documented types of sampling artefacts involves gas-aerosol interactions. Appel et al. (1978) were among the first to note that gaseous nitric acid and particulate nitrate can interconvert on filter media. Depending on particle composition, artefacts can be either positive or negative (Zhang and McMurry 1987). There are numerous methods for minimising these artefacts, however. One of the most rigorous is the use of diffusion denuders (Sickles et al. 1990). These devices first denude the air stream of condensable gases and then sample the aerosols with a filter pack that can collect both the aerosol and any volatilised products from it.

Organic aerosols pose significant sampling problems, which have yet to be resolved (Appel et al. 1989). Many organic compounds have significant vapour pressures, so that exchange between condensed and vapour phases is continuous. Any change in pressure or temperature during sampling will displace this equilibrium, distorting the ambient distribution between the phases. Furthermore, many organic vapours adsorb strongly on filter media, generating positive artefacts. The most common approach to correct for this artefact involves the use of a second filter in series with the first (Novakov et al. 1997). However, this relies on the often-questionable assumption that vapours are only weakly adsorbed, so that the same artefact will be generated on the second filter as on the first. Much remains to be done before organic aerosol observations can be used with confidence.

Another set of problems is associated with the use of inlet tubes (Okazaki et al. 1987). The tendency of particles to continue moving with the ambient flow, rather than entering the flow of sampling inlets, results in decreasing collection efficiency for larger particles. Because both sea salt and mineral dust particles have, for the most part, mass larger than one micrometer, this inertial effect makes it particularly difficult to collect representative samples. Inertial problems are magnified when aerosols are sampled from aircraft, which often fly at 100 m s^{-1} or faster. The deceleration of air within inlet systems generates intense turbulence that propels many of the particles to the walls of the inlet diffuser cone. This results in transmission efficiencies for supermicrometer particles of only 5 to 20%, whereas submicrometer particles can be collected with 50 to 100% efficiency (Huebert and Lee 1990; Sheridan and Norton 1998). Recent measurements during the ACE-2 experiment (Schmid et al. 2000) demonstrated the impact of this on measurements of the extinction of radiation by dust and sea salt: a nephelometer (sampling from an inlet) underestimated the former by about 85% and the latter by about half relative to inlet-less approaches. Similarly, a comparison of a total aerosol sampler with a conventional curved-tube inlet during the PEM-Tropics experiment found that MSA and non-sea salt sulphate were underestimated by a factor of 20 to 30 by filters behind the conventional inlet. The result is that many of the published concentrations of aerosols above the surface are likely to be underestimates of the true ambient concentrations. Users of these data must be aware of this pervasive artefact.

There are new technologies being developed, however, which will improve the situation. A new total aerosol sampler (TAS) has been demonstrated to collect all aerosols that enter its tip, without artefact. The disadvantage of TAS is that all size information is lost in the extraction process. To enable size-dependent sampling, a new laminar-flow inlet developed at Denver University by Seebaugh and Lafleur was tested in flight in 2000. Another solution to the calibration problem might lie in the use of identical, non-aspirated, particle sizing probes both outside the aircraft and inside the sample plumbing (see, e.g., Noone et al. 1992a).

Aerosol sampling problems have no doubt compromised much of the data now in the refereed literature. However, as awareness of these issues grows and technological improvements are realised, much more representative values should be obtained.

5.4.8 Summary of Future Needs

A number of important aerosol parameters are measured poorly or cannot be measured at all using current techniques. As mentioned above, improvements in the measurement of hygroscopicity and in the measurement of aerosol optical properties at ambient relative humidity are needed. Although recent developments show promise, current capabilities to measure the absorption of the ambient aerosol are limited and improvements in sensitivity and calibration are necessary. Problems with inlet sampling affect a wide range of aerosol measurements and improved sampling techniques for organic aerosols are needed. Although physical techniques for the measurement of the elemental composition of single particles have shown great progress, capabilities to determine the molecular composition of particles are still limited, especially with respect to organic compounds. Better techniques are needed for readily quantifying cloud-forming particles and for determining the size-resolved composition of particles.

5.5 Satellite Instruments for Tropospheric Chemistry

5.5.1 Introduction

It is possible to monitor long lived tropospheric compounds adequately with a limited number of measurement stations distributed around the globe. However, to obtain global observations of short lived compounds whose sources are more variable in space and time, satellite-based measurements, using remote sensing techniques, are required. Current techniques are based on the passive sensing of scattered or transmitted sunlight, or emitted thermal or microwave radiation. These passive techniques have employed limb viewing (along horizontal paths), nadir viewing, or a combination of the two to perform the required measurements. Observations from instruments such as TOMS, SBUV, and SAGE since the late 1970s, and from the UARS satellite during the 1990s have provided the scientific community with a comprehensive picture of the chemical composition of the stratosphere and mesosphere. These measurements will be continued during the next decade by the NASA Aura (formerly EOS Chem) and ESA ENVISAT satellites. A selection of past, present, and future satellite instruments designed for sensing atmospheric composition is summarised in Table 5.2.

Initially, the remote sensing of atmospheric composition by satellite sensors was limited to the stratosphere. Satellite remote sensing of trace gases and aerosols in the troposphere is substantially more difficult. Limb viewing techniques have limited utility near and below the tropopause because of frequent obstruction by clouds. Limited vertical resolution, difficulties in separating tropospheric concentration from the total column, and accounting for the effects of clouds and water vapour have been persistent problems with nadir-viewing techniques. Tropospheric chemistry will be advanced in the coming decade by satellite sensors with new capabilities for observing the troposphere: passive sensors with improved spectral coverage; passive sensors in geostationary and L1 orbits providing regional and continental-scale observations with high spatial and temporal resolution; and lidars, active sensors based on lasers, which offer the potential for observations of aerosols and some trace gases with greatly improved vertical resolution, at the cost of restricted geographic coverage. Development of these new sensors has been enabled by technologies that have matured during the 1990s. These new technologies and the advances expected from

Table 5.2. Summary of instruments discussed in Sect. 5.5. This list is not intended to be comprehensive, but gives examples of past, present, and future capabilities. Only measurements of primary significance to atmospheric chemistry are listed

Instrument	Satellite	Type	Relevant Measurements	References
AVHRR	TIROS-N, NOAA-6 through NOAA-14, 1978 to present	Vis-NIR radiometer	Aerosols	Husar et al. 1997
TOMS	Nimbus 7, 1979–1992; Meteor, 1992–1994; ADEOS, 1996–1997; Earth Probe, 1996 to present	UV radiometer	Ozone, aerosols	Heath et al. 1975; Herman and Larko 1994; Herman et al. 1997
SAGE I/II	Atmospheric Explorer, 1979–1981; Earth Radiation Budget Satellite, 1984 to present	Solar occultation	O_3, NO_2, H_2O, and aerosols in stratosphere and upper troposphere	McCormick et al. 1989; Kent et al. 1998
MAPS	Space Shuttle, 1981, 1984, 1994	Gas filter correlation radiometer	CO	Connors et al. 1999
LITE	Space Shuttle, Sept. 1994	Backscatter lidar	Aerosols	Winker et al. 1996
GOME	ERS-2, 1995 to present	UV-vis spectrometer	O_3, NO_2, H_2O, BrO, OClO, SO_2, HCHO, aerosols	Burrows et al. 1999
IMG	ADEOS, 1996–1997	IR FTS	O_3, N_2O, H_2O, CH_4, CO	Ogawa et al. 1994
POLDER	ADEOS, 1996–1997	Wide field, polarization radiometer	Aerosols	Deuzé et al. 1999
MISR	Terra, 12/99 to present	Multiangle radiometer	Aerosols	Diner et al. 1991; Martonchik 1997
MODIS	Terra, 12/99 to present; Aqua, 2001	Vis-NIR-IR radiometer	Aerosols	Kaufman et al. 1997; Tanré et al. 1997
MOPITT	Terra, 12/99 to present	Gas filter correlation radiometer	CO, CH_4	Drummond and Mand 1996; Wang et al. 1999
SCIAMACHY	ENVISAT, 2001	UV-vis-NIR spectrometer	O_3, NO, NO_2, N_2O, H_2O, BrO, OClO, CO, CH_4, SO_2, HCHO, aerosols	Bovensmann et al. 1999
EPIC	Triana	UV-vis radiometer	Ozone, aerosols	Valero et al. 1999
OMI	Aura, 2003	UV-vis spectrometer	O_3, SO_2, NO_2, aerosols	
TES	Aura, 2003	IR FTS	O_3, NO, N_2O, CO, CH_4, HNO_3, etc.	Glavich and Beer 1991
GIFTS	Geostationary satellite, 2003	IR FTS	O_3, CO, H_2O, etc.	Smith et al. 2000
Lidar	CALIPSO, 2004	Polarization lidar	Aerosols	Winker and Wielicki 1999
GeoTRACE	TDRS	UV-vis spectrometer; gas filter correlation radiometer	O_3, CO, aerosols	Neil et al. 2000

them are discussed in this section. Several satellite instruments flown during the 1990s, which are forerunners of the next generation of satellite sensors, are used to illustrate the discussion.

5.5.2 Recent Advances

Satellite measurements of aerosols and molecules in the troposphere are much more difficult than in the upper atmosphere and measurement capabilities from space have been much more limited. The limb viewing geometry is characterised by horizontal resolutions on the order of 400 km, which is inconsistent with the spatial variability of many tropospheric trace compounds. Additionally, tropospheric clouds interfere with limb viewing measurements, so observations of tropospheric composition must rely primarily on nadir- or near-nadir-viewing instruments. Passive nadir-viewing instruments have limited vertical resolution, however, and separation of the tropospheric composition from that of the total column can be very difficult.

5.5.2.1 *UV-Visible Techniques*

Total column ozone has been derived since 1979 from ultraviolet measurements of scattered sunlight by the TOMS instrument. In the late 1980s, a technique was developed to derive tropospheric column of ozone as the residual of the total column measured by TOMS and the stratospheric column from co-located SAGE II measurements (Fishman et al. 1990). The drawback of applying this technique to SAGE II data is that only climatological studies can be done, as SAGE II longitudinal coverage is sparse and full latitudinal coverage is obtained only over about 40 days.

The Global Ozone Monitoring Experiment (GOME) has operated since April 1995 on the ERS-2 satellite, which flies in a 780 km sun-synchronous orbit. GOME is a nadir-viewing spectrometer designed for trace gas measurements by observing reflected sunlight in the ultraviolet and visible (Burrows et al. 1999). In contrast to TOMS and SBUV, which operate at a number of discrete wavelengths, GOME observes the entire spectrum between 240 and 790 nm with a spectral resolution that varies from 0.2 nm in the UV to 0.4 nm in the red. GOME is superior to TOMS and SBUV because of its improved coverage of the ozone spectrum and improved signal to noise ratio. In the troposphere and lower stratosphere, GOME is able to resolve the vertical profile of ozone with an effective resolution of 10 km. For other gases, the tropospheric column is determined by subtracting the stratospheric column from the total column, assuming local longitudinal homogeneity of the stratosphere. Employing this residual technique, GOME has been used to observe tropospheric NO_2, formaldehyde (HCHO) (Fig. 5.3), and SO_2 resulting from pollution and biomass

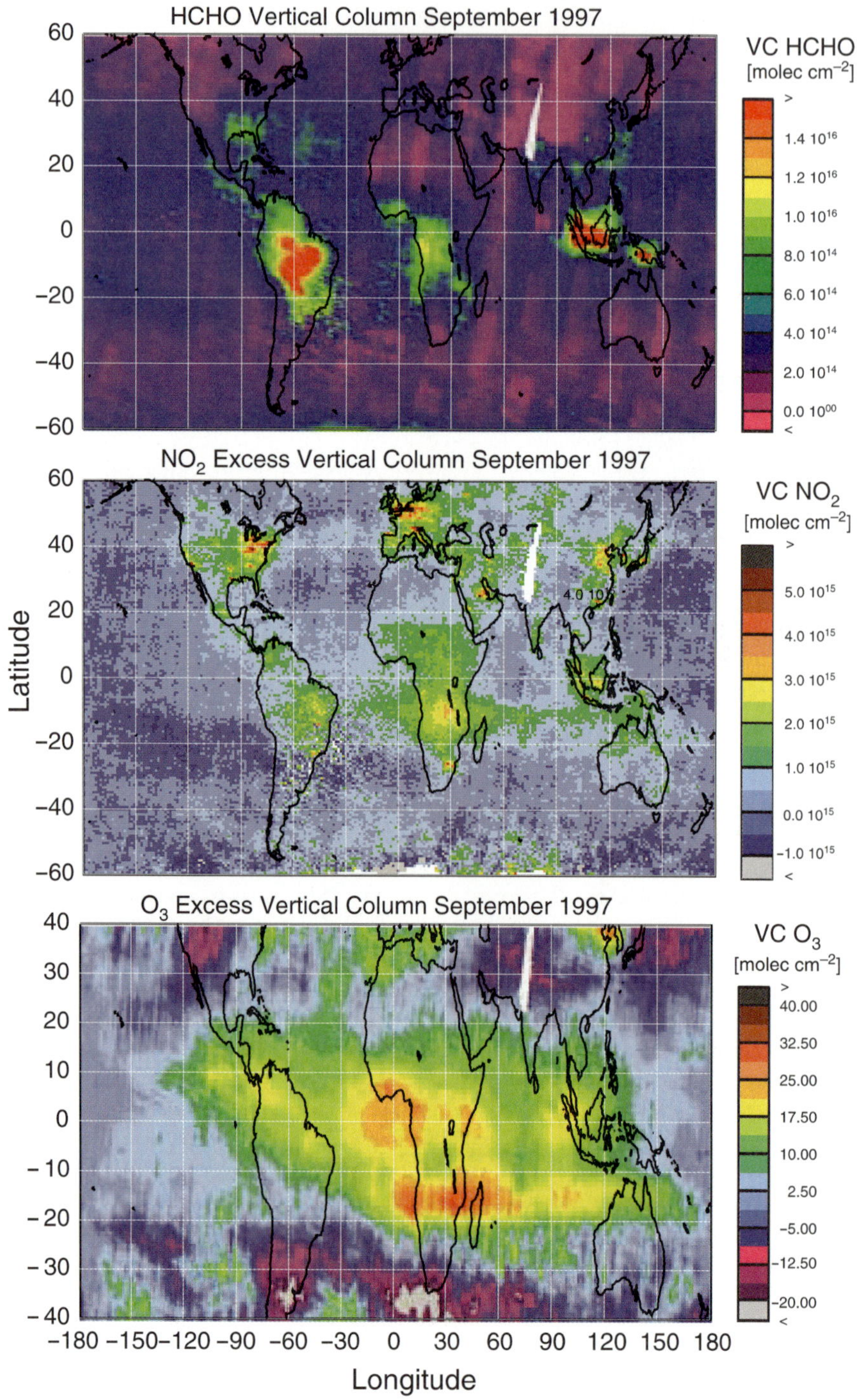

Fig. 5.3. Global tropospheric vertical column (VC) amount (molecules cm^{-2}) of NO_2, HCHO, and O_3 for September 1997 derived from GOME data. Elevated levels of NO_2 can be seen over the industrial areas of Europe, the United States, China, and the Middle East oil fields. The production of NO_2 and HCHO from biomass burning in the Southern Hemisphere is also evident (J. P. Burrows, personal communication, and Burrows 1999)

burning. GOME is the first satellite instrument suitable for studying halogen oxides in the lower troposphere and has been used to observe the global distribution of BrO.

The observations begun by GOME will be continued by three GOME-2 instruments scheduled to be flown on the EUMETSAT/ESA METOP series of satellites, as well as by the conceptually similar OMI instrument, scheduled to be flown on Aura. The capabilities of GOME are extended by SCIAMACHY, now flying on ENVISAT. SCIAMACHY is similar to GOME in design concept but with spectral coverage extended into the near infrared, between 240 and 2 380 nm, and with three viewing geometries: nadir, limb-scanning, and occultation (Bovensmann et al. 1999). Extended spectral coverage and matched limb/nadir measurements allow the tropospheric column amounts of a variety of trace gases to be determined: O_3, NO_2, N_2O, BrO, CO, CH_4, H_2O, and, under polluted conditions, HCHO and SO_2. Additionally, SCIAMACHY will measure vertically resolved profiles of O_3, H_2O, N_2O, and CH_4.

In the last decade, satellites have been used for semi-quantitative global monitoring of aerosol distribution. The Advanced Very High Resolution Radiometer (AVHRR), flying on a series of NOAA satellites, has provided aerosol observations over the oceans since 1979 (Husar et al. 1997). Algorithms using two of the AVHRR wavelengths (0.65 µm and 0.85 µm) have been used to retrieve both aerosol optical depth and Ångström coefficient (the spectral coefficient of aerosol scattering) (Higurashi and Nakajima 1999; Mishchenko et al. 1999)(see example in Fig. 4.2). Operational aerosol products, however, have been limited to ocean regions due to the difficulty of retrieving aerosols over inhomogeneous and/or bright land surfaces. More recently an algorithm to retrieve an aerosol index related to aerosol absorption has been developed (Herman et al. 1997) and applied to TOMS UV measurements to detect absorbing aerosols, primarily dust and smoke, over land and marine surfaces. This recent work has resulted in new insights into the global sources, production mechanisms, and transport of dust and smoke. However, quantitative retrievals of aerosol optical depth require either external information or assumptions regarding the type and height of the aerosol and the nature of the underlying surface (Torres et al. 1998).

POLDER was the first satellite instrument designed expressly to provide quantitative estimates of particle properties, column burden, and radiative forcing. It combines spectral measurements in four channels (0.4–0.9 µm) with polarisation and angular measurements in a geometry similar to a "fish eye camera", observing Earth with a wide range of view angles. The results are global maps of aerosol optical thickness, estimates of refractive index and the wavelength exponent of particulate extinction over the ocean, and an aerosol index over the land (Deuzé et al. 1999). The large size of the foot print, 6 km, restricts the cloud free regions that can be used for aerosol evaluation, however.

The payload of the Terra satellite, launched in December 1999, includes the Moderate Resolution Imaging Spectro-Radiometer (MODIS) and Multi-angle Imaging Spectro-Radiometer (MISR) instruments which represent a significant advance over previous capabilities for aerosol sensing. MODIS is a nadir-viewing imaging filter radiometer with 36 spectral channels and improved calibration relative to AVHRR. Seven of those channels, between 550 and 2 100 nm, are used for retrievals of aerosol optical depth over dark vegetation as well over ocean (Kaufman et al. 1997b; Chu et al. 1998). It is also possible to retrieve the mean aerosol size and the relative contribution to the aerosol optical depth of the accumulation and coarse modes over the ocean (Tanré et al. 1997, 1999). The MISR provides multi-angle observations in four spectral bands between 446 and 866 nm. Each band is imaged with nine view angles spread out along the flight path between 70 degrees forward and 70 degrees aft. These multi-angle measurements provide additional information on aerosol angular scattering characteristics and allow the classification of aerosols into one of about a dozen types, distinguished by size, composition, and shape (spherical-non-spherical), as well as measurements of aerosol optical depth (Kahn et al. 1997).

5.5.2.2 *Infrared Techniques*

Many trace compounds exhibit spectral signatures in the thermal infrared, and infrared techniques hold the potential to observe many compounds which cannot be observed using UV/visible techniques. Satellite measurements of tropospheric gases have been performed using the gas filter correlation technique and by inversion of the radiance spectrum observed at high spectral resolution. In the gas filter correlation technique, an onboard gas cell filled with the target gas provides measurements of a specific gas without the necessity of high spectral resolution and tight spectral requirements. This technique does not require the absorption lines to be resolved and is characterised by high signal to noise ratio. The first satellite measurements of CO were obtained by the MAPS (Measurement of Air Pollution from Satellites) instrument using this technique in the mid-infrared. MAPS flew on the Space Shuttle in 1981, 1984, and 1994, and provided the scientific community with valuable insights into the importance of biomass burning as a source of CO (Connors et al. 1999). The utility of MAPS was limited by the short duration of shuttle missions, however. Global observations of CO on a continuing basis are now being provided by MOPITT on the NASA Terra satellite that was launched in December 1999.

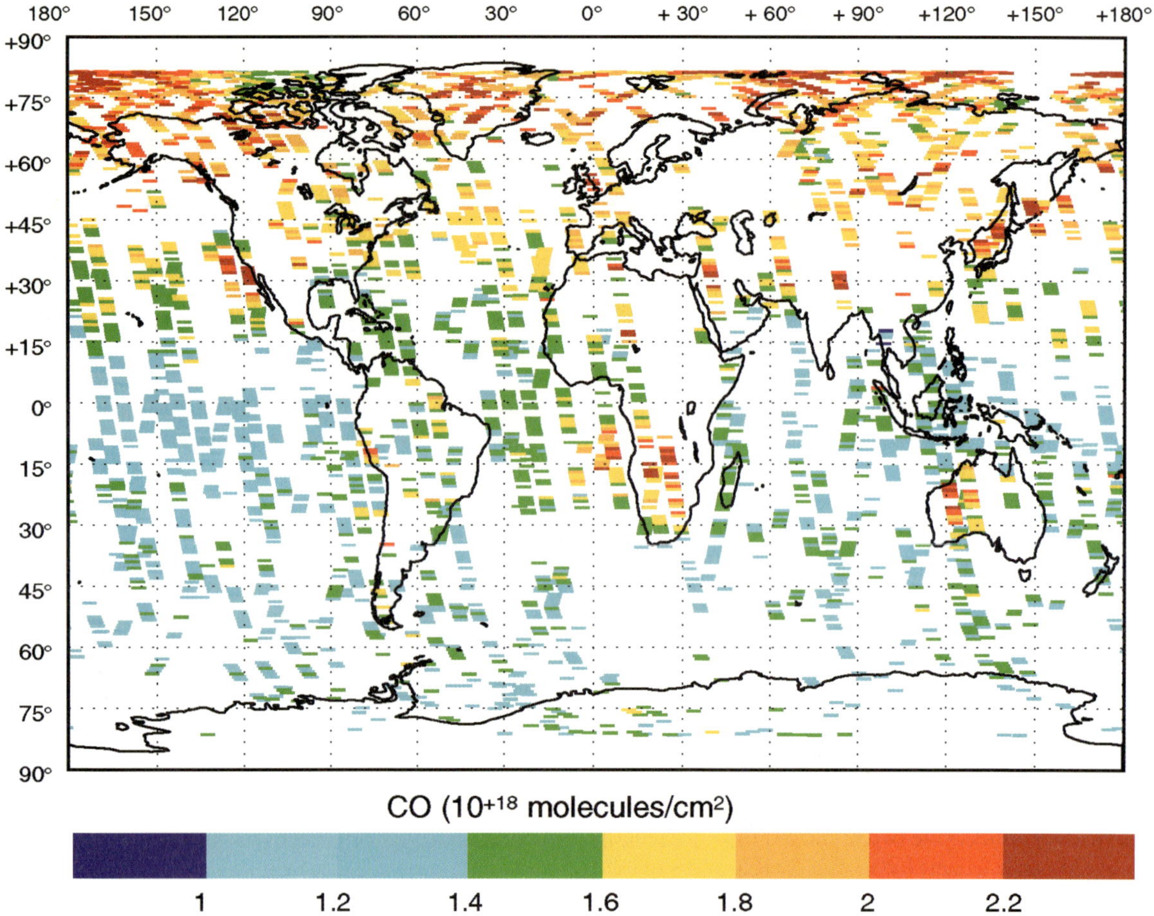

Fig. 5.4. Global distribution of CO total column for June 16–19, 1997 as retrieved from cloud-filtered IMG spectra. *Dots* corresponding to the measured pixels were enlarged on the plot for clarity (source: C. Clerbaux, personal communication 2000)

High spectral resolution infrared spectrometers hold the potential to observe a broader range of compounds than is feasible using the gas filter correlation technique. Limited vertical resolution of CO, CH_4, and O_3 is possible if spectral resolution is high enough to resolve the absorption line structure. The TES (Tropospheric Emission Spectrometer) instrument, part of the Aura payload, is a Fourier transform spectrometer (FTS) observing the spectral region between 3.2 and 15.4 µm at a resolution of 0.025 cm^{-1}. TES is expected to provide tropospheric measurements of O_3, CO, CH_4, HNO_3, NO, and N_2O with vertical resolutions of 2 to 6 km. The IMG (Interferometric Monitor for Greenhouse Gases) instrument, which operated between October 1996 and June 1997 on the Japanese ADEOS satellite, illustrates the potential of high spectral resolution observations from instruments such as TES. IMG was a nadir-viewing FTS operating between 3.3 and 14.0 µm with 0.1 cm^{-1} spectral resolution (Ogawa et al. 1994). The instrument had an $8 \times 8\ km^2$ footprint and 32 km swath. As shown in Fig. 5.4, IMG data have provided global measurements of tropospheric CO, using a neural net retrieval approach (Clerbaux et al. 1999; Hadji-Lazaro et al. 1999).

5.5.3 Future Trends

5.5.3.1 *Active Techniques*

The global distribution of tropospheric aerosol is highly variable due to the variety and variability of sources and the short residence time of aerosols in the atmosphere. Due to the current limited knowledge of aerosol source strengths and production and transformation mechanisms, aerosol distribution cannot be adequately determined by modelling. Improved satellite measurements, especially of the vertical distribution of aerosol, are necessary to obtain a more complete picture of the global aerosol.

The new observing capabilities of the Terra instruments will provide major improvements in knowledge of the global aerosol. None of these instruments, however, is sensitive to aerosols at very low concentrations or provides any information on the vertical profile of aerosols. Active lidar sensors, based on short-pulse lasers, are required.

The Lidar In-space Technology Experiment (LITE) provided the first demonstration of the promise of sat-

Fig. 5.5. Vertical cross-section of aerosol distribution observed by LITE. A deep haze layer (*yellow* and *red*) is seen over the eastern US and extending into the Atlantic Ocean. *Yellow lines* trace wind back-trajectories at the 850 hPa level over the five days previous to the LITE overpass

ellite lidar for aerosol studies (Fig. 5.5). The LITE system used a high-power pulsed laser operating at 355 nm, 532 nm, and 1 064 nm (Winker et al. 1996) and flew on the Space Shuttle in September 1994. Figure 5.5 illustrates the capability of satellite lidar to detect and profile aerosols over both land and ocean. Lidar data also show the spatial relationships between clouds and aerosols, critical information for studies of aerosol-cloud interactions. A study by Kent et al. (1998) combined aerosol data from LITE and SAGE II to investigate the sources, distribution, and optical properties of aerosols in the upper troposphere. This study highlights the potential of combining observations from lidar and passive sensors.

The relatively recent development of rugged, long-life, and efficient diode-pumped solid-state lasers has enabled the development of lidars which can be operated from long-term satellite platforms in low Earth orbit. CALIPSO (formerly PICASSO-CENA) is a recently approved NASA three-year mission planned to begin in 2004 (Winker and Wielicki 1999). CALIPSO will fly in formation with the NASA Aura and Aqua satellites. The CALIPSO instrument suite includes a polarisation-sensitive lidar operating at two wavelengths (532 nm and 1 064 nm). The spectral dependence of the lidar aerosol backscatter provides a height-resolved indication of aerosol particle size. Polarisation sensitivity allows the discrimination of hydrated aerosol droplets from dry (non-spherical) aerosols. These capabilities provide a limited ability to classify aerosols by type. To provide a comprehensive picture of the global distribution and properties of aerosols, data from many diverse sources – satellite, ground-based *in situ*, and remote sensing instruments – will have to be integrated, using assimilation techniques. The CALIPSO satellite, flying in formation with the NASA Aqua satellite, will allowing the combination of MODIS and lidar observations. The OMI instrument on Aura will detect aerosols using the TOMS retrieval approach (Herman et al. 1997). The unknown vertical distribution of aerosol results in significant uncertainties in quantitative retrievals using this algorithm. Near-coincident aerosol profiles from CALIPSO can be used to improve the accuracy of OMI aerosol retrievals.

Technologies required for space-based sensing of trace gases by differential absorption lidar (DIAL) systems are now under development. DIAL systems have not yet been flown in space, but have proven their utility through many years of use in ground-based and airborne measurement campaigns. Laser technology has now progressed to the point where spaceborne DIAL systems are in the early planning stages and it is likely that DIAL systems to measure tropospheric ozone and water vapour will be in orbit by the latter part of this decade. Because active sensors provide limited horizontal coverage, assimilation techniques will need to be developed to merge high vertical resolution data from lidar with broad spatial coverage data from passive sensors.

5.5.3.2 *Observations from Geostationary and Lagrange Orbits*

Instruments in low Earth orbit cannot detect processes which take place on small scales, that is with horizontal spatial resolutions on the order of 10 km or less, and temporal resolutions on the order of fractions of an hour to a few hours. Satellites in low Earth orbit take a day or more to provide complete global coverage and any given region is revisited no more than twice a day. Instruments on satellites in geostationary or Lagrange orbits are required in order to document the spatial distribution, temporal trends, and variability of a number of environmentally significant shorter lived atmospheric compounds. A satellite in geosynchronous orbit – at an altitude of about 36 000 km – orbits Earth once a day and so has a stable and continuous view of a large portion of Earth's surface. Four to five longitudinally spaced satellites in geosynchronous orbit are required to obtain full coverage of Earth up to a latitude of 50–60 degrees. The Lagrange L1 point is between Earth and the sun, about 1.6 million km from Earth. The gravitational pull of both bodies is balanced at this point, so that the satellite maintains a constant distance from each, thereby providing a continuous view of the solar-illuminated Earth disk.

The use of sensors in geosynchronous and L1 orbits to observe tropospheric composition has been enabled by the development of large-format detector arrays and their associated low-noise readout electronics, high bandwidth satellite communications links, and high-volume data handling facilities. These large arrays enable sensors to make quasi-continuous measurements over Earth's disk with high spatial, spectral, and temporal resolution. These new satellite platforms will allow the investigation of a whole new class of atmospheric phenomena, including processes occurring at small spatial scales and the transport of pollutants from local source regions to the global atmosphere. Instruments in these orbits can also give the synoptic context for measurements from ground networks or field campaigns.

GIFTS is a mission planned for launch in 2005 as part of NASA's New Millennium Programme. The programme's objective is to demonstrate new technologies and the potential for high spectral resolution infrared observations from geosynchronous orbit. GIFTS is an imaging Fourier transform spectrometer, which will provide spectra in two bands in the thermal infrared. The spectral resolution is programmable, from 18 cm^{-1} to 0.6 cm^{-1}. The primary focus of GIFTS is to demonstrate an improved capability for temperature and humidity sounding to increase the accuracy of weather forecasts. However, observations at the highest spectral resolution promise to provide vertically resolved profiles of ozone, CO, and other constituents with vertical resolutions ranging from 3 to 11 km. Views of mesoscale-sized regions can be obtained every half hour.

Triana is a NASA mission (originally scheduled to launch in early 2001, but now delayed) planned to place several instruments at the L1 point to provide continuous observations of the sun-illuminated side of Earth from this unique vantage point. A 10-band imaging filter radiometer called EPIC will provide hourly measurements of ozone over the full sunlit disk at a spatial resolution of 8–14 km. This is done using the UV-backscatter technique developed for TOMS.

GeoTrace is an instrument somewhat similar to the Triana EPIC camera, but with higher spectral resolution. GeoTrace has been proposed to the NASA New Millennium Programme to fly in a geosynchronous orbit; it would provide measurements similar to GOME but with continuous full-disk coverage. GeoTrace gives atmospheric column amounts, and is aimed at monitoring distribution, diurnal changes, and transport.

Instruments such as GIFTS, GeoTrace, and Triana will provide unique insights into the temporal behaviour of atmospheric composition and dynamics at small to medium scales. Rapid changes in tropospheric constituents will be observed, providing a means of separating the tropospheric burden from the relatively slowly changing stratospheric burden.

5.5.4 Summary

Global remote sensing of atmospheric constituents is essential to understanding the natural processes that determine the composition of the global atmosphere and to assess the impact of human activities on the atmosphere. Following successes in the development of satellite instruments to study the stratosphere, the next challenge is to improve capabilities for tropospheric measurements. Pathfinder instruments such as GOME, IMG, and LITE have demonstrated the potential for the new generation of instruments to be flown during the next decade. These new satellite sensors will provide major advances in our understanding of the composition of the troposphere on regional and global scales.

However, these new remote sensing capabilities challenge the scientific community if their full potential is to be realised. How will these new satellite measurements be validated and what modelling tools will be required to interpret them? New and improved ground-based measurement capabilities will be required for validation, as well as improved measurement strategies. Even though the new generation of satellite sensors will provide greatly improved observations of trace gases and aerosols, the data will necessarily fall short of full 3-D and temporally continuous coverage. One of the major challenges to the modelling community in the next decade is to develop data assimilation techniques that provide an objective means of estimating the space and time varying composition of the global atmosphere from sparse and irregular satellite observations.

5.6 Long-Term Measurements

5.6.1 Introduction

The last decade has seen many advances in measurement techniques for tropospheric chemistry. At the same time, many advances have been made in the understanding of the methodologies for making measurements. Since the tropospheric system is large and heterogeneous in both space and time over many orders of magnitude, a sound measurement strategy is essential if the small sample of possible measurements is to be meaningful in a global context. There are many aspects to this problem and only a small selection of these will be discussed in detail.

The success of any global network depends on careful calibrations and intercomparisons. There have been sustained, and largely successful, efforts to make comparisons of different instruments which measure the same atmospheric quantity, whether these are multiple copies of the same instrument or instruments using different measurement methodologies. The first category of comparison is an essential ingredient in any network whose data will have scientific meaning. However these comparisons only really address the issue of precision of measurement. It is the second category of comparison which can (but not necessarily will) determine instrumental biases. In some cases the comparisons can be made directly by sampling the same air volume, but in cases where co-location is not an option other techniques, such as the use of standards, are required. The control methodologies have also been strengthened to permit a comparison of techniques without introducing a bias from the comparison itself. Examples of where such strategies have been applied are the NOMHICE (hydrocarbons) and NOHALICE (nitrous oxide and halocarbons) under IGAC, the global network of CO_2 sensors (Francey 1999), and the intercomparison of techniques for measuring OH (Crosley 1997).

The pertinent issues for making long-term measurements to determine trends are also better understood after several decades of trying to make such measurements. Some very obvious logistical problems are the maintenance of the site for such an extended period and changes in the nature of the site, such as the encroachment of an urban environment into a previously rural area. Other issues that have been recognised, if not always understood, are the calibration and quality control of the time series and the necessity of some linkage between the observations that are being made. It has become apparent that a very significant, sustained effort is required to keep the data from a network to a standard that is usable in any comprehensive study of trends over many years. Examples of successful networks are the Global Atmospheric Watch (GAW) ozonesonde and CO_2 networks, and CMDL and AGAGE halocarbon, nitrous oxide, and methane networks.

With the advent of long-term measurements of tropospheric constituents from space, measurement validation is very significant. In this context, comparability of measurements has proved a difficult challenge. Significant aspects are problems of making "simultaneous" measurements at any point on the globe, and of properly accounting for temporal, vertical, and spatial coverage of both measurements. For example, it is very difficult to compare a satellite instrument with effective resolution of 4 km in the vertical, 25 km in the horizontal, and one second temporally with an aircraft sampling experiment of vertical resolution 500 m, spatial resolution of 1 km, and temporal frequency of one hour, even if the measurements are nominally co-located. Nevertheless if the satellite measurements are to become acceptable, these problems must be addressed in a manner that is agreeable to the community.

5.6.2 Global Networks

Issues such as global warming, depletion of the stratospheric ozone layer, acid rain, urban pollution, and atmospheric transport of hazardous material can be understood only by a combination of process studies with long-term monitoring. Within the United Nations system, the World Meteorological Organisation (WMO) through its GAW programme has a continuing responsibility to coordinate long-term ground-based measurement of atmospheric chemistry by maintaining its own operational observation networks (e.g., AGAGE, CMDL) and by fostering links with other active networks on a world-wide basis.

The GAW system is designed to coordinate research with two related objectives: to understand the relationship between changing atmospheric composition and changes of global and regional climate, and to describe the regional and long-range atmospheric transport and deposition of natural and man-made substances.

GAW includes groups with networks measuring ozone (total column, vertical profile, and near the surface), greenhouse gases (CO_2, CFCs, CH_4, N_2O), solar radiation including UV, aerosol characteristics, reactive gases (SO_2, NO_x, CO), chemical composition of rain, radionuclides, and meteorological parameters. To ensure the required quality of data a number of measurement manuals have been and are being prepared and a data quality control plan has been developed. IGAC and GAW have worked closely on a number of these issues.

To collect, process, analyse, and distribute data obtained from the GAW stations, six World Data Centres have been established: on ozone and UV (Toronto, Canada), greenhouse gases (Tokyo, Japan), precipitation chemistry (Albany, New York, US), solar radiation (St. Petersburg, Russia), aerosols (Ispra, Italy), and surface ozone (Kjeller, Norway). The GAW data are published regularly and are available upon request to all organisations, scientific institutions, and individual scientists directly from the Centres.

A most important aspect of the GAW has been the establishment of Quality Assurance/Science Activity Centres (QA/SAC) to oversee the quality of the data produced under GAW. Four such centres are located in Germany, Japan, Switzerland and the United States. The QA/SACs play a major role in training, quality control and establishing protocols for measurements. In coordination with the QA/SACs, a system of World Calibration Centres has been designated for specific measurements. Similar quality control activities are carried out through cooperation between the CMDL and AGAGE networks for chlorocarbons, methane, and nitrous oxide.

Under the GAW, CMDL and AGAGE programmes, close contact is maintained with the scientific community so that when new advances are made in laboratory and field measurements they can eventually be used in the long-term measurement strategy.

5.6.2.1 *Passive Samplers*

A new robust technique for atmospheric chemistry measurements has been developed over the last five years, i.e. passive samplers. These are ion exchange columns that can be exposed in the field from periods of a day to up to a month. After exposure, the sampler is sent to a central laboratory for analysis. Samplers can be configured to measure SO_2, NO_2, VOCs, O_3, and other substances, and they have been tested favourably against continuous measurement techniques. They can be used in both rural and urban environments. The major advantages of the passive sampler systems are that they are inexpensive, robust, and easy to implement in developing countries. Networks have already been established in Asia, Africa, and South America and the first measurements are now being evaluated. The programme is coordinated through the GAW (Carmichael 1998).

5.6.2.2 *Atmospheric Chemistry Measurements from Commercial Aircraft*

Whereas satellites enable global coverage, and research aircraft provide maximum detail about atmospheric composition, commercial aircraft carrying automated analysers and air samplers deliver a little of each. After the first use of commercial airliners in the US (GASP) (Nastrom 1979), three mutually quite different schemes evolved elsewhere and are in fact still operational. Using a Japan Airlines (JAL) Boeing 747, Matsueda and Innoue (1996) measure CO_2, CH_4, and CO monthly, chiefly on routes between Tokyo and Australia. This JAL project is practical and going strong; clean air compressed by the jet engine bypass is collected in titanium flasks for retrospective analysis.

MOZAIC (Marenco et al. 1998) has a more global character; here five aircraft are equipped with water vapour sensors and UV absorption ozone analysers. Well over 10 000 long distance flights have been completed. Funded by the European Commission's environmental programme, MOZAIC has been extended to equip the five aircraft with CO analysers. One Lufthansa MOZAIC aircraft is also equipped with an NO_y analyser.

CARIBIC (Brenninkmeijer et al. 1999) is different again; here one aircraft (Boeing 767, LTU International Airlines) is fitted with an inlet compatible with aerosol and air sampling. A measurement container (1100 kg) is carried in the forward cargo bay (directly above this inlet) on flights from Germany to tropical destinations once to twice a month. Thus, at a lower frequency than in MOZAIC, a rather wide range of data is obtained. CARIBIC records *in situ*: ozone (chemi-luminescence and UV absorption), CO, and aerosols (>5, >12 and >18 nm); it also collects grab samples for trace gas analysis (halocarbons, SF_6, N_2O, CH_4, various hydrocarbons, CO, and CO, CO_2, and CH_4 isotope analysis) and aerosol samples for PIXE analysis. Funded by the European Commission's Environmental 5th Framework programme, NO, NO_y, H_2O, precise CO_2, and aerosol optical properties are targeted as well. CARIBIC is expected to continue and expand (mass spectrometers) for another ten years or more. A typical CARIBIC flight overview is illustrated in Fig. 5.6.

From the success of the above projects that use automated equipment in passenger aircraft it is evident that commercial aircraft can play an important role as a platform in atmospheric chemistry measurements by filling the observational gap between satellite-based, global scale remote sensing and research aircraft intensive operations. Further discussion of measurements from commercial aircraft can be found in Chap. 3.

5.6.2.3 *Monitoring of Trends*

As we gain a better understanding of the processes changing the chemical composition of the atmosphere at one point in time through process studies and intensive campaigns, we must turn our attention next to the changes over longer time periods. Determining long-term trends adds another dimension to the problems associated with measurement of the atmosphere's chemical composition. With process studies the temporal changes may be large with respect to the precision

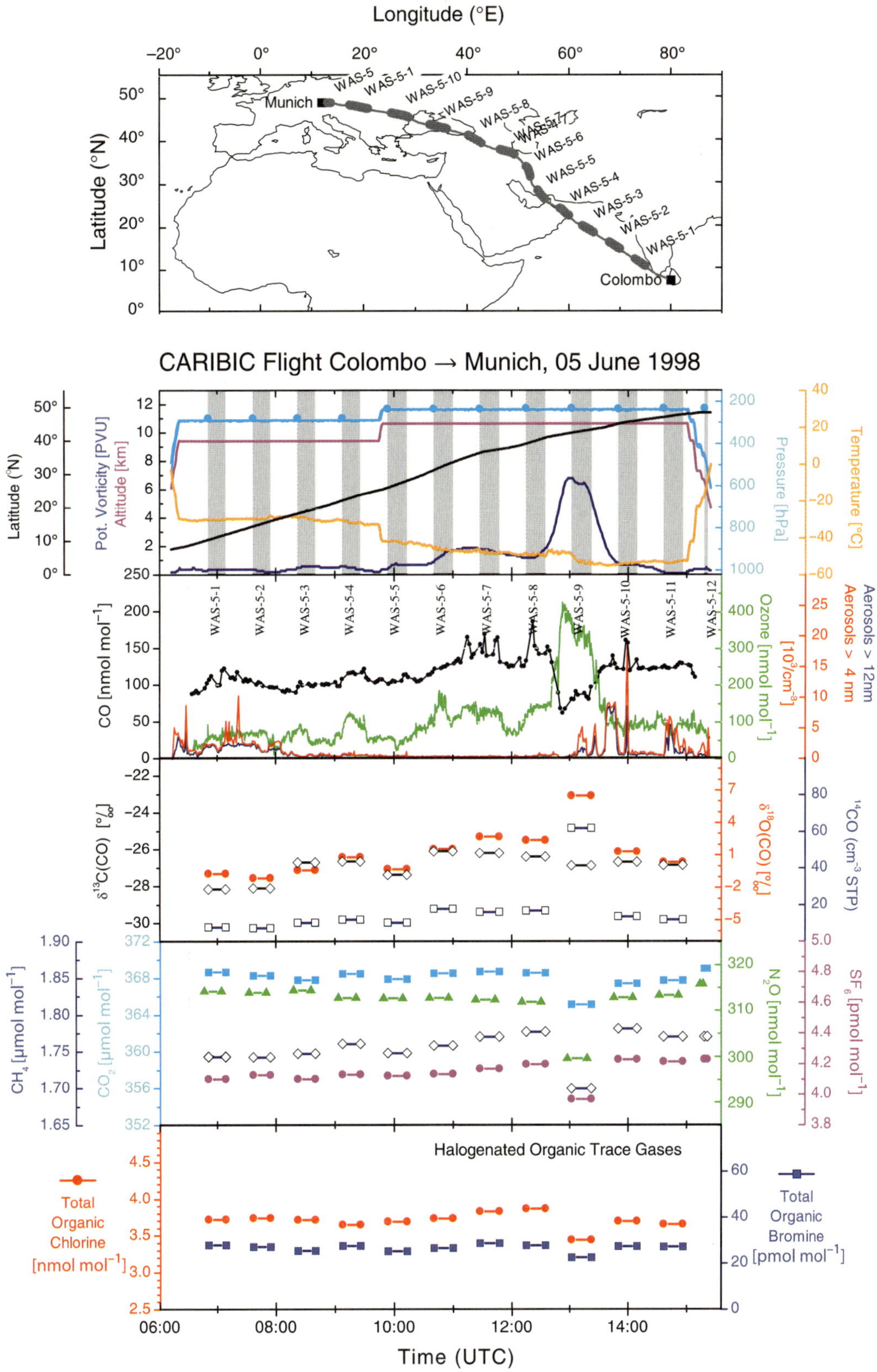

Fig. 5.6. Overview of main results of a CARIBIC flight, June 5, 1998, from Colombo, Sri Lanka, to Munich. The *shaded bars* indicate the collection of the 250 l air samples. For more details about this flight see Zahn et al. (2000) (source: A. Zahn, personal communication 2000)

of the measuring system but with long-term trends the changes often approach the system's precision. It is even more challenging to determine spatial changes. Simpler more robust instrumentation capable of unattended operation needs to be developed so that existing networks, such as the GAW, CMDL, and AGAGE stations, can be augmented for the compounds of interest. With miniaturisation of existing technology, spatial and temporal coverage of the upper troposphere and lower stratosphere could be improved using new instrumentation packages for commercial aircraft deployment on a routine basis. Vertical profiles could be obtained by exploiting existing atmospheric sounding programmes run by national meteorological services. Inherent and critical in any long-term programme are problems of maintaining instrument calibration at each site and inter-calibration between sites. Central or regional calibration facilities may be required, such as GlobalHubs proposed for the global greenhouse gases and associated isotopes by WMO and IAEA (International Atomic Energy Agency).

5.6.2.4 *Gas Chromatographic Measurements*

Gas chromatography is one technique that has been used successfully for delineating long-term trends of many atmospheric trace gases. Networks that have provided invaluable information on the oxidising capability of the troposphere are those of NOAA-CMDL (Montzka et al. 2000), AGAGE (Prinn et al. 2000), and, in the early 1990s, the Eurotrac-TOR project (Kley et al. 1997). Although packed-column chromatography has undergone relatively little scientific refinement in the past 30 years, dramatic technological improvements have resulted in reduced size and weight of these instruments, decreased analysis times, and significantly enhanced reliability and repeatability. These improvements are reflected in several custom-built multi-channel and multi-detector gas chromatographs that operate automatically at remote sampling locations, and onboard aircraft, ships, and balloons (Elkins et al. 1996; Ray et al. 1999; Prinn et al. 2000). At remote stations such custom instruments allow for an air sample to be analysed simultaneously on as many as four different analytical columns and detectors. Each separate "channel" is designed to provide optimum sensitivity and selectivity for one or more trace gases of interest. Close control on temperatures, flows, and pressures allows measurement precision of much less than 1% for gases such as N_2O, CH_4, and CFCs. Chromatographic instruments for operation onboard aircraft or high-altitude balloons have been designed to allow for much higher frequency sampling (Ray et al. 1999; Williams et al. 2000). Rapid sampling and analysis allows one to capture small-scale atmospheric variability in trace gas concentrations. Chromatographic instruments onboard balloons and aircraft have successfully measured CFCs and N_2O with sample frequencies at close to once per minute.

One of the most dramatic improvements in chromatographic techniques over the past 30 years has been in the development of capillary columns to replace packed-columns. Very narrow capillary columns (often called "micro-bore" capillary columns) have the potential to shorten trace analysis times even further and are now being used in the area of atmospheric trace gas analysis. Capillary chromatography techniques are especially powerful when coupled with a mass spectrometer detector (Prinn et al. 2000). Substantial improvements have also been realised in the analysis of air for peroxyacetyl nitrate (PAN) with use of capillary columns and the electron capture detector (Roberts et al. 1989). Capillary columns offer much higher resolving power than packed columns, but generally require pre-concentration of samples before introduction into the instrument. Despite this complication, automated capillary chromatography instruments similar to those described above have been deployed recently at a few remote stations throughout the globe to monitor atmospheric trends in long lived trace gases (Prinn et al. 2000).

5.6.2.5 *Aerosol Optical Properties*

Measurements at ground-based stations can be used to develop instrumental methods, to study temporal variations in aerosol properties over all time scales, and to serve as baseline stations for intensive campaigns. There are presently about a dozen such stations (e.g., Cape Grim, Tasmania baseline station) that conduct routine measurements of aerosol optical properties intended for studying direct climate forcing. Ideally, the number and location of these stations would span the full range of regional aerosol types and the sampling and measurement protocols would be identical at all stations such that their results could be strictly compared. It is important, but difficult, to assess whether *in situ* measurements from fixed surface stations are representative of vertical and regional distributions. One promising approach to this problem involves coupling to co-located lidar measurements.

A large array of ground based autonomous radiometers AErosol RObotic NETwork (AERONET) has been put in place during the last decade with the primary purpose to evaluate and complement satellite information (Holben et al. 1998). These instruments (100 of them worldwide) measure the aerosol spectral optical thickness (0.38–1.02 μm) and the sky spectral radiance. Columnar averages of particle size distribution, estimated refractive index, and single scattering albedo are being derived from the sky data. This information is used to develop an aerosol dynamic climatology, which can be used in models and inversion schemes for the satellite data (Remer and Kaufman 1998; Remer et al. 1998). The information is also used to derive a climatology of the diurnal cycle of the aerosol. The accuracy of the retrievals has recently been evaluated in detail (Dubovik et al. 2000).

5.6.2.6 *Remote Sensing by Lidar*

Lidar has been used to make remote observations of aerosol, water vapour, and ozone vertical profiles for many years, but in the last decade there has been an increasingly routine usage of lidar together with *in situ* and passive remote sensing instruments. The primary advantage of lidar is the ability to provide profile information, with vertical resolutions as small as 10 m. The basic lidar measurement consists of a profile of backscattered signal intensity as a function of range. The derived parameter from simple elastic backscatter lidars is either the aerosol back-scattering cross-section or the ratio of the aerosol back-scattering cross-section to that of the molecular atmosphere. More sophisticated lidars, sensing the Raman-shifted return from a variety of atmospheric compounds or multi-wavelength differential absorption lidars (DIAL, Sect. 5.3.3), are used to profile trace gases, most often water vapour or ozone. Raman lidars are able to measure the vertical profile of aerosol extinction directly (Althausen et al. 2000; Ansmann et al. 1990), and have been used with increasing success in the lower atmosphere (Ansmann et al. 2000).

The development of integrated networks of lidar is just beginning. Compact, autonomous lidars have been added to some of the AERONET sites to provide aerosol profile information in addition to the column properties provided by the AERONET radiometers. An important step for the acquisition of three-dimensional aerosol data was the installation of the European Lidar Network (EARLINET) starting in 2000. Continued progress in the development of robust, automated and less costly lidars will be key to their widespread use. Transition of the more sophisticated systems and methods from occasional to routine observations would be a great asset. More widespread placement of lidars at sun photometer sites would be useful.

5.6.3 Summary and Future Trends

Carefully designed measurement strategies are necessary, given the complex and variable nature of the atmosphere. Direct measurements must be effectively coupled to models and to satellite observations. *In situ* measurements can never sample a significant fraction of the atmosphere, but they offer detailed and, above all, calibrated measurements of properties that cannot be measured by other means. Over the last decade, the atmospheric research community has made substantial efforts to develop and standardise measurement techniques and to better quantify uncertainties (e.g., through closure experiments). Uncertainties can be quantified with minimal assumptions, and often rigorously. One of the major advances of the last decade has been the recognition of the value of carefully integrating modelling studies with *in situ* measurements. In particular, the use of models in planning a field campaign and the availability of regional models running in near-real time have proven to be invaluable (see also Sect. 6.5.2).

In situ data have been acquired from long-term monitoring activities, geographical transects, and intensive field campaigns, with a strong emphasis on the last. The resulting large data sets are rapidly reaching the point where they can be used, in conjunction with models, to characterise important features of the atmosphere on a global scale. Weaknesses do exist, however, that undermine the ability of *in situ* measurements to constrain atmospheric properties for models and satellite retrievals. At the level of experimental design, there is need for more rigorous uncertainty analysis and for more active testing of hypotheses.

Looking ahead to the next decade, a major challenge will be to devise optimal ways to couple *in situ* measurements to satellite observations. A shift is clearly underway (see Sect. 5.4) whereby, for example, satellites will map out aerosol spatial and temporal variations while targeted *in situ* measurements will establish the limits and uncertainties associated with satellite data interpretation. An important step in this direction will be to assess the scales of variability for various gases and aerosols, and future *in situ* measurements should make these tests more definitive.

5.7 Summary and Conclusions

It is interesting to speculate about the future of the measurement techniques. A view of the future can be approached from two directions:

What can we expect through the evolution of current systems and techniques?

Generally miniaturisation is a foreseeable trend, in particular in the electronic components of the instruments, but we will also see applications of micro-mechanical devices, which are presently developing rapidly. Further trends include:

- Wider application of mass spectrometry (e.g., proton transfer, chemical ionisation, time of flight);
- Improvements in optical spectrometry (e.g., DOAS, TDLAS, lidar);
- Miniaturised, automated gas chromatographs;

- Continued development of satellite instrumentation, including active techniques for water vapour and ozone; and
- Application of tomographic techniques to determine the spatial distribution of trace gases.

What do we need?

Future research will require, for instance, more compact, more universal instruments, which are more readily calibrated. Also, the measurement techniques for many free radicals are still not satisfactory (e.g., for RO_2 radicals) or too difficult to use for routine measurements. In addition, modern chemistry-transport models cannot be tested because there are simply no techniques to observe the two- and three-dimensional distributions of trace gases on regional or even global scales. Thus a short list of requirements would include:

- Techniques for high-frequency hydrocarbon (VOC) measurements;
- New flux techniques for gases and aerosols;
- Rapid, reliable measurements of water vapour, NH_3, and NO_2;
- International OH / HO_x instrument intercomparison;
- Instrumentation for unstable N reservoirs (N_2O_5, HNO_4 ...);
- Standardised calibration methods or facilities;
- Development of single-particle instruments with molecular speciation;
- Measurement techniques for reactive halogen compounds (X, XO, OXO, HOX, where X = Cl, Br, I);
- Additional monitoring stations in the Tropics;
- Sondes and lidars to measure vertical profiles of a wider variety of trace gases than is currently possible (particularly reactive gases); and
- Remote sensing (satellite-based) instruments which allow the global observation of three-dimensional trace gas distributions, albeit with limited vertical resolution in the troposphere.

Looking to the future, it is clear that although satellite measurements will greatly supplement our knowledge of tropospheric composition on a global scale and provide a sound basis for environmental policy decisions, satellites are not yet able by themselves to produce a complete picture of the state of the troposphere. There are too many significant compounds that evade quantification from space and too many processes occur on scales that are too short for evaluation from a satellite. There must therefore be a continued emphasis on measurements using platforms and combinations of platforms within the atmospheric system and on the use of modelling to combine these diverse measurements with satellite measurements. These will be needed on time scales from hours to decades. However, if these data are to be useful in a global context, then a global calibration system is required to ensure that the data are indeed comparable. This is a complex exercise involving a combination of funding, determination, and diplomacy to achieve. It is one that will have to be addressed by the community, international organisations such as IGAC, and the various funding agencies.

Chapter 6

Modelling

Lead authors: Claire Granier · Maria Kanakidou · Prasad Kasibhatla
Co-authors: Guy P. Brasseur · Cathy Clerbaux · Frank J. Dentener · Johann Feichter · Sander Houweling · Boris Khattatov Jean-François Lamarque · Mark Lawrence · Sasha Madronich · Natalie Mahowald · Kevin Noone Geoffrey S. Tyndall · Stacy Walters · Chien Wang
Contributors: Carmen Benkovitz · Laura Gallardo · Ivar Isaksen · Kathy Law · Joyce Penner · Dork Sahagian · Will Steffen

6.1 Introduction

The chemistry of the atmosphere is controlled by a large number of complex chemical and physical processes. The study of such a complex system requires the use of numerical models, which have improved substantially over the past ten years. These models are mathematical representations of the main physical and chemical processes controlling the spatial and temporal distributions of trace gases and aerosol particles. The models have been developed to test our understanding of the atmospheric processes, to identify key variables and important interactions, and to interpret local, regional, and global observations. Additionally, they can be used to simulate global distributions of tropospheric compounds, predict the evolution of the chemical state of the atmosphere and of the radiative forcing of climate in response to natural or anthropogenic perturbations, provide an optimal use of satellite data through data assimilation, and help policymakers define emission reduction policies.

Over the last decade, efforts to improve our understanding of atmospheric processes have stimulated the development of a hierarchy of models. Box models and parcel trajectory models continue to be excellent tools to illuminate our understanding of chemical processes. They can now take into account several hundred chemical species linked by a few thousand chemical reactions. Because many tropospheric chemical species are short lived and exhibit large spatial and temporal heterogeneities, the use of regional and global three-dimensional chemistry-transport models (CTMs) for tropospheric chemistry studies has become more common. Considerable progress has been made over the last ten years on the development and application of these sophisticated, global three-dimensional models. Some salient features of several of these models are listed in Table 6.1. These global tropospheric CTMs typically have horizontal resolutions of 150–500 km, and include a reasonably complex representation of chemistry, taking into account in many instances non-methane hydrocarbon chemical schemes or multi-phase and heterogeneous chemistry at varying levels of complexity.

Key advances in the representation of physical processes, numerical techniques, and computer power have benefited tropospheric CTMs over the past decade. These improvements have enabled refined temporal and spatial resolution of the models and more accurate parameterisations of subgrid processes, allowing use of the models for many applications, as seen in Chap. 3 and 4. In this chapter the different types of models used in current studies are described briefly in Sect. 6.2 and model developments over the last decade are discussed in Sect. 6.3. Model evaluations and intercomparisons with regard to specific scientific problems are summarised in Sect. 6.4. A few applications of the models are illustrated in Sect. 6.5 and new model developments, such as coupled Earth system models, data assimilation, inverse modelling, and dynamic aerosol modelling, are discussed in Sect. 6.6.

6.2 Types of Models

Major advances in understanding atmospheric chemistry have resulted from the detailed analysis of individual processes. Zero-dimensional or box models, which consider a box in which the air mass is well mixed, are well suited to this purpose. Box models are used to analyse observations of selected chemical species, and to study tropospheric chemistry under specific conditions. These models are also used to evaluate and parameterise subgrid processes such as the fast chemistry occurring near source areas or heterogeneous processes with time scales shorter than the characteristic transport times in large-scale models.

Eulerian models calculate the distribution of chemical species for fixed spatial grids, while in Lagrangian and trajectory models, the grid box is advected following the air motion and chemistry is calculated along air parcel trajectories. In this type of models, source and sink terms vary as the air parcel moves into different spatial regions. Lagrangian and trajectory models, which allow a detailed consideration of chemistry, are often used for local or regional scale studies. Both types of models have the intrinsic difficulty that they do not correctly include mixing processes with the surrounding

Table 6.1. Salient features of several recent global chemical transport models for oxidant chemistry

Model	Chemistry	Contact person e-mail, web site	References
MOGUNTIA	O_3-CH_4-CO-NO_x-HO_x-NMHC	air@mpch.mainz.mpg.de, mariak@chemistry.uch.gr	Crutzen and Zimmermann 1991; Kanakidou et al. 2000; Poisson et al. 2000
IMPACT	O_3-CH_4-CO-NMHC-NO_x-HO_x	rotman1@llnl.gov	Penner et al. 1998; Rotman et al. 2000
GRANTOUR	NO_x/HNO_3		Penner et al. 1991
GFDL/GCTM	Prescribed NO_x; Parameterized O_3 chemistry		Levy et al. 1997
ECHAM	Online; O_3-CH_4-CO-NO_x-HO_x		Roelofs et al. 1995; 1997
IMAGES	O_3-CH_4-CO-NMHC-NOx-HO_x	muller@oma.be	Muller and Brasseur 1995; Granier et al. 1999
U. Oslo CTM	O_3-CH_4-CO-NMHC-NO_x-HO_x		Berntsen et al. 1997a
TM3	O_3-CH_4-CO-NMHC-NO_x-HO_x		Houweling et al. 1998
GEOS-CHEM	O_3-CH_4-CO-NMHC-NO_x-HO_x	djj@io.harvard.edu http://www.as.harvard.edu/chemistry/trop/geos	Wang et al. 1998a,b,c; Wang and Jacob 1998
MOZART	O_3-CH_4-CO-NMHC-NO_x-HO_x	brasseur@dkrz.de http://www.acd.ucar.edu	Brasseur et al. 1998a; Hauglustaine et al. 1998
TOMCAT	O_3-CH_4-CO-NMHC-NO_x-HO_x	kathy.law@atm.ch.cam.ac.uk	Law et al. 1998; Stockwell et al. 1999; Giannakopoulos et al. 1999
MATCH-MPIC	O_3-CH_4-CO-NO_x-HO_x	http://www.mpch-mainz.mpg.de/~lawrence/MATCH/match_overview.html	Crutzen et al. 1999; Lawrence et al. 1999

air, which makes the validity of these models over periods longer than approximately five days questionable.

One- and two-dimensional models, which do not require large computer power, have been used intensively in the past, as they provide rapid calculations, even for detailed chemical schemes. Over the past decade, these 1-D and 2-D models have been replaced for studies of the troposphere, which presents a very high temporal and spatial variability, by global three-dimensional models. These 3-D models provide more realistic representations of the physical and chemical processes in the troposphere and, presumably, of the distributions of chemical compounds in the troposphere.

The development of global atmospheric models over the past decade was made possible by the remarkable advances in hardware (computational capacity) and software (algorithms) development. There have been algorithm improvements ranging from highly automated solvers using sparse matrix techniques such as SMVGear I and II (Jacobson and Turco 1994; Jacobson 1995). New versions of scientific languages such as Fortran 90 and 95 and modern utilities such as source control software have improved the development process and, ultimately, model reliability. But it is the improvement in raw computer power from the gigaflop (flop = floating-point operation per second) range a decade ago toward the teraflop mark today, together with large increases in storage capacities, that has enabled greatly improved simulations. Another development is the increasing use of multiple processor machines. The CTMs have started to be parallelised, leading to faster throughput for simulations. Not only can higher resolution studies on a global basis be performed regularly but many complex elements such as clouds and sub-grid processes can be simulated with enhanced accuracy at the global scale. This is due to improvements in the understanding of the processes, as described in the previous chapters, and to improvements in the parameterisations of specific processes, as described in the following sections. Today, with computer systems capable of sustaining hundreds of gigaflops, global simulations with regional spatial resolutions are practical. And multi-year runs of models coupling realistic oceans, sea ice, and biogeochemistry with a chemically active atmosphere are on the horizon. This will allow more accurate quantifications of the strong feedbacks between the different components of the Earth system.

6.3 Model Components

Over the past decades, several numerical techniques have been proposed to solve the complex CTM system that includes transport (large-scale advection, vertical mixing, and convection), chemical transformations (gas phase reactions, photolysis), aerosol processes, and deposition (wet removal and dry deposition) processes. The time evolution of the concentration of a chemical species is represented through its continuity equation. In most cases, the system of continuity equations for all chemical species does not yield any analytical solution, and sophisticated numerical methods have to be used in order to obtain accurate solutions. These are described and discussed for example in Brasseur et al. (1999a) and Jacobson (1999). In contemporary models, the set of equations is solved by using an "operator split" formulation in which the different component processes (large-scale transport, subgrid processes, chemical processes, etc.) are isolated and treated sequentially for each time step. Recent advances in the operator splitting technique are discussed in Blom and Verwer (2000). The treatment of these components and different parameterisations commonly used in models are detailed in the following sections.

6.3.1 Treatment of Large-Scale Transport

6.3.1.1 *Advection Schemes*

The simulation of mass transport in CTMs separately accounts for resolved or large-scale advection and subgrid-scale exchanges. These latter processes usually include boundary layer mixing as well as shallow and deep convection. In the real atmosphere, of course, this separation of scales does not exist and the different processes occur simultaneously.

Advection by winds that are resolved by the model spatial grid is theoretically a straightforward process, but deciding how to implement this process numerically in a transport model requires compromise. The qualities desirable in an advection scheme include: computational efficiency, accuracy, local movement of tracers in the downwind direction, monotonic solutions (i.e. no new extrema introduced without divergence in the flow), and mass conservation. So far numerical solutions to the advective transport have had to compromise on one or more of these traits. For example, semi-Lagrangian transport schemes are not conservative by nature, so a "mass fixer" must be introduced, which causes the transport to be slightly non-local (Rasch and Williamson 1990). Second order advection schemes (see details in Jacobson 1999) require a slightly higher memory and computational cost to obtain accurate mass conserving solutions (Russell and Lerner 1981). New formulations of mass flux schemes (Lin and Rood 1996; Rasch and Lawrence 1998) appear to have improved stability and accuracy while being computationally efficient and mass conserving. They have been implemented in various forms in CTMs and general circulation models (GCMs). Prather (1986) recommended a very accurate algorithm for coarse resolution grids. In this formulation, the tracer concentration inside a grid box is represented by second-order polynomials and its spatial distribution by second-order moments. This method, characterised by small numerical diffusion, requires a large memory and computation time. Reviews of the different advection schemes used in CTMs are given, for example, in Rood (1987), Hov et al. (1989), Chock (1991), and Brasseur et al. (1999a).

6.3.1.2 *Subgrid Processes: Convection and Boundary Layer Mixing*

Mixing processes, especially vertical mixing, often occur at scales smaller than those resolved by chemical transport or GCMs. These models generally distinguish between subgrid scale mixing within the planetary boundary layer (referred to as boundary-layer mixing) and subgrid scale mixing between the planetary boundary layer and the free troposphere (referred to as convection). Convective mixing processes have the capability of moving a boundary layer parcel to the tropopause during an intense event, such as a thunderstorm, and the inclusion of this process in models appears to improve the simulated vertical transport significantly (Mahowald et al. 1997a). Currently, there appears to be no optimal way to include cumulus convection in either chemical transport or GCMs. Moist convective mixing parameterisations vary from simple increased diffusion in areas of moist convective instability (Levy et al. 1985) to entraining-plume models (Tiedtke 1989; Feichter and Crutzen 1990; Emanuel 1991) as well as more complicated schemes using the available convective potential energy (Pan and Wu 1995).

The most detailed direct comparison of cumulus convection schemes to date has been performed by Mahowald et al. (1995) who considered seven different moist convection schemes that varied in terms of the criteria used to calculate cloud mass, and also in the details of cloud entrainment and detrainment. The comparisons were carried out in a one-dimensional framework for a variety of convection regimes occurring in the troposphere. While it is difficult to draw definitive conclusions owing to the one-dimensional aspect of the simulations and the lack of observational data to evaluate the schemes, as discussed in Sect. 6.4, this study highlights the variability between the various schemes tested in redistributing tracers vertically (see Fig. 6.1)

Unfortunately, observations that can be used to evaluate unambiguously the parameterisations of cumulus convection and boundary layer mixing in large-scale models are sparse. A recent comparison of ^{222}Rn simulations from a variety of large-scale models concluded that the representation of these processes in models is generally consistent with the limited set of available measurements (Jacob et al. 1997). More extensive observations of tracers whose sources and sinks are well established are needed to further constrain the treatment of tracer transport by moist convection and turbulent mixing in large-scale CTMs (see also Sect. 6.4).

Dry convective mixing occurs when the boundary layer or other layers are statically unstable, as described in Stull (1988); it is often parameterised using a simple diffusive approach based on local instability such as that of Louis (1979). This approach may simulate somewhat

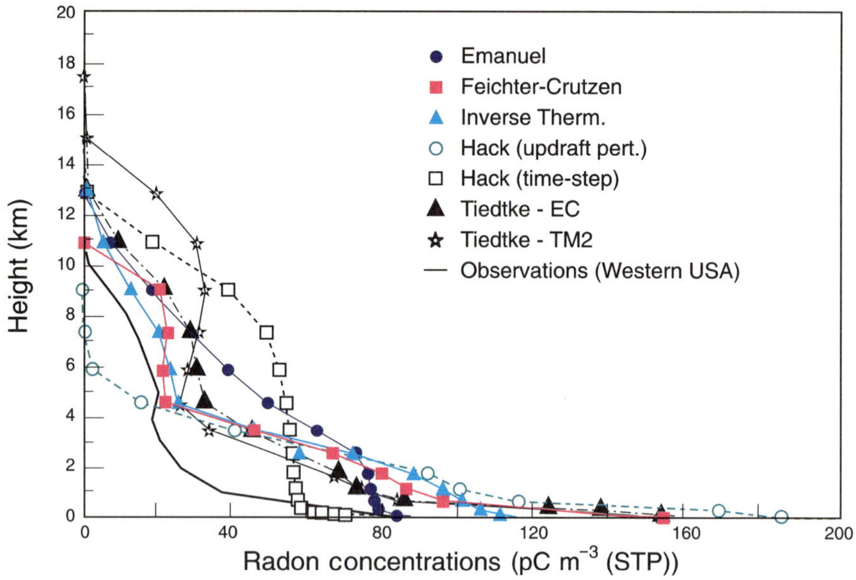

Fig. 6.1.
Simulated average equilibrium vertical profile for a tracer with a one month lifetime (adapted from Mahowald et al. 1995)

too little mixing in the boundary layer under highly unstable (daytime) regimes, which led to the development of non-local schemes such as that of Holtslag and Boville (1993). Comparisons of models such as those presented by Jacob et al. (1997) or Rasch et al. (2000) suggest that the local Louis schemes may produce too little mixing but the non-local schemes may mix slightly too much. Since many chemical species have surface sources, and many observations occur at the surface, improvements of parameterisations for the boundary layer have a high priority. Unfortunately, observations that can be used to evaluate these parameterisations unambiguously are scarce as well.

Most contemporary CTMs are offline models, in which the winds and temperature are input to the model, and have been derived from another model, either a general circulation climate model or a forecast centre model with assimilated meteorological observations. Recently, CTMs have added the ability to treat moisture explicitly in order to capture better the moist convective mixing and precipitation events that control especially soluble chemical species in the atmosphere (e.g., Rasch et al. 1997). In this case, the humidity is predicted (using surface flux and radiative flux fields from the driving model) including the effects of moist convection and stratiform precipitation. These offline models are thus capable of simulating specific events using forecast centre meteorological products such as the NCEP, ECMWF, or DAO winds and accounting for the hydrological cycle (Lawrence et al. 1999a, 2000; Giannakopoulos et al. 1999). Another approach, which allows the simulation of specific events, is to use an online model that is "nudged" to the observed meteorology from forecast centre analyses. The methodology involves adding non-physical relaxation terms to the model equations to force the GCM to simulate specific episodes of interest. Key issues that must be considered carefully in using this technique include the choice of nudging variables and relaxation coefficients, as discussed in Jeuken et al. (1996).

6.3.1.3 *Upper Boundary Conditions*

An important process that must be represented in studies of tropospheric oxidant chemistry is the downward transport of ozone (and to a lesser extent of NO_x and HNO_3) from the stratosphere. To do this accurately, one must realistically simulate stratospheric chemistry, downward transport in the troposphere, and the space-time distribution of stratosphere-troposphere exchange events. The models currently used for global studies of tropospheric chemistry do not include an explicit and rigorous treatment of stratospheric chemistry. In the recent IPCC (1999) report on the impact of aircraft on the atmosphere, 2-D models were used to simulate stratospheric processes, 3-D models were used for the troposphere, and the results were combined to obtain a global picture. This was recognised as a major weakness in the analysis of model results. Models simulating both tropospheric and stratospheric chemistry are currently under development by several groups.

In most current models the concentrations of O_3 and related chemical species are prescribed at some altitude in the stratosphere, usually at or below 10 hPa. An exception is the Harvard model (Table 6.1) which uses an alternative parameterisation of prescribing flux boundary conditions at the tropopause, i.e. cross-tropopause mass fluxes are prescribed in the model based on climatological observations. An additional consideration is that the vertical (and in some cases horizontal) model resolution is often rather coarse (typically 1 km in the vertical and a few degrees in both latitude and longitude) and calculation of the relevant transport processes in the middle atmosphere is missing in most models. This introduces significant uncertainties in model-calculated stratospheric dynamics as well as in the dynamics of stratosphere-troposphere exchanges. For example, using tracer diagnostics from online simulation of tracers, Rind et al. (1999) find that including gravity wave drag in the lower stratosphere and significantly raising the model's top boundary were necessary to simulate stratosphere-troposphere exchanges of tracers in the GISS GCM adequately.

There are indications that the representation of vertical transport by the algorithms discussed in Sect. 6.3.1.1, especially in atmospheric layers with high vertical gradients (e.g., near the tropopause), may be inaccurate, but results can be improved by increasing the vertical resolution of the model. In particular, this limits the ability of global models to estimate accurately the amount of ozone transported from the stratosphere to the troposphere and mid- and highlatitudes.

For example, in an evaluation of five global CTMs using MOZAIC data (see Chap. 3 and 5), good agreement was achieved between calculated and observed distributions in tropopause regions, where our understanding of meteorological processes is good (e.g., over Europe). Agreement was less good in sub-tropical and tropical regions where our knowledge is more limited (e.g., Law et al. 2000). The global tropospheric budget of ozone, and to some extent, the cross-tropopause flux of O_3 are constrained by the fact that over the course of a year, this flux is roughly equal to the difference between net tropospheric ozone chemical production and ozone dry deposition at the surface. However, there is a considerable range in the "simulated" annual-average cross-tropopause O_3 flux in various models. Furthermore, the models differ in terms of the simulated spatial and temporal variability of the cross-tropopause O_3 flux leading to modelled differences in the stratospheric contribution to tropospheric O_3 at particular locations and time periods.

6.3.2 Representation of Chemical Processes

6.3.2.1 *Gas Phase Processes*

The representation of chemical processes in tropospheric models is strongly limited by computing resources. In the global background marine troposphere, it seems reasonable to consider a simplified chemistry scheme based on O_x / NO_x / CH_4 and CO photochemical reactions. However, natural emissions of organic compounds from oceans (mainly alkenes and dimethylsulphide (DMS)) might affect the marine boundary layer chemistry significantly. Halogens may affect the ozone budget in the marine atmosphere too, as discussed by Dickerson et al. (1999). Interactions between N and C and/or S cycles in the polluted marine troposphere may require a more detailed photochemistry.

Over continental areas, both under clean and polluted conditions, chemistry is much more complicated. Consideration of the oxidation chemistry of non-methane hydrocarbons in the troposphere has improved the agreement between model results and observations for chemical species such as carbon monoxide and ozone (Houweling et al. 1998; Wang et al. 1998c; Poisson et al. 2000; Granier et al. 2000). A very large number of chemical reactions and chemical species are involved in the oxidation chains of hydrocarbons. Therefore, a number of techniques were developed in the past years to provide simplified chemistry schemes. Reduction of the chemical reaction schemes introduces inaccuracies in the computation of chemical species, which are evaluated by comparison with: (*1*) field and smog chamber observations; (*2*) other reduced schemes; and (*3*) detailed "master" chemical mechanisms (the best known of the latter are those of NCAR (Madronich and Calvert 1989) and the University of Leeds (Jenkin et al. 1997), which contain several thousand reactions that may be relevant in the troposphere). Unfortunately, most reactions have not been studied because most laboratory data concern the rates and product yields of only the first-generation compounds of interest. Rates of other reactions are estimated, based on reactions involving analogous chemical species. As expected, the level of mechanism reduction is not the same for local and regional pollution models as for global models with resolutions higher than a few degrees in latitude or longitude. Among the best known reduction mechanisms are the Carbon Bond IV (Gery et al. 1988), the Lurmann (Lurmann et al. 1986), the Carter (Carter 1990; Carter and Atkinson 1997), the RADM-RACM (Stockwell et al. 1986), and the EMEP (Simpson et al. 1993) schemes. Although most are based on the same laboratory kinetic data, they differ in the techniques and assumptions used to represent and simplify the organic chemistry.

Two main reduction or "lumping" techniques exist. First is the structural approach, in which the molecular structures or functional groups within the organic molecules are the lumping category (Carbon Bond IV). In the second approach, the numerous emitted organic compounds are grouped and represented by simpler molecules, i.e. a limited number of chemical species is used, each of them representing a certain class of compounds (Lurmann, Carter, and EMEP). Reduced mechanisms can also condense information from both structural and molecular grouping, such as the RADM-RACM mechanisms. The principal requirement for this is that the behaviour of the individual compounds that are lumped must not depart too much from the average behaviour of the category. The disadvantage of the lumped mechanisms is that any change in the reactions or in the reaction rates requires the development and the revaluation of a new mechanism, based on the initial reduction procedure. As expected, the results of models using different chemical mechanisms are not the same, which stresses the importance of evaluating and accounting for uncertainties in the representation of the chemistry. Thus, in recent years, since NMHCs have been shown to be important for regulating oxidant levels, recommended kinetic data have been extended to the still-limited NMHC oxidation chemistry and intercomparison exercises have been performed on an international level (see Sect. 6.4.2).

Recent improvements of chemical schemes in the models include, for example, more detailed hydrocarbon chemistry, with particular attention to second generation products, the incorporation of HONO chemistry and its impact on OH, and the improvement of parameterisations of gas-particle interactions and aqueous phase reactions. New sources of HO_x, PAN, and ozone have been suggested for the upper troposphere as a result of analysis of data from aircraft campaigns (Singh et al. 1998; Jaeglé et al. 1999; Jacob and Prather 1997; Cohan et al. 1999). These mechanisms must also be included in the models, particularly those that emphasise anthropogenic perturbations in the upper troposphere, e.g., aircraft.

Careful evaluation and minimisation of uncertainties and errors in CTMs are expected to enable their application to the study of observed changes in tropospheric ozone. Current models simulate absolute ozone concentrations in the boundary layer within 20 to 50%, as discussed in Chap. 3. Calculated changes in ozone distributions as a result of perturbations, such as changes in surface emissions, may be more accurate. Several recent studies (mainly based on Monte Carlo simulations) focussed on the evaluation of the sensitivity of oxidant levels to various chemical reactions in order to identify crucial processes for error minimisation. Chemical reaction rates are also uncertain: for instance, in the early 1990s, determination of the reaction rates of CH_4 and CH_3CCl_3 with OH suggested that these reactions were about 20% slower than previously measured. Similarly, the reaction between OH and NO_2, which is an important sink for NO_x in the troposphere, has recently been measured at 10–30% lower than previous estimates (Donahue et al. 1997; Brown et al. 1999b; Dransfield et al. 1999). Uncertainties in the temperature dependence of reaction rates can also introduce a 20–40% uncertainty in concentrations of chemical species such as ozone in the troposphere (Stewart and Thompson 1995).

A particularly challenging computational problem arises due to the "stiffness" of the set of differential equations describing the chemical evolution of trace gases. The stiffness arises from the wide range of e-folding times of chemical species that are commonly of interest in tropospheric chemistry. While excellent solution techniques for such problems are available, the challenge in large-scale chemical transport modelling arises from the need for solving these equations over thousands of grid boxes for long time periods. Several innovative approaches have been developed to deal with this challenge. A commonly used practice is to group chemical species that interconvert rapidly into families (Crutzen 1971; Austin 1991; Brasseur and Solomon 1986). The time evolution of family concentration is calculated by solving a differential equation for each family. Concentrations of individual chemical species are then diagnosed using partitioning ratios derived from the inter-species conversion rates. Efficient and more direct approaches for solving the chemical differential equations are also increasingly being used in large-scale CTMs. Such approaches include implicit integration schemes (Stott and Harwood 1993; Carver and Stott 2000) and the multistep implicit-explicit (MIE) method, which combines backward Euler and forward Euler calculations in an iterative manner (Jacobson 1994). Another popular approach is to use the algorithm developed by Gear (Gear 1971), in combination with sparse-matrix and computer optimisation techniques (Jacobson 1994, 1995, 1998; Jacobson and Turco 1994). Jacobson (1999) describes these approaches further. Some packages including different integration methods have been made available to modellers, e.g., the ASAD package (Carver et al. 1997).

6.3.2.2 *Photolysis Rates*

An equally important source of uncertainties in chemical calculations is related to the computation of the rates of photodissociation reactions, which are the driving force of photochemistry in the atmosphere. These rates depend on the actinic flux of incoming solar radiation, the slant column of absorbing gases, solar zenith angle, surface albedo, the presence of clouds and aerosols (size distributions and optical properties), and, in some cases, temperature distribution. Discrepancies in the photolysis rates computed by various methods, as discussed below, result from the use of different parameterisations. Fundamental methods for radiative transfer calculations such as the matrix-operator method require too much computer time to be practical in global CTMs. Therefore, approximate methods have been developed for three-dimensional CTMs. The most commonly used are the two-stream methods, in which the diffuse radiative flux is represented by two components – one upward and one downward. There are a variety of such methods; some of them have been intercompared in the PhotoComp IPCC exercise (Olson et al. 1997), as described in Sect. 6.4. The basic difference among various two-stream methods is the choice of fundamental scattering parameters (Hough 1988). Recently, new approximations for online calculations of photolysis rates in sophisticated CTMs have been developed (Landgraf and Crutzen 1998; Berntsen and Isaksen 1997; Krol and Van

Weele 1997; Wild et al. 2000) to allow computationally efficient consideration of feedbacks between chemistry and photochemical rates in the models. Another approach to overcoming computing limitations is the use of look-up tables wherein photolysis rates are precalculated from accurate radiative transfer models over a range of a limited number of controlling parameters. The accuracy of photolysis rate calculations in both types of approaches is particularly limited by the number of parameters that can be taken into account and by corrections for the impact of clouds and aerosols, whose distributions and optical properties remain poorly documented. The sensitivity of photolysis rates to aerosol properties has been discussed in Dickerson et al. (1997) and He et al. (1999). Furthermore, uncertainties in parameters such as the $O_3 \longrightarrow O(^1D)$ quantum yield and HNO_3 absorption cross section, directly affect any computational approach of the photolysis frequencies, and thus the calculated tropospheric OH and the lifetimes of many tropospheric constituents.

6.3.2.3 *Aerosol and Cloud Chemistry*

Over recent years, growing attention has also been focussed on the importance of heterogeneous and multiphase reactions in the troposphere, and their impact on gaseous compounds such as ozone (Chap. 4). Based on limited laboratory studies and numerous assumptions, the impact on tropospheric chemistry of in-cloud multiphase processes (reactions that can occur in different phases) has been described (Chameides and Davis 1982; Jacob 1986, 2000; Lelieveld and Crutzen 1991). An example of such a multi-phase process is the oxidation of S(IV), which can occur in the gas phase via the $SO_2 + OH$ reaction, and also in the liquid phase via the $HSO_3^- + O_3$ and $HSO_3^- + H_2O_2$ reactions. The impact on tropospheric chemistry of multi-phase processes occuring in clouds has been pointed out by Chameides and Davis (1982), Jacob (1986, 2000) and Lelieveld and Crutzen (1991).

It has also been shown that scavenging of compounds such as N_2O_5 followed by reaction in clouds and/or on aerosols can be crucial for the oxidant levels in the troposphere (Dentener and Crutzen 1993). Other potentially important heterogeneous processes include: activation of chlorine and bromine in sea salt in the marine boundary layer (Vogt et al. 1996; Sander and Crutzen 1996); reactions of HNO_3, NO_2, and O_3 on soot (Lary et al. 1997); and reactions on mineral particles (Zhang et al. 1994b; Dentener et al. 1996; Xiao et al. 1997; Phadnis and Carmichael 2000). Heterogeneous reactions in the troposphere affect sulphur oxidation (Sievering et al. 1991; Chameides and Stelson 1992), production of halogen radicals, and HO_x and NO_x production and destruction with subsequent impact on ozone. Jacob (2000) provides a comprehensive review of heterogeneous chemistry and its influence on ozone.

The interactions between gas-phase chemical species and liquid cloud-aerosol particles involve gas-phase diffusion to and from the surface of the droplet, interfacial mass transfer of chemicals through the gas-liquid phase boundary, and aqueous phase diffusion and reactions in the droplet-aerosol (see reviews by Ravishankara (1997), Seinfeld and Pandis (1998), and Sander (1999)). The uptake of gases on particles is usually described by a mass accommodation coefficient (α), which gives the fraction of molecules colliding with a surface that remains in the particle, and by the reactive uptake coefficient (γ), which is the fraction of molecules that collide and react. Both coefficients depend on surface characteristics, exposure time, and reactivity of the reactant at the surface. Unfortunately, neither laboratory experiments nor models yet provide a comprehensive picture of all processes involved in the uptake of gases and, contrary to the situation for gas phase tropospheric chemistry, a reference scheme for heterogeneous chemistry applicable to realistic atmospheric aerosol conditions does not exist.

Reactions in liquid water clouds may also exert a strong influence on the oxidation of SO_2 and other photochemical processes. Relative to reactions in aerosols, there is more mechanistic knowledge of reactions in these more dilute aqueous systems. The most recent reference mechanisms are provided by Herrmann et al. (2000) and in an unpublished update (H. Herrmann, personal communication), which contains approximately 100 components and 400 reactions. A reduced version of the mechanism is approximately half this size and is even then too large to be included in global models. In this mechanism, rate constants of several important reactions are estimated rather than measured, and for most reactions temperature dependencies are not determined.

Finally, although the type and degree of parameterisation of heterogeneous chemistry vary considerably from model to model, no intercomparison has yet been performed to evaluate their effects on the results provided by current CTMs.

6.3.2.4 *Aerosol Composition and Distribution*

Based on their chemical composition, tropospheric aerosols can be divided into the following six main categories: sulphur aerosols (mainly SO_4^{2-}), black carbon, organic carbon, mineral dust, sea salt, and NO_3^- aerosols. However, in the troposphere, we commonly observe aerosols that are a mixture of the above mentioned chemical components and, thus, have to deal with chemical and optical properties that depend on the specific chemical mixture. For reasons of simplicity, the approach adopted in most earlier studies was to measure and model each aerosol component, as well as the impact of its presence on heterogeneous chemistry separately, namely, by assuming a chemically uniform particle of one of the above mentioned major chemical components (Langner and Rodhe 1991; Tegen and Fung 1994; Chin et al. 1996; Feichter et al. 1996, 1997; Liousse et al. 1996; Cooke and Wilson 1996; Cooke et al. 1999). Distributions, sources, and sinks of these single-component particles were therefore studied individually based on observations and model results. In addition, in many studies particulate matter was treated as a gas, i.e. by ignoring the dynamics governing its size distribution and without evaluating the related inaccuracies in the model simulations of the mass loading of key aerosol constituents. For instance, at the sub-hemispheric scale measured and modelled daily-average concentrations of sulphate typically agree within a factor of two. Considerable effort is now being directed toward developing models of aerosol microphysical processes to gain predictive capability of size-resolved aerosol mass and chemical composition distributions. Size-resolved representation of aerosols is necessary for accurate evaluation of radiative forcing, especially the indirect forcing, in which cloud and aerosol microphysics play the central role (see Chap. 4). Likewise, from the perspective of developing strategies to control local and regional particulate air pollution, it is necessary to develop model-based representations of the processes responsible for aerosol burdens that can confidently relate these burdens to sources of gases and particulate matter. However, in view of the still limited understanding of many of these processes, much effort must be directed toward enhancing understanding before representing them realistically in models.

A major objective of global-scale aerosol modelling is the evaluation of aerosol impacts on climate at present, during the industrial period, and for various possible future emission scenarios. Until recently, such assessments have used temporally-averaged aerosol fields or surface albedo calculated offline, i.e. in transport-transformation models separately from the climate model used to assess the climate influence. It is now recognised that this approach may misrepresent key features of aerosol-climate interactions, for example, correlations between relative humidity and particulate mass that might enhance direct forcing or, more intrinsically, the influence of aerosols on precipitation development and resultant changes in the hydrological cycle. For these reasons computationally efficient size- and composition-dependent aerosol process parameterisations should be developed and incorporated in climate models to allow realistic simulations of aerosol distributions and their climate impact.

Indeed, recent attempts have been made to calculate particle number concentrations using parameterisations of aerosol formation and dynamical processes. Two kinds of such models were developed, using modal or bin schemes. In the modal scheme, one or more modes of the aerosol particle distribution are described by lognormal distributions (Schultz et al. 1998; Wilson et al. 1999). In bin schemes, the particle mass is distributed over different size classes, each bin being characterised by its geometric mean diameter. Such schemes can reproduce changes in size distribution when a large number of size classes is taken into account. They have been applied, for example, to sulphuric acid in the stratosphere (Timreck 2000) and sea salt aerosols (Gong et al. 1997a,b). A third approach to modelling of aerosol dynamics in atmospheric models uses the quadrature method of moments (QMOM) (McGraw et al. 1995; McGraw 1997). This approach simultaneously tracks an arbitrary number of moments of an aerosol particle size distribution and has been implemented in a semi-hemispheric CTM (Wright et al. 2000). The approach includes accurate representations of aerosol nucleation, condensation, coagulation, dry deposition, wet removal, and cloud processes and allows the subsequent calculation of the optical properties needed to study the influence of these aerosols on climate.

Efforts have been made to characterise the chemical composition of aerosols, understand the mechanisms of their formation, and parameterise the internal mixture of the various chemical components of the aerosols. The first attempts of global 3-D modelling of secondary aerosol formation from the oxidation of biogenic nonmethane hydrocarbon emissions (Kanakidou et al. 1999c) and aerosol equilibrium modelling (Adams et al. 1999) have been performed. It is foreseen that in the near future detailed modelling of aerosol dynamics will be possible as indicated in Sect. 6.6.4.

Further work is needed to acquire the prerequisite knowledge for understanding nucleation and condensation processes, and for improved models capable of describing the formation, growth, and hygroscopicity of internally mixed aerosols taking into account their mass and size distributions together with their chemical and optical properties. Such progress will be made in parallel with the improvement of our understanding of heterogeneous reactions in the troposphere.

6.3.2.5 *Physical Removal*

The removal of gases and particles from the atmosphere via scavenging in and below clouds and/or deposition at Earth's surface strongly affects tropospheric chemistry. The description of the hydrological cycle is of great importance for the calculation of the abundance of soluble chemical species and for the OH radical distributions. The amount of H_2O vapour regulates the chemical production of OH, and in- and below-cloud scavenging of soluble gases by condensed H_2O has a strong influence on the levels of oxidants (Crutzen and Lawrence 2000). The parameterisation of these processes is highly variable from one model to another. The simplest parameterisation of washout is based on the use of a reaction rate, and more detailed parameterisations are based on observed or calculated precipitation formation rates and Henry's Law constants for the different chemical species considered. Most models discriminate between large-scale (stratiform) precipitation and convective (sub-grid) precipitation (e.g., Giannakopoulos et al. 1999). Given the large differences in the treatment of cloud schemes in models, it is likely that in- and below-cloud scavenging also differ. The on-line (or coupled) calculation of the distribution of sulphur compounds has been achieved recently in a global model taking into account aerosol interactions with clouds (Lohmann et al. 1999b), and it has been shown that the global sulphur budgets are very sensitive to changes in the cloud cover parameterisations.

The overall effectiveness of wet removal schemes for aerosol components has been tested by comparisons with observations of the radionuclide ^{210}Pb (Balkanski et al. 1993; Guelle et al. 1997). It appears that the altitude dependency of wet removal is a sensitive and poorly constrained parameter. One of the uncertainties that is associated with wet removal processes was demonstrated by Lawrence and Crutzen (1998). They showed that gravitational settling and evaporation of cloud droplets and ice particles can induce a significant downward flux of HNO_3 and H_2O_2 affecting trace compounds such as OH and NO_x. However, the strong uptake of HNO_3 on ice assumed in Lawrence and Crutzen (1998) based on laboratory measurements of Abbatt (1997) and Zondo et al. (1997) was disputed by Tabazadeh et al. (1999). An accurate simulation of the effects of the hydrological cycle remains a challenge in both CTMs and GCMs.

Several gases such as O_3 and CO, as well as nitrogen and sulphur compounds, undergo irreversible absorption at the earth's surface. The importance of dry deposition may be illustrated by the fact that, according to modelling studies, approximately 25–30% of SO_2 emissions are removed by dry deposition (Benkovitz et al. 1994). The downward flux associated with this dry deposition is usually expressed as the product of the surface layer concentration of the chemical species and a quantity called the deposition velocity (Ganzeveld and Lelieveld 1995; Wesely 1989). The deposition velocity is generally parameterised as the inverse of serial resistances as explained in detail in Chap. 2. Dry deposition of ozone and many other gases is controlled by surface cover type and its biological and physical properties (temperature, wetness, stomatal uptake, and amount of biomass). Dry deposition of aerosols is controlled by impaction, by Brownian diffusion, and by sedimentation, all of which depend on aerosol mass and size distributions. It is expected that increased use of parameters measured by satellite (e.g., amount of vegetation, snow-ice cover) may lead to a better estimate of dry deposition velocities. Dry deposition fluxes are, however, difficult to evaluate due to the lack of regular measurements.

6.3.3 Input to Chemistry-Transport Models

6.3.3.1 *Meteorological Fields*

Non-coupled, or offline, CTMs can be driven from data assimilated by weather prediction models at the various centres. These fields can also be provided by GCMs. Forecast centres such as ECMWF (*http://www.ecmwf.int*), NCEP (*http://www.ncep.noaa.gov*), and NASA-DAO (*http://dao.gsfc.nasa.gov*) archive their forecast analyses routinely; these analyses represent a combination of the model predictions and observational data. In addition, re-analyses of observational data from previous years using one numerical weather prediction model have occurred at several centres (ECMWF, NCEP-NCAR (Kalnay et al. 1996), and NASA-DAO). These re-analyses have produced several decadal scale analyses that are available for use in CTMs. General circulation model fields are also used to drive the CTMs, either to provide a more realistic data set for longer climate simulations, or to capture the climatology of chemical species rather than specific events.

Numerical weather prediction has improved over the last one or two decades, in part due to better observations, fewer simplifying assumptions, improved numerical techniques, increased model resolution, improved physical parameterisations, and improved data assimilation methodologies. Bengtsson (1999) discussed the importance of these factors, and concluded that the time scale of useful forecasts over most of the Northern Hemisphere extra-tropics has been extended to more than a week. It should however be noted that, due to the lack of accurate observations of cloud properties and distributions and of parameters describing the planetary boundary layer, it is difficult to validate the data assimilation products concerning these quantities.

Both forecast centre analyses and GCM output have advantages and disadvantages. Analysed wind fields attempt to capture the variability of the weather on specific days and thus can be used to drive offline CTM simulations for comparison against specific field campaigns (e.g., Ramonet et al. 1996; Law et al. 1998). However, these data sets are not without uncertainties. Model studies suggest that short range transport (a few days) driven by ECMWF and NCEP meteorological fields using the same model framework have relatively similar transport, while longer range or inter-hemispheric transport can be quite different (Mahowald et al. 1997b).

These differences in transport that result from using different analysed data sets are consistent with meteorological comparisons of the data sets and the importance of the sub-grid scale mixing parameterisations. The analysed wind fields are largely model output at grid points that are not coincident with meteorological observation sites. Thus, in regions such as the Tropics where observation stations are sparse, there can be large differences among the analysed fields (e.g., Newman et al. 2000; Mo and Wiggins 1996). Transport in regions more densely sampled by meteorological observations should be much more similar among the different data sets. There may be large biases in the analysed fields, especially in the hydrological cycle (e.g., Trenberth and Guillemot 1998). This could have important implications for many trace constituents, such as the hydroxyl radical, soluble gases, and aerosols, although the differences in subgrid scale treatment may lead to larger uncertainties. As an example, Rasch et al. (2000) show that simulations using the same year and source for winds (1990–1991, ECMWF data sets), but different advective and subgrid schemes lead to different vertical mixing and thus transport.

The forecast centre analyses have the disadvantage of not archiving all the mixing coefficients required for calculating the transport of chemical species, and instead employ a forcing term derived from nudging to the observations, which can be non-physical and produce inconsistent fields. This last effect has been reduced substantially with the 3-D assimilation schemes adopted by the analysis agencies (e.g., Kalnay et al. 1996).

General circulation model fields are used for many different CTM purposes, especially for climatological data analysis, or for time periods where there are no observational data sets, or for future climate predictions. General circulation model fields such as cloud water distribution and cloud cover have the advantage of being internally consistent, and usually consistent with offline models that use the meteorological fields. Since subgrid-scale parameterisations are highly important for trace constituent distributions, this consistency represents an important advantage. In addition, on-line tracer calculations are used in GCMs as diagnostics to provide insights into specific model shortcomings (e.g., Rind and Lerner 1996; Rind et al. 1999). As discussed earlier, GCMs have also been used to simulate specific weather episodes using Newtonian relaxation techniques (Dentener et al. 1999; Feichter and Lohmann 1999; Jeuken et al. 1996).

Most models can now use both types of meteorological fields discussed above, i.e. fields calculated using GCMs, or provided by forecast centre analyses. Therefore, the type of meteorological input used in the CTMs is not indicated in Table 6.1.

6.3.3.2 *Thermodynamic and Kinetic Data*

The accuracy of the calculated trace species concentrations relies strongly on the chemical scheme that is adopted in the model, and on the quality of the kinetic data that are determined in the laboratory. The data used in atmospheric chemistry models rely on reviews, assessments, and evaluations of the laboratory data. Most kinetics papers usually include a realistic estimate of the precision of the measurement and a discussion of the potential systematic uncertainties associated with the measurement. Such factors are, of course, best assessed by the authors, who are in a position to understand the complexities of their respective experiments. The combined uncertainty is then usually expressed as either an absolute (plus-minus) or fractional quantity as a function of temperature (for several reasons, uncertainties tend to be lowest at room temperature).

The role of evaluations is to consider all the available data and to make an informed judgement as to the best central value of the rate parameter and the maximum associated uncertainty. This process acknowledges that systematic differences exist between different measurement techniques (and even between experiments conducted using the same technique in different laboratories). Consequently, the uncertainties given in an evaluation are often larger than those in any given study, since they encompass contributions from all the measurements; that is, the recommended uncertainties are not necessarily statistical, but involve some subjective judgement as to the reliability of the combined data set. The most commonly used evaluations are those by the NASA-JPL Panel for Data Evaluation (DeMore et al. 1997; Sander et al. 1997) and the IUPAC Subcommittee for Gas Kinetic Data Evaluation (Atkinson et al. (1997a) and later supplements available at *http://www.iupac-kinetic.ch.cam.ac.uk/*). These provide recommendations for the optimal kinetic, thermodynamic, and photochemical parameters to be used in stratospheric and tropospheric models. Updated evaluations are produced regularly for stratospheric chemistry (including heterogeneous reactions) and a limited number of tropospheric reactions.

To date, no comprehensive evaluation of solubility constants (Henry's Law constants) exists, but a large and evolving compilation of data has been made available on the Internet by R. Sander (*http://www.mpch-mainz.mpg.de/~sander*). The impact of kinetic uncertainties on model results has not been thoroughly assessed. Some studies have been performed with two-dimensional models (Considine et al. 1998), but, due to their computing requirements, this has not been carried out with three-dimensional models.

6.3.3.3 *Emissions*

Chemistry-transport model calculations require specification of the conditions that exist at the boundaries of the spatial domain. These boundary conditions, surface emissions, and deposition represent the impact of the external environment on the domain under consideration and are often representative of complex processes at the interface between different components of the Earth system. Their estimation can be a difficult problem, as these conditions depend on several factors, such as biological processes and human activities, as discussed in more detail in Chap. 2 and 4.

The development of the CTMs has stimulated the development of global inventories of both natural and anthropogenic emissions. The first simulations using CTMs were generally based on emissions databases developed by modellers at the resolution used by each model (Müller 1992). As the number of models and model studies increased, there has been more need for consistent surface emissions, at a high resolution. IGAC-GEIA (Global Emissions Inventory Activity) was created in order to establish a framework for developing and evaluating global emission inventories at a high resolution. A large group of scientists contributed to this tremendous effort, which has resulted in a large set of data (Table 6.2) that is currently available publicly from the internet (*http://www.onesky.umich.edu*). The file format is kept uniform to facilitate use of the different inventories. Most of the data sets correspond to a 1 × 1° grid and to annual averages. Exceptions are the emissions of nitrogen oxides from soils and anthropogenic activities, anthropogenic emissions of sulphur dioxide, natural emissions of volatile organic compounds (isoprene, terpenes, and other hydrocarbons), and black carbon from biomass burning; these are also available on a seasonal or monthly average basis. Similar emission inventories for chemical species such as CO or methane are currently under development.

Surface emissions resulting from large-scale biomass burning, i.e. forest fires or savannah burning (see Chap. 2), are still very difficult to specify in global CTMs. Currently, most models specify average distributions of carbon released during the fires as given by the compilation of Hao and Lin (1994) or by the FAO (Food and Agriculture Organisation of the United Nations). The emission of a particular chemical species is then derived using emission ratios such as the ones given in Table 2.3. Such an approach does not take into account the large year-to-year variations of the fires. Recent work has started to address this issue. An inventory of carbonaceous particles emitted by boreal and temperate forests has been proposed by Lavoue et al. (2000), together with its temporal variability for the 1960–1997 period. Furthermore, it is expected that, in the near future, satellite data will provide more accurate estimates of burnt areas and amount of biomass burned in particular years.

EDGAR (Emission Database for Global Atmospheric Research) (Olivier et al. 1996) is available on line (ftp: *info.rivm.nl*) and provides estimates of the average emissions of several compounds by sector of human activity on a 1 × 1° grid basis for the year 1990. Aircraft emission estimates are given with 1 km altitude resolution. The database has been constructed using the same approach for all chemical species, i.e. emissions were first calculated on a country basis by multiplying activity levels with emission factors per compound. The selection of main source categories and spatial resolution was based on available statistical data and quality and consistency of these data. The list of inventories available within EDGAR, as well as the global source strength for each chemical species, is given in Table 6.3. These inventories are updated regularly.

Aircraft transportation leads to the emissions of gases and particles directly into the troposphere and the lower stratosphere. Three-dimensional comprehensive inventories (altitude, longitude, and latitude) have been developed by the European Commission, under the Abate-

Table 6.2.
Emissions inventories for base year 1985 produced by the IGAC/GEIA activity (*http://weather.engin.umich.edu/ geia/*)

Species	Yearly (Y) or seasonal (S)	Reference
NO_x from soils	S	Yienger and Levy 1995
Anthropogenic NO_x	Y	Benkovitz et al. 1997
Anthropogenic NO_x	S	Voldner et al. (GEIA website)
Anthropogenic SO_2	Y	Benkovitz et al. 1996
Anthropogenic SO_2	S	Voldner et al. (GEIA website)
Volcanic sulfur	Y	Andres and Kasgnoc 1998
Nitrous oxide	Y	Bouwman et al. 1995
Volatile organic compounds	S	Guenther et al. 1995
Reactive chlorine	Y	Graedel and Keene 1999
Black carbon from fossil fuel	Y	Dignon et al. (GEIA website)
Black carbon from biomass burning	Y	Cooke and Wilson 1996

Table 6.3. Inventories available for base year 1990 within EDGAR and estimated global source strength for each chemical species

Species	Global amount emitted (Tg)
CH_4	Tg CH_4 yr^{-1}
Fossil fuel: combustion	4.8
Fossil fuel: production	89.3
Biofuel	14.1
Industrial processes	0.8
Land use/waste treatment	211.4
CO	Tg CO yr^{-1}
Fossil fuel	262.6
Biofuel	181.0
Industrial processes	34.8
Land use/waste treatment	495.9
NO_x	Tg N yr^{-1}
Fossil fuel	21.9
Biofuel	1.3
Industrial processes	1.5
Land use/waste treatment	6.2
NM-VOC	Tg NM-VOC yr^{-1}
Fossil fuel: combustion	41.6
Fossil fuel: non-combustion	27.3
Biofuel	30.7
Industrial processes	33.5
Land use/waste treatment	44.4

ment of Nuisances Caused by Air transport (ANCAT), and by NASA in the US for different periods: 1976, 1984, and 1992. Both these inventories incorporate an extensive air traffic movement database, assumptions about flight operations, and calculations of fuel burned and emissions during the entire flights. The NASA and ANCAT inventories provide emissions of gases and particles at a 1 × 1° in latitude and longitude resolution, and at a 1 km vertical resolution. The 1992 inventories provided by ANCAT and NASA are in good agreement, both for total fuel used by aviation and for total emissions of chemical compounds (Gardner et al. 1997; Baughcum et al. 1996; IPCC 1999).

As detailed in Chap. 2 and 4, inventories of the natural emissions of sea spray, mineral soil, and DMS have been developed for use in global climate models and in regional air quality models (see, for example, Marticorena and Bergametti (1995), Gong et al. (1997a,b), and Kettle et al. (1999)).

These estimates of surface emissions and exchanges of chemical constituents between the surface and the atmosphere rely in most cases on field observations or on laboratory experiments, as detailed in previous chapters (2 and 4). These measurements are generally representative of a relatively small area and, due to the high variability of the processes responsible for the emissions, the extrapolation of these local measurements to broader areas remains a major problem. Methods are needed for extrapolating from leaf, branch, and canopy level measurements of VOC emissions to spatial scales ranging from ecosystem to regional and global. It is impossible to characterise the VOC emission pattern of every plant species of the entire world due to the heterogeneity of plant ecosystems. As a consequence, default emission factors are used for many landscapes, particularly in the Tropics. Empirical light- and temperature-dependent algorithms are used for calculating VOC emission inventories using experimentally determined plant-specific emission rates. These models can account for short term changes (minutes to hours) of some compounds, including isoprene, but are limited in their ability to predict longer term (days to years) variations. Over the past years, the degree of uncertainty surrounding global estimates of emissions has been reduced with modelling exercises but the numbers of VOCs for which emission data are available increased since the 1960s because our ability to measure a much wider range of species in both temperate and tropical regions has increased. Similar problems exist for the determination of natural or anthropogenic emissions of all chemical compounds, and, as a consequence, the uncertainties in global emissions data are large, both for anthropogenic and natural processes.

There are still large uncertainties about the amount of nitric oxide emitted as a result of lightning, and its spatial and temporal distribution. Lightning arises from the breakdown of the charge separation in electrified convective clouds. Estimates of the total emissions range from 1–20 Tg N yr^{-1} (Lawrence et al. 1995; Price et al. 1997). Current parameterisations of the NO production from lightning typically rely on the work of Price and Rind (1992) and Price et al. (1997), who provide estimates for the mean energy per flash in thunderclouds and the average NO yield per unit energy. The effect of NO production from lightning on tropospheric chemical species also depends on the altitude at which NO is released from the thunderstorm. Pickering et al. (1998) defined three standard vertical profiles based on several observations and results from a cloud-resolving transport model. These parameterisations are still very crude, and to improve their accuracy more explicit cloud microphysics parameterisations are required; these are under development (Solomon and Baker 1998). More sophisticated treatment of lightning emissions in CTMs does result in better simulation of free tropospheric NO_x (Stockwell 1999).

Global distributions of lightning flashes are starting to be available at a high spatial and temporal resolution, for example from the Optical Transient Detector

(OTD, on board Orbview) and from the Lightning Images Sensor (LIS, on board TRMM). Both data sets are available from *http://thunder.nsstc.nasa.gov/data*. It is expected that such data sets will be used more and more in CTMs for better parameterisations of lightning emissions.

An additional consideration derives from the need for consistency between emissions of NO_x and the model meteorology. For example, Levy et al. (1996) use simulated moist convection statistics to couple NO_x emissions by lightning to strong convective transport in the model.

Because CTMs are used to simulate both past and future evolutions of atmospheric composition (Sect. 6.5.1), there is currently a major need to develop historical anthropogenic sources inventories as well as future inventories. The emissions of several gases such as sulphur or nitrogen oxides over the last century have already been studied (Dignon and Hameed 1989; Bertnsen et al. 1996) using estimates of the evolution of fossil fuel consumption. Estimates for different future scenarios have been adopted by the WMO and IPCC assessments (WMO 1995; IPCC 2001). However, realistic estimations of past and future emissions of trace gases are still urgently needed.

6.4 Model Evaluation

6.4.1 Comparisons with Observations

One of the necessary and ultimate tests of a model is to compare its predictions with observations. This may seem to be a trivial exercise; however, there are a number of significant challenges involved. These include determining which data to use in the comparison, reconciling the different spatial and temporal scales of the model results and measurements, and dealing with the fact that, in many cases, data are not available at all.

Even in the cases where published data are available for comparison, the limitations imposed by data quality and time coverage are important issues. There are many instances in which central values of measurements are presented, but little or no information about the variability of the data is available. In such cases it is difficult to assess the precision of the data, or the representativeness of the central value. Data can be biased by limitations in measurement techniques themselves, or by the fact that many measurement systems use various criteria to accept or reject data when, for example, local pollution sources or circulation patterns could influence the accuracy of the measurements. It is necessary for data users to understand the way measurements were made so they are able to assess the appropriateness of a given data set.

Model results are produced for finite grid volumes that have very different spatial and temporal scales than those on which measurements are made (and often not at high enough vertical or horizontal resolution). Ground-based measurements can be made over long periods of time, but they are limited to a fixed location. Measurements from moving platforms such as ships or aircraft provide data at different locations, but for limited time periods at each, and often only in a very limited number of grid volumes. Often, measurements are not co-located with model grid volumes, requiring interpolation of one of the fields to match the other.

Thus, the most significant limiting factor in further model evaluation is the lack of reference data sets of observations covering the whole globe. This implies that new parameterisation schemes or other developments in global models can be compared more accurately with other model results than with observations. While model intercomparisons are both necessary and extremely useful, the ultimate goal of model development is to better describe and predict the behaviour of real systems. In some cases, box models are more appropriate to evaluate the accuracy of individual process parameterisations by comparisons with spatially and temporally well-defined observations.

The issue of the scale mismatch, though difficult, may be the most approachable of the three challenges discussed here. The issues of data availability and quality are more difficult to resolve. Increasingly, strategies will be developed with the explicit goal of making sure that the measurements are useful for comparisons with model results to the largest extent possible. This was done during the recent measurement campaigns such as MOZAIC and POLINAT (Flatoy et al. 2000; Lawrence et al. 2000; see also Chap. 3).

As described in Chap. 5, global measurements of tropospheric compounds from space are starting to become available; examples include CO and CH_4 columns from the IMG-ADEOS experiment, NO_2 and CH_2O tropospheric columns from the GOME satellite, and CO concentrations at different tropospheric levels from the MOPITT experiment. Such data will become more and more accurate, and will cover longer time periods, which will allow more detailed and comprehensive evaluation of global CTMs.

There is a clear need for continued direct collaboration and cooperation between the measurement and modelling communities. Cooperative efforts allow assessment of the accuracy and precision of the data as well as selection of data that are appropriate for comparisons between model results and observations. Input from the modelling community is necessary to help plan new measurement campaigns, for determining observational strategies and quantities to measure. Significant progress on such collaborations has been made in the last decade, particularly as a result of IGAC-sponsored activities.

6.4.2 Model Intercomparisons

The accuracy and consistency of models used to simulate tropospheric chemistry and perform future predictions have been tested over the past five years through a number of intercomparison exercises, as further described below.

Parameterisations of convection were tested by comparing model-simulated results and observed distributions of ^{222}Rn. Twenty global 2-D and 3-D models were included in that exercise, which demonstrated that deep convection is not well represented by most models. Upper troposphere simulations appear to be very sensitive to the treatment of moist convection and to the scale of deep convection (Jacob et al. 1997).

Twenty-four photochemical codes used in CTMs of various scales were isolated from the transport parameterisations and were compared in the 1994 IPCC "Photocomp" intercomparison exercise (Olson et al. 1997; WMO 1995). The results show that model-to-model differences of 30% or more exist in the calculation of ozone and OH concentrations. These differences result from differences in the calculated photolysis rates, kinetic data, numerical methods used to solve the equations, and the set of photochemical reactions adopted to represent tropospheric photochemistry. In particular, the ozone photodissociation rates differ by ±20%, whereas the best agreement of ±5% was obtained for the NO_2 photolysis rates, the other rates agreeing within ±15%. The largest differences among models were obtained when non-methane hydrocarbon chemistry was included.

The large dispersion of results from several 2-D and 3-D models of ozone changes resulting from an imposed 20% increase in CH_4 concentrations showed that the ability of the models to predict tropospheric ozone changes induced by methane perturbations is not satisfactory (WMO 1995).

A more recent intercomparison by Kuhn et al. (1998) of tropospheric chemical models for atmospheric conditions over Europe showed differences in oxidant levels up to 40% due to different representations of the gas phase chemistry.

Following the recent rapid progresses in the development of global 3-D CTMs, an intercomparison exercise of this new generation of models was performed in 1997–1998, under the initiative of the IGAC Global Integration and Modelling (GIM) Activity. The objectives of the Tropospheric Ozone Global Model Intercomparison Exercise were: (*1*) to evaluate systematically the capabilities of the current generation of 3-D global models to simulate tropospheric ozone and their precursor gases; and (*2*) to identify key areas of uncertainty in our understanding of the tropospheric ozone budget (Kanakidou et al. 1999). Thirteen global CTMs were included in this exercise, which used different inputs, meteorological fields, surface emissions, and chemical schemes. Results differed significantly among models although all of them captured the general patterns in the global distribution of carbon monoxide, nitrogen oxides, and ozone. Comparison of model results and observations

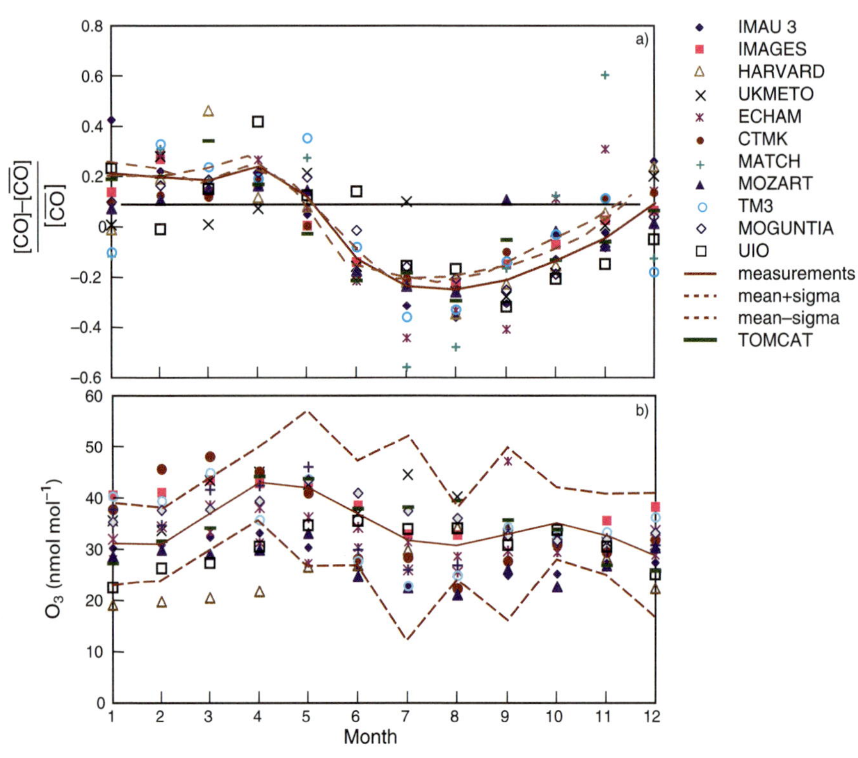

Fig. 6.2. Comparison between concentrations derived from different global chemical transport models and observations. **a** Seasonal variation of CO mixing ratio at Mace Head, Ireland, normalized to the annual mean, $[\overline{CO}]$; **b** seasonal variation of surface O_3 mixing ratio at Mace Head (Kanakidou et al. 1999)

at selected stations indicated that the models deviate from the observed annual mean CO concentrations by about ±50% (Fig. 6.2). The main features of the ozone distribution differed among models at some locations by up to a factor of three. Maximum deviations of models from surface ozone observations are of the order of 50% whereas the models deviated much more from observations in the free troposphere, particularly close to the tropopause where the ozone concentrations are sensitive to the parameterisation of cross-tropopause transport (Fig. 6.3). Interestingly, the photochemical lifetime of methane in the troposphere up to 300 hPa (or to the closest model level) was calculated to be 7.5 years globally and 8.6 years in the Southern Hemisphere against 6.8 years in the Northern Hemisphere (as the medians of model results). Model results varied within 30% of these values. The north/south asymmetry that these results show is the opposite of that inferred by Brenninkmeijer et al. (1992), Montzka et al. (2000), and Prinn et al. (2001) from atmospheric observation, and the asymmetry in the computed OH distribution requires further investigation.

The Comparison of Large-scale Sulphate Aerosol Models study (COSAM), also co-sponsored by IGAC-GIM, compared the performance of atmospheric models with each other and with observations. It involved: (*1*) design of a standard model experiment for the world wide web; (*2*) ten model simulations of the cycles of sulphur and ^{222}Rn / ^{210}Pb conforming to the experimental design; (*3*) assemblage of the best available observations of atmospheric sulphate, SO_2, and MSA; and (*4*) a workshop in Halifax, Canada to analyse model perform-

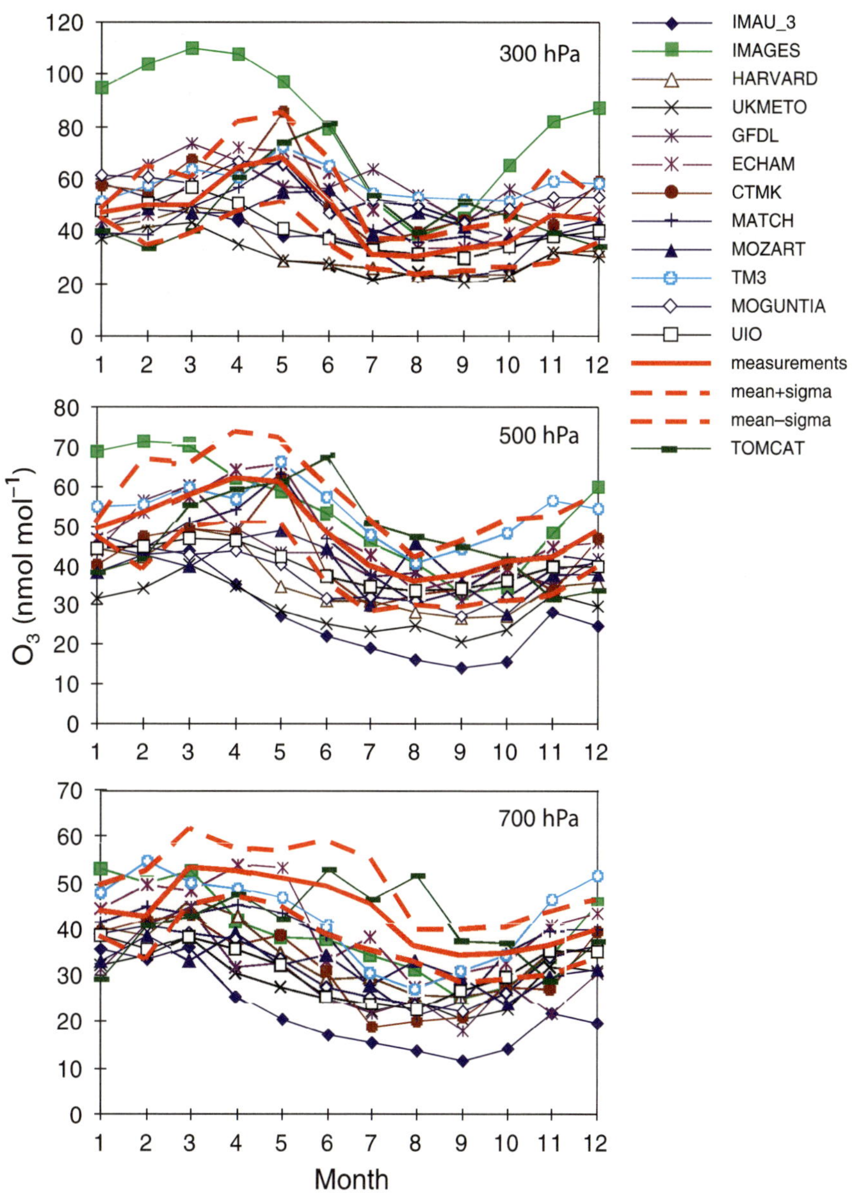

Fig. 6.3.
Comparison of monthly mean O_3 model results (nmol mol^{-1}) to observations at three different pressure levels: 300, 500, and 700 hPa at Hilo, Hawaii. The *thick red line* corresponds to the observations and the *dashed red line* to their standard deviation (Kanakidou et al. 1999)

ance and future model development needs. The analysis examined the variance between models and observations, discussed the sources of that variance, and suggested ways to improve models. Variations between models in the export of SO_x from Europe or North America were not sufficient to explain an order of magnitude variation in spatial distributions of SO_x downwind in the Northern Hemisphere. On average, models predicted surface level seasonal mean sulphate aerosol mole fractions better (most within 20%) than SO_2 mole fractions (over-prediction by factors of two or more). Results suggested that vertical mixing from the planetary boundary layer into the free troposphere in source regions is a major cause of the uncertainty in predicting the global distribution of sulphate aerosols in climate models today. For improvement, it is essential that globally coordinated research efforts continue to address emissions of all atmospheric species that affect the distribution and optical properties of ambient aerosols in models and that a global network of observations be established that will ultimately produce a global climatology of aerosol chemistry.

Intercomparisons between three 3-D CTMs were reported by Thakur et al. (1999) together with comparisons between calculated and observed distributions of reactive nitrogen species. The three models underpredicted NO, and significantly overpredicted HNO_3 and PAN, particularly in the upper troposphere.

The complexity of model intercomparison, and the comparisons of models with data, indicates the need for continuing and expanding these exercises. For the stratosphere, the Models and Measurements exercises (Prather and Remsberg 1993; Park et al. 2000) have led to substantial improvements of the models through a detailed assessment of the difference between models and the development of a database used for model/data comparison. Such an exercise would be more complex in the troposphere, but is worth pursuing in the near future.

6.5 Model Applications

6.5.1 Evolution of the Composition of the Troposphere

The evolution of the composition of the troposphere from the pre-industrial period up to about 2050 has been simulated over the last decade by different groups. Changes in the distributions of ozone and OH have been simulated, for example, by Crutzen and Zimmerman (1991), Lelieveld and van Dorland (1995), Levy et al. (1997), Roelofs et al. (1997), Berntsen et al. (1997), Wang and Jacob (1998), Stevenson et al. (1998), and Brasseur et al. (1998b). All of these studies, looking at the changes from the pre-industrial to the current period, compared the results of two sets of simulations, one in which all anthropogenic emissions (fossil fuel combustion, industry, fertiliser) were set to zero, and one in which emissions typical of current conditions were used. The two simulations were performed using the same meteorological fields to calculate the transport of the chemical species. The model results suggest that between 1850 and 1990 tropospheric ozone burden increased by about 80%. These models generally overestimate pre-industrial surface ozone concentrations as reconstructed from very sparse historical data.

Evaluations of the evolution of the atmospheric composition depend critically on the accuracy of the chemical species emission changes. However, it was not until the 1970s that reliable emissions estimates for the pre-industrial era became available, particularly for biomass burning emissions, which cannot be related easily to population or energy use changes. Previous model studies have assumed that pre-industrial biomass burning emissions were 10 to 30% of the present-day value.

Emissions rates from Europe and North America are available from the EMEP programme (Cooperative Program for Monitoring and Evaluation of the Long Range Transmission of Air Pollutants in Europe) (EMEP 1998). Estimates of fossil fuel consumption are available for the rest of the world from the British Petroleum Statistical Reports (BP 1994, 1998). It should be noted that past and even current emission estimates are scarce for developing countries, though the largest changes in emissions have already been noticed in these countries and are expected to increase even more in the future.

Most simulations of the evolution of the chemical composition of the troposphere are currently performed with constant-year meteorology. Reanalysis of the past 40 years of meteorological data is currently under way (ECMWF ERA-40 project) and similar products are available for shorter periods (from the ECMWF and NCEP meteorological centres). Such data sets are starting to be used in long-term simulations using global CTMs, which will give better indications of the impact of year-to-year variability of meteorological fields on changes in the distributions of tropospheric compounds.

Estimates of emissions during the 21st century have been proposed by different groups, based on scenarios developed by the Intergovernmental Panel on Climate Change (IPCC 1992, 2000). The scenarios account for World Bank and United Nations estimates of population changes, energy consumption trends, transport development, and agriculture development. As stated in the previous paragraph, possible climate changes will also have to be taken into account; for example, Johnson et al. (1999) found that the impact of future changes in meteorology is likely to be as important as changes in emissions during the next century.

6.5.2 Use of Photochemical Models for Supporting Field Campaigns

Photochemical models can play an important role in field campaigns, both to support the planning and execution of experiments and to evaluate the observations. Constrained box models have been particularly useful in helping to point out possible problems with measurement techniques (Crawford et al. 1996) and to point towards possible incomplete knowledge of photochemical processes such as a missing sink for OH in the marine boundary layer (Eisele et al. 1996), or a missing source of HO_x in the upper troposphere (Wennberg et al. 1999; Jaeglé et al. 1999). Recently, more complex CTMs have also come into the picture. For instance, MOZART (Table 6.1) was used to examine the O_3 photochemical budget over the Pacific Ocean based on data from MLOPEX (Brasseur et al. 1996). Another example of the use of 3-D CTMs for interpreting campaign data is the study of Singh et al. (1998), which concluded that transport of Asian NO_y into the PEM-West A region served as a major source of NO_y during the campaign. Further, Lawrence et al. (1999c) concluded that while convective pumping of low-ozone marine boundary layer air is the main process responsible for producing O_3 minima in the upper troposphere, their severity may be influenced by a missing O_3 sink, e.g., halogens. Finally, models have been employed to evaluate the sampling done during a campaign. Ehhalt et al. (1997) used a CTM with surrogate tracers to demonstrate that the sampling during PEM-West A and B was likely sufficient to capture the mean in the region sampled. In contrast, Lawrence et al. (1999a) used O_3 simulations to conclude that a single cruise through the CEPEX region was likely insufficient to capture the mean, although they did point out that extremely low levels of ozone occur in the upper troposphere, which had not been previously observed.

Model estimates of expected concentrations and gradients of various chemical species are often used in planning campaigns. During the last few years, photochemical models have been used increasingly in forecasting during the intensive field phases of campaigns. Regional forecasts were described by Lee et al. (1997), who forecasted stratospheric constituents in the polar vortex for the ASHOE-MESA and SESAME campaigns, and by Flatøy et al. (2000, and references therein), who provided tropospheric chemistry forecasts for the POLINAT campaign. Global tropospheric chemistry and aerosol forecasts were carried out by Lawrence et al. (2000), Rasch et al. (2000), and Collins et al. (2001) for the INDOEX campaign. In all cases, the forecasts were highly valuable for the field investigators. Despite inaccuracies in the model predictions the forecasts were able to help point towards several events that were of interest to observe.

6.5.3 Climate Assessments

Calculations of the radiative forcing due to changes in the distribution of well-mixed gases such as CO_2, CH_4, N_2O, and the CFCs have been performed over the last decade using radiative models. As seen in the previous chapters, ozone is more and more recognised as a greenhouse gas with an estimated contribution to the enhanced greenhouse effect since the pre-industrial period of about 20%, as indicated in Chap. 1. Accurate calculations of the radiative forcing of ozone require knowledge of the spatial and temporal changes of the ozone distribution in the troposphere. As seen above, the development of the CTMs has allowed the calculation of the three-dimensional changes in ozone from the pre-industrial to the present period (Crutzen and Zimmermann 1991; Berntsen et al. 1997; Roelofs et al. 1997; Brasseur et al. 1998c). Previous studies have shown that changes in ozone in the vicinity of the tropopause might be the most important in terms of radiative forcing. The evolution of the ozone vertical distribution is known for the past few decades from ozone sonde measurements in a few locations (Logan et al. 1999), and for additional locations over the past few years from the MOZAIC measurements (Marenco et al. 1998). However, no data concerning the evolution of ozone at this altitude go back a whole century; also, this is a region where most models have a rather low vertical resolution, and calculated concentrations are sensitive to errors in the parameterisation of stratosphere-troposphere exchanges.

6.6 Current Developments and Future Challenges

6.6.1 Further Development of Current CTMs

Future CTMs of the atmosphere will include more comprehensive formulations of chemical and photochemical processes and specifically of oxidation mechanisms. The schemes describing the degradation of volatile organic compounds which are still uncertain, as discussed in Chap. 3, will become more comprehensive and will account more accurately for the effects of primary and oxygenated hydrocarbons of natural and anthropogenic origin. This is important for improving our understanding of the ozone budget. Oxidation processes by halogen compounds will also have to be taken into account due to their important and documented role in the destruction of ozone in the polar marine boundary layer and perhaps at other latitudes. Models focussing on gas-phase oxidation processes and on aerosol chemistry and microphysics will continue to be merged into a single interactive framework. Emphasis will be put on cloud-chemistry interactions.

At the same time, the formulation of transport by resolved grid-scale winds as well as by sub-grid scale transport will be improved. An important requirement for advective models is that they are mass conserving and accurate in regions such as the tropopause, where vertical gradients are strong. Improvements are also expected regarding the parameterisation of boundary layer exchanges with the free troposphere, vertical transport in shallow and deep convective systems, wet scavenging, and dry deposition, by optimal use of detailed cloud physics parameterisations in chemistry-transport models.

As computers become more powerful, the spatial and temporal resolution of CTMs will be considerably enhanced. It is desirable to increase their horizontal resolution to represent features with a spatial scale of roughly 20–100 km (e.g., regions with large emissions and/or high photochemistry activity, tropopause). This will reduce inaccuracies in the computation of the distribution of chemical species due to non-linearities in the photochemistry and enable the capture of regional features in global models. Non-uniform grids with enhanced resolution in regions of interest will probably be used. By using regional "windows" within global models, long-range influences can be accounted for, and the problem arising from the choice of lateral boundary conditions in regional models can be avoided. The layered structure of the atmosphere, which has a strong effect on the vertical exchanges of chemical compounds and on the rates of chemical conversion, will be best reproduced if the number of model levels is increased. Future global CTMs will probably typically include 30–100 levels in the vertical (vertical resolution in the troposphere of 500 m or less). One objective for the next decade will be to develop regional and even global non-hydrostatic models at high spatial resolution with interactive dynamics, microphysics, and chemistry.

The upper boundary of transport models will be moved to higher altitudes in order to account better for the influence of the middle atmosphere on the troposphere, and to study the behaviour of chemical compounds in the stratosphere and even in the mesosphere or thermosphere. This will allow, for example, the quantification of the impact of stratospheric ozone changes on the composition of the troposphere, both as a result of changes in the tropospheric UV field and of changes in stratosphere-troposphere transport of ozone. Not only will the vertical domain of the model be extended but, in addition, the physics included in the model will have to account for the specific processes observed in these higher regions.

Although many CTMs will be coupled on-line with existing atmospheric GCMs and used in a prognostic mode (to predict future climate changes, for example), these models also will be used increasingly in a diagnostic mode with analysed (observed) meteorology to interpret observational data (e.g., satellite, aircraft measurements, etc.).

The greatest impact on future CTMs will likely come from the seemingly ever increasing computational power. Looking out over the next decade it seems highly probable that scientific computing machines will be composed of multiple nodes with each node having several vector and scalar processing units. Furthermore, the basic functional units will all reside on a single chip, perhaps a few square centimetres in size. The nodes will be interconnected with high-speed networks. A typical CTM could run on several hundred nodes comprising a total of thousands of CPUs. Although an imprecise measure, such a computing platform may well deliver tens of teraflops of genuine throughput compared to present performance of 10–100 gigaflops at best. However, it remains an open question whether the development of CTMs for use on massively parallel machines is fruitful or if vectorised transport parameterisations are more beneficial for tropospheric chemistry-climate simulations.

6.6.2 Chemical Data Assimilation

A large amount of observations of meteorological parameters, distributions of atmospheric chemical species, ecosystems, etc., has been collected over the last two decades, and the new generation of Earth satellite observations will provide even much larger data sets. However, it is not possible to measure all atmospheric quantities simultaneously in time and space, and the retrieval of satellite data will generally give access to sparse, heterogeneous, and irregular distributions of atmospheric quantities. Objective approaches to combine our *a priori* knowledge about the physical system under consideration with these usually sparse and irregular observations are often referred to as data assimilation. Up to now, data assimilation techniques have mostly been used in numerical weather prediction (Krishnaumurti et al. 1991), data retrievals from remote sensing experiments, and inverse modelling. Some interesting pioneering work on assimilating observations of chemical species in the atmosphere using photochemical models has been done recently, as discussed below, and the results demonstrate the feasibility of the approach and promise important benefits.

The mathematical basis of data assimilation is estimation theory or inverse problem theory. In a conventional forward problem one uses a set of *a priori* parameters to predict the state of the physical system. In the inverse or estimation problem one attempts to use available observations of the state of the system to estimate poorly known model parameters and/or the state itself.

Very broadly, commonly used data assimilation methods can be divided into variational and sequential techniques. In variational data assimilation, one attempts to find optimal parameters (e.g., optimal initial conditions) that minimise a discrepancy between model results and measurements for a chosen analysis period. The variational data assimilation technique can be thought of as a constrained least-squares fit to a set of observations distributed over some period of time. The constraints are given by the model equations. In the sequential method, observations are blended with model simulations with certain weights as they become available to form new initial conditions for the model for the next time step. The optimal weights are obtained from estimates of the model errors, errors of the current observations, and errors of observations incorporated into the model previously.

Lyster et al. (1997) were the first to use a two-dimensional transport model on isentropic surfaces and the Kalman filter technique to assimilate CLAES- and HALOE-measured methane data from the UARS satellite. Although very computationally expensive, their pioneering approach allows production of synoptic maps from irregularly distributed satellite measurements. Fisher and Lary (1996) used variational data assimilation for assimilating and mapping the UARS-CLAES observations of O_3, NO_2, and HNO_3 with a fairly simple (6 chemical species, 19 reactions) photochemical box model in conjunction with a trajectory model. This was the first application of data assimilation techniques for analysis of photochemically active chemical species in the stratosphere. Elbern et al. (1997) extended this variational approach to assimilation of various tropospheric gases. Khattatov et al. (1999) applied the variational approach as well as the extended Kalman filter for assimilation of satellite data for chemical species measured in the stratospheric using a relatively sophisticated box model and a trajectory model. They also showed that concentrations of a number of chemical species that were not observed could be derived successfully from available data. Lamarque et al. (1999) and Clerbaux et al. (1999) applied a similar method to assimilate CO data obtained from space with a 3-D CTM.

6.6.3 Inverse Modelling

The emissions of trace gases specified at the surface are generally poorly quantified, as they depend on complex processes related to meteorological conditions and human factors (see Chap. 2). As more and more observations of chemical species become available, it is possible in principle, using inverse modelling techniques, to optimise surface emissions estimates, which also improves the agreement between observed and calculated concentrations. Different methods have been developed over the past few years. They have been used to optimise surface emissions or lifetimes of CFCs (Cunnold et al. 1983; Hartley and Prinn 1993; Mahowald et al. 1997a), CO_2 (Enting and Mansbridge 1991; Bousquet et al. 1999), CH_4 (Hein et al. 1997; Houweling et al. 1999), and CO and its isotopes (Bergamaschi et al. 1999). These powerful techniques were the subject of an IGAC-GIM workshop in 1998 and a subsequent monograph (Kasibhatla et al. 2000). The use of such methods is currently limited because the number of observation sites is relatively small. Furthermore, optimisation of surface emissions using inverse modelling requires an accurate treatment of the mixing between the boundary layer and the free troposphere. In the near future, more and more observations of tropospheric chemical species will become available from satellite observations, which may considerably improve the results and possibly lead to new insights into the distribution of surface emissions of chemical species and on their spatial and temporal variability.

A systematic transport model intercomparison effort, performed under the TransCOM sub-project (*http://transcom.colostate.edu*) of the GAIM (Global Analysis, Interpretation and Modelling) Activity of IGBP, is currently underway within the carbon cycle modelling community. Its purpose is to characterise the present uncertainty in carbon sinks that arises from the uncertainty in CTM transport and to prioritise improvements in both the models and the observing system that might reduce this uncertainty. Several research groups are now using global 3-D tracer models to perform time-dependent inverse calculations of the atmospheric carbon budget. The primary objective of the on-going phase of TransCom is to determine the effect of differences in simulated tracer transport among atmospheric models on the results of carbon cycle inversions. Secondary objectives are to diagnose the mechanisms that produce these differences and to suggest improvements to models and the global observing system to produce more robust inversions. These developments will provide support for current efforts to develop inverse modelling methodologies that can be applied to more reactive chemical compounds.

6.6.4 Dynamic Aerosol Modelling

The representation of the climatic effect of aerosols in global models initially used static simplified spatial distributions derived from observations and Mie-calculations to assign optical parameters. Most model studies of the direct and indirect effect of aerosols and climate used these static climatologies. Since the pioneering study by Langner and Rodhe (1991), who used a coarse resolution chemical transport model based on climatological meteorology, the complexity of parameterisations of aerosol precursor chemistry and dry and wet removal has increased dramatically. Several models now treat aerosol precursor chemistry and evolution of particle mass interactively with meteorology with account taken of the complex interactions between cloud processes, heterogeneous chemistry, and wet removal (Benkovitz et al. 1994; Benkovitz and Schwartz 1997; Feichter et al. 1997; Roelofs et al. 1998; Rasch et al. 2000; Koch et al. 1999; Lohmann et al. 1999a). However, the physical and optical properties of the aerosol components are prescribed. Such an approach allows the representation of the high variability of the mass distributions of the particles and is used more particularly when aerosol-cloud interactions are studied.

Most global climate and chemistry transport models have so far only considered the bulk aerosol mass of some specific components rather than the size spectra of externally and internally mixed aerosols and their size-dependent chemical composition. However, the size distribution and the chemical composition of the particles control the optical and deliquescence properties, the activation of particles to cloud condensation nuclei, and their subsequent formation to cloud droplets. Currently, work is under way using thermodynamic equilibrium models to improve the understanding of the presence of nitrate and organic aerosols. In particular, the description of the state of mixture is a challenging task for future research (see Chap. 4).

6.6.5 Nesting-Variable Resolution

With the computer power currently available, global CTMs with extensive chemistry can incorporate horizontal grids of ~250 km or even ~100 km. It can be important to increase the resolution of a model over a region of interest, since transport and chemistry (including sources and sinks) are expected to be simulated more accurately with increasing resolution. Furthermore, comparisons of measurements and model results are more straightforward and significant in a high-resolution model. Two different approaches can achieve the increase in resolution: (*1*) place a high-resolution grid inside a coarse resolution grid where greater resolution or accuracy is desired (over a fixed region or following the path of a particular phenomenon); and (*2*) create a model with variable resolution.

Both approaches have pros and cons and have been used extensively in meteorological forecasting but to a lesser extent in studies involving chemistry. This section will describe both techniques briefly. It is important to remember that CTMs require knowledge of meteorological fields (including cloud properties) to calculate the transport and transformation of chemical species. A meteorological model simulation over the area of interest provides this information (Brasseur et al. 1999 and references therein), which is then used off-line.

6.6.5.1 *Nesting Approach*

The nesting approach consists of inserting a finer resolution model inside a coarse resolution model. There are basically two approaches to limited-area modelling, which can be characterised as one- and two-way interacting (Phillips and Shukla 1973). In the one-way approach, time-dependent conditions are specified at the boundaries of a limited area and the model is then integrated at high resolution over this area. The larger scales of the flow that cannot be simulated on the smaller domain of the fine grid are allowed to affect the fine grid solution, but the fine grid solution does not feed back to the coarse grid. The two-way interactive nested grid approach consists of integrating the fine grid along with the coarse grid. The lateral boundary conditions for the fine grid are taken from the coarse grid solution. The solution on the coarse grid is then updated with the fine grid solution at any coarse grid location where the two grids overlap (Skamarock et al. 1989).

In a meteorological forecast model, boundary conditions are set on all the prognostic variables (wind, temperature, humidity, and cloud properties). In a CTM, boundary conditions pertain also to the chemical species' distributions. Lateral boundary conditions (values or fluxes) are calculated by interpolating the necessary fields from the coarse grid solution. This interpolation introduces errors and noise. Because short lived chemical species adjust quickly to the ambient levels of long lived ones, only the latter need to be calculated at the boundary. Also, long lived chemical species usually exhibit a smoother distribution that tends to decrease the interpolation errors. The short lived chemical species are then calculated explicitly for the conditions defined by the long lived ones. This ensures that the chemical system is balanced with the set of equations defining the chemical mechanism, both on the coarse and the fine grids. As far as we know only the one-way nesting approach has been used in CTMs (Chang et al. 1987; Ebel et al. 1991; Hess et al. 2000).

6.6.5.2 *Variable Resolution Approach*

Even in an ideal situation where all the problems related to the specification of boundary conditions are solved, there is still the practical difficulty of having two systems (global and nested) to maintain (Côté et al. 1993). It would therefore be interesting to have a single model in which the resolution could be varied such that high resolution is focussed over an area of interest, and therefore provides both global and regional simulations.

Courtier and Geleyn (1988) described the first application of a conformal transformation that creates a region of high resolution with a continuous transition from coarse to high. This variable resolution model, with the pole rotated over the area of interest, leaves the governing equations almost unchanged (except for the multiplication of a few terms by a local map factor). Because there is a variation in model resolution, problems similar to loss of information and wave reflection at the boundaries in the nesting approach will have counterparts in this method but with a less damaging intensity (Courtier and Geleyn 1988). There are nevertheless severe limitations, mostly associated with the use of a spectral model (Côté et al. 1993). Côté et al. have therefore applied the same map transformation but with a finite-element discretisation (and other numerical specificities) with considerable success.

The meteorological fields required for variable-resolution CTMs can be derived by interpolating results from regular-grid meteorological models. This can, first, be a cumbersome task and, second, introduce noise in the dynamical fields, which might translate into spurious vertical velocities. Such an approach has nevertheless been applied to the case study of the MLOPEX data (Ginoux 1997). In that case, a triangular grid was used, with the equations of transport solved by the finite element method. Although the variable resolution approach is conceptually appealing, the limited number of models that can create the necessary fields for driving the CTM has hampered its use. This situation is changing as evidenced by the recent development of the Canadian Global Environmental Multiscale (GEM) weather forecast model.

6.6.6 Cloud-Resolving Chemical Models

In recent years there has been significant progress in the development of mesoscale chemical models that resolve the dynamics and physics of convection. These processes are sub-grid scale in the global models where they must instead be simulated using parameterisations.

Deep convection influences tropospheric chemistry not only through vertical transport but also by changing UV fluxes (and thus the photochemical rates), by scavenging soluble chemical species including aerosols, and by producing NO molecules through lightning (which then change the main chemical pathways). Obviously, these complicated physical and chemical processes cannot be realistically simulated using models with coarse spatial resolution.

In recent years, non-hydrostatic cloud-resolving models, e.g., two- or three-dimensional models with horizontal resolution $\leq$3–4 km and relatively detailed cloud dynamics and microphysics, have been used to study tropospheric chemistry in deep convection. The chemistry modules in these models range from tracer models (e.g., Flossmann and Pruppacher 1988; Garstang et al. 1988; Wang et al. 1995; Thompson et al. 1997; Skamarock et al. 2000) to aqueous chemistry and transport models (e.g., Tremblay and Leighton 1986; Niewiadomski 1989; Wang and Chang 1993a; Barth 1994; Flossmann 1994; Kreidenweis et al. 1997; Andronache et al. 1999; Audiffren et al. 1999). While the majority of the chemistry models listed above are developed to run offline, a fully interactive model including dynamics, aerosol and cloud microphysics, radiation, lightning parameterisation, and gaseous and aqueous chemistry has also been reported recently (Wang and Prinn 2000). High-resolution models have revealed the complicated features of convective transport including strong time dependency of transport patterns (Wang and Chang 1993b; Skamarock et al. 2000; Wang and Prinn 2000), small scale ozone intrusion from the stratosphere (e.g., Wang et al. 1995; Stenchikov et al. 1996; Wang and Prinn 2000), and the signatures of lower tropospheric chemical sources in detrained air in the upper troposphere (e.g., Scala et al. 1990; Wang and Chang 1993b; Thompson et al. 1997; Skamarock et al. 2000). Interestingly, few of these studies support the popular "chimney" concept model, i.e. where chemically reactive compounds freshly emitted in the air undergo reactions favoured in the highly polluted environment of the air parcel where they are emitted, which is relatively isolated from the surrounding air, before being mixed. Various cloud-resolving models with aqueous chemistry have suggested that aqueous phase reactions are far less efficient in convective clouds than the theoretical estimates, caused by much lower solubility of SO_2 relative to H_2O_2 and the conversion of water to ice phase particles, which terminates the aqueous chemistry (Wang and Chang 1993c; Barth 1994; Wang and Crutzen 1995; Flossmann and Wobrock 1996; Kreidenweis et al. 1997). The most recent study further indicates that the limited coverages and lifetimes of the liquid phase portions of the convective clouds provide another reason for expecting low aqueous production rates of sulphate in convective clouds (Wang and Prinn 2000). Several studies have also found that nucleation scavenging (scavenging of gases due to nucleation) is more important than impact scavenging (i.e. physical removal following the flow of a scavenger) and in-cloud oxidation of SO_2 in producing aqueous sulphates (e.g., Wang and Chang 1993c; Flossmann 1994; Kreidenweis et al. 1997). In addition, the more complicated issue of changes in the gas phase chemistry induced by deep convection has also been studied using cloud-resolving models. Some early work using embedded one-dimensional photochemical models (Pickering et al. 1990, 1991; Thompson et al. 1997) suggests that ozone production in the downwind side of the cloud anvil may be enhanced. A recent study using a more comprehensive model (Wang and Prinn 2000) indicates that gas-phase production of some chemical species can be changed by more than a factor of two due to deep convection. The most significant, but uncertain, issue in this respect is probably the NO production rate by lightning as well as its subsequent impact on regional tropospheric chemistry. It is expected that the development of more accurate convective schemes and lightning emission parameterisations, which might be based on satellite data as discussed in Sect. 6.3.3.3, will help reduce current uncertainties.

Compared to the current global CTMs, cloud-resolving models can interpret detailed nonlinear physics and chemistry and thus are far better platforms for evaluating theoretical assumptions and laboratory experimental data. They are also potentially better models for interpreting field (e.g., aircraft, ship) measurement results, and helping the design of field experiments. However, the problem of specifying the boundary conditions for these mesoscale models needs to be solved (e.g., by imbedding them in global CTMs based on observed winds) before their full potential is reached. With a careful numerical experimental design, these models could also be used to derive parameterisations of deep convection for large-scale models. This could make the use of this type of model in tropospheric chemistry studies increasingly more practical in the future.

6.6.7 Coupled Earth System Models

The rationale for developing comprehensive, three-dimensional Earth system models is to simulate the coupled processes and feedbacks that exist within and among the atmosphere, the oceans, and the continental biosphere and then, for example, to assess how chemical processes in the atmosphere affect these coupling mechanisms. Significant progress has been made during the past decade in modelling the separate elements of the Earth system and in recent years some of these elements have been coupled to address specific issues. Climate models, for instance, couple the ocean and the atmosphere and also include some formulation of the physical interactions between the atmosphere and terrestrial vegetation. Studies of the hydrological cycle also require the use of coupled models. Looking ahead, comprehensive, integrated Earth system models must account not only for physical interactions among components, but also must represent biogeochemical cycles in some detail. A major challenge for the next decade will be to couple within elaborated models the physical and biogeochemical systems and to assess these interactions at different time scales.

The long time scales involved in the study of the Earth system as a whole and the need to address uncertainties (e.g., through ensembles of model integrations) limit our ability to couple models that account for the full complexity of the system, given present conceptual constraints and computer power limitations. There is a need therefore to develop Earth system models of intermediate complexity, or EMICs (e.g., Alcamo et al. 1994; Prinn et al. 1994; *http://www.pik-postdam.de/projects/climber/welcome.html*). The goal of these models is to describe most of the processes included in comprehensive models, but in a more highly parameterised form. The challenge will be to determine the minimum model complexity necessary to account for the critical causes, effects, and interactive nonlinearities that affect global feedbacks. The validation of such coupled models, using information provided by complex models and by observations, will represent a challenge in the future.

Even with all our present computational power, the development of highly coupled models is a daunting task that requires compromises to achieve century-scale simulations. The spatial resolution of the experiments is often coarse, and therefore even parameterisations of important processes have to be greatly simplified. We can expect that future developments leading to more accurate model components as well as future advances in computing will allow us to build efficient, interactive, coupled models capable of century-scale simulations.

Chapter 7

An Integrated View of the Causes and Impacts of Atmospheric Changes

Lead author: Guy P. Brasseur
Co-authors: Paulo Artaxo · Leonard A. Barrie · Robert J. Delmas · Ian Galbally · Wei Min Hao · Robert C. Harriss
Ivar S. A. Isaksen · Daniel J. Jacob · Charles E. Kolb · Michael Prather · Henning Rodhe · Dieter Schwela
Will Steffen · Donald J. Wuebbles

7.1 Introduction

The preceding chapters have presented a topical synthesis of results obtained by the international scientific community after a decade of intensive research efforts focussing on the chemistry of the global atmosphere. Areas highlighted in this book are the relations between atmospheric composition and biospheric processes, the budget of atmospheric photooxidants including tropospheric ozone, and the importance of aerosols for the chemical composition of the atmosphere and for the climate system. The perturbing role of human activities has been stressed.

The data gathered from numerous field experiments (often organised under IGAC auspices), the analyses of observations performed with a hierarchy of numerical models, and the fundamental investigations carried out in the laboratory have provided a wealth of new and important information. The scientific progress made regarding our understanding of the processes that govern the observed state of the atmosphere allows us to provide today improved answers to some of the important questions that decision-makers often pose. In this last chapter, we present some responses to the societally important questions raised in Chap. 1. Specifically, we highlight the key factors that determine the chemical composition of the natural and anthropogenically perturbed atmosphere; we discuss the perturbations in global biogeochemical cycles and in the oxidising power of the atmosphere resulting from human-induced activities; we summarise the impacts of atmospheric changes on the biosphere, food production, and human health; and, we consider the possible future evolution of the atmospheric chemical composition with special attention to potential abrupt and irreversible changes. Finally, we suggest some general directions for future research, recognising that more detailed scientific priorities for the next 10 to 15 years, a subject beyond the scope of the present book, remain to be carefully established by the international community.

7.2 What Determines the Chemical Composition of the Atmosphere?

One of the remarkable discoveries of the past 50 years has been the increasing recognition of the role that biological processes play in determining the chemical composition of the atmosphere and the habitability of Earth (e.g., Schlesinger 1997; IGBP 1994; Matson and Harriss 1995; Daily 1997; NRC 1999a). Geological evidence indicates that the biosphere and atmosphere have co-evolved over billions of years. Complex Earth system feedback relationships are hypothesised to maintain the composition of the atmosphere within limits that allow for the habitability of life on Earth (Lovelock 1979, 1988). This hypothesis remains, however, a matter of debate within the scientific community.

Today, there is no pristine natural atmosphere remaining – the chemical fingerprint of human civilisation can be found in the most remote regions of both the lower and upper atmosphere. Grand challenges for atmospheric chemists are to attempt to derive the composition of the pre-industrial atmosphere from fossil air samples and to understand the consequences of the increasing human perturbation of atmospheric composition. The intimate relationship between the chemical composition of the atmosphere, human actions, and the undisturbed natural biosphere may mean that changes in trace gas concentrations are the best "early warning" indicator of changes in the status and health of Planet Earth over a range of local to global scales.

Trace chemicals in the atmosphere respond to the solar, biological, geological, chemical, and human forces that vary on a variety of time and space scales. At the 10 to 100 km scale, dramatic variations in the chemical composition of the atmosphere typically reflect intense emission sources of gaseous or particulate materials (e.g., volcanoes, biomass burning, or industrial pollutants). Beginning with the industrial revolution the sources, chemistry, and human health impacts of urban air pollution have been observed and studied (NRC 1991). The combustion of fossil fuels is the major anthropogenic source of gases such as carbon monoxide, nitrogen oxides, and hydrocarbons that subsequently react with other gases and particles in the presence of sunlight to reduce visibility and produce enhanced concentrations of ozone. Gaseous sulphur and nitrogen pollutants react with cloud droplets to acidify rain regionally (NRC 1986). The spatial scale of urban and industrial air pollution is determined by the magnitude of emission sources, meteorological dispersion of the air pollutants, and chemical processes that transform and remove pollutants (Chameides and Perdue 1997). The detrimental environmental and economic consequences of air pollution have been well documented. Unfortunately, the policy basis for regulating local and regional air quality remains incomplete due to the complexity of integrating competing economic, political, social, technological, and scientific factors into a robust risk management framework (NRC 1999b). In addition, regulatory measures, believed to be beneficial can have negative side effects. For example, reductions in emissions of SO_2 are beneficial for health but can augment greenhouse gas-driven climate warming by lowering sulphate aerosol concentrations.

At locations remote from human activities, variability in atmospheric trace gas composition is typically dominated by the physiology of the undisturbed natural landscape. Plant photosynthesis and respiration, microbial decomposition of organic matter, and other biological processes can be detected in near-surface variations in air chemistry (Matson and Harriss 1995). The concentrations of carbon, nitrogen, and sulphur gases are especially sensitive to variations in the ecology of the natural landscape.

The cumulative impact of human activities on the composition of the global atmosphere is clearly reflected by increasing concentrations of a variety of trace gases that have long atmospheric lifetimes such as carbon dioxide (CO_2), methane (CH_4), and nitrous oxide (N_2O). Studies of concentration trends of these and other gases using ancient air trapped in ice, along with direct measurements in recent decades, indicate that current levels are unprecedented for at least the past 400 000 years. The growth and accumulation of these gases primarily reflect increased global energy production, land-use change, food production, and waste generation over the past century (NRC 1999a). The basic trends in carbon dioxide, methane, and nitrous oxide, and their relation to human activities, are undisputed. However, there are significant variations in the rates of growth and accumulation of certain trace gases in the global atmosphere, which are poorly understood. The large-scale ecosystem sources and removal mechanisms for carbon and nitrogen trace gases are intimately connected to variability in the climate system and to climate-biosphere interactions. A new generation of Earth-observing satellites and *in situ* measurement technologies will soon resolve some of the vexing questions related to human activities and the future of the atmosphere.

It is estimated that human activities are now the dominant biological force shaping the landscape of Planet Earth (Vitousek et al. 1997), so the name "Anthropocene" has been suggested (Crutzen and Stoermer 2000) to characterise the past 200 to 300 years of the Holocene. The IGAC Project reflects the urgent need to understand how the human appropriation and transformation of nature will influence the dynamical and chemical properties of the future atmosphere. The scales of interest range from microbial production and consumption of atmospheric gases to tropical deforestation and global food production. The composition of the atmosphere will continue to serve as the most effective indicator of the pulse of Planet Earth. Biology rules the composition of critical atmospheric trace gases, but today and in the future the human species will be in control.

7.3 How Have Human Activities Altered Atmospheric Composition?

When the Industrial Revolution changed much of the world from a rural to an industrial society, the chemical composition of the atmosphere began to change as well. Changes were accelerated by world population growth. Ironically, societies that are in the process of modernisation are causing atmospheric change and, at the same time, being adversely affected by it. The most important causes of these changes include the use of specialty chemicals, fossil fuel combustion, land clearance for agriculture, ranching, and urbanisation, as well as the use of fertilisers. These human activities result in an increase in concentrations of many trace gases in the atmosphere that subsequently destroy ozone in the stratosphere, change the chemistry of the troposphere, and affect global climate. In particular, the concentration of greenhouse gases in the atmosphere has increased substantially since the pre-industrial period. The radiative forcing of anthropogenic origin (about 1% of the natural one) caused by these gases is now equivalent to the effect of a glacial-interglacial climate change.

The use of specialty chemicals significantly damages the global environment through a wide range of halogenated compounds, specifically chlorofluorocarbons (CFCs), halons, chlorocarbons, and bromocarbons. While inert, or very nearly so, in the troposphere these compounds are broken apart by short-wavelength UV light in the stratosphere. Halogen atoms are released, which react with and destroy ozone especially over Antarctica and the Arctic. The depletion of polar stratospheric ozone has spread to the midlatitude stratosphere. Decreases in the ozone layer and concomitant increases in ultraviolet radiation (UV-B) can cause sunburn, skin cancer, and eye damage in humans. Concentrations of organic chlorine contained in long and short lived chlorocarbons have increased since the late 1970s. After reaching a maximum value of about 3.7 nmol mol^{-1} in the mid-1990s, they have begun to decrease slowly in the troposphere as a result of the implementation of the 1987 Montreal Protocol on Substances that Deplete the Ozone Layer and its amendments (WMO 1999). Concentrations of organic bromine are, however, still increasing in the troposphere, largely because of the growth of Halon-1211. Chlorofluorocarbons are not only reactive with ozone in the stratosphere, they themselves are potent greenhouse gases.

Fossil fuel burning, deforestation, and cement production are the largest net sources behind the recent increases of atmospheric carbon dioxide, which is the most important greenhouse gas after water vapour. During the pre-industrial period, carbon dioxide concentrations were estimated to be 275–284 µmol mol^{-1}, based on ice core and firn measurements (Etheridge et al. 1996). The emission of carbon dioxide from fossil fuel burning has increased exponentially at a nearly constant rate of 4.3% per year between 1850 and the early 1970s, except during the Great Depression and World Wars I and II. Carbon dioxide concentrations increased from about 285 µmol mol^{-1} to 325 µmol mol^{-1} in the same period. In the past three decades, the growth rate of fossil fuel combustion has slowed due to the 1973 energy crisis, increased nuclear generating capacity, and energy conservation in some countries. Atmospheric CO_2 concentrations increased at a rate of about 1.5 µmol mol^{-1} yr^{-1} from 1980 to 1989; the present level is about 370 µmol mol^{-1}.

The radiative forcing of atmospheric methane is presently one third that of CO_2. The average CH_4 concentrations in Antarctic ice cores were shown to be around 670 nmol mol^{-1} between A.D. 1000 and A.D. 1800 (Etheridge et al. 1998). The anthropogenic increase started in the Northern Hemisphere during the second half of the 18th century. The current level of atmospheric CH_4 is about 1700–1750 nmol mol^{-1}, i.e. more than twice the pre-industrial CH_4 concentrations. The growth rate of CH_4 increased from 5 nmol mol^{-1} yr^{-1} in 1945 to its peak of 17 nmol mol^{-1} yr^{-1} in 1981. It has declined in the late 1990s to its pre-1945 rate. Microbial processes in rice cultivation, animal waste, and landfills, along with fossil fuel and vegetation combustion, are the major human activities that may have caused the increase of atmospheric concentrations. Uncertainties are associated with the extent of each source contributing to the increase of atmospheric CH_4 concentrations. It is also not known what has caused the declining growth rates in the 1990s, although fluctuations of the emissions, rather than of the sinks, seem to play a dominant role in the observed changes.

The utilisation of synthetic fertilisers and breeding of high-yield crops are two major trends in modern agriculture. These activities also have changed the chemical composition of atmospheric nitrous oxide, nitric oxide (NO), and CH_4 since the 19th century. Another important greenhouse gas, nitrous oxide, is produced during nitrification and denitrification of nitrogen fertilisers. Atmospheric N_2O concentrations have risen steadily from about 275 nmol mol^{-1} before the Industrial Revolution to 299 nmol mol^{-1} by 1976 and to 311–312 nmol mol^{-1} by 1996 (Elkins et al. 1998). Nitrous oxide increased at a rate of approximately 0.25% yr^{-1} during the 1980s, but more slowly in the 1990s. Although the causes of changes in atmospheric N_2O concentrations are not clear, agriculture is probably a major contributor to the increase. Nitrous oxide emissions from agricultural land have increased almost four times from 1900 to 1994 (Kroeze et al. 1999). Agricultural and industrial emissions now rival natural emissions in quantity.

In addition to CO_2, carbon monoxide (CO), sulphur dioxide (SO_2), and the higher nitrogen oxides (NO plus nitrogen dioxide, NO_2) are the other major pollutants emitted by fossil fuel burning. Approximately 20 to 40% of atmospheric CO is produced by fossil fuel burning. There is, however, only limited information on trends in global atmospheric CO concentrations. Measurements in polar ice cores suggest that CO concentrations may have increased in higher latitudes of the Northern Hemisphere since 1850, but not in the Southern Hemisphere (Haan et al. 1996). Sulphate and nitrate concentrations in the Northern Hemisphere have shown marked increases since the beginning of the 20th century, based on Greenland ice core measurements (Mayewski et al. 1990; Legrand et al. 1997). The sulphate concentrations in the late 1960s are four times those of the pre-industrial levels, but have begun to decline since about 1980 because of stringent regulation of pollutant emissions in North America and western Europe. Approximately half of global nitrogen oxides are produced by fossil fuel burning. Nitrate concentrations doubled between 1950 and 1989, but have started to decline in recent years because of the use of catalytic converters in motor vehicles and regulation curbing industrial pollution. Sulphate and nitrate are responsible for the acid precipitation phenomenon affecting several regions of North America and Eurasia.

Biomass burning, a unique source of many atmospheric trace gases and particulates, has been and continues to be a common land management practice, especially in the Tropics. Fires are associated with deforestation, shifting cultivation, grazing in savannas, clearing agricultural residue, and fuel wood. The amount of carbon monoxide produced by biomass burning is as great as that produced by fossil fuel combustion; it accounts for about a quarter of the total sources of atmospheric CO (WMO 1999). Biomass burning also contributes from 5 to 10% of atmospheric methane. Nitric oxide and ammonia are the major nitrogen-containing compounds produced by human-caused fires. The amount of NO emitted from biomass burning is about 30% of that produced from industrial sources. In addition, significant amounts of methyl chloride (CH_3Cl) and methyl bromide (CH_3Br) are emitted by vegetation fires. As with other halogenated compounds, both CH_3Cl and CH_3Br can react with ozone in the stratosphere. Most of the carbon- and nitrogen-containing compounds produced by biomass burning have short lifetimes, are chemically reactive, and are precursors of tropospheric ozone. Thus, inter-annual and seasonal trends of tropospheric ozone concentrations observed by the Nimbus 7 and TOMS satellites between 25° N and 25° S correspond to the seasonal cycle and extent of biomass burning in tropical Africa, Latin America, and Asia (Thompson et al. 1999). European ice records show that atmospheric ammonium concentrations started to grow before sulphate and nitrate (Döscher et al. 1996).

Little information is available regarding trends in tropospheric ozone since the pre-industrial era. Obser-

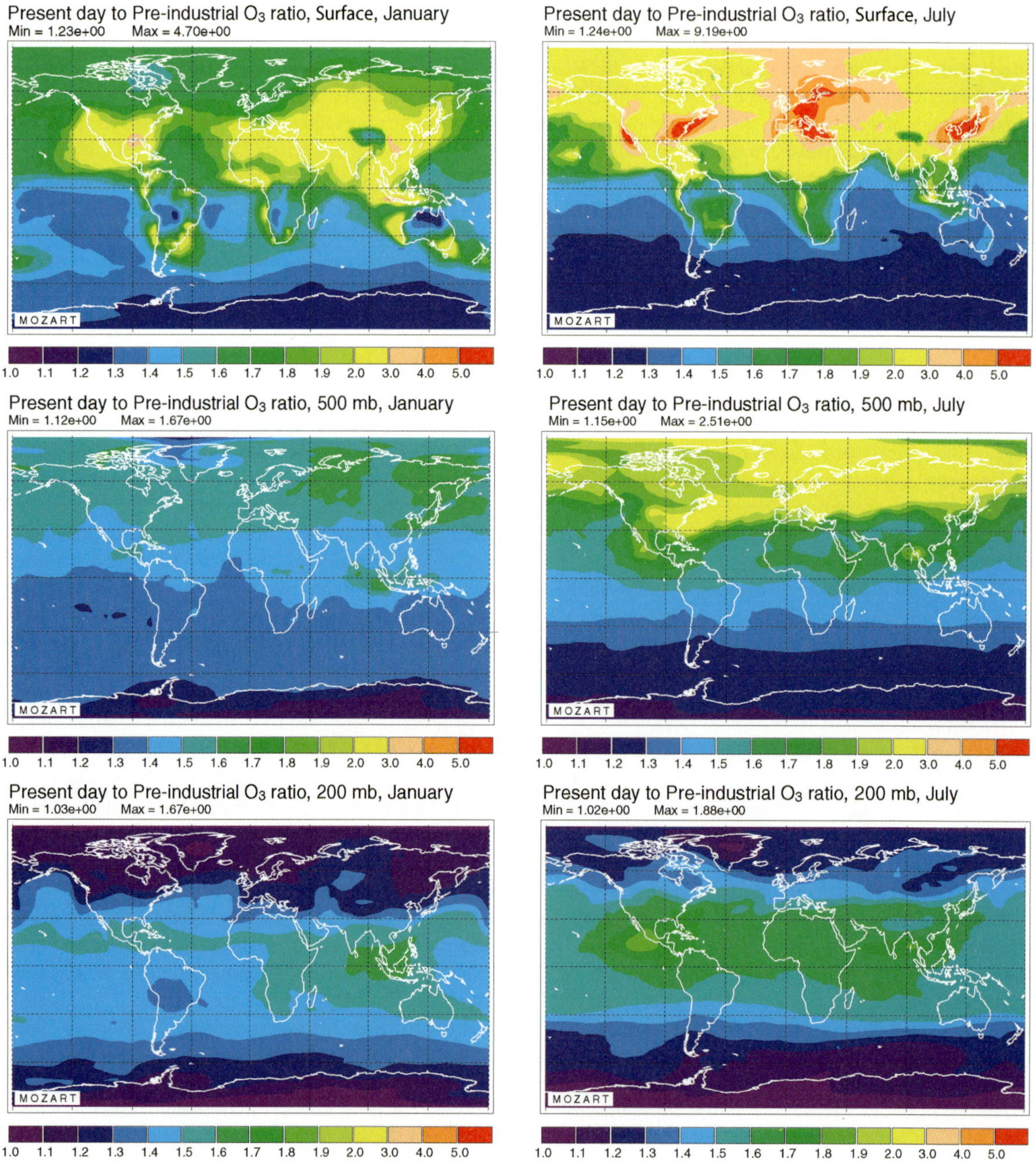

Fig. 7.1. Estimated changes (ratio) in the ozone concentration since the pre-industrial era at three different altitudes (surface, 500 hPa, and 200 hPa) in January (*left*) and July (*right*). The changes are expressed as ratio versus the concentration estimated for year 1850 (Hauglustaine and Brasseur 2001)

vations performed at the end of the 19th century or at the beginning of the 20th in different parts of the world (Volz and Kley 1988; Wang and Jacob 1998; Hauglustaine and Brasseur 2001) suggest that ozone mixing ratios at the surface were typically 15 to 20 nmol mol^{-1}, i.e. a factor two to three lower than current summertime values. These early measurements based on the technique established by Schönbein (Volz and Kley 1988; Marenco et al. 1994) must, however, to be viewed with caution because they can be biased by other environmental factors such as the relative humidity of the air. Global chemical transport models that account for changes in emissions of ozone precursors by fossil fuel, fertiliser usage, and biomass burning, also show substantial increases in the concentration of tropospheric ozone (Fig. 7.1) and its precursors (Fig. 7.2). The changes are most pronounced during summertime in the Northern Hemisphere, especially over industrialised and urbanised regions of North America, Europe, and parts of Asia. These estimates remain uncertain, however, primarily due to the limited information available regarding past changes in biomass burning frequency, especially in the Tropics.

Concentrations of carbonaceous particles, heavy metals (e.g., zinc, copper, cadmium, and lead), hydrogen peroxide, and acetic acid in glacial ice are several times higher now than they were in the 1800s. For instance, from pre-industrial (1755–1890) to modern times (1950–1975), carbonaceous particle concentrations at a high-Alpine glacier site have increased by factors of 3.7, 3.0, 2.5, and 2.6, for black carbon, elemental carbon, organic carbon, and total carbon, respectively (Lavanchy et al. 1999). In Greenland ice, the ratio between pre-industrial levels and the last 20 years is significantly higher for metals than for sulphate or nitrate: factors of 2.3, 2.7, and 9, for instance, for Zn, Cu, and Cd, respectively. The ratio is considerably higher for Pb (namely, 250) due to the use of organic lead derivates in combustion engine fuels.

Long-term time series of atmospheric dust over the Atlantic indicate a secular increase presumably associated with desertification of northern Africa. Dust is a major contributor to aerosol optical depth, scattering solar radiation and absorbing terrestrial radiation. In addition, deposition of dust to the ocean supplies nutrients (iron and other metals) that increase the efficiency of CO_2 uptake in high nutrient, low chlorophyll areas. Increasing desertification over the next 100 years is expected as a result of agricultural practices and irrigation pressure in the Tropics and subtropics. A resulting increase in atmospheric dust could have important implications for climate change, both directly by radiative forcing and indirectly by fertilisation of the oceans (see Chap. 2).

Historical records of the composition of reactive oxidants and their precursors in the atmosphere are scarce and limited mostly to high latitude regions. No instrumental record longer than 15 years exists. Hydrogen peroxide (H_2O_2) is directly related to the oxidising power of the atmosphere. The most recent Greenland atmospheric, snow pit, and firn core studies lead to an estimate of an overall H_2O_2 increase of 60 ±12% since pre-industrial time. The concentration profile of H_2O_2 departed from natural background fluctuations in the middle of the 19th century, i.e. long before that of nitrate. In the most recent years, the increasing trend seems to be more pronounced and the seasonal amplitude has tripled since 1970 (Anklin and Bales 1997). Formaldehyde measurements in Greenland snow are scarce and still uncertain. They suggest a doubling of pre-industrial concentration (about 2 ng g^{-1}) since that era, with an accelerated increase in the most recent decades.

Light carboxylic acids in the atmosphere are linked to natural emissions by vegetation and biomass burning. Anthropogenic activities are also potential contributors, directly or indirectly, to their global carboxylic acid budget. Interestingly, excluding sporadic biomass burning spikes, a decreasing temporal trend in formate concentration is observed over the last century in Greenland snow, suggesting that the anthropogenic source is not dominant for this chemical species.

It can be expected that the composition of the atmosphere will continue to change in the future. For several long lived chemical species, the trends will be hard to reverse. On the other hand, for short lived pollutants, stringent measures could rapidly be profitable. Green-

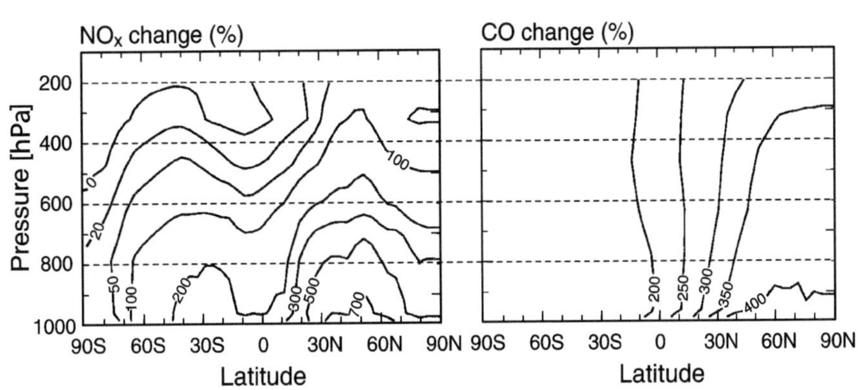

Fig. 7.2.
Percentage change in the yearly and zonally averaged concentration of NO_x and CO from pre-industrial times to present (Wang and Jacob 1998)

house gas emissions, mainly associated with fossil-fuel burning, will still increase. In the Northern Hemisphere, methane and tropospheric ozone could become more important in radiative forcing. Industrialisation of developing countries and changes in land use in tropical countries will have a significant impact on trace gas and particulate concentrations in the atmosphere.

7.4 How Have Human Activities Changed the Global Atmospheric Budgets of Carbon, Nitrogen, and Sulphur?

Global budgets are an instructive way of summarising our understanding of the occurrence and fluxes of the key chemical species in the atmosphere. Such budgets enable one to compare the magnitude of the various fluxes and to find out whether estimates of the sources are balanced by estimates of the sinks. The relative importance of natural and anthropogenic sources can be seen easily. But there are also shortcomings associated with the global budget approach. Many chemical species, including several compounds of nitrogen and sulphur, are not long lived enough to have a uniform global distribution, implying that their concentrations vary from region to region depending on the geographical distributions of the emissions. In such cases global totals may not be representative of the actual situation in specific places. Budgets over specific regions could then be more illustrative.

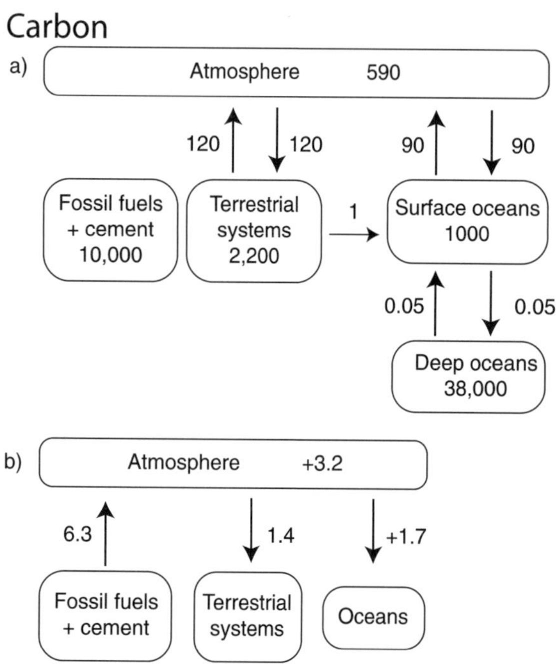

Fig. 7.3. a Schematic representation of the global carbon cycle for pre-industrial times (fluxes in Pg C yr^{-1} and burdens are in Pg C); and **b** present-day perturbations in the cycle in response to human activities (fluxes in Pg C yr^{-1}). Numerical values should be regarded only as illustrative of recent estimates

In the following paragraphs, we provide estimates of the budgets of carbon, nitrogen, and sulphur for pre-industrial and present-day times, respectively. It should be stressed that, in many cases, large uncertainties remain on the estimated fluxes and their evolution with time. The numerical values that are provided in Fig. 7.3 to 7.10 should therefore be regarded as illustrative only. Budget estimates provided by different authors are available in IPCC (2002).

Carbon. The main carbon-containing compound in the atmosphere is carbon dioxide (CO_2). It is of fundamental importance to the climate system because of its strong absorption of infrared radiation giving rise to the greenhouse effect. For a detailed discussion of the global carbon cycle, the sources and sinks of atmospheric CO_2, and the human induced perturbation of these fluxes, see the assessment report by IPCC (1996).

The global fluxes of CO_2 to and from the atmosphere, as well as the atmospheric burden during pre-industrial times and during the past decade, are shown in Fig. 7.3. Present-day emissions from fossil fuel combustion, cement manufacturing, and deforestation amount to only about 5% of the natural (pre-industrial) emissions. Nevertheless, the atmospheric burden has increased by 30% because of these anthropogenic emissions. The main reason is that the natural exchanges between the atmosphere on the one hand, and the oceans and the terrestrial ecosystems on the other, are gross fluxes that essentially balance over a year, whereas the human induced emissions represent a net input. The atmospheric burden during pre-industrial times – corresponding to an atmospheric concentration of about 280 $\mu mol\ mol^{-1}$ – is the result of a complex interplay between volcanic CO_2 emissions, burial of organic sediments, and storage of carbon in terrestrial biota and in the oceans.

The nominal turnover time of CO_2 in the atmosphere, obtained as the ratio of the pre-industrial burden and the total (gross) removal rate in Fig. 7.3a, is about three years. This means that, on average, a CO_2 molecule in the atmosphere spends about three years in the atmosphere before it is taken up by terrestrial biota or by the oceans. But since most of the CO_2 molecules will re-enter the atmosphere within a few years, the effective lifetime, before eventual sequestration in the deep oceans, is of the order of 100 years (1 000 years for the return of carbon from the deep ocean to the atmosphere). This 100-year time scale also represents the time scale of adjustment of atmospheric CO_2 to changes in emissions. If all anthropogenic emissions were halted, it would take hundreds of years before the concentration approached steady state again. This steady concentration would be lower than the current level but still not quite as low as that during the pre-industrial period. Because of the long turnover time of CO_2 in the atmosphere, its concentration is relatively uniform worldwide and the upward trend appears wherever the measurements are made.

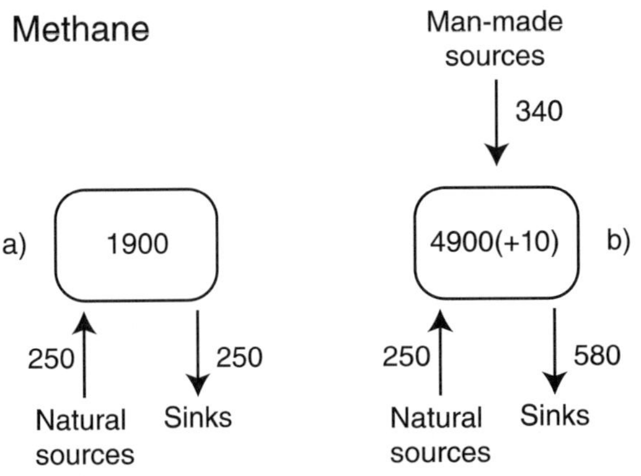

Fig. 7.4. a Schematic representation of the global atmospheric methane budget for pre-industrial times, and **b** the present-day. Fluxes are in Tg CH_4 yr^{-1} and burdens are in Tg CH_4. Numerical values should be regarded only as illustrative of recent estimates

The second most abundant carbon-containing compound is methane (CH_4). This gas plays an important role in atmospheric photochemistry (see Chap. 3), and it is also a major contributor to the greenhouse effect. The methane budget is summarised in Fig. 7.4. In this case, the anthropogenic emissions are more than twice as large as the natural emissions and the atmospheric burden has increased by the same proportion, i.e. by some 140% since pre-industrial times. As with CO_2, the pre-industrial burden of CH_4 has been estimated from measurements of air trapped in ice cores (Dlugokencky et al. 1994). The most important anthropogenic CH_4 sources include rice cultivation, exhalations from domestic animals, biomass burning, and coal mining. The atmospheric turnover time of CH_4 is around nine years. This is long enough for it to be reasonably well mixed around the globe but short enough for the atmospheric concentration to respond within a few years to changes in emissions. This means that a reduction in emissions will be followed within a few years by an approximately corresponding reduction in atmospheric concentration.

In absolute terms, the human induced increase in CH_4 is much less than that of CO_2: 1 vs. 90 µmol mol^{-1}. Despite this, the contribution of the CH_4 increase to the current greenhouse effect is as large as 30% of the corresponding contribution of CO_2. This is because CH_4, per molecule, is about 40 times more efficient as a greenhouse gas than is CO_2.

Carbon monoxide (CO) is another key carbon-containing molecule in the atmosphere. This gas plays an important role as a precursor to ozone formation in the troposphere, and as a reactant with the hydroxyl (OH) radical (c.f., Chap. 3). The atmospheric content of carbon monoxide has also increased substantially because of human activities, see Fig. 7.5. The primary anthropogenic fluxes include combustion processes (mainly traffic, forest clearing, and savannah burning) and oxidation of CH_4 derived from human induced sources. In the case of CO, no direct estimate is available of the pre-industrial burden; the number in Fig. 7.5a is scaled from the present burden by the ratio of natural to total emissions. The turnover time of CO is about two months, indicating a less uniform distribution around the globe than CO_2 and CH_4. The annual and latitudinal average concentration of CO in surface air shows a strong inter-hemispheric gradient with higher values in the Northern Hemisphere due to anthropogenic emissions (c.f., Chap. 3, Fig. 3.8). A secondary maximum in the Tropics is associated with the biomass burning.

In addition to CO_2, CH_4, and CO, there are small but chemically significant concentrations of a number of gaseous and particulate carbon compounds, many of them affected by human induced emissions (see Chap. 2 and 4). On the global scale, biogenic nonmethane hydrocarbon emissions (e.g., isoprene) are substantially larger than anthropogenic sources; however, natural sources can be reduced significantly as a result of deforestation. No attempt is made here to summarise their budgets.

Nitrogen. Nitrous oxide (N_2O) is important both as a greenhouse gas and because of its influence on ozone concentration in the stratosphere. Its anthropogenic sources include emissions from cultivated soils (fertilised with nitrogen), biomass burning, and various industrial processes, several of them not well quantified. They sum up to be almost as large as the natural emissions (c.f., Fig. 7.6). The turnover time of N_2O is around 100 years. The current rate of increase in its atmospheric concentration is about 0.2% per year, indicating an imbalance between sources and sinks.

NO and NO_2 – the sum of them is often referred to as NO_x – are key components of the atmosphere with a strong influence on the concentration of ozone (c.f., Chap. 3, Sect. 3.3.5). Their oxidation product (nitric acid,

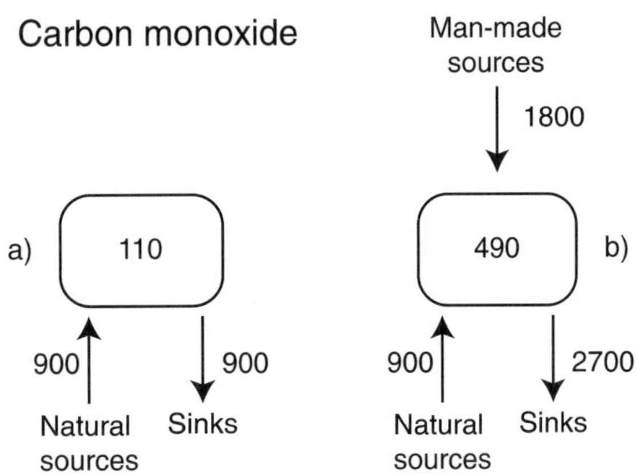

Fig. 7.5. a Schematic representation of the global atmospheric carbon monoxide budget for pre-industrial times, and **b** the present-day. Fluxes are in Tg CO yr^{-1} and burdens are in Tg CO. Numerical values should be regarded only as illustrative of recent estimates

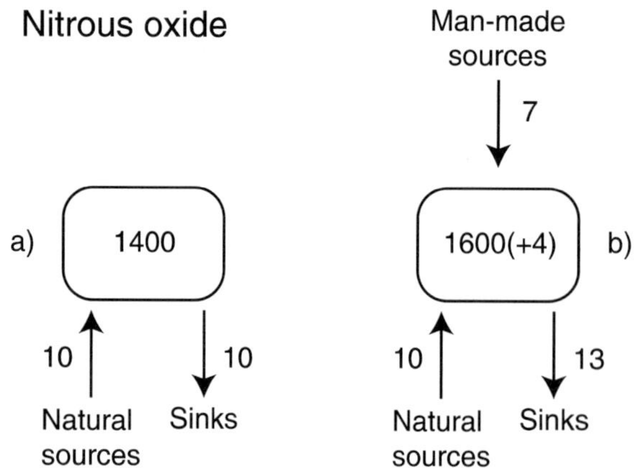

Fig. 7.6. a Schematic representation of the global atmospheric nitrous oxide budget for pre-industrial times, and **b** the present-day. Fluxes are in Tg N yr^{-1} and burdens are in Tg N (+ annual increase). Numerical values should be regarded only as illustrative of recent estimates

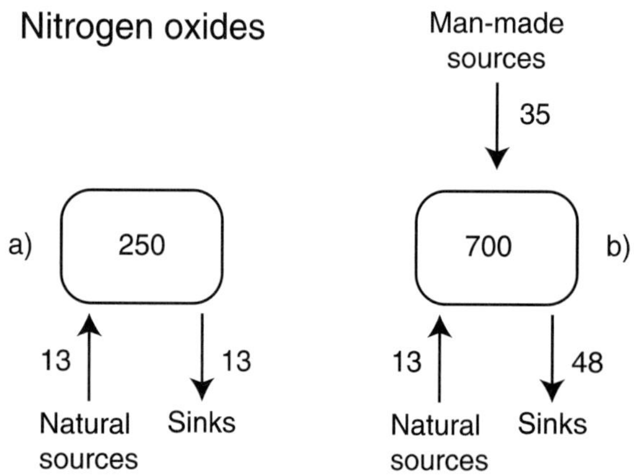

Fig. 7.7. a Schematic representation of the global atmospheric nitrogen oxides ($NO_x = NO + NO_2$) budget for pre-industrial times; and **b** the present-day. Fluxes are in Tg N yr^{-1} and burdens are in Mg N. Numerical values should be regarded only as illustrative of recent estimates

HNO_3) contributes to the acidity of rainwater and, when deposited to the ground, also provides a nutrient that is limiting in many terrestrial and marine ecosystems. Furthermore, NO_x is a toxic pollutant with adverse health effects at concentrations that may occur mainly in urban areas. The global budget of NO_x is presented in Fig. 7.7. The current anthropogenic emissions are dominated by fossil fuel combustion and biomass burning. A fraction of the emission from soils is also likely to be affected by human activities.

Since the atmospheric turnover time of NO_x is limited to a few days, the geographical distribution of its atmospheric concentration and rate of deposition is patchy, with higher values concentrated in and around the most industrialised regions (c.f., Chap. 3, Fig. 3.4). The short turnover time also implies that any future change in emission will be immediately seen as corresponding changes in local concentration and deposition rates.

The global budget of ammonia/ammonium (NH_x) is shown in Fig. 7.8. Ammonia is the only important gaseous alkaline component in the atmosphere. It has a strong influence on the acid-base balance of aerosol particles, cloud droplets, and rain water and is a major component of many aerosol particles. When deposited it can have both positive and negative effects on ecosystems. Most emissions related to human activities are due to domestic animals, biomass burning, and fertilisers. Taken together the human induced emissions account for about 75% of the total. In addition to Europe and North America, India (with its large cattle population) shows up as a major source of NH_3. The increasing use of nitrogen fertilisers worldwide points towards a continued increase in anthropogenic emissions.

Aerosols and their precursors. Emissions of sulphur compounds (mainly SO_2) associated with human activities have a profound impact on the acid-base status of aerosols, clouds, and precipitation. High concentrations of SO_2 may also lead to harmful effects on human health, plant growth, and corrosion of building material. In addition, the aerosol sulphate resulting from these emissions exerts a strong influence on climate (c.f., Chap. 4, Sect. 4.3.2.3). On a global basis, about 70% of the current emissions of gaseous sulphur compounds derive from anthropogenic sources, mainly fossil fuel combustion (c.f., Fig. 7.9). The short atmospheric turnover time of SO_2 and aerosol sulphate (only a few days), together with the uneven geographical distribution of the human induced emissions, leads to a pronounced patchiness in the distribution of these sulphur compounds. In the most polluted regions in North America, Europe, and China, anthropogenic emissions represent more than 90% of total emissions, implying a dramatic impact on the chemical climate. The concern regarding environmental effects, including acid deposition, has led to a substantial reduction in SO_2 emissions in some parts of the world (mainly Europe and North America) during the 1980s and 1990s. In other regions, including South and East Asia, emissions show a strong positive trend.

There are several other environmentally important aerosol components, including organic compounds, soot, and soil dust, whose concentrations have been substantially affected by human activities. However, neither the anthropogenic nor the natural fluxes have yet been well quantified for many of these components. The global budget of soot (black carbon) is shown in Fig. 7.10 as an example of a primary aerosol component. Natural emissions of soot are believed to be small compared to those from fossil fuel combustion and biomass burning, but uncertainties are large.

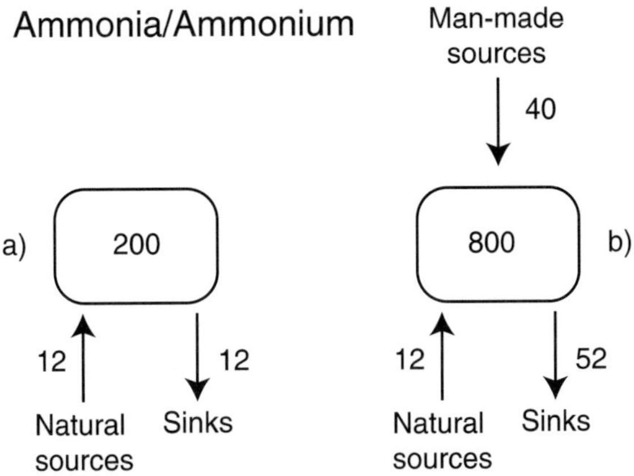

Fig. 7.8. **a** Schematic representation of the global atmospheric ammonia/ammonium (NH_x) budget for pre-industrial times; and **b** the present-day. Fluxes are in Tg N yr^{-1} and burdens are in Mg N. Numerical values should be regarded only as illustrative of recent estimates

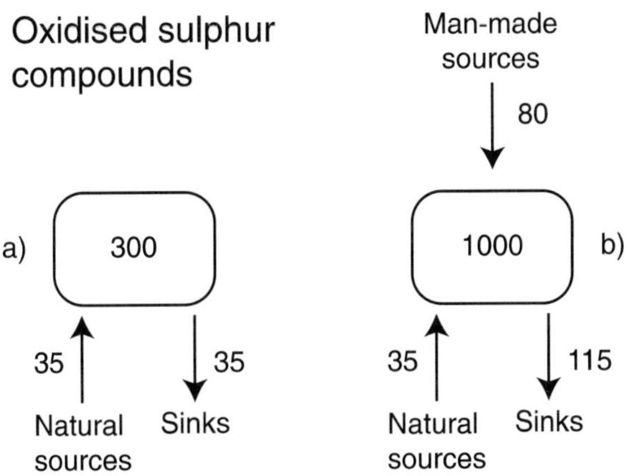

Fig. 7.9. **a** Schematic representation of the global atmospheric budget of oxidised sulphur compounds (SO_X) for pre-industrial times, and **b** the present-day. Fluxes are in Tg S yr^{-1} and burdens are in Mg S. Numerical values should be regarded only as illustrative of recent estimates

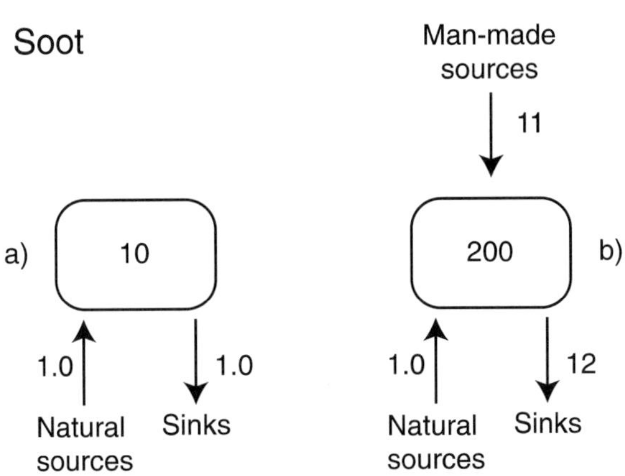

Fig. 7.10. **a** Schematic representation of the global atmospheric budget of soot carbon for pre-industrial times, and **b** the present-day. Fluxes are in Tg C yr^{-1} and burdens are in Mg C. Numerical values should be regarded only as illustrative of recent estimates

7.5 What Controls Tropospheric Ozone?

Ozone in the troposphere (the lowest part of the atmosphere, extending from the surface to 10–15 km altitude) is a chemical species of central environmental interest (see Chap. 3 for more details). Photolysis of ozone in the presence of water vapour produces the OH radical, which is the main atmospheric oxidant responsible for removal of a large number of important gases including methane, other hydrocarbons, CO, NO_x, and hydrochlorofluorocarbons. At the earth's surface, ozone is toxic to humans and vegetation; it is one of the principal harmful components of smog. In the middle and upper troposphere, ozone is a major greenhouse gas.

Only about 10% of total atmospheric ozone resides in the troposphere. The rest resides in the stratosphere above, where it is produced by photolysis of molecular oxygen at UV wavelengths shorter than 240 nm. Such strong UV radiation does not penetrate into the troposphere. Until the 1970s it was thought that tropospheric ozone was mainly supplied by transport from the stratosphere, and removed by deposition involving reactions with organic materials at Earth's surface. Research over the last two decades has shown that tropospheric ozone is in fact largely controlled by chemical production and loss within the troposphere.

The general mechanism for ozone production in the troposphere is by photochemical oxidation of hydrocarbons and CO by OH in the presence of nitrogen oxides ($NO_x = NO + NO_2$). In the simplest case of oxidation of CO, the mechanism is as follows:

$$O_3 + h\nu \longrightarrow O_2 + O(^1D) \tag{7.1}$$

$$O(^1D) + H_2O \longrightarrow 2OH \tag{7.2}$$

$$CO + OH \longrightarrow CO_2 + H \tag{7.3}$$

$$H + O_2 + M \longrightarrow HO_2 + M \tag{7.4}$$

$$HO_2 + NO \longrightarrow OH + NO_2 \tag{7.5}$$

$$NO_2 + h\nu \longrightarrow NO + O \tag{7.6}$$

$$O + O_2 + M \longrightarrow O_3 + M \tag{7.7}$$

The mechanism involves oxidation of CO by O_2 catalysed by hydrogen oxide radicals ($HO_x = OH + H + HO_2$) and by NO_x. It can be viewed as a reaction chain initiated by generation of HO_x radicals from photolysis of ozone (Reactions 7.1–7.2), and propagated by cycling of HO_x and NO_x radicals through Reactions 7.3–7.7. The stoichiometric net effect of Reactions 7.3–7.7 is

$$CO + 2O_2 \longrightarrow CO_2 + O_3 \tag{7.8}$$

The critical element in the chain propagation is the cycling of NO_x radicals between NO and NO_2, leading to ozone production following NO_2 photolysis. Oxidation mechanisms for hydrocarbons are more complicated but follow the same general schematic as for CO.

Laboratory and field studies over the past two decades have provided considerable evidence for ozone production by the mechanism described above. Sensitive instrumentation for measuring HO_x and NO_x radicals has been developed that allows quantitative estimates of radical concentrations and associated ozone production rates in different regions of the troposphere. Chemical loss of ozone, which takes place principally by Reactions 7.1–7.2 and by reactions with HO_x radicals, can similarly be estimated. Rapid progress in global 3-D models of tropospheric chemistry has provided a tool for placing these ozone production and loss rates in a global context and coupling them to transport processes. Considerable uncertainties, however, remain regarding a quantitative understanding of stratosphere-troposphere mass exchange.

As a result of these different advances, it is now well established that chemical production of ozone within the troposphere dominates over transport from the stratosphere as a source of tropospheric ozone. Similarly, it is also well established that chemical loss dominates over deposition as the major sink of tropospheric ozone. In this manner, ozone in the global troposphere is largely determined by a balance between chemical production and chemical loss. Because the lifetime of ozone is relatively long (typically several weeks), this balance is strongly modulated by transport on global scales. Net ozone production takes place in polluted regions with strong UV radiation, and in the upper troposphere; net ozone loss takes place in remote regions of the lower troposphere.

Building on the above result, much recent work has been done to better understand the sources of tropospheric ozone precursors (hydrocarbons, CO, NO_x). Sources of hydrocarbons and CO include natural biogenic emissions, combustion, and industrial activities. Sources of NO_x include lightning, biogenic processes in soils, and combustion. Research over the past two decades has established that ozone production is largely limited by the supply of NO_x and not by the supply of hydrocarbons or CO.

Aircraft campaigns have built a large database for concentrations of NO_x and related chemical species in various regions of the world. Loss of NO_x takes place on a time scale of a day by atmospheric oxidation and hydrolysis to nitric acid (HNO_3) and to organic nitrates including, in particular, peroxyacetyl nitrate (PAN). This short lifetime limits the extent to which NO_x can be transported directly from continental sources to the remote troposphere. There is now good evidence that NO_x in many remote regions of the troposphere is maintained by thermal decomposition of PAN transported over long distances (thousands of kilometres or more). PAN's efficacy as a reservoir for NO_x greatly increases the scale over which anthropogenic emissions from combustion can increase NO_x and hence ozone production in the troposphere.

There remains much uncertainty regarding the sources of NO_x in the remote troposphere. In the Tropics, where production of ozone is of particular importance as a source of OH, the dominant sources of NO_x appear to be lightning, soils, and seasonal biomass burning. Present estimates of the global lightning source vary by an order of magnitude, and estimates of emissions from soils and seasonal biomass burning vary several fold. In the upper troposphere, where ozone is a major greenhouse gas, recent field studies indicate that emissions from aircraft, lightning, and deep convective transport of combustion products all make significant contributions to NO_x levels; the relative contribution from each source remains a matter of much uncertainty. Some field observations have suggested that NO_x levels in the upper troposphere must be maintained by fast chemical recycling of HNO_3 through an unknown mechanism, but the interpretation of these observations is controversial. Better understanding of the factors controlling tropospheric NO_x is critical for future assessments of the effects of human activity on tropospheric ozone.

7.6 Is the "Cleansing Efficiency" of the Atmosphere Changing?

The capacity of our atmosphere to remove a chemical injected into it is conveniently expressed as the inverse of the lifetime of the chemical. This lifetime over which a molecule, atom, free radical, or aerosol particle remains in the atmosphere is determined in turn by a number of atmospheric chemical and physical processes. For example, certain pollutants are rapidly removed from the atmosphere either because their chemical reactivity is high or because they are highly soluble in water. Others that are chemically stable and relatively insoluble in water (such as the chlorofluorocarbons) may remain in the atmosphere for years, decades, or even centuries. In this case, they are transported to remote polar regions and to the middle atmosphere where they can play important roles in chemistry (such as that of ozone depletion).

The atmosphere is an oxidising medium. Almost all hydrogen-containing gases of natural and anthropogenic origin are oxidised by the hydroxyl (OH) free radical, a chemical species often referred to as the "detergent" of the atmosphere. Other oxidation processes include reactions with ozone (O_3), hydrogen peroxide (H_2O_2), the nitrogen trioxide free radical (NO_3), and halogen species. Hydrogen peroxide, for example, converts sulphur dioxide to sulphuric acid in cloud and rain droplets and hence contributes to the formation of acid precipitation. The NO_3 free radical plays a significant role at night, primarily in polluted areas.

The rate at which oxidation occurs varies widely from chemical species to chemical species. For example, the oxidation of methane typically takes nine years, while that of isoprene takes less than a day. Thus, even though the magnitude of the surface emissions of these two hydrocarbons is similar, their atmospheric burdens differ by several orders of magnitude.

The rate at which chemical compounds are eliminated from the atmosphere is very often not determined by the precipitation rate, but rather by the rate at which they are oxidised into a water-soluble form. This oxidation rate therefore plays a central role in many global environmental problems. An important issue is to quantify any potential change in the oxidising power or "cleansing efficiency" of the atmosphere that may have resulted from human activities (Thompson 1992).

Over the past decade and century, the oxidising power of the atmosphere may have changed. For example, the increasing penetration of solar UV light in response to stratospheric ozone depletion and the increased emissions of NO_x and nonmethane hydrocarbons from combustion tend to increase ozone and OH concentrations in the troposphere. Conversely, the increase in carbon monoxide emissions, which has also resulted from increased biomass and fossil fuel burning since the pre-industrial era, tends to reduce OH abundance. Similarly, increases in the methane abundance due to increases in cattle, rice, and natural gas usage should have contributed to a decrease in the global OH concentration and therefore in the oxidising power of the atmosphere.

Since there are presently no long-term regional to global scale observations of tropospheric OH, the change in the oxidising power of the atmosphere cannot be directly monitored. Estimates must therefore rely primarily either on model calculations or deductions using measurements of compounds destroyed by OH. Model calculations suggest that global tropospheric levels of OH have probably decreased significantly since the pre-industrial era. The quantitative value of this decrease varies from model to model, however, with figures ranging from 3% in Lelieveld and Van Dorland's IS model (1995), to 22% in that of Roelofs et al. (1997). An intermediate value of 9% was provided by Wang and Jacob (1998) and a range of 5–10% was estimated by Crutzen and Zimmermann (1991). Earlier estimates based on one-dimensional models suggested a decrease ranging from 10 to 30% (Thompson 1992). The general consensus from modelling studies is that, despite the large increase in the emissions of carbon monoxide, hydrocarbons, and nitrogen oxides, and the stratospheric ozone depletion during the 20th century, the partially offsetting effects of these factors implies that the globally averaged concentration of OH has remained relatively constant. Decreases probably do not exceed 20% (Wang and Jacob 1998).

The analysis of the methylchloroform (CH_3CCl_3) observations performed since 1978 at several surface monitoring stations of the Atmospheric Lifetime Experiment (ALE), Global Atmospheric Gases Experiment (GAGE), and the Advanced Global Atmospheric Gases Experiment (AGAGE) (Prinn et al. 2000 and references therein) provides an indirect method to derive trends in the global concentration of tropospheric OH. Methylchloroform's sources, which are purely anthropogenic, are believed to be known very accurately. By combining the estimates of these sources with the CH_3CCl_3 measurements, little change in the OH density was deduced by Prinn et al. (1995) for the period 1978–1994 (specifically 0.0 ±0.2% per year, 1σ error range). A similar study by Krol et al. (1998) suggested a small increase in the OH density (0.46 ±0.6% per year, full error range) from 1978–1993. A more recent analysis of the methylchloroform data for 1978–2000 by Prinn et al. (2001) suggests that the global level of OH increased between 1978 and 1988, but then decreased between 1988 and 2000. Overall, the global trend in the average OH abundance between 1978 and 2000 was estimated to be $-0.64 \pm 0.60\%\ yr^{-1}$, with the most pronounced decrease in the Southern Hemisphere.

In summary, recent analyses suggest that the oxidising power of the atmosphere has remained relatively constant since the pre-industrial era, with a decrease in the OH concentration, and hence in the lifetime of major primary pollutants (or greenhouse gases such as methane) that has not exceeded 20%. On shorter time scales, the abundance of OH may have changed rather substantially, however; for example, in response to rapid ozone depletion in the stratosphere, fluctuations in tropospheric water vapour, or in the emissions of nitrogen oxides or carbon monoxide (see, e.g., Madronich and Granier 1992; Granier et al. 1996). Future changes in the oxidising power of the atmosphere will depend on future anthropogenic emissions and hence on future economic development and population growth.

7.7 How Does Atmospheric Chemistry Affect the Biosphere and Food Production?

The changes in atmospheric composition and chemistry described earlier in this volume and elsewhere have numerous direct impacts on both natural and managed ecosystems. One of the best known cases is damage caused by emission of SO_2 and NO_x from the industrial and transport sectors, especially in northern Europe and parts of northeastern North America. These emissions cause acid deposition, which reduces the alkalinity of lakes and streams and also impact on soils and terrestrial ecosystems directly. The effects on freshwater ecology, forest health, and soil functioning have been relatively well characterised. Although sulphur emissions in Europe and North America have been declining, in many other regions of the world, e.g., parts of Asia, sulphur emissions are growing rapidly, causing concern about future acidification problems.

The "acid rain" saga in Europe and North America was the first well-documented regional-scale air pollution phenomenon. The number and type of atmospheric impacts are now growing, extending to other regions, and connecting across regions to become global-scale phenomena.

7.7.1 Impact on Terrestrial and Marine Ecosystems

Ecosystems are affected by atmospheric change in many direct ways, including fertilisation effects, toxicity, calcification, and radiation damage, as well as indirectly through impacts on biodiversity. In addition, the interaction between atmospheric change and ecosystem response is having some profound and surprising feedbacks to the functioning of the Earth System.

7.7.2 Direct Effects

Fertilisation. Enhanced atmospheric CO_2 has long been known to cause increased growth in terrestrial and marine ecosystems, and there is some evidence that the deposition of fixed N compounds (NO_3^-, NH_4^+) also has a fertilisation effect. The recent rates of regrowth of some European forests, more vigorous than in the historical past, have been attributed to this effect. In some cases atmospheric deposition can provide a substantial portion of required nutrients. For example, a study of aerosol deposition in southern Africa indicates that in most of the ecosystems of the Okavango Delta wetlands, aerosols supply up to 52% of phosphates and 30% of nitrates needed by the vegetation (Garstang et al. 1998).

Toxicity. The same fixed N compounds that fertilise ecosystems in lower doses can become toxic in high concentrations. This is particularly likely when other nutrients are strongly limiting and N compounds are present in excess.

Large emissions of SO_2 and NO_x from industrial processes and transportation, and the subsequent deposition of these compounds and their associated acids, have led to widespread damage to sensitive ecosystems especially in northern Europe and parts of northeastern North America. This type of impact has been well documented for freshwater ecosystems. Acid deposition reduces the alkalinity of lakes and streams. In waters with low buffering capacity, the pH can be reduced to levels that cause acute and chronic impacts associated with increased aluminium levels that accompany the lowered pH. Such acidification can also affect crops, water resources, and structures (corrosion). The effects of acidification on soils and terrestrial ecosystems are complex. Sulphur deposition has increased concentrations of absorbed sulphate in soils and caused a depletion of base cations, especially magnesium and potassium, leading to a nutrient deficiency. In parts of northern Europe, the acidity of forest soils has increased considerably (pH has decreased by 0.5–1.0 units) during the past 30–60 years, at least partly as a result of acidification. This has led to reduced root distributions of forest trees. An additional stress to forests is caused by the increased mobilisation of metals (e.g., Al, Cd). The combined effect of the various pollutant-related stresses to forest ecosystems is not yet well characterised.

Although the sulphur emissions in Europe and North America are now declining – as a result of policy actions – sulphate deposition is still well above the so-called "critical load" above which acidification damage to sensitive ecosystems will occur. This means that soils in many areas continue to deteriorate as base cations are leached out of the ground.

In other regions of the world – e.g., southern and eastern Asia – sulphur emissions are growing rapidly and causing concern about future acidification problems. Widespread distribution of alkaline soil dust in the atmospheres of some parts of these regions – northern parts of China and a large part of India – has contributed to maintaining rainwater pH above 5.0 or 6.0. However, this buffering effect is likely to be exhausted if emissions of SO_2 and NO_x continue to grow.

Deposition of nitrogen compounds (NO_3^-, NH_4^+) has led to the fertilisation of many terrestrial ecosystems and, following nitrification of NH_4^+, to increased nitrate leaching to groundwater and runoff. Even more important may have been the use of nitrogen fertilisers. This fertilisation effect has caused changes in the functioning and stability of many sensitive ecosystems (e.g., heathlands and bogs) in the industrialised parts of the world. In very polluted areas (e.g., The Netherlands) the drinking water standard for nitrate has been exceeded. On the positive side, the addition of nitrogen in soils may have led to increased biomass and hence to net uptake of CO_2 from the atmosphere.

Elevated deposition of nitrogen compounds, as well as high runoff of such compounds from adjacent land areas, has contributed to eutrophication of lakes and coastal waters in polluted regions. Episodes of high wet deposition of nitrogen are also suspected to affect more remote marine ecosystems occasionally.

In some polluted regions there is evidence for O_3 impacts, including visible leaf damage on sensitive plant species and even reductions in the growth of some forest tree species (e.g., beech). Whereas direct damage due to elevated concentrations of SO_2 and NO_2 has been documented mainly in or around urban areas and close to large pollution sources, damage due to elevated O_3 levels is known to occur regularly during the summer season in extended regions in northern Europe, North America, and some parts of the Middle East and Asia. The damage due to O_3 produces visible leaf injury on sensitive plant species and causes a decrease in the yields of economically important crops (see Table 7.1). These effects also could have implications for forest production on broad geographical and time scales. There is an urgent need to increase our understanding of the current – and potential future – extent of O_3 damage, especially in those parts of the world (e.g., China and India) where emissions of O_3 precursors (NO_x, hydrocarbons, and CO) are rising. The possibility that harmful O_3 levels may occur in connection with biomass burning in tropical and subtropical regions should also be investigated.

Biodiversity. Experimental research on grassland plots has shown that N addition leads to changes in plant species composition and an overall reduction in diversity. Similar effects have been noted in the field in Europe, where heath species have been eliminated from some heathlands in The Netherlands and where increasing dominance by grasses is occurring in many heath, meadow, and forest ecosystems. In general, the impacts are most pronounced in nutrient poor systems, where N deposition enhances growth of the most responsive species, which often outcompete and eliminate rare species that occupy N-deficient habitats (Mooney et al. 1999; Sala et al. 1999).

Calcification. Changing atmospheric composition also impacts on marine ecosystems. Rising CO_2 reduces oceanic alkalinity and hence the calcium carbonate saturation state of seawater; this leads to decreased rates of calcification (growth) of organisms which form coral reefs. A recent study estimates that a doubling of atmospheric CO_2 may reduce the calcification rate by as much as 30% (Kleypas et al. 1999).

UV damage. Increased levels of UV radiation at the earth's surface, caused by stratospheric ozone depletion over recent decades, have likely resulted in damage to terrestrial organisms including plants and microbes. However, the impact may have been reduced by the protection and repair process capabilities in many organisms. The major concern regarding effects on ecosystems seems to be focussed on marine phytoplankton, especially in polar areas where ozone-related UV increases are the greatest. Macroalgae, sea grasses, sea urchins, and corals have also been found to be sensitive to UV radiation. For some of such populations even current levels of UV may be a limiting factor (UNEP 1998).

7.7.3 Earth System Feedbacks

Carbon dioxide and N fertilisation of terrestrial ecosystems. The fertilisation effects noted above also have global scale implications. It is likely that both increasing atmospheric CO_2 concentration and increasing deposition of fixed N compounds have individually and interactively increased significantly the global uptake of atmospheric carbon by terrestrial ecosystems (Schimel 1995). The magnitude and longevity of these effects are still open to question, however (Scholes et al. 1999).

Table 7.1. Agricultural optimisation model (AOM) estimates of 1990 total change in welfare in the markets for eight major US crops (corn, soybeans, wheat, alfalfa hay, cotton, grain sorghum, rice, and barley) due to a reduction in ozone pollution (billions of 1990 dollars). Case I is a 100% reduction in all anthropogenic emissions of VOCs and NO_X. Cases IIA and IIB are a 10% and a 100% reduction in motor-vehicle related emissions of VOCs and NO_X, respectively (adapted from Murphy et al. 1999)

	Case I		Case IIA				Case IIB			
			Direct emissions[a]		Direct + indirect[a]		Direct emissions[a]		Direct + indirect[a]	
	Low	High	Low	High	Low	High	Low	High	Low	High
Change in producer surplus	4.71	12.07	0.27	0.58	0.31	0.61	2.74	6.25	3.10	6.59
Change in deficiency payments[b]	2.49	7.85	0.14	0.37	0.16	0.39	1.33	3.91	1.53	4.11
Change in consumer surplus	0.63	1.53	0.03	0.08	0.04	0.08	0.39	0.80	0.44	0.86
Change in total welfare	2.84	5.76	0.17	0.28	0.19	0.30	1.80	3.14	2.01	3.34

[a] Direct emissions are tailpipe and evaporative emissions from motor vehicles. Indirect emissions include emissions from the production of motor fuels, the servicing of motor vehicles, the production of crude oil used to make motor fuel, the production of motor vehicles, etc. (see Delucchi 1996 for details).

[b] Deficiency payments were the result of a voluntary federal crop price support program that guaranteed growers a minimum price for all acreage enrolled in the program. If the market price fell below a target price, these payments were the difference between the target price and the actual market price. This program no longer exists, but it was in effect during 1990, the year of the analysis.

Iron fertilisation of the oceans. The Vostok ice core record hints that deposition of Fe-laden aerosols from continents might affect ocean carbon uptake through fertilisation of phytoplankton growth. The recent SOIREE experiment near New Zealand demonstrated that such an effect is plausible. The most striking evidence, however, comes from the recent START (Global Change System for Analysis, Research and Training) synthesis (Piketh et al. 2000), which shows that aerosols derived from the southern African continent are very likely increasing carbon uptake downwind in the Indian Ocean. The START work documents the emission, atmospheric transport, and deposition of an aerosol mix derived from both industrial and terrestrial sources, which recirculates around the southern African region in an anticyclonic system, exits the continent off the southeast coast of South Africa, and comes to ground several thousand kilometres out in the Indian Ocean, precisely where independent measurements have shown strong marine uptake of atmospheric CO_2.

Trace gas emissions. In addition to fertilising and toxifying effects, the research described in Chap. 2 of this volume shows that deposition of N compounds on ecosystems already N-saturated can lead to the emission back to the atmosphere of a mix of N trace gases, with the ratio dependent on soil temperature, N content, and water status.

7.7.4 Impacts on Agricultural Production Systems

Carbon dioxide fertilisation. A recent synthesis of the research on elevated CO_2 impacts on agricultural production indicates that modest enhancement of yield is likely occurring, with a doubling of atmospheric CO_2 concentration leading to a 10–15% increase in yield under normal field conditions (Gregory et al. 1999). In semi-arid regions enhancements of growth and yield may also be caused by CO_2-induced increases in water use efficiency, leading to enhanced soil moisture.

Tropospheric O_3. There is some evidence that increased tropospheric O_3 in the northern midlatitudes is leading to significant reductions in crop yields. For example, it is estimated that for wheat there is a 30% yield reduction for a seasonal seven-hour O_3 daily mean of 80 nmol mol^{-1}, a concentration level that has been found in parts of the US and Europe (Fuehrer 1996). Economic losses in the US alone may amount to several billion dollars annually (c.f., Table 7.1). It is difficult to generalise, however, as responses vary considerably among species, and even between varieties in a single species. In addition, there is some evidence that increasing atmospheric CO_2 may partially ameliorate the O_3 damage through an increase in stomatal closure.

Aerosols. Increasing aerosol loading might have indirect impacts on regional agricultural production through a reduction of incident solar radiation. An assessment of the effect of atmospheric aerosols and regional haze on the yields of rice and winter wheat in China suggests that this effect is currently depressing optimal yields of ca. 70% of these crops grown in China by at least 5–30% (Chameides et al. 1999).

Thus, in regions where tropospheric O_3 and aerosol concentrations are significantly enhanced, their negative impacts on crop yields are of approximately the same size as or larger than the beneficial effects due to CO_2 fertilisation.

7.8 How Does Atmospheric Chemistry Affect Human Health?

Impacts of air pollution on human health have been studied carefully in relation to the smog episodes in London in the 1950s and the photochemical smog in Los Angeles in the 1960s and 70s. More recently, health effects due to atmospheric changes have become more subtle and more widespread as atmospheric change has grown from a local to a global phenomenon. The present situation with respect to human health effects of air pollution can be summarised as follows (Schwela 2000).

Sulphur dioxide. Most information on the acute effects of SO_2 comes from controlled chamber experiments on volunteers exposed to SO_2 concentrations above 500 µg m^{-3} for periods ranging from a few minutes up to a few hours (WHO 1999a, 2000; Romieu 1999). Normal healthy individuals present upper respiratory symptoms, while asthmatics and persons with respiratory hyperreactivity exhibit acute responses. Acute responses occur within the first few minutes after inhalation begins. Effects in sensitive individuals include reductions in the mean forced expiratory volume over one second (FEV_1) and forced vital capacity (FVC), increases in specific airway resistance, and symptoms such as wheezing or shortness of breath. These effects are enhanced by exercise that increases the volume of air inspired, as it allows SO_2 to penetrate farther into the respiratory tract.

Several studies showed a significant increase in total mortality, cardiovascular mortality and respiratory mortality with increase in SO_2 (Touloumi et al. 1994; Xu et al. 1994; Spix and Wichmann 1996; Sunyer et al. 1996; Touloumi et al. 1996; Vigotti et al. 1996; Wietlisbach et al. 1996; Zmirou et al. 1996, 1998). Other recent studies could not establish a significant increase in total mortality with increase in SO_2 in the presence of other pollutants, except under special seasonal conditions (Schwartz and Dockery 1992; Spix et al. 1993; Verhoeff et al. 1996; Dab et al. 1996; Anderson et al. 1996b).

Nitrogen dioxide. There are indications of weak associations between short-term NO_2 exposure due to gas cooking and respiratory symptoms and decrement in lung function parameters in children. This association is not found consistently in exposed women. A number of outdoor studies indicate that children with long-term exposure exhibit increased respiratory symptoms, decreases in lung function, and increase of the incidences of chronic cough, bronchitis, and conjunctivitis. Other long-term studies provide little evidence of such effects of NO_2 in adults. In multi-pollutant time-series studies of the association between NO_2 and mortality and morbidity, either no association was found or a single-pollutant association disappeared when other pollutants were included in the analysis. There is still not enough evidence from epidemiological studies to establish a causal relationship of NO_2 and observed health effects (Ackermann-Liebrich and Rapp 1999).

In addition to the direct effects from elevated levels of SO_2 and NO_x (Lübkert-Alcamo and Krzyzanowski 1995), indirect health effects due to increased dissolution of metals in soils (e.g., Al, Cd), and in water pipes (Cu), in relation to the acidification of air and of precipitation are also suspected. Whereas the direct effects of SO_2 and NO_x are normally confined to urban and industrial areas, the indirect effects may also occur farther away from the sources. An encouraging development – establishment of emission controls – has taken place in many European and North American cities during the past two decades. As a result, concentrations have been reduced several fold.

Carbon monoxide. CO absorbed in the lungs binds with haemoglobin to form carboxyhaemoglobin (COHb), which reduces the oxygen-carrying capacity of the blood and impairs the release of oxygen from haemoglobin. Potential health effects of CO include hypoxia, neurological deficits, neurobehavioral changes, and increases in daily mortality and hospital admissions for cardiovascular diseases. Several studies showed small, statistically significant relationships between CO and daily mortality. However, other pollutants and other environmental variables were also shown to be significant. It is still a question whether the demonstrated associations are causal or whether CO might be acting as a proxy for particulate matter. Some studies appear to show that the association between ambient CO and mortality and hospital admissions due to cardiovascular diseases persists even at very low CO levels, indicating no threshold for the onset of these effects. It is possible that ambient CO may have more serious health consequences than the COHb formation, and at lower levels than mediated through elevated COHb levels (USEPA 1993; Bascom et al. 1996; WHO 1999a,b, 2000; Maynard and Waller 1999; Romieu 1999).

Tropospheric ozone. Surface layer levels of O_3 can have adverse health effects, even for short exposure times and relatively low concentrations. Such concentrations commonly occur during sunny periods over large regions in Europe and North America and in other parts of the world where emissions of O_3 precursors are large.

Short-term acute effects of O_3 include pulmonary function decrements, increased airway responsiveness and airway inflammation, aggravation of pre-existing respiratory diseases such as asthma, increases in daily hospital admissions and emergency department visits for respiratory causes, and excess mortality (USEPA 1993; WHO 1999a, 2000; Thurston and Ito 1999). Exposure-response relationships appear to be non-linear for the associations between O_3 concentration and the FEV_1, inflammatory changes, and changes in hospital admissions, respectively. A linear relationship was established for the association between the percent change in symptom exacerbation among adults and asthmatics. In studies on the relationships between ozone exposure and daily mortality and hospital admissions for respiratory diseases, single-pollutant associations between O_3 and these health effects remained statistically significant even in multi-pollutant models.

Controlling O_3 pollution is difficult because of the complex interactions between the various precursors, which originate from both natural and anthropogenic sources, and the inter-continental transport of some O_3 precursors. Chameides et al. (1999), for example, have shown that in urban areas where trees emit large amounts of hydrocarbons ozone control must concentrate on NO_x.

Stratospheric ozone depletion. Halogenated hydrocarbons (including the CFCs), while not harmful as such, contribute to depletion of the stratospheric ozone layer. The resulting increase in UV-B radiation at Earth's surface is estimated to cause adverse health effects including eye cancer and cataract, skin cancer (in fair skinned populations), and a depression of the immune system for certain tumours and infectious diseases (UNEP 1998).

Suspended particulate matter (SPM). SPM denotes particles of various sizes and varying chemical composition. In a public health context, particle size distributions are divided into coarse (mostly larger than 2.5 μm in aerodynamic diameter), fine (mean aerodynamic diameter between 0.1 and 2.5 μm), and ultrafine fractions (mean aerodynamic diameter below 0.1 μm) (Lippmann and Maynard 1999; Morawska et al. 1998). A broad variety of processes produce SPM in the ambient air. Coarse particles usually contain Earth crustal materials and fugitive dust from roads, while fine and ultrafine particles contain aerosols formed secondarily, i.e. combustion particles and recondensed organic and metal vapours (see Chap. 4). Air pollution by particulate matter has been considered to be primarily an urban phenomenon. In many areas of developed countries, however, urban-rural differences in PM_{10} (particles with aerodynamic diameter less than 10 µm) are small or even absent, indicating that SPM exposure is widespread.

Increasing SPM concentrations can cause adverse health effects related primarily to the respiratory system. Although the specific mechanisms are not known, recent evidence suggests that impacts are more closely related to the fine and ultrafine particles. Regional scale impacts of aerosol loading can be significant. The extensive fires on the Indonesian islands of Kalimantan and Sumatra during 1997 caused a smoke haze episode that had major health impacts. An estimated 20 million people in Indonesia alone suffered from respiratory problems, mainly asthma, upper respiratory tract illness, and skin and eye irritation during the episode. About 210 000 persons were treated clinically, with nearly four times as many acute respiratory illnesses as normal reported in South Sumatra and a two- to threefold increase in outpatient visits in Kuching, Borneo (Kalimantan) (Heil and Goldammer 2001).

An extensive body of experimental and epidemiological literature (WHO 1999; Bascom et al. 1996; Wilson and Spengler 1996; Romieu 1999; Pope and Dockery 1999; Ghio and Samet 1999) demonstrates significant associations between daily fluctuations in PM_{10} or $PM_{2.5}$ (particles with aerodynamic diameter less than 2.5 µm) levels and the rates of mortality and morbidity in the human population. The effects include increased daily mortality, increased rates of hospital admissions for exacerbation of respiratory disease, fluctuations in the prevalence of bronchodilator use, cough, and peak flow reductions (see also Chap. 1, Table 1.3). The current time-series epidemiological studies do not indicate a threshold below which no effects occur. Rather they suggest that even at low levels of SPM, short-term exposure is associated with daily mortality, daily hospital admissions, exacerbation of respiratory symptoms, bronchodilator use, incidence of cough, and reduction of peak expiratory flow.

Evidence is also emerging that long-term exposure to low concentrations of SPM in air is associated with mortality and other chronic effects, such as increased rates of bronchitis, and reduced lung function (WHO 1999a, 2000; Pope and Dockery 1999; Zemp et al. 1999). Several population-based and cohort-based mortality studies conducted in the US suggest that life expectancy may be two to three years shorter in communities with high SPM than in communities with low SPM (Pope and Dockery 1999). The results showed that long-term average exposures to low SPM levels, starting at about 10 $\mu g\ m^{-3}$ of fine particulate matter, were associated with a reduction in life expectancy (Brunekreef 1997).

Lead and other heavy metals. The level of lead (Pb) in blood is the best available indicator of current and recent past environmental Pb exposure. The biological effects of Pb can therefore be related to blood Pb levels as an indicator of internal exposure. Potential effects of Pb in adults and children include encephalopathic signs and symptoms, symptoms in the central nervous system, cognitive effects, increase in blood pressure, and a reduction in measures of child intelligence. While lowest observed effect levels are established for most of these effects, the question of increase in blood pressure is still controversial. That low levels of Pb exposure may result in mental deficit has also been controversially debated recently (WHO 1995, 1999a, 2000; Romieu 1999; Wadge 1999). During the last decades, atmospheric lead has decreased substantially in many areas of the world due to the removal of this compound from petrol.

The mobilisation and transport of other heavy metals, such as mercury (Hg), is a growing concern. The deposition of Hg over large regions combined with leaching of mercury from acidified soils is leading to toxic levels in fish in some lakes in Scandinavia and Canada.

7.9 What is the Connection between Atmospheric Composition and Climate?

Atmospheric composition plays a crucial role in determining our planet's climate. On average, the net incoming radiation from the Sun must be in balance with the outgoing radiation to space. It is this balance that largely determines the climate on Earth. Some of the incoming solar radiation (about 31%) is reflected back to space, largely by clouds and by aerosols in Earth's atmosphere, with about 9% of the 31% due to the reflectivity, or albedo, of Earth's surface. As it passes through the atmosphere a fraction of the solar radiation (about 20%) is absorbed by clouds and gases such as molecular oxygen and ozone. The rest of the solar radiation is absorbed by Earth's surface and then is reradiated into the atmosphere, largely at infrared wavelengths. Energy from the surface is also released as latent heat. The absorption and reemission of this infrared radiation by greenhouse gases and by clouds in the atmosphere reduce the radiative energy emitted to space and increase the amount of energy maintained in the atmosphere-Earth system. This "greenhouse effect" process is crucial to the existence of life on Earth, at least as we know it. Without the existence in the atmosphere of water vapour, carbon dioxide, ozone, methane, and other greenhouse gases, the planet would be much colder and largely uninhabitable. As we have seen throughout this volume, most of the gases and aerosols that play such a large role in determining climate also greatly affect and are affected by atmospheric chemistry.

Changes in atmospheric composition can also lead to changes in climate. A perturbation to the atmospheric concentration of an important greenhouse gas, or the distribution of aerosols, induces a radiative forcing that can affect climate. In keeping with the definition adopted by the Intergovernmental Panel on Climate Change, radiative forcing of the surface-troposphere system is defined as the change in net radiative flux at the tropopause due to a change in either solar or infrared radiation. Generally, this net flux is calculated after allowing for stratospheric temperatures to re-adjust to radiative equilibrium. A positive radiative forcing tends on average to warm the surface; a negative radiative forcing tends to cool it. This definition is based on earlier climate modelling studies, which indicated an approximately linear relationship between the global mean radiative forcing at the tropopause and the resulting global mean surface temperature change. However, recent studies of greenhouse gases indicate that the climatic response can be sensitive to the altitude, latitude, and nature of the forcing.

There is now strong evidence of a discernible influence on the climate over recent decades as a result of human-induced changes in atmospheric composition. In 1896, Svante Arrhenius was the first to hypothesise that a doubling of the natural concentration of carbon dioxide could have a significant warming impact on climate. He thought such a change in CO_2 concentrations possible due to rapid increases in industry and burning of fossil fuels, but he had no clear evidence that such a change in atmospheric composition was occurring. We now know that atmospheric concentrations of CO_2 and other greenhouse gases and aerosols have increased since pre-industrial times, largely as a result of human activities. Meanwhile, globally-averaged temperatures have increased by about 0.65 °C over the last century, with about half of the warming occurring in the last few decades (Jones et al. 1997). Recent globally-averaged temperatures are greater than those of the past 1 000 years (Mann et al. 1999). The available evidence strongly suggests that these changes in climate are connected tightly to the human-related changes in atmospheric composition.

Concentrations of carbon dioxide have increased by 30% since the beginning of the Industrial Era, from about 280 μmol mol^{-1} to about 370 μmol mol^{-1} in 2001. Such a CO_2 concentration has probably never been exceeded in the last 420 000 years. Carbon cycle analyses indicate that this increase is largely the result of fossil fuel burning and, to a lesser extent, deforestation and other land use changes. Atmospheric methane concentrations have increased from 0.7 μmol mol^{-1} to 1.75 μmol mol^{-1}, or by 150%, over the last several hundred years. Concentrations of nitrous oxide have increased by 16%, to 315 nmol mol^{-1}, over this time period. Although the exact changes are less certain, concentrations of sulphate aerosols have increased by 40% or more. Concentrations of other aerosols have also increased.

As seen in Fig. 1.6, analyses of the direct radiative forcing due to the changes in greenhouse gas concentrations since the late 1700s give an increase of about 2.4 W m^{-2}. To put this into perspective, a doubling of CO_2 from pre-industrial levels would correspond to about 4 W m^{-2}. Approximately 0.5 W m^{-2} of the increase has occurred within the last decade (Hansen et al. 1998). By far the largest effect on radiative forcing has been the increasing concentration of carbon dioxide, accounting for about 64% of the total change in forcing (Hansen et al. 1998). Of the greenhouse gases, methane is the second largest contributor to the changes in radiative forcing. The response of the climate system is expected to be amplified by a feedback involving water vapour. Also evident from Fig. 1.6 is that the changes in greenhouse gases have not been the only forcing on climate since the pre-industrial era.

Changes in tropospheric and stratospheric ozone also affect climate, but the increase in tropospheric ozone over the last century and the decrease in stratospheric ozone over recent decades have had a relatively small combined effect on radiative forcing compared to CO_2. The radiative forcing from the changes in amount of stratospheric ozone that have occurred over the last few decades, primarily as a result of human-related emissions of halogenated chemical species containing chlorine and bromine, is generally well understood. In contrast, the changes in concentrations of tropospheric ozone over the last century, and the resulting radiative forcing, are much less well understood. The chemistry and changing emissions affecting ozone in the troposphere are discussed in Chap. 3.

Changes in the amounts of sulphate, nitrate, and carbonaceous aerosols induced by natural and human activities have all contributed to changes in radiative forcing over the last century. The direct effect on climate from sulphate aerosols occurs primarily through the scattering of solar radiation. This scattering produces a negative radiative forcing, and has resulted in a cooling tendency on Earth's surface that counteracts some of the warming effect from the greenhouse gases. A recent estimate of the geographical distribution of the direct climate forcing due to sulphate aerosols of human origin is shown in Fig. 7.11 (Kiehl et al. 2000). In the global average, increases in amounts of carbonaceous aerosols, which absorb solar and infrared radiation, have likely counteracted some of the effect of the sulphate aerosols. However, the aerosol forcing estimates shown in Fig. 1.7 are not strictly comparable to the forcings of the greenhouse gases; the forcing from aerosols can differ greatly in different regions. Aerosols can also produce an indirect radiative forcing by acting as condensation nuclei for cloud formation. There is large uncertainty in determining the extent of

Fig. 7.11.
Derived geographical distribution of the climate forcing (both direct and indirect) due to sulphate aerosols of human origin (Kiehl et al. 2000)

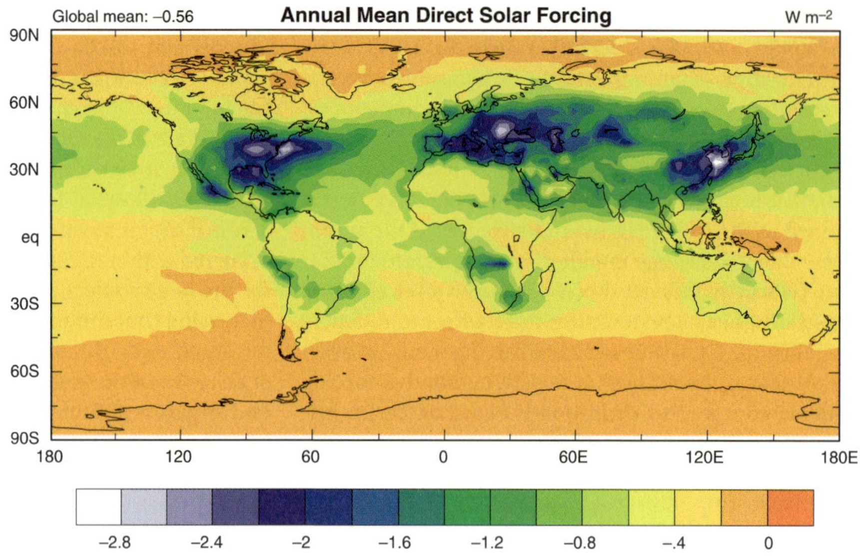

radiative forcing that has resulted from this indirect effect, as indicated in Fig. 1.7. If the upper (most negative) end of the uncertainty range applies, the negative forcing due to aerosols may have balanced essentially all of the positive forcing due to greenhouse gases. On the other hand, if the aerosol forcing is near the small end of the range, greenhouse warming should have been the dominating effect during the past century. It is obvious that research aiming at reducing the uncertainty of aerosol forcing is of utmost importance for the climate change issue. For example, it is likely, but not proven, that some climate warming may have been delayed by the increasing aerosol load in Earth's atmosphere. Because the residence time of aerosol is much shorter than that of greenhouse gases (and hence aerosols will not accumulate), the warming effect of the greenhouse gases may dominate (in relative terms) over long periods of time. The processes affecting aerosol loading in the troposphere are discussed in Chap. 4.

Changes in the solar energy output reaching Earth are an important external forcing on the climate system. The Sun's output of energy is known to vary by small amounts over the 11-year cycle associated with sunspots, and there are indications that the solar output may vary by larger amounts over longer time periods. Slow variations in Earth's orbital parameters, over time scales of multiple decades to thousands of years, have varied the solar radiation reaching Earth and have affected the past climate. Solar variations over the last century are thought to have had a small but important effect on the climate, but are not important in explaining the large increase in temperatures over the last few decades (e.g., Stott et al. 2000).

As indicated by Fig. 1.6, evaluation of the radiative forcing from all of the different sources since pre-industrial times indicates that the net globally-averaged radiative forcing on climate has increased. Although a number of recent climate modelling studies have treated the combined radiative forcing from greenhouse gases in terms of equivalent CO_2 forcing, it should be noted that the radiative forcing from the important greenhouse gases actually varies appreciably with latitude and season (Jain et al. 2000). Also, because of the hemispheric and other inhomogeneous variations in concentrations of aerosols, the overall change in radiative forcing is much greater or much smaller at specific locations over the globe.

Future changes in radiative forcing, and thus changes in climate, will depend on future emissions and concentrations of greenhouse gases and aerosols. Projections of future greenhouse gas concentrations depend on assumptions made about the factors governing future emissions from human-related sources, on any projected changes in atmospheric sinks, and on any potential changes in the biogeochemical cycles that affect atmospheric concentrations. However, a series of scenarios developed for the IPCC suggests that, without major policy actions, concentrations of CO_2 and other greenhouse gases are likely to continue to increase throughout this century. In these business-as-usual scenarios, concentrations of CO_2 are likely to double or more than double those of the pre-industrial atmosphere. Concentrations of methane and nitrous oxide are projected to continue to increase. However, growing recognition of the direct impacts of sulphate aerosols on humans, crops and vegetation appears likely to result in controls that will decrease SO_2 emissions gradually over the century. The net result is an increase in radiative forcing of 2.6 to 7.9 W m^{-2} from 1990 to 2100 for these scenarios. This, in turn, leads to an estimated globally averaged temperature increase of about 1.4 to 5.8 °C over this time period (accounting for uncertainties in the climate feedbacks from clouds and other processes). Such changes in temperature and other climate variables (e.g., precipitation) could have significant impacts on humanity and ecosystems. As with the climate change itself, there are expected to be large regional variations in the resulting impacts. However, these impacts include concerns about human health, ecosystems, water resources, agriculture, and local air quality. The possibility of changes in the frequency of severe events, like floods, droughts, and intense storms, could add to these concerns.

Changes in climate could also have important implications for atmospheric chemistry. For example, a warmer climate is expected to enhance surface evaporation, and hence the tropospheric abundance of water vapour. In addition to its amplifying effects on greenhouse warming, water also provides starting material for the formation of the hydroxyl radical (OH) and hence affects the atmospheric lifetime of many chemical compounds (including greenhouse gases such as methane). The distribution of ozone (the other starting material for OH formation) in the troposphere is also expected to change with potential effects on the related climate forcing.

Climate change could also result in more intense convective activity. As a consequence, the formation rate of nitrogen oxides by lightning and hence the concentration of ozone in the free troposphere would probably increase, producing an amplification of radiative forcing.

The response of natural emissions from important biogeochemical cycles could provide an important feedback on atmospheric composition, with potential significant implications on the actual changes in climate. Greenhouse gases such as CO_2, CH_4, and N_2O could be affected by the response of biogenic sources to a changing climate. Surface emissions of other gases with biogenic sources (including nitrogen oxides by soils, isoprene and other hydrocarbons by vegetation, dimethylsulphide by the ocean) are expected to increase on a warmer planet. These compounds are all precursors of tropospheric ozone and/or aerosols, and hence could indirectly impact Earth's climate. Changes in soil moisture associated with changes in precipitation patterns would affect the intensity and the geographical locations of savannah and forest fires, with significant consequences on biomass burning emissions, and hence on the concentration of several chemical compounds, including organic and black carbon particles, carbon monoxide, nitrogen oxides, and ozone. Finally, changes in precipitation rates associated with climate change would modify the rate at which soluble gases and aerosol particles are removed from the atmosphere, with consequences on the atmospheric residence times of many compounds, leading to potential feedbacks on the climate system.

7.10 How Might Chemical Composition Evolve in the Future?

Future changes in the chemical composition of the atmosphere will be determined principally by changes in primary emissions of chemical compounds such as CH_4, CO, NO_x, VOC, and SO_2. The emissions of these compounds will change in response to human activities, specifically with the level of fossil fuel consumption, with the intensity of industrial activities, and with the extent and nature of land-use changes. In addition, emissions will produce changes in the atmospheric sinks of secondary compounds like O_3, OH, and H_2O_2. The future evolution of atmospheric composition will also be affected by climate change through changes in the hydrological cycle (water vapour abundance, precipitation rates, cloud processes), temperature, and atmospheric circulation. Finally, the rate at which chemical compounds are destroyed in the atmosphere will be affected by changes in short-wave solar radiation caused by potential changes in stratospheric ozone and in tropospheric cloud or aerosol distributions.

Several global 3-D chemical transport models have been used to predict changes in the chemical composition of the atmosphere over the next century. These studies have been conducted as part of the 2001 IPCC assessment, using projected future emission scenarios from the IPCC Special Report on Emission Scenarios (IPCC 2000); the scenarios make different assumptions regarding population growth, economic development, energy use, etc. Four principal scenarios have been established. In the most optimistic of them (rapid technological development, use of renewable energy sources, global economic convergence) the fossil fuel combustion source of CO_2 will peak at 10.4 Pg C yr^{-1} in 2060, a 70% increase over present, and then decline by 2100 to its present-day value. In the worst case scenario the fossil fuel source of CO_2 will gradually increase to four times its present-day value by 2100. Estimates of future emissions for NO and CO tend to roughly follow those for CO_2. In the case of SO_2, emissions are expected to decrease by 2100 (after reaching a peak in the mid-21st century in two of the scenarios) due to emission controls and transition away from coal.

Based on the selected emission scenarios for primary compounds such CH_4, CO, VOC, and NO_x, a wide range of concentrations for all compounds is predicted for year 2100. When adopting the scenario with the largest emission increases (IPCC scenario A2), the global methane concentration more than doubles by 2100 relative to 1990. At the same time, the globally averaged tropospheric ozone concentration increases by 30 nmol mol^{-1} (with most of the increase occurring in the Northern Hemisphere) and the globally averaged OH concentration decreases by approximately 12% (due to large increases in NO_x emissions). These models suggest (see Fig. 7.12) that the largest increase in the ozone concentration will be observed over such regions as south and southeast Asia, as well as over areas of South America and Africa. Thus, while large ozone increases occurred in Europe and North America during the 20th century, it is likely that ozone pollution events will now become more fre-

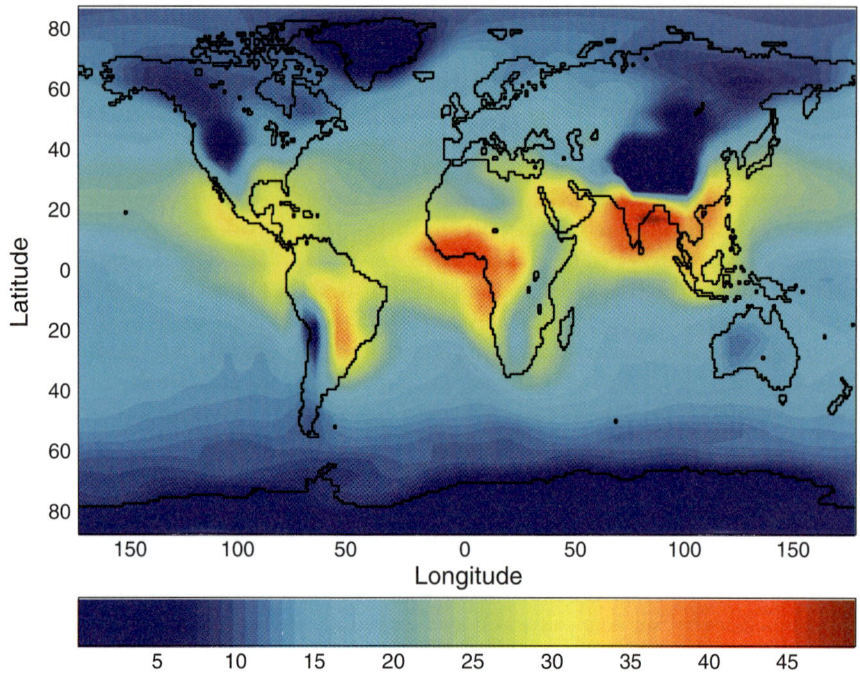

Fig. 7.12.
Predicted annual mean difference in the ozone mole fraction (nmol mol^{-1}) between years 2000 and 2100 averaged over a 2 km layer above the surface. These predictions, based on SRES scenario A2, are provided by the chemical-transport model of the University of Oslo, and account for estimated changes in tropospheric as well as stratospheric chemistry (courtesy of M. Gauss, University of Oslo, Norway)

quent in the rapidly developing nations of the Tropics and subtropics. Radiative forcing will be greatly augmented as a consequence of these large changes in atmospheric composition. In addition, the ability of the atmosphere to cleanse itself of trace pollutants will be reduced, so that air quality in most regions will be degraded, even if local emissions are unchanged.

It is important to note that the atmospheric destruction rate of many chemical compounds will be changed as a result of increasing surface emissions. As already stated in the response to Question 6 in this chapter, the release of compounds like carbon monoxide, methane, non-methane hydrocarbons, and nitrogen oxides tends to modify the concentration of the hydroxyl radical (OH) and hence the oxidising power of the atmosphere. Model studies suggest that recent changes in the methane oxidation rate have been influenced by increases in NO emissions at low latitudes (Gupta et al. 1998; Karlsdottir and Isaksen 2000). The decline in the methane growth observed over the last one to two decades (Dlugokencky et al. 1998) could therefore have been the result of enhanced loss rate, caused by increases in OH concentrations (Krol et al. 1998; Karlsdottir and Isaksen 2000). Past and future changes in the yearly and zonal mean OH field, based on the model predictions of Lelieveld and Dentener (2000) are shown in Fig. 7.13. It is believed, however, that most of the observed slowdown in methane growth is due to reduced methane emission rates, as documented by the ALE/GAGE measurements (Prinn et al. 1995).

Future changes in the chemical composition of the atmosphere could also be produced by changes in the climate system. Climate changes could, for instance, modify natural ecosystems and, hence, the emission and deposition rates of many trace constituents. However, only a few studies have assessed how the chemical composition of the atmosphere could be affected by climate-induced changes. Model studies show that global ozone loss and OH production should increase in a warmer and wetter future climate. This leads to a reduced atmospheric abundance and reduced radiative forcing for ozone and methane compared to a case with the same emission increase but with unchanged climate. Wang et al. (1996b) calculated for 2050 a reduction of approximately 10% and 16% in tropospheric total ozone and methane levels, respectively, as a result of changes in temperature and water vapour. The corresponding reduction in radiative forcing is of the order of −0.2 W m^{-2} for both compounds. Similar calculations by Brasseur et al. (1998b), using the NCAR CCM3 model, provide a 7% increase in the OH density and a 5% decrease in tropospheric O_3. In a recent study a 3-D tropospheric chemistry model was coupled to the Hadley Centre coupled climate model in an experiment using the high emission (A2) SRES scenario (Stevenson et al. 2000). This model, which was run for the period 1990 to 2100, suggests that the increases in global CH_4 and in Northern Hemispheric ozone concentrations were approximately three-quarters and one half, respectively, of the values obtained in a model run without climate-chemistry coupling.

Most studies of future changes in aerosol distributions have been performed for fixed current climate conditions as part of the 2001 IPCC assessment. The models suggest an approximate linear relation between

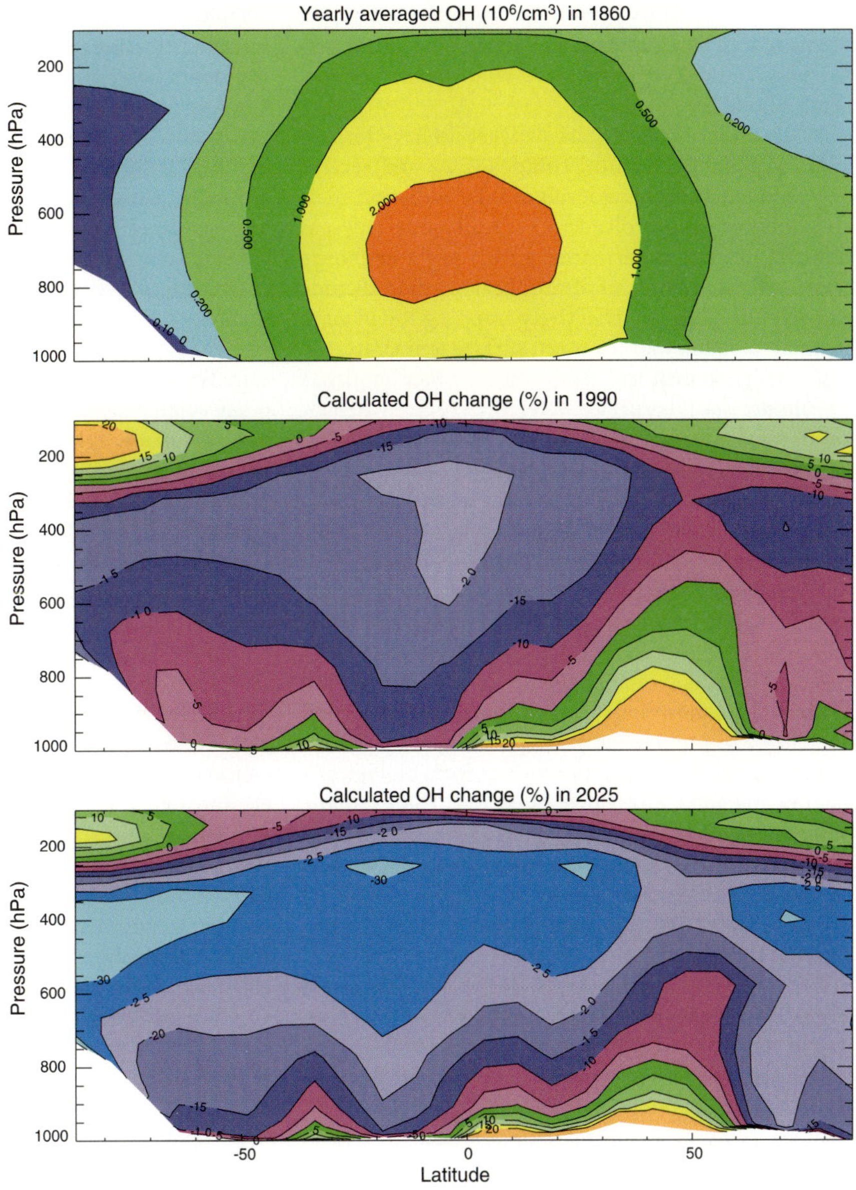

Fig. 7.13. Yearly and zonally averaged OH field calculated for year 1860, and estimated percentage changes in 1990 and in 2025. Calculations performed by the TM3 model of Lelieveld and Dentener (2000). *White areas* denote orography (courtesy of M. Krol, University of Utrecht, The Netherlands)

long-term changes in anthropogenic aerosol abundance and emissions for current meteorological conditions. Thus, predictions of future changes in the atmospheric aerosol load will be determined primarily by the changes in surface emissions. The SRES scenarios for SO_2 emissions for the time period 1990 to 2100 are significantly different from the scenarios used in previous IPCC assessments (e.g., IPCC 1996) in that there is a marked drop in the emissions after 2020–2030. This difference leads to a reduction in the calculated sulphate burden at later dates, and a similar reduction in the negative radiative forcing (less cooling) toward the end of the 21st century. Predictions of the burden of other anthropogenic aerosols suggest significant changes toward the end of the 21st century. Finally, expected future climate change should affect the emissions of sea salt particles and of mineral dust, through changes in wind velocity near the surface. Due to the marked reduction in predicted future sulphur emissions, it is likely that the role of sulphate aerosols for radiative forcing will decrease relative to other anthropogenic and natural aerosols.

7.11 Are There Risks of Abrupt Changes and/or Irreversible Changes in Atmospheric Composition?

It is important to realise that the current chemical state of Earth's atmosphere is the dynamic product of interactions among a wide range of geochemical, biological, and anthropogenic processes. The clearest illustration of this is seen by a comparison between the current atmospheric states of three terrestrial planets: Venus, Earth, and Mars.

While all three planets are believed to have formed with comparable chemical compositions and receive only moderately different levels of solar radiation, their atmospheres have evolved to very disparate current states. Venus has experienced a runaway greenhouse effect with high levels of atmospheric carbon dioxide and sulphur oxides so effectively trapping the planet's thermal radiation that the mean surface temperature is high enough to melt lead. Mars, on the other hand, has lost almost all of its surface water, thus severely depleting its atmosphere of water vapour, an extremely effective greenhouse gas, thereby helping to produce a mean surface temperature well below the freezing point of water. Only Earth has managed to retain levels of atmospheric greenhouse gases that maintain a mean planetary surface temperature compatible with the abundant levels of liquid water which are believed to be necessary for life to thrive. In return, biological processes help control the buildup of atmospheric carbon dioxide and produce the atmospheric oxygen that supports animal life. They also provide the atmospheric molecular oxygen and ozone that shield the surface from biologically detrimental solar ultraviolet radiation. The dramatic differences among the current states of these three planetary atmospheres suggest that there is good reason to be concerned that our planet's atmosphere might be driven to a very different, possibly irreversible, composition in the future.

Earth's paleoclimate records, particularly those retrieved from Greenland and Antarctic glacial ice cores, clearly correlate changes in major atmospheric greenhouse gases, including methane and carbon dioxide, with significant changes in mean surface temperatures, strongly tying atmospheric composition to climate (c.f., Chap. 1, Fig. 1.5). The fact that atmospheric composition and climate are closely connected is not surprising (see Question 9, "What is the Connection Between Atmospheric Composition and Climate?"), but the rapid changes in atmospheric composition and correlated surface temperature surrogates are. It appears that natural processes can cause significant changes in atmospheric composition (and therefore climate) on time scales significantly shorter than a thousand years, and perhaps on time scales of less than a century (Severinghaus and Brooke 1999). In the context of geological time scales these ice core-recorded atmospheric changes are abrupt indeed.

Finally, the impact of human activities on atmospheric change must be considered. We have known for a long time that sustained emissions from factories, commercial establishments, and homes, coupled with certain meteorological conditions, can lead to very unhealthful atmospheres in urban and industrial environments. London's killer fogs during the 1930s to 1950s, Los Angeles' photochemical smogs during the 1950s to 1970s, and current heavy photochemical oxidant and aerosol loadings in many developing megacities, such as Beijing, Bangkok, and Mexico City, are among the best known examples. Measured, globally-averaged atmospheric levels of a wide variety of radiatively and/or chemically active pollutants, including carbon dioxide, methane, nitrous oxide, hydrogen, carbon monoxide, and tropospheric ozone, have been growing dramatically. Growing anthropogenic emissions of methane and other volatile organic compounds, nitrogen oxides, carbon monoxide, sulphur dioxide, and ammonia are expected to have significant effects on the production of photochemical oxidants and aerosols in Earth's future atmosphere (see Question 10, "How Might Chemical Composition Evolve in the Future?").

While we believe that the poor air quality found in many urban and industrial regions can be reversed if emissions of primary pollutants are significantly reduced, we do need to be concerned that some combination of changing climate and atmospheric composition may lead to rapid and possibly irreversible change. This concern is buttressed by analyses of atmospheric models and by experience. Several research groups have published analyses of mathematical models of both stratospheric and tropospheric photochemistry which demonstrate that relatively moderate variations in trace pollutant concentrations can lead to divergent atmospheric photochemical states (Prather et al. 1979; White and Dietz 1984; Hess and Madronich 1997; Field et al. 2001). These analyses indicate that the current photochemical state of the atmosphere may be metastable, and that variations in emission fluxes and climate parameters could push the current atmospheric state into a new regime.

While there is no guarantee that the instabilities found in idealised models of atmospheric photochemistry, with their less-than-realistic boundary conditions, will be manifested in the real atmosphere, one recent experience with a sudden and dramatic shift in atmospheric chemical content certainly lends credence to the possibility. The development each austral spring of a continent-sized "hole" in the stratospheric ozone layer over Antarctica is a dramatic change of state for a large portion of the atmosphere, which was not present before 1980. Subsequent analyses have revealed that the Antarctic ozone hole occurs when inorganic halogen (chlorine and bromine) compound concentrations exceed NO_x concentrations (WMO 1999). Under these conditions unanticipated heterogeneous reactions on polar stratospheric cloud particles formed in the very cold Antarctic polar vortex release highly reactive halogen species into the gas phase while sequestering nitrogen oxide species (which would ordinarily scavenge reactive halogen forming less reactive chemical species) in the cloud particles. At the onset of austral spring the reactive halogen species catalyse the photochemical destruction of ozone, removing most of the stratospheric ozone from a huge volume of polar air. We now anticipate that large Antarctic ozone holes will form for approximately 50 more years, until controls on manmade halogen source compounds reduce the stratospheric inorganic halogen levels back below levels found before 1980 (WMO 1999). The elimination of the Antarctic ozone hole as a persistent seasonal feature of our atmosphere will perhaps be delayed by global tropospheric warming, which also results in colder stratospheric temperatures that may well lead to more abundant polar stratospheric cloud particles which trigger the chemistry leading to severe ozone depletion.

We do not know if other unpleasant atmospheric composition surprises, like the Antarctic ozone hole, lie in our future. However, if we continue to increase our emissions of radiatively and chemically active pollutants the odds are greatly increased that we will encounter sudden and potentially irreversible changes in the composition of the atmosphere.

7.12 What Should the Research Strategy Be to Address Unresolved Questions?

Ten years ago, scientific programs addressing global tropospheric chemistry were in their infancy. Almost no observations of tropospheric composition on a large scale were available, many chemical transformation mechanisms were unknown, and global atmospheric chemistry models were rather crude. Global atmospheric chemistry research blossomed during the last decade of the 20th century. We have learned much about the global cycles (sources, transformation, and sinks) of most important atmospheric chemical species. Existing satellite observations have provided a wealth of data regarding the chemical composition of the stratosphere, and new satellite instruments probing the troposphere have recently been or are about to be launched. Multi-platform process studies of atmospheric chemical processes have been conducted on an unprecedented scale. Global chemical transport models can now simulate with some success the distribution of key tropospheric chemical species, and are capable of simulating future global atmospheric composition scenarios. Furthermore, short lived, radiatively-active substances such as ozone and aerosols are now incorporated as active constituents in most global climate models.

As scientific understanding of the elements of atmospheric chemistry has been developed, the necessity of understanding the linkage between atmospheric composition and other components of the Earth system has been realised more explicitly. Ten years ago, the "Earth system" was a rather abstract idea. Today we are on the threshold of a more quantitative understanding of the role of atmospheric chemistry in Earth system processes and of developing strategies to integrate that knowledge into a predictive capability. As IGAC is completing a decade of activities, new overall objectives of global atmospheric chemistry research in the next decade could be:

1. To provide a fundamental understanding of the processes that govern the behaviour of chemical compounds in the atmosphere and their impact on climate;
2. To contribute to the development of an integrating framework describing the physical, chemical, and biological interactions among the different components of the Earth system;
3. To improve the ability to predict atmospheric composition and budgets for the coming decades through models that integrate knowledge gained from process studies and global observations; and
4. To address policy issues through application of atmospheric chemistry research that includes the human dimensions.

After a decade of interdisciplinary cooperation focussed on understanding atmospheric chemistry on a global scale, it is now time to begin to integrate atmospheric chemistry into Earth system science while focussing upon the environmental aspects of the solutions to socio-economic problems. Integration should take place from at least two perspectives.

Integration of Earth system components. In the next ten years, a focus on the chemical and biological interactions between atmosphere and land as well as the atmosphere and ocean should be of high priority. Just as global climate models move toward the physical coupling of atmosphere and ocean, atmospheric chemistry must move toward chemically coupling the atmosphere to components of the earth's surface.

Integration of knowledge over broad time and space scales. Scales are a very important concept in the study of atmospheric chemistry and environmental science. Scale links processes in space and time. Focussed continental- and ocean-basin-scale regional studies will help understanding the Earth system in a global perspective. Opportunities for truly integrated studies between atmosphere, hydrosphere, biosphere, and human dimensions can be developed most easily on a regional scale.

In order to understand the coupling between atmospheric composition and other components of the Earth system, much closer linkage of chemistry with physics and biology is necessary. Furthermore, advances in mathematics and computer science must be promptly incorporated into theory and numerical modelling. Integration of science disciplines, and specifically of issues related to the biogeochemical system, the climate system, and the socio-economic system could take place through new international efforts involving a strong collaboration between the International Geosphere-Biosphere Programme (IGBP), the World Climate Research Programme (WCRP), and the International Human Dimension Programme (IHDP). Therefore, in order to accomplish its goals in the next decade, the atmospheric chemistry community will have to strengthen old partnerships and explore new ones. Closer collaboration between major national research programmes that are global in scope with the international research enterprise must take place.

References

Abbatt JP (1997) Interaction of HNO_3 with water ice surfaces at temperatures of the free troposphere. Geophys res lett 24:1479–1482

Abbatt JPD, Demerjian KL, Anderson JG (1990) A new approach to free radical kinetics: Radially and axially resolved high-pressure discharge flow with results for OH + (C_2H_6, C_3H_8, n-C_4H_{10}, n-C_5H_{12}) ∅ products at 297 K. J chem phys 94:4566–4575

Abdul-Razzak H, Ghan SJ (2000) A parameterization of aerosol activation 2. Multiple aerosol types. J geophys res 105:6837–6844

Ackermann-Liebrich U, Rapp R (1999) Epidemiological effects of oxides of nitrogen, especially NO_2. In: Holgate ST, Samet JM, Koren HS, Maynard RL (eds) Air pollution and health. Academic Press, London, pp 559–584

Adams PJ, Seinfeld JH, Koch DM (1999) Global concentrations of tropospheric sulfate, nitrate, and ammonium aerosol simulated in a general circulation model. J geophys res 104:13791–13823

Agarwal JK, Sem GJ (1980) Continuous flow, single-particle-counting condensation nucleus counter. J aerosol sci 11:343–357

Akimoto H, Nakane H, Matsumoto Y (1994) The chemistry of oxidant generation: Tropospheric ozone increase in Japan. In: Calvert JG (ed) Chemistry of the atmosphere: Its impact on global change. Blackwell Science, pp 261–273

Albrecht BA (1989) Aerosols, cloud microphysics, and fractional cloudiness. Science 245:1227–1230

Alicke B, Hebestreit K, Stutz J, Platt U (1999) Iodine oxide in the marine boundary layer. Nature 397:572–573

Aliwell SR, Jones RL (1998) Measurements of tropospheric NO_3 at midlatitude. J geophys res 103:5719–5727

Allan BJ, Burgess RA, Carslaw N, Coe H, Plane JMC (1999) Observations of the nitrate radical in the marine boundary layer. J atmos chem 33:129–154

Allan BJ, McFiggans G, Plane JMC, Coe H (2000) Observations of iodine monoxide in the remote marine boundary layer. J geophys res 105:14363–14369

Allen D, Pickering K, Stenchikov G, Thompson A, Kondo Y (2000) A three-dimensional total odd nitrogen (NO_y) simulation during SONEX using stretched-grid chemical transport model. J geophys res 105(3):3851–3876

Althausen D, Müller D, Ansmann A, Wandinger U, Hube H, Clauder E, Zörner S (2000) Scanning 6-wavelength 11-channel aerosol lidar. J atmos ocean tech 17:1469–1482

Andersen KA, Armenguaud A, Genthon C (1998) Atmospheric dust under glacial and interglacial conditions. Geophys res lett 25:2281–2284

Anderson HR, de Leon P, Bland JM, Bower JS, Strachan DP (1996b) Air pollution and daily mortality in London: 1987–1992. Brit med J 312:665–669

Anderson TL, Ogren JA (1998) Determining aerosol radiative properties using the TSI 3563 integrating nephelometer. Aerosol sci tech 29:57–69

Anderson TL, Covert DS, Marshall SF, Laucks ML, Charlson RJ, Waggoner AP, Ogren JA, Caldow R, Holm RL, Quant FR, Sem GJ, Wiedensohler A, Ahlquist NA, Bates TS (1996a) Performance characteristics of a high-sensitivity, three-wavelength, total scatter/backscatter nephelometer. J atmos ocean tech 13:967–986

Anderson TL, Covert DS, Wheeler JD, Harris JM, Perry KD, Trost BE, Jaffe DJ, Ogren JA (1999) Aerosol backscatter fraction and single scattering albedo: Measured values and uncertainties at a coastal station in the Pacific Northwest. J geophys res 104:26793–26807

Andreae MO (1983) Soot carbon and excess fine potassium: Long-range transport of combustion-derived aerosols. Science 220:1148–1151

Andreae MO (1993) The influence of tropical biomass burning on climate and the atmospheric environment. In: Oremland RS (ed) Biogeochemistry of global change: Radiatively active trace gases. Chapman and Hall, New York, pp 113–150

Andreae MO (1995) Climatic effects of changing atmospheric aerosol levels. In: Henderson-Sellers A (ed) Future Climates of the World. Elsevier, Amsterdam, pp 347–398

Andreae MO (1997) Emissions of trace gases and aerosols from savanna fires. In: van Wilgen BW, Andreae MO, Goldammer JG, Lindesay JA (eds) Fire in the Southern African Savanna: Ecological and environmental perspectives. Witwatersrand University Press, Johannesburg, South Africa, pp 161–183

Andreae MO, Barnard WR (1984) The marine chemistry of dimethylsulfide. Mar chem 14:267–279

Andreae MO, Crutzen PJ (1997) Atmospheric aerosols: Biogeochemical sources and role in atmospheric chemistry. Science 276:1052–1058

Andreae MO, Ferek RJ (1992) Photochemical production of carbonyl sulfide in seawater and its emission to the atmosphere. Global biogeochem cy 6:175–183

Andreae MO, Merlet P (2001) Emission of trace gases and aerosols from biomass burning. Global biogeochem cy 15(4):955–966

Andreae MO, Raemdonck H (1983) Dimethylsulfide in the surface ocean and the marine atmosphere: A global view. Science 221:744–747

Andreae MO, Schimel DS (1989) Exchange of trace gases between terrestrial ecosystems and the atmosphere. J. Wiley, New York

Andreae MO, Talbot RW, Berresheim H, Beecher KM (1990) Precipitation chemistry in central Amazonia. J geophys res 95:16987–16999

Andreae MO, Fishman J, Lindesay J (1996) The Southern Tropical Atlantic Region Experiment (STARE): Transport and atmospheric chemistry near the Equator-Atlantic (TRACE-A) and Southern African Fire/Atmosphere Research Initiative (SAFARI): An introduction. J geophys res 101:23519–23520

Andreae MO, Andreae TW, Annegarn H, Beer J, Cachier H, Le Canut P, Elbert W, Maenhaut W, Salma I, Wienhold FG, Zenker T (1998) Airborne studies of aerosol emissions from savanna fires in southern Africa: 2. Aerosol chemical composition. J geophys res 103:32119–32128

Andreae MO, Artaxo P, Fischer H, Fortuin J, Gregoire J, Hoor P, Kormann R, Krejci R, Lange L, Lelieveld J, Longo K, Peters W, Reus MD, Scheeren B, Silva-Dias MD, Strøm J (2001) Transport of biomass burning smoke to the upper troposphere by deep convection in the equatorial region. Geophys res lett 28:951–954

Andreas EL (1998) A new sea-spray generation function for wind speeds up to 32 m s^{-1}. J phys oceanogr 28:2175

Andreas EL, Edson JB, Monahan EC, Rouault MP, Smith SD (1995) The spray contribution to net evaporation from the sea: A review of recent progress. Bound-lay meteorol 72:3–52

Andres RJ, Kasgnoc AD (1998) A time-averaged inventory of subaerial volcanic sulfur emissions. J geophys res 103:25251–25261

Andronache C, Donner LJ, Seman CJ, Ramaswamy V, Hemler RS (1999) Atmospheric sulfur and deep convective clouds in tropical Pacific: A model study. J geophys res 104:4005–4024

Anklin M, Bales RC (1997) Recent increase in H_2O_2 concentration at Summit, Greenland. J geophys res 102:19099–19104
Ansmann A, Riebesell M, Weitkamp C (1990) Measurement of atmospheric aerosol extinction profiles with a raman lidar. Opt lett 15:746–748
Ansmann A, Mattis I, Wandinger U, Wagner F, Reichardt J, Deshler T (1997) Evolution of the Pinatubo aerosol: Raman Lidar observations of particle optical depth, effective radius, mass and surface area over central Europe at 53.4° N. J atmos sci 54:2630–2641
Ansmann A, Althausen D, Wandinger U, Franke K, Müller D, Wagner F, Heintzenberg J (2000) Vertical profiling of the Indian aerosol plume with six-wavelength lidar during INDOEX: A first case study. Geophys res lett 27:963–966
Antia N, Harrison P, Oliveira L (1991) The role of dissolved organic nitrogen in phytoplankton nutrition, cell biology and ecology. Phycologia 30:1–89
Apel EC, Calvert JG, Fehsenfeld FC (1994) The non-methane hydrocarbon intercomparison experiment (NOMHICE): Tasks 1 and 2. J geophys res 99:16651–16664
Apel EC, Calvert JG, Riemer D, Pos W, Zika R, Kleindienst T, Lonneman W, Fung K, Fujita E, Shepson P, Starn T, Roberts PT (1998) Measurements comparison of oxygenated volatile organic compounds at a rural site during the 1995 SOS Nashville Intensive. J geophys res 103:22295–22316
Apel EC, Calvert JG, Gilpin TM, Fehsenfeld FC, Parrish DD, Lonneman WA (1999) The nonmethane hydrocarbon intercomparison experiment (NOMHICE): Task 3. J geophys res 104:26069–26086
Appel BR, Wall SM, Tokiwa Y, Haik M (1978) Interference effects in sampling particulate nitrate in ambient air. Atmos environ 13:319–325
Appel BR, Cheng W, Salaymeh F (1989) Sampling of carbonaceous particles in the atmosphere II. Atmos environ 23:2167–2175
Appenzeller C, Holton JR, Rosenlof KH (1996) Seasonal variation of mass transport across the tropopause. J geophys res 101:15071–15078
Arimoto R, Duce DLSRA, Prospero JM (1992) Trace elements in aerosol particles from Bermuda and Barbados: Concentrations, sources and relationships to aerosol sulfate. J atmos chem 14:439–457
Arimoto R, Gao Y, Zhou MY, Lee DS, Chen L, Gu D, Wang Z, Zhang X (1997) Atmospheric deposition of trace elements to the Western Pacific Basin. In: Baker JE (ed) Atmospheric Deposition of Contaminants to the Great Lakes and Coastal Waters. SETAC Press, Pensacola, FL, pp 209–225
Arking A (1991) The radiative effects of clouds and their impact on climate. B am meteorol soc 72:795–813
Arnott WP, Moosmüller H, Abbott RE, Ossofsky MD (1995) Thermoacoustic enhancement of photoacoustic spectroscopy: Theory and measurements of the signal to noise ratio. Rev sci instrum 66:4827–4833
Arnott WP, Moosmüller H, Rogers CF, Jin T, Bruch R (1999) Photoacoustic spectrometer for measuring light absorption by aerosol: Instrument description. Atmos environ 33:2845–2852
Artaxo P, Storms FBH, van Grieken R, Maenhaut W (1988) Composition and sources of aerosols from the Amazon Basin. J geophys res 93:1605–1615
Artaxo P, Maenhaut HSW, van Grieken R (1990) Aerosol characteristics and sources for the Amazon Basin during the wet season. J geophys res 95:16971–16985
Artaxo P, Gerab F, Yamasoe MA, Martins JV (1994) Fine mode aerosol composition at three long-term atmospheric monitoring sites in the amazon basin. J geophys res 99:22857–22868
Asman W (1994) Emission and deposition of ammonia and ammonium. Nova Acta Leopoldina 70:263–297
Asman WAH, Harrison RM, Ottley CJ (1994) Estimation of the net air-sea flux of ammonia over the southern bight of the north sea. Atmos environ 28:3647–3654
Atkinson R (1997a) Gas-phase tropospheric chemistry of volatile organic compounds: 1. Alkanes and alkenes. J phys chem ref data 26:215–290
Atkinson R (1997b) Atmospheric reactions of alkoxy and *b*-hydroxy radicals. Int J chem kinet 29:99–111
Atkinson R, Arey J (1998) Atmospheric chemistry of biogenic organic compounds. Acc Chem Res 31(9):574–583
Atkinson R, Aschmann SM, Arey J, Shorees B (1992) Formation of OH radicals in the gas phase reactions of O_3 with a series of Terpenes. J geophys res 97:6065–6073
Atkinson R, Baulch DL, Cox RA, Hampson Jr. RF, Kerr JA, Rossi MJ, Troe J (1997) Evaluated kinetic and photochemical data for atmospheric chemistry, Supplement 5. J phys chem-us ref data 26:521
Atlas EL, Ridley BA (1996) The Mauna Loa Observatory Photochemistry Experiment: Introduction. J geophys res 101:14531–14541
Audiffren N, Cautenet S, Chaumerliac N (1999) A modeling study of the influence of ice scavenging on the chemical composition of liquid-phase precipitation of a cumulonimbus cloud. J appl meteorol 38:1148–1160
Austin JF, Follows MJ (1991) The ozone record at Payerne – An assessment of the cross-tropopause flux. Atmos environ 25(9):1873–1880
Ayers GP, Gillett RW (1988) Acidification in Australia. In: Rodhe H, Herera R (eds) Acidification in tropical countries. J. Wiley, New York, pp 347–402
Ayers GP, Larson TV (1990) Numerical study of droplet size dependent chemistry in oceanic wintertime stratus cloud at southern mid-latitudes. J atmos chem 11:143–167
Ayers GP, Ivey JP, Gillett RW (1991) Coherence between seasonal cycles of dimethyl sulphide, methanesulphonate and sulphate in marine air. Nature 349:404–406
Ayers GP, Penkett SA, Gillett RW, Bandy B, Galbally IE, Meyer CP, Elsworth CM, Bentley ST, Forgan BW (1992) Evidence for photochemical control of ozone concentrations in unpolluted marine air. Nature 360:446–448
Ayers GP, Bentley ST, Ivey JP, Forgan BW (1995) Dimethylsulfide in marine air at Cape Grim, 41° S. J geophys res 100:21013–21021
Ayers G, Gillet R, Hara H (1996a) Acidic deposition in east Asia and Oceania. In: World Meteorological Organization (ed) Global acid deposition assessment. WMO, Geneva, Switzerland, pp 107–134
Ayers GP, Penkett SA, Gillett RW, Bandy BJ, Galbally IE, Meyer CP, Elsworth CM, Bentley ST, Forgan BW (1996b) Annual cycle of peroxides and ozone in marine air at Cape Grim, Tasmania. J atmos chem 23:221–252
Ayers GP, Graneck H, Boers R (1997) Ozone in the marine boundary layer at Cape Grim: Model simulation. J atmos chem 27:179–195
Baker JE, Poster DL, Clark CA, Church TM, Scudlark JR, Ondov JM, Dickhut RM, Cutter G (1997) Loadings of atmospheric trace elements and organic contaminants to the Chesapeake Bay. In: Baker JE (ed) Atmospheric deposition of contaminants to the Great Lakes and coastal waters. SETAC Press, Pensacola, FL, pp 171–194
Baker JM, Reeves CE, Nightingale PD, Penkett SA, Gibb SW, Hatton AD (1999) Biological production of methyl bromide in the coastal waters of the North Sea and open ocean of the northeast Atlantic. Mar chem 64:267–285
Baldocchi D, Guenther A, Harley P, Klinger L, Zimmerman P, Lamb B, Westberg H (1995) The fluxes and air chemistry of isoprene above a deciduous hardwood forest. Philos t roy soc a 350:279–296
Balkanski YJ, Jacob GMGDJ, Graustein WC, Turekian KK (1993) Transport and residence times of tropospheric aerosols inferred from a global three-dimensional simulation of ^{210}Pb. J geophys res 98:20573–20586
Ball SM, Hancock G, Murphy IJ, Rayner SP (1993) The relative quantum yields of $O_2(a^1(\Delta)$ from the photolysis of ozone in the wavelength range 270 nm $\leq \lambda \leq$ 329 nm. Geophys res lett 20:2063–2066
Ball SM, Hanson DR, Eisele FL, McMurry PH (1999) Laboratory studies of particle nucleation: Initial results for H_2SO_4, H_2O, and NH_3 vapors. J geophys res 104:23709–23718
Baltensperger U, Gäggeler HW, Jost DT, Lugauer M, Schwikowski M, Weingartner E, Seibert P (1997) Aerosol climatology at the high-alpine site Jungfraujoch, Switzerland. J geophys res 102:19707–19715
Bange HW, Bartell UH, Rapsomanikis S, Andreae MO (1994) Methane in the Baltic and North Seas and a reassessment of the marine emission of methane. Global biogeochem cy 8:465–480

Barbosa PM, Perieira JMC, Gregoire J-M (1998) Compositing criteria for burned area assessment using multitemporal low resolution satellite data. Remote sens environ 65:38–49

Barnard WR, Andreae MO, Watkins WE, Bingemer H, Georgii H-W (1982) The flux of dimethylsulfide from the oceans to the atmosphere. J geophys res 87(C11):8787–8793

Barnard WR, Andreae MO, Iverson RL (1984) Dimethylsulfide and *Phaeocystis poucheti* in the southeastern Bering Sea. Cont shelf res 3:103–113

Barrie L, Platt U (1997) Arctic tropospheric chemistry: An overview. Tellus 49B:449–454

Barrie LA, Bottenheim JW, Schnell RC, Crutzen PJ, Rasmussen RA (1988) Ozone destruction and photochemical reactions at polar sunrise in the lower Arctic atmosphere. Nature 334:138–141

Barrie LA, Gregor D, Hargrave B, Lake R, Muir D, Shearer R, Tracey B, Bidleman T (1992) Arctic contaminants: Sources, occurrence and pathways. Sci total environ 122:1–74

Barrie LA, Yi Y, Leaitch WR, Lohmann U, Kasibhatla P, Roelofs G-J, Wilson J, McGovern F, Benkovitz C, Mélières MA, Law K, Prospero J, Kritz M, Bergmann D, Bridgeman C, Chin M, Christensen J, Easter R, Feichter J, Land C, Jeuken A, Kjellström E, Koch D, Rasch P (2001) A comparison of large scale atmospheric sulphate aerosol models (COSAM): Overview and highlights. Tellus 53B:615–645

Barth MC (1994) Numerical modeling of sulfur and nitrogen chemistry in a narrow cold-frontal rainband: The impact of meteorological and chemical parameters. J appl meteorol 33:855–868

Barth MC, Rasch PJ, Kiehl JT, Benkovitz CM, Schwartz SE (2000) Sulfur chemistry in the National Center for Atmospheric Research Community Climate Model: Description, evaluation, features and sensitivity to aqueous chemistry. J geophys res 105:1387–1415

Bascom R, Bromberg PA, Costa DA, Devlin R, Dockery DW, Frampton MW, Lambert W, Samet JM, Speizer FE, Uttell M (1996a) Health effects of outdoor air pollution, Part 1. Am J resp crit care 153:3–50

Bascom R, Bromberg PA, Costa DA, Devlin R, Dockery DW, Frampton MW, Lambert W, Samet JM, Speizer FE, Uttell M (1996b) Health effects of outdoor air pollution, Part 2. Am J resp crit care 153:477–498

Bates TS (1999) Preface to special section: First Aerosol Characterization Experiment (ACE-1) Part 2. J geophys res-atmos 104:21645

Bates TS, Cline JD (1985) The role of the ocean in a regional sulphur cycle. J geophys res 90:9168–9172

Bates TS, Cline JD, Gammon RH, Kelly-Hansen S (1987) Regional and seasonal variations in the flux of oceanic dimethylsulfide to the atmosphere. J geophys res 92:2930–2938

Bates TS, Lamb BK, Guenther A, Dignon J, Stoiber RE (1992) Sulfur emissions to the atmosphere from natural sources. J atmos chem 14:315–337

Bates TS, Kelly KC, Johnson JE (1993) Concentrations and fluxes of dissolved biogenic gases (DMS, CH_4, CO, CO_2) in the equatorial Pacific during the SAGA-3 experiment. J geophys res 98:16969–16978

Bates TS, Kiene RP, Wolfe GV, Matrai PA, Chavez FP, Buck KR, Blomquist BW, Cuhel RL (1994) The cycling of sulphur in surface seawater of the northeast Pacific. J geophys res 99:7835–7843

Bates TS, Kelly KC, Johnson JE, Gammon RH (1995) Regional and seasonal variations in the flux of oceanic carbon monoxide to the atmosphere. J geophys res 100:23093–23101

Bates TS, Kelly KC, Johnson JE, Gammon RH (1996) A reevaluation of the open-ocean flux of methane to the atmosphere. J geophys res 101:6953–6962

Bates TS, Huebert BJ, Gras JL, Griffiths FB, Durkee PA (1998) International Global Atmospheric Chemistry (IGAC) project's first aerosol experiment (ACE-1): Overview. J geophys res 103:16297–16318

Battle M, Bender M, Sowers T, Tans PP, Butler JH, Elkins JW, Ellis JT, Conway T, Zhang N, Lang P, Clarke AD (1996) Atmospheric gas concentrations over the past century measured in air from firn at the South Pole. Nature 383:231–235

Baughcum SL, Tritz TG, Henderson SC, Pickett DC (1996) NASA contractor report 4700. Scheduled civil aircraft emissions inventories for 1992: Database development and analysis, US National Aeronautics and Space Administration, Washington, D.C.

Baumann MEM, Brandini FP, Staubes R (1994) The influence of light and temperature on carbon-specific DMS release by cultures of *Phaeocystis antarctica* and three Antarctic diatoms. Mar chem 45:129–136

Beauford W, Barber J, Barringer AR (1975) Heavy metal release from plants into the atmosphere. Nature 256:35–37

Beauford W, Barber J, Barringer AR (1977) Release of particles containing metals from vegetation into the atmosphere. Science 195:571–573

Beck J, Ammann C, Rummel U, Meixner FX, Andreae MO, Helas G (2000) Aerosol particle flux measurements over the rainforest of Amazonia. In: Sixth International Conference on Air-Surface Exchange of Gases and Particles, Edinburgh, Scotland

Becker K-H (1996) The European photoreactor EUPHORE: Design and technical development of the European photoreactor and first experimental results. EC-Project, Wuppertal, Germany

Beekmann M, Ancellet G, Blonsky S, DeMuer D, Ebel A, Elbern H, Hendricks J, Kowol J, Mancier C, Sladkovic R, Smit HGJ, Speth P, Trickl T, Van Haver P (1997) Regional and global tropopause fold occurrence and related ozone flux across the tropopause. J atmos chem 28(1–3):29–44

Beier C, Gundersen P, Rasmussen L (1992) A new method for estimation of dry deposition of particles based on throughfall measurements in a forest edge. Atmos environ 26A:1553–1559

Bekki S (1997) On the possible role of aircraft-generated soot in the middle latitude ozone depletion. J geophys res 102:10751–10758

Belviso S, Morrow R, Mihalopolous N (2000) An Atlantic meridional transect of surface water dimethyl sulfide concentrations with 10–15 km horizontal resolution and close examination of ocean circulation. J geophys res 105D:14423–14431

Bender M, Conrad R (1994) Methane oxidation activity in various soils and freshwater sediments: Occurrence, characteristics, vertical profiles, and distribution on grain size fractions. J geophys res 99:16531–16540

Bengtsson L (1999) From short range barotropic modelling to extended range global weather prediction: A 40 year perspective. Tellus 51(A-B):13–32

Benkovitz CM, Schwartz SE (1997) Evaluation of modeled sulfate and SO_2 over North America and Europe for four seasonal months in 1986–1987. J geophys res 102:25305–25338

Benkovitz CM, Easter RC, Nemesure S, Wagener R, Schwartz SE (1994) Sulfate over the North Atlantic and adjacent continental regions: Evaluation for October and November 1986, using a three-dimensional model driven by observation-derived meteorology. J geophys res 99:20725–20756

Benkovitz CM, Scholtz MT, Pacyna J, Tarrasón L, Dignon J, Voldner EV, Spiro PA, Logan JA, Graedel TE (1996) Global gridded inventories of anthropogenic emissions of sulfur and nitrogen. J geophys res 101D:29239–29253

Berg OH, Swietlicki E, Krejci R (1998) Hygroscopic growth of aerosol particles in the marine boundary layer over the Pacific and Southern Oceans during the First Aerosol Characterization Experiment (ACE-1). J geophys res 103:16535–16545

Bergamaschi P (1997) Seasonal variations of stable hydrogen and carbon isotope ratios in methane from a Chinese rice paddy. J geophys res 102:25383–25393

Bergamaschi P, Brenninkmeijer CAM, Hahn M, Röckmann T, Scharffe DH, Crutzen PJ, Elansky NF, Belikov IB, Trivett NBA, Worthy DEJ (1998a) Isotope analysis based source identification for atmospheric CH_4 and CO sampled across Russia using the Trans-Siberian railroad. J geophys res 103:8227–8235

Bergamaschi P, Lubina C, Konigstedt R, Fischer H (1998b) Stable isotopic signatures (^{13}C, δD) of methane from European landfill sites. J geophys res 103:8251–8265

Bergamaschi P, Hein MHR, Crutzen PJ (1999) Inverse modeling of CO mixing ratios. J geophys res 105:1909–1928

Bergamaschi P, Schupp M, Harris GW (1994) High-precision direct measurements of $^{13}CH_4/^{12}CH_4$ and $^{12}CH_3D/^{12}CH_4$ ratios in atmospheric methane sources by means of a long-path tunable diode laser absorption spectrometer. Appl optics 33:7704–7716
Bergan T, Gallardo L, Rodhe H (1999) Mercury in the global troposphere: A three-dimensional model study. Atmos environ 33:1575–1585
Berkowitz CM, Daum PH, Spicer CW, Busness KM (1996) Synoptic patterns associated with the flux of excess ozone to the western North Atlantic. J geophys res 101:28923–28933
Berntsen T, Isaksen ISA (1997) A global 3-D chemical transport model for the troposphere;1. Model description and CO and ozone. J geophys res 102:21239–21280
Berntsen T, Isaksen ISA, Wang W-C, Liang X-Z (1996) Impacts of increased anthropogenic emissions in Asia on tropospheric ozone and climate: A global 3-D model study. Tellus 48B:13–32
Berntsen T, Isaksen ISA, Myhre G, Fuglestvedt JS, Stordal F, Larsen TA, Freckleton RS, Shine KP (1997) Effects of anthropogenic emissions on tropospheric ozone and its radiative forcing. J geophys res 102:28101–28126
Bethan S, Vaughan G, Gerbig C, Volz-Thomas A, Richer H, Tiddeman DA (1998) Chemical air mass differences near fronts. J geophys res 103:13413–13434
Bey I, Aumont B, Toupance G (2001a) A modeling study of the nighttime radical chemistry in the lower continental troposphere: 1. Development of a detailed chemical mechanism including nighttime chemistry. J geophys res 106(D9):9959–9990
Bey I, Aumont B, Toupance G (2001b) A modeling study of the nighttime radical chemistry in the lower continental troposphere: 2. Origin and evolution of HO_x. J geophys res 106(D9):9991–10001
Bigg EK, Leck C (2001) Cloud-active particles over the central Arctic Ocean. J geophys res-atmos 106:32139–32154
Bigg EK, Leck C, Nilsson ED (1996) Sudden changes in Arctic atmospheric aerosol concentrations during summer and autumn. Tellus 48B:254–271
Bigg EK, Leck C, Nilsson ED (2001) Sudden changes in aerosol and gas in the central arctic marine boundary layer: Causes and consequences. J geophys res 106:32167–32186
Birmili W, Stratmann F, Wiedensohler A (1999a) Design of a DMA-based size spectrometer for a large particle size range and stable operation. J aerosol sci 30:549–553
Birmili W, Yuskiewicz B, Wiedensohler A, Stratmann F, Choularton TW, Bower KN (1999b) Climate-relevant modification of the aerosol size distribution by processes associated with orographic clouds. Atmos res 50:241–263
Blake DR, Rowland FS (1988) Continuing worldwide increase in tropospheric methane, 1978 to 1987. Science 239:1129–1131
Blake NJ, Blake DR, Chen T-Y, Collins Jr. JE, Sachse GW, Anderson BE, Rowland FS (1997) Distribution and seasonality of selected hydrocarbons and halocarbons over the western Pacific basin during wintertime. J geophys res 102:28315–28333
Blanchard DC (1963) The electrification of the atmosphere by particles from bubbles in the sea. Prog oceanogr 1:71–202
Blanchard DC (1983) The production, distribution and bacterial enrichment of the sea-salt aerosol. In: Liss PS, Slinn WGS (eds) Air-sea exchange of gases and particles. D. Reidel, Dordrecht, pp 407–454
Blanchard DC, Woodcock AH (1957) Bubble formation and modification in the sea and its meteorological significance. Tellus 9:145–158
Blitz M, Pilling MJ, Robertson SH, Seakins PW (1999) Direct studies on the decomposition of the tert-butoxy radical and its reaction with NO. Phys chem chem phys 1:73–80
Blom JG, Verwer JG (2000) A comparison of integration methods for atmospheric transport-chemistry models. J comput appl math 126:381–396
Blunier T, Chappellaz JA, Schwander J, Barnola J-M, Desperts T, Stauffer B, Raynaud D (1993) Atmospheric methane record from a Greenland ice core over the last 1000 years. Geophys res lett 20:2219–2222
Blunier T, Chappellaz JSJ, Stauffer B, Raynaud D (1995) Variations in atmospheric methane concentration during the Holocene epoch. Nature 374:46–49
Bluth GJ, Schnetzler CC, Krueger AJ (1993) The contribution of explosive volcanism to global atmosphere sulfphur dioxide concentrations. Nature 366:327–329
Bodelier PLE, Roslev P, Henckel T, Frenzel P (2000) Stimulation by ammonium-based fertilizers of methane oxidation in soil around rice roots. Nature 403:421–424
Bodhaine BA (1989) Barrow surface aerosol: 1976–1986. Atmos environ 23:2357–2369
Bodhaine BA, DeLuisi JJ (1985) An aerosol climatology of Samoa. J atmos chem 3:107–122
Bodhaine BA, Deluisi JJ, Harris M, Houmere P, Bauman S (1986) Aerosol measurements at the South Pole. Tellus 38B:223–235
Boeckx P, van Cleemput O (1996) Flux estimates from soil methanogenesis and methanotrophy: Landfills, rice paddies, natural wetlands and aerobic soils. Environ monit assess 42:189–207
Boers R, Mitchell RM (1994) Absorption feedback in stratocumulus clouds influence on cloud top albedo. Tellus 46A:229–241
Bogner JE, Sass RL, Walter BP (2000) Model comparisons of methane oxidation across a management gradient: Wetlands, rice production systems and landfill. Global biogeochem cy 14(4):1021–1033
Bond TC, Charlson RJ, Heintzenberg J (1998) Quantifying the emission of light-absorbing particles: Measurements tailored to climate studies. Geophys res lett 25:337–340
Bond TC, Anderson TL, Campbell D (1999) Calibration and intercomparison of filter-based measurements of visible light absorption by aerosols. Aerosol sci tech 30:582–600
Bopp L, Monfray P, Aumont O, Orr JC, Dufresne J-L, Treut Hl (2000) Potential impact of climate change on marine export production. In: SOLAS Open Science Conference, Damp, Germany
Bottenheim JW, Barrie LA, Atlas E, Heidt LE, Niki H, Rasmussen RA, Shepson PB (1990) Depletion of lower tropospheric ozone during Arctic spring – The polar sunrise experiment. J geophys res 95:18535–18568
Boucher O, Lohmann U (1995) The sulphate-CCN-cloud albedo effect. Tellus 47B:281–300
Boucher O, Schwartz SE, Ackerman TP, Anderson TL, Bergstrom B, Bonnel B, Chylek P, Dahlback A, Fouquart Y, Fu Q, Halthore RN, Haywood JM, Iversen T, Kato S, Kinne S, Kirkevåg A, Knapp KR, Lacis A, Laszlo I, Mishchenko MI, Nemesure S, Ramaswamy V, Roberts DL, Russell P, Schlesinger ME, Stephens GL, Wagener R, Wang M, Wong J, Yang F (1998) Intercomparison of models representing direct shortwave radiative forcing by sulfate aerosols. J geophys res 103:16979–16908
Bousquet P, Ciais P, Peylin P, Ramonet M, Monfray P (1999) Inverse modeling of annual atmospheric CO_2 sources and sinks. 1. Method and control inversion. J geophys res 104:26161–26178
Bouwman AF (1996) Direct emissions of nitrous oxide from agricultural soils. Nutr cycl agroecosys 46:53–70
Bouwman AF (1998) Nitrogen oxides and tropical agriculture. Nature 392:866–867
Bouwman AF, van der Hoek K, Olivier JGJ (1995) Uncertainties in the global sources distribution of nitrous oxide. J geophys res 100:2785–2800
Bouwman AF, Lee DS, Asman WAH, Dentener FJ, van der Hoek KW, Olivier JGJ (1997) A global high-resolution emission inventory for ammonia. Global biogeochem cy 51:561–587
Bovensmann H, Burrows JP, Buchwitz M, Frerick J, Noël S, Rozanov VV, Chance KV, Goede APH (1999) SCIAMACHY – Mission objectives and measurements modes. J atmos sci 56:127–150
Bower KN, Choularton TW, Gallagher MW, Colvile RN, Wells M, Beswick KM, Wiedensohler A, Hansson H-C, Svenningsson B, Swietlicki E, Wendisch M, Berner A, Kruisz C, Laj P, Facchini MC, Fuzzi S, Bizjak M, Dollard G, Jones B, Acker K, Wieprecht W, Preiss M, Sutton MA, Hargreaves KJ, Storeton-West RL, Cape JN, Arends BG (1997) Observations and modelling of the processing of aerosol by a hill cap cloud. Atmos environ 31:2527–2543
Boyd PW, Watson AJ, Law CS, Abraham ER, Trull T, Murdoch R, Bakker DCE, Bowie AR, Buesseler KO, Chang H, Charette M, Croot P, Downing K, Frew R, Gall M, Hadfield M, Hall J, Harvey M, Jameson G, Laroche J, Liddicoat M, Ling R, Maldonado MT, Mckay RM, Nodder S, Pickmere S, Pridmore R, Rintoul S, Safi K, Sutton P, Strzepek R, Tanneberger K, Turner S, Waite A, Zeldis J (2000) A mesoscale phytoplankton bloom in the polar Southern Ocean stimulated by iron fertilization. Nature 407:695–702

Bradshaw J, Davis D, Crawford J, Chen G, Shetter R, Müller M, Gregory G, Sachse G, Blake D, Heikes B, Singh H, Mastromarino J, Sandholm S (1999) Photofragmentation two-photon laser-induced fluorescence detection of NO_2 and NO: Comparison of measurements with model results based on airborne observations during PEM-Tropics A. Geophys res lett 26:4471–4474

Bradshaw J, Davis D, Grodzinsky G, Smyth S, Newell R, Sandholm S, Liu S (2000) Observed distributions of nitrogen oxides in the remote free troposphere from the NASA Global Tropospheric Experiment Program. Rev geophys 38:61–116

Brandenburger U, Brauers T, Dorn H-P, Hausmann M, Ehhalt D (1998) *In situ* measurements of tropospheric hydroxyl radicals by folded long-path laser absorption during the field campaign POPCORN. J atmos chem 31:181–204

Brasseur GP, Granier C (1992) Mount Pinatubo aerosols, chlorofluorocarbons, and ozone depletion. Science 257:1239–1242

Brasseur GP, Solomon S (1986) Aeronomy of the middle atmosphere. D. Reidel, Dordrecht, Holland

Brasseur GP, Hauglustaine DA, Walters S (1996) Chemical compounds in the remote Pacific troposphere: Comparisons between MLOPEX measurements and chemical transport model calculations. J geophys res 101:14795–14813

Brasseur GP, Cox RA, Hauglustaine D, Isaksen I, Lelieveld J, Lister DH, Sausen R, Schumann U, Wahner A, Wiesen P (1998a) European scientific assessment of the atmospheric effects of aircraft emissions. Atmos environ 32(13):2329–2418

Brasseur GP, Hauglustaine DA, Walters S, Rasch PJ, Muller J-F, Granier C, Tie XX (1998b) MOZART, a global chemical transport model for ozone and related chemical tracers: 1. Model description. J geophys res 103:28265–28289

Brasseur GP, Kiehl JT, Muller JF, Schneider T, Granier C, Tie XX, Hauglustaine D (1998c) Past and future changes in global tropospheric ozone. Geophys res lett 25:3807–3810

Brasseur GP, Orlando JJ, Tyndall GS (1999a) Atmospheric chemistry and global change. Oxford University Press, Cambridge

Brasseur GP, Khattatov B, Walters S (1999b) Modeling. In: Brasseur G, Orlando JJ, Tyndall G (eds) Atmospheric chemistry and global change. Oxford University Press, Oxford

Brauers T, Aschmutat U, Brandenburger U, Dorn H-P, Hausmann M, Hebling M, Hofzumahaus A, Holland F, Plass-Dülmer C, Ehhalt DH (1996) Intercomparison of tropospheric OH radical measurements by multiple folded long-path laser absorption and laser induced fluorescence. Geophys res lett 23:2545–2548

Brauers T, Hausmann M, Bister A, Kraus A, Dorn H-P (2001) OH radicals in the boundary layer of the Atlantic Ocean: 1. Measurements by long-path absorption spectroscopy. J geophys res 106:7399–7414

Bregman A, Arnold F, Burger V, Fischer H, Lelieveld J, Scheeren BA, Schneider J, Siegmund PC, Strom J, Waibel A, Wauben WMF (1997) *In situ* trace gas and particle measurements in the summer lower stratosphere during STREAM II: Implications for O_3 production. J atmos chem 26(3):275–310

Brenguier J-L, Chuang PY, Fouquart Y, Johnson DW, Parol F, Pawlowska H, Pelon J, Schüller L, Schröder F, Snider J (2000) An overview of the ACE-2 CLOUDYCOLUMN closure experiment. Tellus 52B:815–827

Brenninkmeijer CAM, Manning MR, Lowe DC, Wallace GW, Sparks RJ, Volz-Thomas A (1992) Interhemispheric asymmetry in OH abundance inferred from measurements of atmospheric ^{14}CO. Nature 356:50–52

Brenninkmeijer CAM, Crutzen PJ, Fischer H, Güsten H, Hans W, Heinrich G, Heintzenberg J, Hermann M, Immelmann T, Kersting D, Maiss M, Nolle M, Pitscheider A, Pohlkamp H, Scharffe D, Specht K, Wiedensohler A (1999) CARIBIC: Civil aircraft for global measurement of trace gases and aerosols in the tropopause region. J atmos ocean tech 16:1373–1383

Broadgate WJ, Liss PS, Penkett SA (1997) Seasonal emissions of isoprene and other reactive hydrocarbons gases from the ocean. Geophys res lett 24:2675–2678

Brocard D, Lacaux J-P (1998) Domestic biomass combustion and atmospheric emission associated in Western Africa. Global biogeochem cy 12:127–139

Brock CA, Hamill P, Wilson JC, Jonsson HH, Chan KR (1995) Particle formation in the upper tropical troposphere: A source of nuclei for the stratospheric aerosol. Science 270:1650–1653

Broecker WS, Ledwell JR, Takahashi T, Weiss R, Merlivat L, Memery L, Peng T-H, Jähne B, Munnich KO (1986) Isotopic versus micrometeorologic ocean CO_2 fluxes: A serious conflict. J geophys res 91:10517–10527

Broecker WS, Peng T-H, Ostlund G, Stuiver M (1995) The distribution of bomb radiocarbon in the ocean. J geophys res 90:6953–6970

Brook EJ, Sowers T, Orchardo J (1996) Rapid variations in atmospheric methane concentrations during the past 110 000 years. Science 273:1087–1091

Brook JR, Zhang L, Li Y, Johnson D (1999a) Description and evaluation of a model of deposition velocities for routine estimates of air pollutant dry deposition over North America. Part II: Review of past measurements and model results. Atmos environ 35:5053–5070

Brook JR, Zhang L, Padro J (1999b) Description and evaluation of a model of deposition velocities for routine estimates of air pollutant dry deposition over North America. Part I: Model development. Atmos environ 33:5037–5051

Browell EV, Fenn MA, Butler CF, Grant WB, Clayton ME, ishman J, Bachmeier AS, Anderson BE, Gregory GL, Fuelberg HE, Bradshaw JD, Sandholm ST, Blake DR, Heikes BG, Sachse GW, Singh HB, Talbot RW (1996a) Ozone and aerosol distributions and air mass characteristics over the South Atlantic basin during the burning season. J geophys res 101:24043–24068

Browell EV, Fenn MA, Butler CF, Grant WB, Merrill JT, Newell RE, Bradshaw JD, Sandholm ST, Anderson BE, Bandy AR, Bachmeier AS, Blake DR, Davis DD, Gregory GL, Heikes BG, Kondo Y, Liu SC, Rowland FS, Sachse GW, Singh HB, Talbot RW, Thornton DC (1996b) Large-scale air mass characteristics observed over Western Pacific during summertime. J geophys res 101:1691–1712

Browell EV, Ismail S, Grant WB (1998) Differential Absorption Lidar (DIAL) measurements from air and space. Appl phys b 67:399–410

Browell EV, Ismail S, Hall WM, Moore AS, Kooi SA, Brackett VG, Clayton MB, Barrick J, Schmidlin FJ, Higdon NS, Melfi SH, Whiteman DN (1997) LASE validation experiment. In: Ansmann A, Neuber R, Rairoux P, Wandinger U (eds) Advances in atmospheric remote sensing with lidar. Springer-Verlag, Berlin, pp 289–295

Browell EV, Fenn MA, Butler CF, Grant WB, Ismail S, Ferrare RA, Kooi SA, Brackett VG, Clayton MB, Avery MA, Barrick JDW, Fuelberg HE, Maloney JC, Newell RE, Zhu Y, Mahoney MJ, Anderson BE, Blake DR, Brune WH, Heikes BG, Sachse GW, Singh HB, Talbot RW (2001) Large-scale air mass characteristics observed over the remote tropical Pacific Ocean during March–April 1999: Results from PEM Tropics B field experiment. J geophys res 106:32481–32501

Brown CW, Yoder JA (1994) Coccolithophorid blooms in the global ocean. J geophys res 99(C4):7467–7482

Brown J, Colling A, Park D, Phillips J, Rothery D, Wright J (1989) Ocean circulation. Open University/Pergamon Press, New York

Brown SS, Talukdar RK, Ravishankara AR (1999a) Reconsideration of the rate constant for the reaction of hydroxyl radicals with nitric acid. J chem phys 103:3031–3037

Brown SS, Talukdar RK, Ravishankara AR (1999b) Rate constants for the reaction $OH + NO_2 + M \rightarrow HNO_3 + M$ under atmospheric conditions. Chem phys lett 299:277–284

Browning KA (1990) Organization of clouds and precipitation in extratropical cyclones. In: American Meteorological Society (ed) Extratropical cyclones: The Erik Palmen memorial volume. American Meteorological Society

Brugger A, Slezak IOD, Herndl GJ (1998) Photolysis of dimethylsulfide in the northern Adriatic Sea: Dependence on substrate concentration, irradiance and DOC concentration. Mar chem 59:321–331

Brune WH, Faloona IC, Tan D, Weinheimer AJ, Campos T, Ridley BA, Vay SA, Collins JE, Sachse GW, Jaeglé L, Jacob DJ (1998) Airborne *in situ* OH and HO_2 observations in the cloud-free troposphere and lower stratosphere during SUCCESS. Geophys res lett 25:1701–1704

Brune WH, Tan D, Faloona IF, Jaegle L, Jacob DJ, Heikes BG, Snow J, Kondo Y, Shetter R, Sachse GW, Anderson B, Gregory GL, Vay S, Singh HB, Davis DD, Crawford JH, Blake DR (1999) OH and HO_2 chemistry in the North Atlantic free troposphere. Geophys res lett 26:3077–3080

Brunekreef B (1997) Air pollution and life expectancy: Is there a relationship? Occup environ med 54:781–784
Brunner D, Staehelin J, Jeker D (1998) Large-scale nitrogen oxide plumes in the tropopause region and implications for ozone. Science 282(5392):1305–1309
Brussaard CPD, Kempers RS, Kop AJ, Riegman R, Heldal M (1996) Virus-like particles in a summer bloom of *Emiliania huxleyi* in the North Sea. Aquat microb ecol 10:105–113
Burrows JP (1999) Current and future passive remote sensing techniques used to determine atmospheric constituents. In: Bouwman AF (ed) Approaches to scaling of trace gas fluxes in ecosystems. Elsevier, Berlin
Burrows JP, Weber M, Buchwitz M, Rozanov VV, Ladstätter-Weißenmayer A, Richter A, De RBeek, Hoogen R, Bramstedt K, Eichmann KU (1999) The Global Ozone Monitoring Experiment (GOME): Mission concept and first scientific results. J atmos sci 56:151–175
Butler JH (1994) The potential role of the ocean in regulating atmospheric CH_3Br. Geophys res lett 21:185–188
Butler JH, Battle M, Bender ML, Montzka SA, Clarke AD, Saltzman ES, Sucher CM, Severinghaus JP, Elkins JW (1999) A record of atmospheric halocarbons during the twentieth century from polar firn air. Nature 399:749–755
Butler TJ, Likens GE (1995) A direct comparison of throughfall plus stemflow to estimates of dry and total deposition for sulfur and nitrogen. Atmos environ 29:1253–1265
Cabalo J, Zelenyuk A, Baer T, Miller RE (2000) Two-color laser induced evaporation dynamics of liquid aerosols probed by time-of-flight mass spectrometry. Aerosol sci tech 33:3–19
Cachier H, Buat-Ménard P, Fontugne M (1985) Source terms and source strengths of the carbonaceous aerosol in the Tropics. J atmos chem 3:469–489
Cahoon DR Jr., Stocks BJ, Levine JS, Cofer WR III, Pierson JM (1994) Satellite analysis of the severe 1987 forest fires in northern China and southeastern Siberia. J geophys res 99:18627–18638
Caldeira K (1989) Evolutionary pressures on planktonic production of atmospheric sulphur. Nature 337:732–734
Callander BA (1995) Scientific aspects of the Framework Convention on Climate Change and national greenhouse gas inventories. Environ monit assess 38:129–140
Calvert JG, Lazrus AL, Kok GL, Heikes BG, Walega JG, Lind JA, Cantrell CA (1985) Chemical mechanisms of acid generation in the troposphere. Nature 317:27–35
Cammas J-P, Jacoby-Koaly S, Suhre K, Rosset R, Marenco A (1998) Atlantic subtropical vorticity barrier as seen by Measurements of Ozone by Airbus In-Service Aircraft (MOZAIC) flights. J geophys res 103:25681–25693
Campolongo F, Saltelli A, Jensen NR, Wilson J, Hjorth J (1999) The role of multiphase chemistry in the oxidation of dimethylsulfide (DMS). A latitude dependent analysis. J atmos chem 32:327–356
Cantrell CA, Stedman DH (1982) A possible technique for the measurement of atmospheric peroxy radicals. Geophys res lett 9:846–849
Cantrell CA, Lind JA, Shetter RE, Calvert JG, Goldan PD, Kuster W, Fehsenfeld FC, Montzka SA, Parrish DD, Williams EJ, Buhr MP, Westburg HH, Allwine G, Martin R (1992) Peroxy radicals in the ROSE experiment: Measurement and theory. J geophys res 97:20671–20686
Cantrell CA, Shetter RE, Gilpin TM, Calvert JG (1996a) Peroxy radicals measured during Mauna Loa Observatory Photochemistry Experiment 2: The data and first analysis. J geophys res 101:14643–14652
Cantrell CA, Shetter RE, Gilpin TM, Calvert JG, Eisele FL, Tanner DJ (1996b) Peroxy-radical concentrations measured and calculated from trace-gas measurements in the Mauna Loa Observatory Photochemistry Experiment 2. J geophys res 101:14653–14644
Cantrell CA, Shetter RE, Calvert JG (1996c) Peroxy-radical chemistry during FIELDVOC 1993 in Brittany, France. Atmos environ 30(23):3947–3957
Cantrell CA, Shetter RE, Calvert JG (1996d) Dual-inlet chemical amplifier for atmospheric peroxy-radical measurements. Anal chem 68:4194–4199
Cantrell CA, Shetter RE, Calvert JG, Eisele FL, Williams E, Baumann K, Brune WH, Stevens PS, Mather JH (1997a) Peroxy radicals from photostationary-state deviations and steady-state calculations during the Tropospheric OH Photochemistry Experiment at Idaho Hill, Colorado, 1993. J geophys res 102:6369–6378
Cantrell CA, Shetter RE, Calvert JG, Eisele FL, Tanner DJ (1997b) Some considerations of the origin of night-time peroxy radicals observed in MLOPEX-2c. J geophys res 102:15899–15913
Cantrell CA, Stedman DH, Wendel GJ (1984) Measurement of atmospheric peroxy radicals by chemical amplification. Anal chem 56:1496–1502
Cao M, Marshall S, Gregson K (1996) Global carbon exchange and methane emissions from natural wetlands: Application of a process based model. J geophys res 101(D9):14399–14414
Capaldo KP, Pandis SN (1997) Dimethylsulfide chemistry in the remote marine atmosphere: Evaluation and sensitivity analysis of available mechanisms. J geophys res 102:23251–23267
Carlson TN (1991) Mid-latitude Weather Systems. Routledge, London, New York
Carmichael G (1998) Report on passive samplers for atmospheric chemistry measurements and their role in GAW
Carmichael GR, Zhang Y, Chen LL, Hong MS, Ueda H (1996) Atmospheric seasonal variation of aerosol composition at Cheju Island, Korea. Atmos environ 30:2407–2416
Carmichael GR, Hong MS, Ueda H, Chen LL, Murano K, Park JK, Kang C, Shim S (1997) Aerosol composition at Cheju Island, Korea. J geophys res 102:6047–6061
Carpenter LJ, Monks PS, Bandy BJ, Penkett SA, Galbally IE, Meyer CP (1997) A study of peroxy radicals and ozone photochemistry at coastal sites in the northern and southern hemispheres. J geophys res 102:25417–25427–14652
Carpenter LJ, Sturges WT, Penkett SA, Liss PS, Alicke B, Hebestreit K, Platt U (1999) Short-lived alkyl iodides and bromides at Mace Head, Ireland: Links to biogenic sources and halogen oxide production. J geophys res 104(D1):1679–1689
Carrico CM, Rood MJ, Ogren JA (1998) Aerosol light scattering properties at Cape Grim, Tasmania during the First Aerosol Characterization Experiment (ACE-1). J geophys res 103:16565–16574
Carsey TP, Churchill DD, Farmer ML, Fischer CJ, Pszenny AAP, Ross VB, Saltzman ES, Springer-Young M, Bonsang B (1997) Nitrogen oxides and ozone production in the North Atlantic marine boundary layer. J geophys res 102:10653–10665
Carslaw N, Carpenter LJ, Plane JMC, Allan BJ, Burgess RA, Clemitshaw KC, Coe H, Penkett SA (1997a) Simultaneous measurements of nitrate and peroxy radicals in the marine boundary layer. J geophys res 102:18917–18933
Carslaw N, Plane JMC, Coe H, Cuevas E (1997b) Observations of the Nitrate Radical in the Free Troposphere at Izaña de Tenerife. J geophys res 102:10613–10622
Carslaw N, Creasey DJ, Heard DE, Lewis AC, McQuaid JB, Pilling MJ, Monks PS, Bandy BJ, Penkett SA (1999a) Modeling OH, HO_2, and RO_2 radicals in the marine boundary layer: 1. Model construction and comparison with field measurements. J geophys res 104:30241–30255
Carslaw N, Jacobs PJ, Pilling MJ (1999b) Modelling OH, HO_2 and RO_2 radicals in the marine boundary layer: 2. Mechanism reduction and uncertainty analysis. J geophys res 104:30257–30273
Carslaw N, Creasey DJ, Harrison D, Heard DE, Hunter MC, Jacobs PJ, Jenkin ME, Lee JD, Lewis AC, Pilling MJ, Saunders SM, Seakins PW (2001) OH and HO_2 radical chemistry in a forested region of north-western Greece. Atmos environ 35:4725–4737
Carson PG, Johnston MV, Wexler AS (1997) Laser desorption/ionization of ultrafine aerosol particles. Rapid commun mass sp 11:993–996
Carter WPL (1990) A detailed mechanism for the gas-phase atmospheric reactions of organic compounds. Atmos environ 24A:481–518
Carter WPL, Atkinson R (1997) Development and evaluation of a detailed mechanism for the atmospheric reactions of isoprene and NO_x. Int J chem kinet 28:497–530
Carver GD, Stott PA (2000) IMPACT: An implicit time integration scheme for chemical species and families. Ann geophys 18:337–346

Carver GD, Brown PD, Wild O (1997) The ASAD atmospheric chemistry integration package and chemical reactions database. Comput phys commun 105:197–215

Cember R (1989) Bomb radiocarbon in the Red Sea: A medium-scale gas exchange experiment. J geophys res 94:2111–2123

Cess RD, Zhang MH, Ingram WJ, Potter GL, Alekseev V, Barker HW, Cohen-Solal E, Colman RA, Dazlich DA, Del ADGenio, Dix MR, Dymnikov V, Esch M, Fowler LD, Fraser JR, Galin V, Gates WL, Hack JJ, Kiehl JT, Le Treut H, Lo KK-W, McAvaney BJ, Meleshko VP, Morcrette J-J, Randall DA, Roeckner E, Royer J-F, Schlesinger ME, Sporyshev PV, Timbal B, Volodin EM, Taylor KE, Wang W, Wetherald RT (1996) Cloud feedback in atmospheric general circulation models: An update. J geophys res 101:12791–12794

Chamberlain AC, Chadwick RC (1953) Deposition of airborne radioiodine vapour. Nucleonics 11:22

Chameides WL, Davis DD (1982) The free radical chemistry of cloud droplets and its impact upon the composition of rain. J geophys res 87:4863–4877

Chameides WL, Perdue EM (1997) Biogeochemical cycles: A computer-interactive study of Earth system science and global change. Oxford University Press, Oxford

Chameides WL, Stelson AW (1992) Aqueous-phase chemical processes in deliquescent sea-salt aerosols: A mechanism that couples the atmospheric cycles of S and sea-salt. J geophys res 97:20565–20580

Chameides WL, Walker JCG (1973) A photochemical theory for tropospheric ozone. J geophys res 78:8751–8760

Chameides WL, Yu H, Liu SC, Bergin M, Zhou X, Mearns L, Wang G, Kiang CS, Saylor RD, Luo C, Huang Y, Steiner A, Giorgi F (1999) Case study of the effects of atmospheric aerosols and regional haze on agriculture: An opportunity to enhance crop yields in China through emission controls? P natl acad sci usa 96:13626–13633

Chang JS, Brost RA, Isaksen ISA, Madronich S, Middleton P, Stockwell WR, Walcek CJ (1987) A three-dimensional Eulerian acid deposition model: Physical concepts and formulation. J geophys res 92:14681–14700

Chang LY, Liu HY, Lam KS, Wang T (1998) Analysis of the seasonal behavior of tropospheric ozone at Hong Kong. Atmos environ 32:159–168

Chappellaz J, Barnola J-M, Korotkevich YS, Lorius C (1990) Ice-core record of atmospheric methane over the past 160 000 years. Nature 345:127–131

Chappellaz J, Blunier DRT, Barnola J-M, Schwander J, Stauffer B (1993) Synchronous changes in atmospheric CH_4 and Greenland climate between 40 and 8 kyr BP. Nature 366:443–445

Chappellaz J, Blunier SKT, Dällenbach A, Barnola J-M, Schwander J, Raynaud D, Stauffer B (1997) Changes in the atmospheric CH_4 gradient between Greenland and Antarctica during the Holocene. J geophys res 102:15987–15997

Charlson RJ, Lovelock JE, Andreae MO, Warren SG (1987) Oceanic phytoplankton, atmospheric sulphur, cloud albedo and climate. Nature 326:655–661

Charlson RJ, Langner J, Rodhe H (1990) Sulphate aerosol and climate. Nature 348:22

Charlson RJ, Langner J, Rodhe H, Leovy CB, Warren SG (1991) Perturbation of the northern hemisphere radiative balance by backscattering of anthropogenic sulphate aerosols. Tellus 43AB:152–163

Charlson RJ, Schwartz SE, Hales JM, Cess RD, Coakley JA Jr., Hansen JE, Hofmann DJ (1992) Climate forcing by anthropogenic aerosols. Science 256:423–430

Charlson RJ, Anderson TL, Rodhe H (1999) Direct climate forcing by anthropogenic aerosols: Quantifying the link between atmospheric sulfate and radiation. Contributions to atmospheric physics 72(1):79–94

Chatfield RB (1994) Anomalous HNO_3/NO_x ratio of remote tropospheric air: Conversion of nitric acid to formic acid and NO_x? Geophys res lett 21:2705–2708

Chatfield RB, Crutzen PJ (1984) Sulfur dioxide in remote oceanic air: Cloud transport of reactive precursors. J geophys res 89:7111–7132

Chatfield RB, Vastano HBSJA, Sachse G (1996) A general model of how fire emissions and chemistry produce African/Oceanic plumes (O_3, CO, PAN, smoke) in TRACE-A. J geophys res 101:24279–24306

Chen G, Davis PKD, Bandy A, Thornton D, Blake D (1999) A mass-balance/photochemical assessment of DMS sea-to-air flux as inferred from NASA GTE PEM-West A and B observations. J geophys res 104:5471–5482

Chin M, Davis DD (1993) Global sources and sinks of OCS and CS_2 and their distributions. J geophys res 7:321–337

Chin M, Jacob DJ (1996) Anthropogenic and natural contributions to atmospheric sulfate: A global model analysis. J geophys res 101:18691–18699

Chin M, Jacob DJ, Munger JW, Parrish DD, Doddridge BG (1994) Relationship of ozone and carbon monoxide over North America. J geophys res 99:14565–14573

Chin M, Jacob DJ, Gardner GM, Foreman-Fowler MS, Spiro PA (1996) A global three-dimensional model of tropospheric sulphate. J geophys res 101:18667–18690

Chock DP (1991) A comparison of numerical methods for solving the advection equation. Atmos environ 25A:853

Christaki U, Belviso S, Dolan JR, Corn M (1996) Assessment of the role of copepods and ciliates in the release to solution of particulate DMSP. Mar ecol-prog ser 141:119–127

Christian RR, Boyer JN, Stanley DW (1991) Multi-year distribution patterns of nutrients within the Neuse River Estuary. Mar ecol-prog ser 71:259–274

Chu DA, Kaufman YJ, Remer LA, Holben BN (1998) Remote sensing of smoke from MODIS airborne simulator during the SCAR-B experiment. J geophys res 103:31979–31988

Chuang CC, Penner JE, Taylor KE, Grossman AS, Walton JJ (1997a) An assessment of the radiative effects of anthropogenic sulphate. J geophys res 102:3761–3778

Chuang PY, Charlson RJ, Seinfeld JH (1997b) Kinetic limitations on droplet formation in clouds. Nature 390:594–596

Ciccioli P, Brancaleoni E, Frattoni M, Di Palo V, Valentini R, Tirone G, Seufert G, Bertin N, Hansen U, Csiky O, Lenz R, Sharma M (1999) Emission of reactive terpene compounds from orange orchards and their removal by within-canopy processes. J geophys res 104:8077–8094

Cielisk S (1998) Energy and ozone fluxes in the atmospheric surface layer observed in Southern Germany highlands. Atmos environ 32:1273

Cieslik S, Labatut A (1996) Ozone surface fluxes and stomatal activity. In: Borrell PM, Borrell P, Kelly K, Cvitas T, Seiler W (eds) Proceedings of the EUROTRAC symposium '96: Transport and transformation of pollutants in the troposphere. Computational Mechanics Publications, Garmisch-Partenkirchen, Germany

Cieslik S, Labatut A (1997) Ozone deposition on various surface types. In: Slanina S (ed) Biosphere-atmosphere exchange of pollutants and trace substances. Springer-Verlag, Berlin, pp 225–243

Cinderby S, Cambridge HM, Herrera R, Hicks WK, Kuylenstierna JCI, Murray F, Olbrich K (1998) Global assessment of ecosystem sensitivity to acidic deposition. SEI, Stockholm

Claquin T, Schulz M, Balkanski YJ (1999) Modeling the mineralogy of atmospheric dust sources. J geophys res 104:22243–22256

Clark JS, Cachier H, Goldammer JG, Stocks B (1997) Sediment records of biomass burning and global change. Springer-Verlag, Berlin-Heidelberg

Clarke AD (1993) Atmospheric nuclei in the Pacific mid-troposphere: Their nature, concentration and evolution. J geophys res 98:20633–20647

Clarke AD, Porter JN (1993) Pacific marine aerosol, 2: Equatorial gradients in chlorophyll, ammonium, and excess sulfate during SAGA 3. J geophys res 98:16997–17010

Clarke AD, Li Z, Litchy M (1996) Aerosol dynamics in the equatorial Pacific marine boundary layer: Microphysics, diurnal cycles and entrainment. Geophys res lett 23:733–736

Clarke AD, Uehara T, Porter JN (1997) Atmospheric nuclei and related aerosol fields over the Atlantic: Clean subsiding air and continental pollution during ASTEX. J geophys res 102:25281–25292

Clarke AD, Davis D, Kapustin VN, Eisele F, Chen G, Paluch I, Lenschow D, Bandy AR, Thornton D, Moore K, Mauldin L, Tanner D, Litchy M, Carroll MA, Collins J, Albercook G (1998) Particle nucleation in the tropical boundary layer and its coupling to marine sulfur sources. Science 282:89–92

Clarke AD, Eisele F, Kapustin VN, Moore K, Tanner D, Mauldin L, Litchy M, Lienert B, Carroll MA, Albercook G (1999a) Nucleation in the equatorial free troposphere: Favorable environments during PEM-Tropics. J geophys res 104:5735–5744

Clarke AD, Kapustin VN, Eisele FL, Weber RJ, McMurry PH (1999b) Particle production near marine clouds: Sulfuric acid and predictions from classical binary nucleation. Geophys res lett 26:2425–2428

Clausen HB, Hammer CU, Hvidberg CS, Dahl-Johsen D, Steffensen JP, Kipstuhl J, Legrand M (1997) A comparison of the volcanic records over the past 4000 years from the Greenland Ice Core Project and Dye 3 Greenland ice cores. J geophys res 102:26707–26723

Clerbaux C, Hadji-Lazaro J, Payan S, Camy-Peyret C, Megie G (1999) Retrieval of CO columns from IMG/ADEOS spectra. Ieee t geosci remote 37:1657–1661

Climate Monitoring, Diagnostics Laboratory (2001) N_2O data, halocarbons and other atmospheric trace species group. Climate Monitoring and Diagnostics Laboratory (HATS/CMDL), Boulder, Colorado, US, *http://www.cmdl.noaa.gov/hats* or anonymous *ftp://ftp.cmdl.noaa.gov/hats/n2o/*

Coale KH, Johnson KS, Fitzwater SE, Gordon RM, Tanner S, Chavez FP, Ferioli L, Sakamoto C, Rogers P, Millero F, Steinberg P, Nightingale P, Cooper D, Cochlan W, Landry MR, Constantinou J, Rollwagen G, Trasvina A, Kudela R (1996) A massive phytoplankton bloom induced by an ecosystem-scale iron fertilization experiment in the equatorial Pacific Ocean. Nature 383:495–501

Cohan DS, Schultz MG, Jacob DJ, Heikes BG, Blake DR (1999) Convective injection and photochemical decay in the tropical upper troposphere: Methyl iodide as a tracer of marine convection. J geophys res 104:5717–5724

Collins DR, Jonsson HH, Seinfeld JH, Flagan RC, Gasso S, Hegg DA, Russell PB, Schmid B, Livingston JM, Öström E, Noone KJ, Russell LM, Putaud JP (2000) *In situ* aerosol size distributions and clear column radiative closure during ACE-2. Tellus 52B:498–525

Collins W, Rasch P, Eaton B, Khattatov B, Lamarque J-F, Zender C (2001) Simulating aerosols using a chemical transport model with assimilation of satellite aerosol retrievals: Methodology for INDOEX. J geophys res 106:7313–7336

Conen F, Dobbie KE, Smith KA (2000) Predicting N_2O emissions from agricultural land through related soil parameters. Glob change biol 6:417–426

Connors VS, Flood M, Jones T, Gormsen B, Nolf S, Reichle HG Jr. (1996) Global distribution of biomass burning and carbon monoxide in the middle troposphere during early April and October 1994. In: Levine JS (ed) Biomass burning and global change. MIT Press, Cambridge, Mass., pp 99–106

Connors VS, Gormsen BB, Nolf S, Reichle HG (1999) Spaceborne observations of the global distribution of carbon monoxide in the middle troposphere during April and October 1994. J geophys res 104:21455–21470

Conny JM, Currie LA (1996) The isotopic characterization of methane, non-methane hydrocarbons and formaldehyde in the troposphere. Atmos environ 30:621–638

Conrad R (1996) Soil microorganisms as controllers of atmospheric trace gases (H_2, CO, CH, OCS, N_2O, NO). Microbiol rev 60:609–640

Conrad R, Dentener FJ (1999) The application of compensation point concepts in scaling of fluxes. In: Bouwman AF (ed) Approaches to scaling of trace gas fluxes in ecosystems. Elsevier, Berlin, pp 205–216

Considine DB, Dessler AE, Jackman CH, Rosenfield JE, Meade PE, Schoeberl MR, Roche AE, Waters JW (1998) Interhemispheric asymmetry in the 1 mbar O_3 trend: An analysis using an interactive zonal mean model and UARS data. J geophys res 103:1607–1618

Cooke WF, Wilson JJ (1996) A global black carbon aerosol model. J geophys res 101:19395–19409

Cooke WF, Koffi B, Gregoire J-M (1996) Seasonality of vegetation fires in Africa from remote sensing data and application to a global chemistry model. J geophys res 101:21051–21065

Cooke WF, Liousse C, Cachier H, Feichter J (1999) Construction of a 1×1° fossil fuel emission dataset for carbonaceous aerosol and implementation and radiative impact in the ECHAM-4 model. J geophys res 104:22137–22162

Cooper DJ, Watson AJ, Ling RD (1998a) Variation of pCO_2 along a North Atlantic shipping route (UK to the Caribbean): A year of automated observations. Mar chem 60:147–164

Cooper OR, Moody JL, Davenport JC, Oltmans SJ, Johnson BJ, Chen X, Shepson PB, Merrill JT (1998b) The influence of springtime weather systems on vertical ozone distributions over three North American sites. J geophys res 103:22001–22103

Cooper OR, Moody JL, Parrish DD, Trainer M, Ryerson TB, Holloway JW, Hübler G, Fehsenfeld FC, Oltmans SJ, Evans MJ (2001) Trace gas signatures of the airstreams within North Atlantic cyclones: Case studies from the North Atlantic Regional Experiment (NARE '97) aircraft intensive. J geophys res 106(D6):5437–5456

Copeland BJ, Gray J (1991) Status and trends report of the Albemarle-Pamlico Estuary. In: Steel J (ed) Albemarle-Pamlico Estuarine study. NC Dept. of Environ. Health & Nat. Resources, Raleigh

Cornell S, Rendell A, Jickells T (1995) Atmospheric inputs of dissolved organic nitrogen to the oceans. Nature 376:243–246

Correll D, Ford D (1982) Comparison of precipitation and land runoff as sources of estuarine nitrogen. Estuar coast shelf s 15:45–56

Côté J, Roch ASM, Fillion L (1993) A variable-resolution semi-lagrangian finite-element global model of the shallow-water equations. Mon weather rev 121:232–243

Côté J, Desmarais J-G, Gravel S, Méthot A, Patoine A, Roch M, Staniforth A (1998a) The operational CMC-MRB Global Environmental Multiscale (GEM) model: Part II – Results. Mon weather rev 126:1397–1418

Côté J, Gravel S, Méthot A, Patoine A, Roch M, Staniforth A (1998b) The operational CMC-MRB Global Environmental Multiscale (GEM) model: Part I – Design considerations and formulation. Mon weather rev 126:1373–1395

Courtier P, Geleyn J-F (1988) A global numerical weather prediction model with variable resolution: Application to shallow-water equations. Q j roy meteor soc 114:1321–1346

Covert DS, Charlson RJ, Ahlquist NC (1972) A study of the relationship of chemical composition and humidity to light scattering by aerosols. J appl meteorol 11:968–976

Covert DS, Kapustin VN, Quinn PK, Bates TS (1992) New particle formation in the marine boundary layer. J geophys res 97:20581–20589

Covert DS, Kapustin VN, Bates TS, Quinn PK (1996a) Physical properties of marine boundary layer aerosol particles of the mid-Pacific in relation to sources and meteorological transport. J geophys res 101:6919–1930

Covert DS, Wiedensohler A, Aalto P, Heintzenberg J, McMurry PH, Leck C (1996b) Aerosol number size distributions from 3 to 500 nm diameter in the Arctic marine boundary layer during summer and autumn. Tellus 48B:197–212

Covert DS, Gras JL, Wiedensohler A, Stratmann F (1998) Comparison of directly measured CCN with CCN modeled from the number-size distribution in the marine boundary layer during ACE-1 at Cape Grim, Tasmania. J geophys res 103:16597–16608

Cowling EB, Chameides WL, Kiang CS, Fehsenfeld FC, Meagher JF (1998) Introduction to special section: Southern Oxidants Study Nashville/Middle Tennessee ozone study. J geophys res 103(D17):22209–22212

Cowling EB, Chameides WL, Kiang CS, Fehsenfeld FC, Meagher JF (2000) Introduction to special section: Southern Oxidants Study Nashville/Middle Tennessee ozone study, Part 2. J geophys res 105(D7):9075–9077

Cox RA (1999) Ozone and peroxy radical budgets in the marine boundary layer: Modelling the effect of NO_x. J geophys res 104:8047–8056

Cox RA, Bloss W, Rowley DM, Jones RL (1999) OIO and the atmospheric chemistry of iodine. Geophys res lett 26:1857–1860

Crawford J, Davis D, Chen G, Bradshaw J, Sandholm S, Gregory G, Sachse G, Anderson B, Collins J, Blake D, Singh H, Heikes B, Talbot R, Rodriguez J (1996) Photostationary state analysis of the NO_2-NO system based on airborne observations from the western and central North Pacific. J geophys res 101:2053–2072

Crawford J, Davis DD, Chen G, Bradshaw J, Sandholm S, Kondo Y, Merrill J, Liu S, Browell E, Gregory G, Anderson B, Sachse G, Barrick J, Blake D, Talbot R, Pueschel R (1997a) Implications of large scale shifts in tropospheric NO_x levels in the remote tropical Pacific. J geophys res 102:28447–28468

Crawford J, Davis D, Chen G, Bradshaw J, Sandholm S, Kondo Y, Liu S, Browell E, Gregory G, Anderson B, Sachse G, Collins J, Barrick J, Blake D, Talbot R, Singh H (1997b) An assessment of ozone photochemistry in the extratropical western North Pacific: Impact of continental outflow during the late winter/early spring. J geophys res 102:28469–28487

Cros BR, Nganga D, Manga A, Fishman J, Brackett V (1992) Distribution of tropospheric ozone at Brazzaville, Congo, determined from ozone sonde measurements. J geophys res 97:12869–12875

Crosley DR (1997) Tropospheric OH photochemistry experiment: A summary and perspective. J geophys res 102:6495

Crutzen PJ (1971) Ozone production rates in an oxygen-hydrogen-nitrogen oxide atmosphere. J geophys res 81:7411–7327

Crutzen PJ (1973) Photochemical reactions initiated by and influencing ozone in unpolluted tropospheric air. Tellus 26:47–57

Crutzen PJ (1976) The possible importance of OCS for the sulfate layer of the stratosphere. Geophys res lett 3:73–76

Crutzen PJ (1979) The role of NO and NO_2 in the chemistry of the troposphere and the stratosphere. Annu rev earth pl sc 7:443–472

Crutzen PJ (1998) Atmospheric aerosols: To the forefront of research in atmospheric chemistry. IGACtivities Newsletter 11:2

Crutzen PJ, Andreae MO (1990) Biomass burning in the Tropics: Impact on atmospheric chemistry and biogeochemical cycles. Science 250:1669–1678

Crutzen PJ, Brühl C (1993) A model study of atmospheric temperatures and the concentrations of ozone, hydroxyl, and some other photochemically active gases during the glacial, the preindustrial holocene and the present. Geophys res lett 20:1047–1050

Crutzen PJ, Fishman J (1977) Average concentrations of OH in the troposphere and the budgets of CH_4, CO, H_2, and CH_3CCl_3. Geophys res lett 4:321–324

Crutzen PJ, Lawrence MG (2000) The impact of precipitation scavenging on the transport of trace gases: A 3-dimensional study. J atmos chem 37:81–112

Crutzen PJ, Stoermer EF (2000) The "Anthropocene". Global change newsletter (IGBP) 41:12–13

Crutzen PJ, Zimmermann PH (1991) The changing photochemistry of the troposphere. Tellus 43AB:136–151

Crutzen PJ, Heidt LE, Krasnec JP, Pollock WH, Seiler W (1979) Biomass burning as a source of atmospheric gases CO, H_2, N_2O, NO, CH_3Cl, and COS. Nature 282:253–256

Crutzen PJ, Fall IGR, Lindinger W (1999a) Parameters for global ecosytem models. Nature 399:535

Crutzen PJ, Lawrence MG, Poschl U (1999b) On the background photochemistry of tropospheric ozone. Tellus 51A-B:123–146

Csiky O, Seufert G (1999) Terpenoid emissions of Mediterranean oaks and their relation to taxonomy. Ecol appl 9:1138–1146

Dab W, Medina S, Quenel P, Le Moullec Y, Le Tertre A, Thelot B, Moteil C, Lameloise P, Pirard P, Momas I, Ferry R, Festy B (1996) Short-term respiratory health effects of ambient air pollution: Results of the APHEA project in Paris. J epidemiol commun h 50(Suppl. 1):S30–S35

Dacey JWH, Wakeham SG (1986) Oceanic dimethylsulfide: Production during zooplankton grazing. Science 238:1314–1316

Dacey JWH, Howse FA, Michaels AF, Wakeham SG (1998) Temporal variability of dimethylsulfide and dimethylsulfoniopropionate in the Sargasso Sea. Deep-sea res 45:2085–2104

Dacey JWH, Edson JB, Holland PH, McGillis WR (1999) *In situ* estimation of air-sea gas transfer using DMS. AMS Boundary Layer Turbulence, 425–426

Dahneke BE, Cheng YS (1979) Properties of continuum source particle beams, 1: Calculation methods and results. J aerosol sci 10:257–274

Daily GC (1997) Nature's services: Societal dependence on natural ecosystems. Island Press, Washington, D.C.

Danielsen EF (1968) Stratospheric-tropospheric exchange based on radioactivity, ozone and potential vorticity. J atmos sci 25:502–518

Davidson CI, Tang W, Finger S, Etyemezian V, Striegel MF, Sherwood SI (2000a) Soiling patterns on a tall limestone building: Changes over 60 years. Environ sci technol 34:560–565

Davidson EA, Matson PA, Vitousek PM, Riley R, Dunkin K, Garcia-Mendez G, Maass JM (1993) Processes regulating soil emissions of NO and N_2O in a seasonally dry tropical forest. Ecology 74(1):130–139

Davidson EA, Kingerlee W (1997) A global inventory of nitric oxide emissions from soils. Nutr cycl agroecosys 48:37–50

Davidson EA, Keller M, Erickson HE, Verchot LV, Veldkamp E (2000b) Testing a conceptual model of soil emissions of nitrous and nitric oxides. Bioscience 50(8):667–680

Davies CW (1962) Ion association. Butterworth, London

Davies WE, Vaughan G, O'Connor FM (1998) Observation of near-zero ozone concentrations in the upper troposphere at mid-latitudes. Geophys res lett 25:1173–1176

Davis DD, Crawford J, Chen G, Chameides W, Liu S, Bradshaw J, Sandholm S, Sachse G, Gregory G, Anderson B, Barrick J, Bachmeier A, Collins J, Browell E, Blake D, Rowland S, Kondo Y, Singh H, Talbot R, Heikes B, Merrill J, Rodriguez J, Newell RE (1996) Assessment of ozone photochemistry in the western North Pacific as inferred from PEM-West A observations during the fall 1991. J geophys res 101:2111–2134

Davis DD, Crawford J, Lui S, McKeen S, Bandy A, Thornton D, Rowland FS, Blake D (1997) Potential impact of iodine on tropospheric levels of ozone and other critical oxidants. J geophys res 101:2135–2147

Davis DD, Chen G, Bandy A, Thornton D, Eisele F, Mauldin L, Tanner D, Lenschow D, Fuelberg H, Huebert B, Heath J, Clarke A, Blake D (1999) Dimethyl sulfide oxidation in the equatorial Pacific: Comparison of model simulations with field observations for DMS, SO_2, H_2SO_4(g), MSA(g), MS, and NSS. J geophys res 104:5765–5784

De Felice TP, Cheng RJ (1998) On the phenomenon of nuclei enhancement during the evaporative stage of a cloud. Atmos res 47/48:15–40

De Haan DO, Brauers T, Oum K, Stutz J, Nordmeyer T, Finlayson-Pitts BJ (1999) Heterogeneous chemistry in the troposphere: Experimental approaches and applications to the chemistry of sea salt particles. Int rev phys chem 18:343–385

DeBaar HJW, Boyd PW (2000) The role of iron in plankton ecology and carbon dioxide transfer of the global ocean. In: Hanson RB, Ducklow HW, Fields JG (eds) The changing ocean carbon cycle. Cambridge University Press, Cambridge, pp 61–140

Debye P, McAulay J (1925) Das elektrische Feld der Ionen und die Neutralsalzwirkung. Z phys 26:22–29

del Grosso SJ, Parton WJ, Mosier AR, Ojima DS, Potter CS, Borken W, Brumme R, Butterbach-Bahl K, Crill KPM, Dobbie E, Smith KA (2000) General CH_4 oxidation model and comparisons of CH_4 oxidation in natural and managed systems. Global biogeochem cy 14:999–1020

Delany AC, Fitzjarrald DR, Lenschow DH, Pearson R, Wendel GJ, Woodruff B (1986) Direct measurements of nitrogen oxides over grassland. J atmos chem 4:429

Delmas R, Druilhet A, Cros B, Durand P, Delon C, Lacaux J, Brustet J, Serça D, Affre C, Guenther A, Greenberg J, Baugh W, Harley P, Klinger L, Ginoux P, Brasseur G, Zimmerman P, Grégoire J, Janodet E, Tournier A, Perr P (1999) Experiment for Regional Sources and Sinks of Oxidants (EXPRESSO): An overview. J geophys res 104:30609–30624

DeMore WB, Sander SP, Golden DM, Hampson RF, Kurylo MJ, Howard CJ, Ravishankara AR, Kolb CE, Molina MJ (1997) Chemical kinetics and photochemical data for use in stratospheric modeling. NASA Jet Propulsion Laboratory 97-4, Pasadena, CA

Dennis DR (1997) Using the Regional Acid Deposition Model to determine the nitrogen deposition airshed of the Chesapeake Bay watershed. In: Baker JE (ed) Atmospheric deposition of contaminants to the Great Lakes and coastal waters. SETAC Press, Pensacola, FL, pp 393–413

Dentener FJ, Crutzen PJ (1993) Reaction of HNO_5 on tropospheric aerosols: Impact on the global distributions of NO_x, O_3, and OH. J geophys res 98:7149–7163

Dentener FJ, Crutzen PJ (1994) A three-dimensional model of the global ammonia cycle. J atmos chem 19:331–369

Dentener FJ, Carmichael GR, Zhang Y, Lelieveld J, Crutzen PJ (1996) Role of mineral aerosol as a reactive surface in the global troposphere. J geophys res 101:22869–22889

Dentener FJ, Feichter J, Jeuken A (1999) Simulation of transport of ^{222}Rn using on-line and off-line global models at different horizontal resolutions: A detailed comparison with measurements. Tellus 51B:573–602

Derwent RG, Curtis AR (1977) Two dimensional model studies of some trace gases and free radicals in the troposphere. HMSO, London

Derwent RG, Simmonds PG, Collins WJ (1994) Ozone and carbonmonoxide measurements at a remote maritime location, Mace Head, Ireland, from 1990 to 1992. Atmos environ 28:623–2637

Deuzé JL, Herman M, Goulub P, Tanré D, Marchand A (1999) Characterization of aerosols over ocean from POLDER/ADEOS-1. Geophys res lett 26:1421–1424

Devolder P, Fittschen C, Frenzel A, Hippler H, Poskrebyshev G, Striebel F, Viscolz B (1999) Complete falloff curves for the unimolecular decomposition of i-propoxy radicals between 330 and 408 K. Phys chem chem phys 1:675–668

Diab RD, Thompson AM, Zunckel M, Coetzee GJR, Combrink J, Bodeker GE, Fishman J, Sokolic F, McNamara DP, Archer CB, Nganga D (1996) Vertical ozone distribution over southern Africa and adjacent oceans during SAFARI-92. J geophys res 101:23823–23833

Dias-Lalcaca P, Brunner D, Imfeld W, Moser W, Staehelin J (1998) An automated system for the measurement of nitrogen oxides and ozone concentration from a passenger aircraft: Instrumentation and first results of the NOXAR project. Environ sci technol 32:3228–3236

Dickerson RR, Huffman GJ, Luke WT, Nunnermacker LJ, Pickering KE, Leslie ACD, Lindsey CG, Slinn WGN, Kelly TJ, Daum PH, Delany AC, Greenberg JP, Zimmerman PR, Boatman JF, Ray JD, Stedman DH (1987) Thunderstorms: An important mechanism in the transport of air pollutants. Science 235:460–465

Dickerson RR, Kondragunta S, Stenchikov G, Civerolo KL, Doddrige BG, Holben BN (1997) The impact of aerosols on solar ultraviolet radiation and photochemical smog. Science 278:827–830

Dickerson RR, Rhoads KP, Carsey TP, Oltmans SJ, Burrows JP, Crutzen PJ (1999) Ozone in the remote marine boundary layer: A possible role for halogens. J geophys res 104:21385–21395

Dietz RN, Wieser RF (1983) Sulfate formation in oil-fired power plant plumes. Volume 1: Parameters affecting primary sulfate emissions and a model for predicting emissions and plume opacity. Electric Power Research Institute

Dignon J, Hameed S (1989) Global emissions of nitrogen and sulfur oxides from 1960 to 1980. Japca J air waste ma 39:180–186

Dignon J, Eddleman HE, Penner JE (2001) A black carbon emission data base for atmospheric chemistry and climate studies. IGAC/GEIA Web.

Dimmer C, Simmonds P, Nickless G (1999) Production of gaseous hydrocarbons from soil ecosystems. EOS 80:67

Diner DJ, Bruegge CJ, Martonchik JV, Bothwell GW, Danielson ED, Floyd EL, Ford VG, Hovland LE, Jones KL, White ML (1991) A multi-angle imaging spectroradiometer for terrestrial remote sensing from the Earth Observing System. Int J imag syst tech 3:92–107

Dlugokencky EJ, Steele LP, Lang PM, Masarie KA (1994) The growth rate and distribution of atmospheric methane. J geophys res 99:17021–17043

Dlugokencky EJ, Masaire KA, Lang PM, Tans PP (1998) Continuing decline in the growth rate of the atmospheric methane burden. Nature 393:447–450

Dlugokencky EJ, Walter BP, Lang PM, Kasischke ES (2000) Measurements of an anomalous global methane increase during 1998. EOS 81:F76

Dobbie KE, McTaggart IP, Smith KA (1999) Nitrous oxide emissions from intensive agricultural systems: Variations between crops and seasons; key driving variables; and mean emission factors. J geophys res 104:26891–26899

Dockery DW, Pope CA III, Xu X, Spengler JD, Ware JH, Ray ME, Ferris BG Jr., Speizer FE (1993) An association between air pollution and mortality in six US cities. New engl j med 329:1753–1759

Doherty S, Anderson TL, Charlson RJ (1999) Measurement of the lidar ratio for atmospheric aerosols using a 180°-backscatter nephelometer. Appl optics 38:1823–1832

Dolske KE, Gatz GF (1985) A field intercomparison of methos for measurement of particle and gas dry deposition. J geophys res 90:2076–2084

Dome-F Ice Core Research Group (1998) Preliminary investigation of paleoclimate signals recorded in the ice core from Dome Fuji station, East Dronning Maud Land, Antarctica. Ann glaciol 27:338–342

Dominé F, Murrells TP, Howard CJ (1990) Kinetics of the reactions of NO_2 with CH_3S, CH_3SO, CH_3SS, and CH_3SSO at 297 K and 1 Torr. J chem phys 94:5839–5847

Donahue NM, Prinn RG (1993) *In situ* nonmethane hydrocarbon measurements on SAGA 3. J geophys res 98:16915–16932

Donahue NM, Dubey MK, Mohrschaldt R, Demerjian KL, Anderson JG (1997) High-pressure flow study of the reaction $OH + NO_x \rightarrow HO\,NO_x$: Errors in the falloff region. J geophys res 102:6159–6168

Dorn H-P, Brandenburger U, Brauers T, Hausmann M, Ehhalt DH (1996) *In situ* detection of tropospheric OH radicals by folded long-path laser absorption. Results from the POPCORN field campaign in August 1994. Geophys res lett 23:2537–2540

Döscher A, Gäggeler HW, Schotterer U, Schwikowski M (1996) A historical record of ammonium concentrations from a glacier in the Alps. Geophys res lett 23:2741–2744

Dransfield TJ, Perkins KK, Donahue NM, Anderson JG, Sprengnether MM, Demerjian KL (1999) Temperature and pressure dependent kinetics of the gas-phase reaction of the hydroxyl radical with nitrogen dioxide. Geophys res lett 26:687–690

Drummond JR, Mand GS (1996) The Measurement Of Pollution In The Troposphere (MOPITT) instrument: Overall performances and calibration requirements. J atmos ocean tech 13:314–320

Dubovik O, Smirnov A, Holben BN, King MD, Kaufman YJ, Eck TF, Slutzker I (2000) Accuracy assessments of aerosol optical properties retrieved from Aerosol Robotic Network (AERONET) sun and sky radiance measurements. J geophys res 105:9791–9806

Duce R (1986) The impact of atmospheric nitrogen, phosphorus, and iron species on marine biological productivity. In: Buat-Menard P, Reidel D (eds) The role of air-sea exchange in geochemical cycling. Norwell, Mass., pp 497–529

Duce R (1991) Chemical exchange at the air-coastal sea interface. In: Mantoura R, Martin J, Wollast R (eds) Ocean margin processes in global change. J. Wiley & Sons, Chichester, pp 91–110

Duce RA (1995) Source, distributions and fluxes of mineral aerosols and their relationship to climate. In: Charison RJ, Heintzenberg J (eds) Aerosol forcing of climate. J. Wiley, New York, pp 43–72

Duce RA, Tindale NW (1991) Atmospheric transport of iron and its deposition in the ocean. Limnol oceanogr 36:1715–1726

Duce RA, Liss PS, Merril JT, Atlas EL, Buat-Menard P, Hicks BB, Miller JM, Prospero JM, Arimoto R, Church TM, Ellis W, Galloway JN, Hansen L, Jickells TD, Knap AH, Reinhardt KH, Schneider B, Soudine A, Tokos JJ, Tsunogai S, Wollast R, Zhou M (1991) The atmospheric input of trace species to the world ocean. Global biogeochem cy 5:193–259

Duderstadt KA, Carroll MA, Sillman S, Wang T, Albercook GM, Feng L, Parrish DD, Holloway JS, Fehsenfeld FC, Blake DR, Blake NJ, Forbes G (1998) Photochemical production and loss rates of ozone at Sable Island, Nova Scotia during the North Atlantic Regional Experiment (NARE) 1993 summer intensive. J geophys res-atmos 103:13531–13555

Durkee PA, Nielsen KE, Smith PJ, Russell PB, Schmid B, Livingston JM, Holben BN, Tomasi C, Vitale V, Collins D, Flagan RC, Seinfeld JH, Noone KJ, Öström E, Gasso S, Hegg D, Russell LM, Bates TS, Quinn PK (2000a) Regional aerosol optical depth characteristics from satellite observations: ACE-1, TARFOX and ACE-2 results. Tellus 52B:484–497

Durkee PA, Noone KJ, Bluth RT (2000b) The Monterey Area Ship Track (MAST) Experiment. J atmos sci 57(16):2523–2541

Dye JE, Ridley BA, Skamarock W, Barth M, Venticinque M, Defer E, Blanchet P, Thery C, Laroche P, Baumann K, Hubler G, Parrish DD, Ryerson T, Trainer M, Frost G, Holloway JS, Matejka T, Bartels D, Fehsenfeld FC, Tuck A, Rutledge SA, Lang T, Stith J, Zerr R (2000) An overview of the Stratospheric-Tropospheric Experiment: Radiation Aerosols, and ozone – deep convection experiment with results from the July 10, 1996 storm. J geophys res 105:10023–10045

Ebel A, Hass H, Jakobs HJ, Laure M, Memmesheimer M, Oberreuter A, Geiss H, Kuo Y-H (1991) Simulation of ozone intrusion caused by a tropopause fold and cut-off low. Atmos environ 25A:2131–2144

Echalar F, Artaxo P, Gerab F, Yamasoe MA, Martins JV, Longo KM, Maenhaut W, Holben BN (1998) Aerosol composition and variability in the Amazon basin. J geophys res 103:31849–31866

Ehhalt D, Rohrer F, Kraus AB, Prather MJ, Blake DR, Rowland FS (1997) On the significance of regional trace gas distributions as derived from aircraft campaigns in PEM West A and B. J geophys res 102:28333–28351

Ehhalt D, Prather M, Dentener F, Derwent RG, Dlugokencky EJ, Holland E, Isaksen I, Katima J, Kirchoff V, Matson P, Midgley P, Wang M (2001) Atmospheric chemistry and greenhouse gases. In: Houghton JT, Ding Y, Griggs DJ, Noguer M, van der Linden PJ, Dai X, Maskell K, Johnson CA (eds) Climate change 2001: The scientific basis. Contribution of working group I to the third assessment report of the Intergovernmental Panel on Climate Change (IPCC). Cambridge University Press, Cambridge

Ehhalt DH (1999) Gas phase chemistry of the troposphere. In: Zellner (ed) Global aspects of atmospheric chemistry. Springer-Verlag, New York, pp 21–109

Ehhalt DH, Rohrer F, Wahner A (1992) Sources and distribution of NO_x in the upper troposphere at northern midlatitudes. J geophys res 97(D4):3725–3738

Eisele FL, Hanson DR (2000) First measurement of prenucleation molecular clusters. J phys chem a 104:1715–1719

Eisele FL, Mount GH, Fehsenfeld FC, Harder J, Marovich E, Parrish DD, Roberts J, Trainer M, Tanner DJ (1994) Intercomparison of tropospheric OH and ancillary trace gas measurements at Fritz Peak Observatory, Colorado. J geophys res 99:18605–18626

Eisele FL, Tanner D, Cantrell C, Calvert J (1996) Measurements and steady state calculations of OH concentrations at Mauna Loa Observatory. J geophys res 101:14665–14679

Eisele FL, Mount GH, Tanner D, Jefferson A, Shetter R, Harder JW, Williams EJ (1997) Understanding the production and interconversion of the hydroxyl radical during the Tropospheric OH Photochemistry Experiment. J geophys res 102:6457–6465

Elbern H, Schmidt H, Elbert A (1997) Variational data assimilation for tropospheric chemistry modeling. J geophys res 102:15967

Eliassen A (1978) The OECD Study of long range transport of air pollutants: Long range transport modelling. Atmos environ 12:479–487

Eliassen A, Saltbones J (1983) Modelling of long-range transport of sulphur over Europe: A two-year model run and some model experiments. Atmos environ 17:1457–1473

Elkins JW, Fahey DW, Gilligan JM, Dutton GS, Baring TJ, Volk CM, Dunn RE, Myers RC, Montzka SA, Wamsley PR, Hayden AH, Butler JH, Thompson TM, Swanson TH, Dlugokencky EJ, Novelli PC, urst DF, Lobert JM, Ciciora SJ, McLaughlin RJ, Thompson TL, Winkler RH, Fraser PJ, Steele LP, Lucarelli MP (1996) Airborne gas chromatograph for *in situ* measurements of long-lived species in the upper troposphere and lower stratosphere. Geophys res lett 23:347–350

Elkins JW, Butler JH, Hurst DF, Montzka SA, Moore FL, Thompson TM (1998) Nitrous Oxide and Halocompounds Group/Climate Monitoring and Diagnostics Laboratory (NOAA/CMDL) web site (*http://www.cmdl.noaa.gov/noal/*), Boulder, CO, updated data available on anonymous ftp site (*file://ftp.cmdl.noaa.gov/noah/*)

Elliott S, Rowland FS (1995) Methyl halide hydrolysis rates in natural waters. J atmos chem 20:229–236

Elliott S, Liu E, Rowland FS (1989) Rates and mechanisms for the hydrolysis of carbonyl sulfide in natural waters. Environ sci technol 23:458–461

Elvidge CD, Kroehl HW, Kihn RA, Baugh KE, Davis ER, Hao WM (1996) Algorithm for the retrieval of fire pixels from DMSP Operational Linescan System data. In: Levine JS (ed) Biomass burning and global change. MIT Press, Cambridge, Mass., pp 73–85

Emanuel K (1991) A scheme representing cumulus convection in large-scale models. J atmos sci 48:2313

Emerson S, Quay P, Stump C, Wilbur D, Knox M (1991) O_2, Ar, N_2, and Rn-222 in surface waters of the subarctic ocean: Net biological production. Global biogeochem cy 5:49–69

Emmons LK, Carroll MA, Hauglustaine DA, Brasseur GP, Atherton C, Penner J, Sillman S, Levy H II, Rohrer F, Wauben WMF, van Velthoven PFJ, Wang Y, Jacob DJ, Bakwin P, Dickerson R, Doddridge B, Gerbig C, Honrath R, Hubler G, Jaffe D, Kondo Y, Munger JW, Torres A, Volz-Thomas A (1997) Climatologies of NO_x and NO_y: A comparison of data and models. Atmos environ 31:1851–1904

Emmons LK, Hauglustaine DA, Muller JF, Carroll MA, Brasseur GP, Brunner D, Staehelin J, Thouret V, Marenco A (2000) Data composites of airborne observations of tropospheric ozone and its precursors. J geophys res 105(D16):20497–20538

Engel T, Priesack E (1993) Expert-N, a building block system of nitrogen models as a resource for advice, research, water management and policy. In: Eijsackers HJP, Hamers T (eds) Integrated soil and sediment research: A basis for proper protection. Kluwer Academic, Dordrecht, pp 503–507

Enting IG, Mansbridge JV (1991) Latitudinal distributions of sources and sinks of CO_2: Results of an inversion study. Tellus 43B:1560170

Erickson DJ III, Oglesby RJ, Marshal S (1995) Climate response to indirect anthropogenic sulfate forcing. Geophys res lett 22:2017–2020

Erisman JW, Draaijers G, Duyzer J, Hofschreuder P, van Leeuwen N, Römer F, Ruijgrok W, Wyers P (1997) The aerosol project: Introduction and some background information. Atmos environ 31:315–319

Esler GJ, Tan DGH, Haynes PH, Evans MJ, Law KS, Plantevin P-H, Pyle JA (2001) Stratosphere-troposphere exchange: Chemical sensitivity to mixing. J geophys res 106(D5):4717–4731

Esler MB, Wilson SR, Griffith DWT, Steele LP (1999) Baseline trace gas monitoring using fourier transform infrared (FTIR) spectroscopy. Baseline 96:24–35

Esler MB, Griffith DWT, Wilson SR, Steele LP (2000) Precision trace gas analysis by FT-IR spectroscopy. 2. The $^{13}C/^{12}C$ isotope ratio of CO_2. Anal chem 72:216–221

Etheridge DM, Pearman GI, Fraser PJ (1992) Changes in tropospheric methane between 1841 and 1978 from a high accumulation-rate Antarctic ice core. Tellus 44B:282–294

Etheridge DM, Steele LP, Langenfelds RL, Francey RJ, Barnola J-M, Morgan VI (1996) Natural and anthropogenic changes in atmospheric CO_2 over the last 1000 years from air in Antarctic ice and firn. J geophys res 101:4115–4128

Etheridge DM, Steele LP, Francey RJ, Langenfelds RL (1998) Atmospheric methane between A.D. 1000 and present: Evidence for anthropogenic emissions and climate variability. J geophys res 103:15979–15993

Evans MJ, Shallcross DE, Law KS, Wild JOF, Simmonds PG, Spain TG, Berrisford P, Methven J, Lewis AC, MacQuaid JB, Pilling MJ, Bandy BJ, Penkett SA, Pyle JA (2000) Evaluation of a Lagrangian box model using field measurements from EASE (Eastern Atlantic Summer Experiment) 1996. Atmos environ 34(23):3843–3863

Fabian P, Pruchniewicz PG (1977) Meridional distribution of ozone in the troposphere and its seasonal variations. J geophys res 82:2063–2073

Facchini MC, Fuzzi S, Zappoli S, Andracchio A, Gelencser A, Kiss G, Krivacsy Z, Mészaros E, Hansson HC, Asberg T, Zebühr Y (1999) Partitioning of the organic aerosol component between fog droplets and interstitial air. J geophys res 104:26821–26832

Fairall CW, Hare JE, Edson JB, McGillis WR (2000) Parameterization and micrometeorological measurement of air-sea gas transfer. Bound-lay meteorol 96:63–105

Falkowski PG, Kim Y, Kolber Z, Wilson C, Wirick C, Cess R (1992) Natural versus anthropogenic factors affecting low level cloud albedo over the North Atlantic. Science 256:1311–1313

Fall R (1999) Biogenic emissions of volatile organic compounds from higher plants. In: Hewitt CN (ed) Reactive hydrocarbons in the atmosphere. Academic Press, New York, pp 41–96

Fan SM, Wofsy SC, Bakwin PS, Jacob DJ, Anderson SM, Kebabian PL, McManus JB, Kolb CE, Fitzjarrald DR (1992) Micrometeorological measurements of CH_4 and CO_2 exchange between the atmosphere and subarctic tundra. J geophys res 97:16627–16643

Fehsenfeld F (1995) Measurement of chemically reactive trace gases at ambient concentrations. In: Matson P, Harriss R (eds) Biogenic trace gases: Measuring emissions from soil and water. Blackwell Science

Fehsenfeld FC, Liu SC (1993) Tropospheric ozone distribution and sources. In: Hewitt, Sturges (eds) Global atmospheric chemical change. Elsevier Applied Science, New York, pp 169–233

Fehsenfeld FC, Drummond JW, Roychowdhury UK, Galvin PJ, Williams EJ, Buhr MP, Parrish DD, Hübler G, Langford AO, Calvert JG, Ridley BA, Grahek F, Heikes BG, Kok GL, Shetter JD, Walega JG, Elsworth CM, Norton RB, Fahey DW, Murphy PC, Hovermale C, Mohnen VA, Demerjian KL, Mackay GI, Schiff HI (1990) Intercomparison of NO_2 measurement techniques. J geophys res 95:3579–3597

Fehsenfeld FC, Trainer M, Parrish DD, Volz-Thomas A, Penkett SA (1996) North Atlantic Regional Experiment 1993 summer intensive: Forward. J geophys res 101:28869–28875

Feichter J, Crutzen PJ (1990) Parameterization of vertical tracer transport due to deep cumulus convection in a global transport model and its evaluation with ^{222}radon measurements. Tellus 42B:100–117

Feichter J, Lohmann U (1999) Can a relaxation technique be used to validate clouds and sulphur species in a GCM. Q j roy meteor soc 125:1277–1294

Feichter J, Kjellstrom E, Rodhe H, Dentener F, Lelieveld J, Roelofs G-J (1996) Simulation of the tropospheric sulfur cycle in a global climate model. Atmos environ 30:1693–1707

Feichter J, Lohmann U, Schult I (1997) The atmospheric sulfur cycle in ECHAM-4 and its impact on the shortwave radiation. Clim dynam 13:235–246

Feingold G, Kreidenweis SM (2000) Does heterogeneous processing of aerosol increase the number of cloud droplets? J geophys res 105:24351-24361

Fenn MA, Browell EV, Butler CF, Grant WB, Kooi SA, Clayton MB, Gregory GL, Newell RE, Zhu Y, Dibb JE, Fuelberg HE, Anderson BE, Bandy AR, Blake DR, Bradshaw JD, Heikes BG, Sachse GW, Sandholm ST, Singh HB, Talbot RW, Thornton DC (1999) Ozone and aerosol distributions and air mass characteristics over the South Pacific during the burning season. J geophys res 104:16197–16212

Ferek RJ, Andreae MO (1984) Photochemical production of carbonyl sulfide in marine surface waters. Nature 307:148–150

Ferek RJ, Hegg DA, Hobbs PV, Durkee P, Nielsen K (1998) Measurements of ship-induced tracks in clouds off the Washington coast. J geophys res 103:23199-23206

Ferek RJ, Garrett T, Hobbs PV, Strader S, Johnson DW, Taylor JP, Nielsen K, Ackerman AS, Kogan Y, Liu Q, Albrecht BA, Babb D (2000) Drizzle suppression in ship tracks. J atmos sci 57:2707–2728

Ferretti DF, Lowe DC, Martin RJ, Brailsford GW (2000) A new GC-IRMS technique for high precision, N_2O-free analysis of $\delta^{13}C$ and $\delta^{18}O$ in atmospheric CO_2 from small air samples. J geophys res 105:6709–6718

Field RJ, Hess PG, Kalachev LV, Madronich S (2001) Characterization of oscillation and period-doubling transition to chaos reflecting dynamic instability in a simplified model of tropospheric chemistry. J geophys res 106:7553–7565

Finlayson-Pitts BJ, Hemminger JC (2000) Physical chemistry of airborne sea salt particles and their components. J phys chem a 104(49):11463–11477

FIRESCAN Science Team (1996) Fire in ecosystems of boreal Eurasia: The Bor Forest Island Fire Experiment Fire Research Campaign Asia-North (FIRESCAN). In: Levine JS (ed) Biomass burning and global change. MIT Press, Cambridge, Mass., pp 848–873

Firestone MK, Davidson EA (1989) Microbiological basis of NO and N_2O production and consumption in soils. In: Andreae MO, Schimel DS (eds) Exchange of trace gases between terrestrial ecosystems and the atmosphere. J. Wiley, New York, pp 7–21

Fischer H, Wagenbach D, Kipfstuhl J (1998) Sulphate and nitrate firn concentrations on the Greenland ice sheet: 2. Temporal and anthropogenic deposition changes. J geophys res 103:21935–21942

Fischer H, Wienhold FG, Hoor P, Bujok O, Schiller C, Siegmund P, Ambaum M, Scheeren HA, Lelieveld J (2000) Tracer correlations in the northern high latitude lowermost stratosphere: Influence of cross-tropopause mass exchange. Geophys res lett 27(1):97–100

Fisher D, Oppenheimer M (1991) Atmospheric nitrogen deposition and the Chesapeake Bay estuary. Ambio 20:102–108

Fisher M, Lary DJ (1996) Lagrangian four dimensional variational data assimilation of chemical species. Q j roy meteor soc 121:1681

Fishman J, Balok AE (1999) Calculation of daily tropospheric ozone residuals using TOMS and empirically improved SBUV measurements: Application to an ozone pollution episode over the eastern United States. J geophys res 104(D23):30319–30340

Fishman J, Brackett VG (1997) The climatological distribution of tropospheric ozone derived from satellite measurements using version 7 Total Ozone Mapping Spectrometer and Stratospheric Aerosol and Gas Experiment data sets. J geophys res 102(D15):19275–19278

Fishman J, Solomon S, Crutzen PJ (1979) Observational and theoretical evidence in support of a significant *in situ* photochemical source of tropospheric ozone. Tellus 31(5):432–446

Fishman J, Watson CE, Larson JC, Logan JA (1990) Distribution of tropospheric ozone determined from satellite data. J geophys res 95:3599–3617

Fitzgerald JW (1991) Marine aerosols: A review. Atmos environ 25A:533–545

Flagan RC, Friedlander SK (1978) Particle formation in pulverized coal combustion: Review. In: Shaw DT (ed) Recent developments in aerosol science. J. Wiley, New York, pp 25–59

Flatøy F, Hov Ø, Schlager H (2000) Chemical forecasts used for measurement flight planning during POLINAT 2. Geophys res lett 27(7):951–954

Flessa H, Dorsch P, Beese F (1995) Seasonal variation of N_2O and CH fluxes in differently managed arable soils in southern Germany. J geophys res 100:23115–23123

Flöck OR, Andreae MO (1996) Photochemical and non-photochemical formation and destruction of carbonyl sulfide and methyl mercaptan in ocean waters. Mar chem 54:11–26

Flöck OR, Andreae MO, Dräger M (1997) Environmentally relevant precursors of carbonyl sulfide in aquatic systems. Mar chem 59:71–85

Flossmann AI (1994) A 2-D spectral model simulation of the scavenging of gaseous and particulate sulfate by a warm marine cloud. J atmos sci 32:255–268

Flossmann AI, Pruppacher HR (1988) A theoretical study of the wet removal of atmospheric pollutants, part III. J atmos sci 45:1857–1871

Flossmann AI, Wobrock W (1996) Venting of gases by convective clouds. J geophys res 101:18639–18649

Forster PMD, Freckleton RS, Shine KP (1997) On aspects of the concept of radiative forcing. Clim dynam 13(7-8):547–560

Francey R (1998) Report of the ninth WMO meeting of experts on carbon dioxide concentration and related tracer measurement techniques. In: Francey R (ed) Global atmosphere watch. World Meteorological Organization, Geneva

Francey RJ, Manning MR, Allison CE, Coram SA, Etheridge DM, Langenfelds RL, Lowe DC, Steele LP (1999) A history of $\delta^{13}C$ in atmospheric CH_4 from the Cape Grim air archive and Antarctic firn air. J geophys res 104:23631–23643

French NHF, Kasischke ES, Johnson RD, Bourgeau-Chavez LL, Frick AL, Ustin S (1996) Estimating fire-related carbon flux in Alaskan boreal forests using multisensor remote-sensing data. In: Levine JS (ed) Biomass burning and global change. MIT Press, Cambridge, Mass., pp 808–826

Frew NM (1997) The role of organic films in air-sea gas exchange. In: Liss PS, Duce RA (eds) The sea surface and global change. Cambridge University Press, Cambridge, pp 121–172

Frew NM, Goldman JC, Dennett MR, Johnson AS (1990) Impact of phytoplankton-generated surfactants on air-sea gas exchange. J geophys res 95:3337–3352

Fricke W, Kaminski U, Gilge S (1997) Der Hohenpeißenberg im GAW-Programm der WMO. Promet 26(1/2):53–60
Fried A, Henry BE, Calvert JG, Mozurkewich M (1994) The reaction probability of N_2O_5 with sulfuric acid aerosols at stratospheric temperatures and compositions. J geophys res 99:3517–3532
Fried A, Lee Y-N, Frost G, Wert B, Henry B, Drummond JR, Hübler G, Jobson T (2002) Airborne CH_2O measurements over the North Atlantic during the 1997 NARE campaign: Instrument comparisons and distributions. J geophys res 107
Friedlander SK (1978) A review of the dynamics of sulphate containing aerosols. Atmos environ 12:187–195
Frolking S, Crill PM (1994) Climate controls on temporal variability of methane flux from a poor fen in southeastern New Hampshire: Measurement and modeling. Global biogeochem cy 8:385–397
Frolking SE, Mosier AR, Ojima DS, Li C, Parton WJ, Potter CS, Priesack E, Stenger R, Haberbosch C, Dörsch P, Flessa H, Smith KA (1998) Comparison of N_2O emissions from soils at three different agricultural sites: Simulations of year-round measurements by four models. Nutr cycl agroecosys 52:77–105
Fruekilde P, Hjorth J, Jensen NR, Kotzias D, Larsen B (1998) Ozonolysis at vegetation surfaces: A source of acetone, 4-oxopentanal, 6-methyl-5-hepten-2-one and geranyl acetone. Atmos environ 32:1893–1902
Fuehrer J (1996) The critical level for the effect of ozone on crops. In: Karenlampi L, Skarby L (eds) Critical levels for ozone in Europe: Testing and finalising the concepts. UNECE Workshop Report, Kuopio, Finland, pp 27–43
Fuentes J, Lerdau M, Atkinson R, Baldocchi D, Bottenheim J, Ciccioli P, Lamb B, Geron C, Gu L, Guenther A, Sharkey T, Stockwell W (2000) Biogenic hydrocarbons in the atmospheric boundary layer: A review. B am meteorol soc 81:1537–1575
Fung I, John J, Lerner J, Matthews E, Prather M, Steele LP, Fraser PJ (1991) Three-dimensional model synthesis of the global methane cycle. J geophys res 96:13033–13065
Fung K, Grosjean D (1981) Determination of nanogram amounts of carbonyls as 2,4-dinitro-phenylhydrazones by high-performance liquid chromatography. Anal chem 53:168–171
Fuzzi S, Facchini MC, Orsi G, Lind JA, Wobrock W, Kessel M, Maser R, Jaeschke W, Enderle KH, Arends BG, Berner A, Solly I, Kruisz C, Reischl G, Pahl S, Kaminski U, Winkler P, Ogren JA, Noone KJ, Hallberg A, Fierlinger-Oberlinninger H, Puxbaum H, Marzorati A, Hansson H-C, Wiedensohler A, Svenningsson IB, Martinsson BG, Schell D, Georgii HW (1992) The Po Valley fog experiment 1989 – An overview. Tellus 44B:448–468
Gabric A, Murray N, Stone L, Kohl M (1993) Modelling the production of dimethylsulfide during a phytoplankton bloom. J geophys res 98:22805–22816
Gabric A, Matrai PA, Vernet M (1999) Modelling the production and cycling of DMSP and DMS during the vernal bloom in the Barents Sea. Tellus 51B:919–937
Gabric AJ, Whetton PH, Boers R, Ayers GP (1998) The impact of simulated climate change on the air-sea flux of dimethylsulphide in the subantarctic Southern Ocean. Tellus 50B:388–399
Gage D, Rhodes D, Nolte K, Hicks W, Leustek T, Cooper A, Hanson A (1997) A new route for the synthesis of dimethylsulfoniopropionate in marine algae. Nature 387:891–894
Galbally I (1989) The International Global Atmospheric Chemistry (IGAC) Programme. CACGP
Galbally IE (1971) Ozone profiles and ozone fluxes in the atmospheric surface layer. Q j roy meteor soc 97:18
Galbally IE, Roy CR (1980) Destruction of ozone at the earth surface. Q j roy meteor soc 106:599
Galbally IE, Meyer CP, Bentley ST, Ye Y (1996) Studies of ozone, NO_x and VOCs in near-surface air at Cape Grim, 1996. In: Gras JL, Derek N, Tindale NW, Dick AL (eds) Baseline atmospheric program, Australia 1996. Bureau of Meteorology, and CSIRO Division of Atmospheric Research, Melbourne, Australia, pp 103–104
Galenter M, Levy H, Carmichael GR (2000) Impacts of biomass burning on tropospheric CO, NO_x and O_3. J geophys res 105:6633–6653
Gallagher MW, Beswick KM, Duyzer J, Westrate H, Choularton TW, Hummelshøj P (1997) Measurements of aerosol fluxes to Speulder Forest using a micrometeorological technique. Atmos environ 31:359–373
Galloway J, Levy H, Kasibhatia P (1994) Consequences of population growth and development on deposition of oxidized nitrogen, Year 2020. Ambio 23:120–123
Galloway JN, Likens GE, Keene WC, Miller JM (1982) The composition of precipitation in remote areas of the world. J geophys res 87:8771–8786
Galloway JN, Schlesinger WH, Levy H II, Michaels A, Schnoor JL (1995) Nitrogen fixation: Anthropogenic enhancement-environmental response. Global biogeochem cy 9:235–252
Galmarini S, Vila-Guerau J, Duyzer J (1997) Fluxes of chemically reactive species inferred from mean concentration measurements. Atmos environ 31:2371
Galy C, Modi AI (1998) Precipitation chemistry in the Sahelian savanna of Niger, Africa. J atmos chem 30:319–343
Gan J, Yates SR, Papiernik SK, Crowley D (1998) Application of organic amendments to reduce volatile pesticide emission from soil. Environ sci technol 32:3094–3098
Ganzeveld LN, Lelieveld J (1995) Dry deposition parameterization in the climate/chemistry model ECHAM and its influence on the distribution of chemically reactive trace gases. J geophys res 100:20999–21012
Gao W, Wesely ML (1995a) Modeling gaseous dry deposition over regional scales with satellite observations. I. Model development. Atmos environ 29:727–737
Gao W, Wesely ML (1995b) Modeling gaseous dry deposition over regional scales with satellite observations. II. Deriving surface conductances from AVHRR data. Atmos environ 29:739–747
Gao W, Wesely ML, Doskey PV (1993) Numerical modeling of the turbulent diffusion and chemistry of NO_x, O_3, isoprene, and other reactive trace gases in and above a forest canopy. J geophys res 98:18339–18353
Gardner RM, Adams TCK, Deidewig F, Ernedal S, Falk R, Feuti E, Herms E, Johnson CE, Lecht M, Lee DS, Leech M, Lsiter D, Masse B, Metcalfe M, Newton P, Schmitt A, Venderbergh C, Van Drimmelen R (1997) ANCAT/EC (The Abatment of Nuisances Caused by Air Transport) global inventory of NO_x emissions from aircraft. Atmos environ 31:1751–1766
Garland J (2000) On the size dependence of particle deposition velocities. In: Sixth International Conference on Air-Surface Exchange of Gases and Particles. Edinburgh, Scotland
Garstang M, Scala J, Greco S, Harriss R, Beck S, Browell E, Sachse G, Gregory G, Hill G, Simpson J, Tao W-K, Torres A (1988) Trace gas exchanges and convective transports over the Amazonian rain forest. J geophys res 93:1528–1550
Garstang M, Tyson PD, Swap R, Edwards M, Kallberg P, Lindesay JA (1996) Horizontal and vertical transport of air over southern Africa. J geophys res 101:23721–23736
Garstang M, Ellery WN, McCarthy TS, Scholes MC, Scholes RJ, Swap RJ, Tyson PD (1998) The contribution of aerosol- and water-borne nutrients to the Okavango Delta ecosystem, Botswana. S afr J sci 94:223–229
Gasche R, Papen H (1999) A 3-year continous record of nitrogen trace gas fluxes from untreated and limed soil of a N-saturated spruce and beech forest ecosystem in Germany. 2: NO and NO_2 fluxes. J geophys res 104(D15):18505–18520
Gear GW (1971) Numerical initial value problems in ordinary differential equations. Prentice Hall, New York
George LA, Hard TM, O'Brien RJ (1999) Measurement of free radicals OH and HO_2 in Los Angeles smog. J geophys res 104:11643–11655
Georgii HW, Warneck P (1999) Chemistry of the tropospheric aerosol and of clouds. In: Zellner R (ed) Global aspects of atmospheric chemistry. Steinkopff Verlag, Darmstadt, pp 111–169
Gershey RM (1983) Characterization of seawater organic matter carried by bubble-generated aerosols. Limnol oceanogr 28:309–319
Gery MW, Whitten GZ, Killus JP, Dodge MC (1988) Development and testing of the CBM-IV for urban and regional modeling, final report. Atmospheric Sciences Research Laboratory, US
GESAMP Joint Group of Experts on the Scientific Aspects of Marine Environmental Protection (1995) The sea-surface microlayer and its role in global change. World Meteorological Organization, Geneva

Ghan SJ, Leung LR, Easter RC, Abdul-Razzak H (1997) Prediction of cloud droplet number in a general circulation model. J geophys res 102:21777–21794

Ghio AJ, Samet JM (1999) Metals and air pollution particles. In: Holgate ST, Samet JM, Koren HS, Maynard RL (eds) Air pollution and health. Academic Press, London, pp 635–672

Giannakopoulos C, Chipperfield MP, Law KS, Pyle JA (1999) Validation and intercomparison of wet and dry deposition schemes using ^{210}Pb in a global three-dimensional off-line chemical transport model. J geophys res 104:23761–23784

Gibb SW, Fauzi R, Mantoura C, Liss PS (1999) Ocean-atmosphere exchange and atmospheric speciation of ammonia and methylamines in the region of the NW Arabian Sea. Global biogeochem cy 13:161–178

Gieray R, Wieser P, Engelhardt T, Swietlicki E, Hansson H-C, Mentes B, Orsini D, Martinsson B, Svenningsson B, Noone K, Heintzenberg J (1997) Phase partitioning of aerosol constituents in cloud based on single-particle and bulk analysis. Atmos environ 31:2491–2502

Gillet RW, Ayers GP, Noller BN (1990) Rainwater acidity at Jabiru, Australia in the wet season 1983/84. Sci total environ 92:122–144

Gillette DA, Adams J, Endo A, Smith D, Kihl R (1980) Threshold velocities for input of soil particles into the air by desert soils. J geophys res 85:5621–5630

Gilpin T, Apel E, Fried A, Wert B, Calvert J, Genfa Z, Dasgupta P, Harder J, Heikes B, Hopkins B, Westberg H, Kleindienst T, Lee Y-N, Zhou X, Lonneman W, Sewell S (1997) Intercomparison of six ambient [CH_2O] measurement techniques. J geophys res 102:21161–21188

Ginoux P (1997) L'impact de l'activité humaine sur la composition chimique de la troposphère au- dessus de l'Océan Pacifique: Développement d'un modèle telescopique de chimie et de transport atmosphériques et interprétation des résultats de la campagne de mesures MLOPEX. Ph.D. thesis, Université Libre de Bruxelles (Belgium)

Glantz P, Noone KJ (2000) A physically-based mechansim for converting aerosol mass to cloud droplet number. Tellus 52B:1216–1231

Glavich TA, Beer R (1991) Tropospheric emission spectrometer for the Earth Observing System. In: The International Society for Optical Engineering, Society of Photo-Optical Instrumentation Engineers (eds) Infrared technology XVII. Bellingham, Washington, San Diego, pp 148–159

Global Atmospheric Watch (1994) Global atmospheric watch guide. World Meteorological Organization

Glove DM, Reeburgh WS (1987) Radon-222 and radium-222 in southeastern Bering Sea shelf waters and sediment. Cont shelf res 5:433–456

Goldammer JG (1993) Feuer in Waldökosystemen der Tropen und Subtropen. Birkhäuser-Verlag, Basel

Goldammer JG (1999a) Forests on fire. Science 284:1782–1783

Goldammer JG (1999b) Public policies affecting forest fires in Europe and boreal/temperate Asia, FAO Meeting on Public Policies Affecting Forest Fires. FAO Forestry Paper 138:113–164

Goldammer JG, Price C (1998) Potential impacts of climate change on fire regimes in the Tropics based on MAGICC and a GISS GCM-derived lightning model. Climatic change 39:273–296

Goldman JC, Dennett MR, Frew NM (1988) Surfactant effects on air-sea gas exchange under turbulent conditions. Deep-sea res 35:1953–70

Gong SL, Barrie LA, Prospero JM, Savoie DL (1997a) Modeling sea-salt aerosols in the atmosphere, 2: Atmospheric concentrations and fluxes. J geophys res 102:3819–3830

Gong SL, Barrie LA, Blanchet J-P (1997b) Modeling sea-salt aerosols in the atmosphere: I, model development. J geophys res 102:3805–3818

Gong SL, Walmsley JL, Barrie LA, Hopper JF (1997c) Mechanisms for surface ozone depletion and recovery during polar sunrise. Atmos environ 31:969–981

Gong W, Ménard S, Lin X (1997) Uncertainties in regional oxidant modelling: Experience with a comprehensive regional model. In: Gryning S, Chaumerliac N (eds) Air pollution modelling and its applications. Plenum Press

González JM, Kiene RP, Moran MA (1999) Transformations of sulphur compounds by an abundant lineage of marine bacteria in the subclass of the class proteobacteria. Appl environ microb 65:3810–3819

Graedel TE, Crutzen PJ (1993) Atmospheric change: An Earth system perspective. W. H. Freeman and Company, New York

Graedel TE, Keene WC (1999) Preface, reactive chlorine emissions inventory. J geophys res 104:8331–8332

Graedel TE, Bates TS, Bouwman AF, Cunnold D, Dignon J, Fung I, Jacob DJ, Lamb BK, Logan JA, Marland G, Middleton P, Pacyna JM, Placet M, Veldt C (1993) A compilation of inventories of emissions to the atmosphere. Global biogeochem cy 7:1–26

Graf H-F, Kirchner I, Robock A, Schult I (1993) Pinatubo eruption winter climate effects: Model versus observations. Clim dynam 9:81–93

Graf H-F, Feichter J, Langmann B (1997) Volcanic sulfur emissions: Estimates of source strength and its contribution to the global sulphate distribution. J geophys res 102:10727–10738

Gran HH (1931) On the conditions for the production of plankton in the sea. Rapports et Process verbeaux des Reunions Conseil International pour l'exploration de la Mer 75:37–46

Granier C, Müller J-F, Madronich S, Brasseur GP (1996) Possible causes for the 1990–1993 decrease in global tropospheric CO abundances: A three-dimensional sensitivity study. Atmos environ 30:1673–1682

Granier C, Müller J-F, Brasseur G (1999) The impact of biomass burning on the global budget of ozone and ozone precursors. In: Proceedings of the Wengen Conference on Global Change Research: Biomass burning and its inter-relationships with the climate system

Granier C, Pétron G, Muller J-P, Brasseur G (2000) The impact of natural and anthropogenic hydrocarbons on the tropospheric budget of carbon monoxide. Atmos environ 34:5255–5271

Gras JL (1995) CN, CCN and particle size in southern ocean air at Cape Grim. Atmos res 35:233–251

Gravel S, Staniforth A (1992) Variable resolution and robustness. Mon weather rev 120:2633–2640

Gregory GL, Hoell JM Jr., Ridley BA, Singh HB, Gandrud B, Salas LJ, Shetter JD (1990) An intercomparison of airborne PAN measurements. J geophys res 95:10077–10087

Gregory GL, Davis DD, Beltz N, Bandy AR, Ferek RJ, Thornton DC (1993) An intercomparison of aircraft instrumentation for tropospheric measurements of sulfur dioxide. J geophys res 98:23325–23352

Gregory GL, Merrill JT, Shipham MC, Blake DR, Sachse GW, Singh HB (1997) Chemical characteristics of tropospheric air over the Pacific Ocean as measured during PEM-West B: Relationship to Asian outlfow and trajectory history. J geophys res 102:28275–28286

Gregory PJ, Ingram JSI, Campbell B, Goudriaan J, Hunt LA, Landsberg JJ, Linder S, Stafford MSmith, Sutherst RW, Valentin C (1999) Managed production systems. In: Walker B, Steffen W, Canadell J, Ingram J (eds) The terrestrial biosphere and global change. Implications for natural and managed ecosystems. Cambridge University Press, Cambridge, pp 229–270

Groffman PM, Brumme R, Butterbach-Bahl K, Dobbie KE, Mosier AR, Ojima D, Papen H, Parton WJ, Smith KA, Wagner-Riddle C (2000) Evaluating annual nitrous oxide fluxes at the ecosystem scale. Global biogeochem cy 14:1061–1070

Gröne T, Kirst GO (1992) The effect of nitrogen deficiency, methionine and inhibitors of methionine metabolism on the DMSP contents of *Tetraselmis subcordiformis* (Stein). Mar biol 112:497–503

Groszko W, Moore RM (1998) Ocean-atmosphere exchange of methyl bromide: Results from NW Atlantic and Pacific Ocean studies. J geophys res 103:16737–16741

Guelle W, Balkanski YJ, Dibb JE, Schulz M, Dulac F (1997) Wet deposition in a global size-dependent aerosol transport model: 1. Comparison of a 1 year ^{210}Pb simulation with ground measurements. J geophys res 103:11429–11445

Guenther A (1999) Modeling biogenic volatile organic compound emissions to the atmosphere. In: Hewitt C (ed) Reactive hydrocarbons in the atmosphere. Academic Press, New York, pp 97–118

Guenther A, Hills A (1998) Eddy covariance measurement of isoprene fluxes. J geophys res 103:13145–13152

Guenther A, Zimmerman PHP, Monson R, Fall R (1993) Isoprene and monoterpene emission rate variability: Model evaluation and sensitivity analysis. J geophys res 98:12609–12617

Guenther A, Newitt CN, Erickson D, Fall R, Geron C, Graedel T, Harley P, Klinger L, Lerdau M, McKay W, Pierce T, Scholes B, Steinbrecher R, Tallamraju R, Taylor J, Zimmermann P (1995) A global model of natural volatile organic compound emissions. J geophys res 100:8873–8892

Guenther A, Baugh W, Davis K, Hampton G, Harley P, Klinger L, Zimmerman P, Allwine E, Dilts S, Lamb B, Westberg H, Baldocchi D, Geron C, Pierce T (1996a) Isoprene fluxes measured by enclosure, relaxed eddy accumulation, surface-layer gradient, mixed-layer gradient, and mass balance techniques. J geophys res 101:18555–18568

Guenther A, Otter L, Zimmerman P, Greenberg J, Scholes R, Scholes M (1996b) Biogenic hydrocarbon emissions from southern African savannas. J geophys res 101:25859–25865

Guenther A, Baugh B, Brasseur G, Greenberg J, Harley P, Klinger L, Serca D, Vierling L (1999) Isoprene emission estimates and uncertainties for the Central African EXPRESSO study domain. J geophys res 104:30625–30639

Guenther A, Geron C, Pierce T, Lamb B, Harley P, Fall R (2000) Natural emissions of non-methane volatile organic compounds, carbon monoxide, and oxides of nitrogen from North America. Atmos environ 34:2205–2230

Guesten H, Heinrich G (1996) On-line measurements of ozone fluxes. Atmos environ 30:897

Gupta M, Elliott S, Cicerone R (1998) Perturbation to global tropospheric oxidizing capacity due to latitudinal redistribution of surface sources of CH_4, CO, and NO_x. Geophys res lett 25:3931–3934

Haagen-Smit AJ (1952) Chemistry and physiology of Los Angeles smog. Indust eng chem 44:1342

Haan D, Raynaud D (1998) Ice core record of CO variations during the last two millennia: Atmospheric implications and chemical interactions within the Greenland ice. Tellus 50B:253–262

Haan D, Martinerie P, Raynaud D (1996) Ice core data of atmospheric carbon monoxide over Antarctica and Greenland during the last 200 years. Geophys res lett 23(17):2235–2238

Hadji-Lazaro J, Clerbaux C, Thuria S (1999) An inversion algorithm using neural network to retrieve atmospheric CO concentrations from high resolution nadir radiances. J geophys res 104:23841–23854

Hahn DW (1998) Laser-induced breakdown spectroscopy for sizing and elemental analysis of discrete aerosol particles. Appl phys lett 72:2960–2962

Haines TK, Martinez J, Cleaves DA (1998) Influences on prescribed burning activity in the United States National Forest System. Int For Fire News 43–46

Hales JM (1990) Precipitation scavenging. In: Irving PM (ed) Atmospheric process research and process model development. National Acid Precipitation Assessment Program, Washington, DC

Hall SJ, Matson PA (1999) Nitrogen oxide emissions after nitrogen additions in tropical forests. Nature 400:152–155

Hallberg A, Ogren JA, Noone KJ, Heintzenberg J, Berner A, Solly I, Kruisz C, Reischl G, Fuzzi S, Facchini MC, Hansson H-C, Wiedensohler A, Svenningsson IB (1992) Phase partitioning for different aerosol species in fog. Tellus 44B:545–555

Hallquist M, Stewart DJ, Baker J, Cox RA (2000) Hydrolysis of N_2O_5 on submicron sulfuric acid aerosols. Phys chem chem phys 104:3984–3990

Hamill P, Jensen EJ, Russell PB, Bauman JJ (1997) The life cycle of stratospheric aerosol particles. B am meteorol soc 78:1395–1410

Hammer CU, Clausen HB, Dansgaard W, Neftel A, Kristinsdottor P, Johnson E (1985) Continuous impurity analysis along the Dye 3 deep ice core. In: Langway JCC, Oeschger H, Dansgaard W (eds) Greenland ice core: Geophysics, geochemistry and environment. Amer. Geophys. Union, Washington, D.C., pp 90–94

Han Q, Rossow WB, Lacis AA (1994) Near-global survey of effective droplet radii in liquid water clouds using ISCCP data. J climate 7:465–497

Han Q, Rossow WB, Chou J, Welch RM (1998) Global survey of the relationships of cloud albedo and liquid water path with droplet size using ISCCP. J climate 7:1516–1528

Hanisch F, Crowley JN (2001) Heterogeneous reactivity of gaseous nitric acid on Al_2O_3, $CaCO_3$, and atmospheric dust samples: A Knudsen cell study. J phys chem a 105:3096–3106

Hansen ADA, Rosen H, Novakov T (1982) An instrument for the real-time measurement of the absorption coefficient of aerosol particles. Appl optics 21:3060–3062

Hansen J, Sato MKI, Ruedy R, Lacis A, Asamoah K, Borenstein S, Brown E, Cairns B, Caliri G, Campbell M, Curran B, de Castro S, Druyan L, Fox M, Johnson C, Lerner J, McCormick MP, Miller R, Minnis P, Morrison A, Pandolfo L, Ramberran I, Zaucker F, Robinson M, Russell P, Shah K, Stone P, Tegen I, Thomason L, Wilder J, Wilson H (1996) A Pinatubo climate modeling investigation. In: Fiocco G, Fua D, Visconti G (eds) The Mount Pinatubo eruption: Effects on the atmosphere and climate. Springer-Verlag, Heidelberg, Germany, pp 233–272

Hansen J, Sato M, Ruedy R (1997a) Radiative forcing and climate response. J geophys res 102:6831–6864

Hansen J, Sato M, Ruedy R, Lacis A, Asamoah K, Beckford K, Borenstein S, Brown E, Cairns B, Carlson B, Curran B, de Castro S, Druyan L, Etwarrow P, Ferede T, Fox M, Gaffen D, Glascoe J, Gordon H, Hollandsworth S, Jiang X, Johnson C, Lawrence N, Lean J, Lerner J, Lo K, Logan J, Luckett A, McCormick MP, McPeters R, Miller R, Minnis P, Ramberran I, Russell G, Russell P, Stone P, Tegen I, Thomas S, Thomason L, Thompson A, Wilder J, Willson R, Zawodny J (1997b) Forcings and chaos in interannual to decadal climate change. J geophys res 102:25679–25720

Hansen J, Sato M, Lacis A, Ruedy R, Tegen I, Matthews E (1998) Climate forcing in the industrial era. P natl acad sci usa 95:12753–12758

Hanson DR, Eisele F (2000) Diffusion of H_2SO_4 in humidified nitrogen: Hydrated H_2SO_4. J phys chem a 104:1715–1719

Hansson ME (1994) The Renland ice core: A Northern Hemisphere record of aerosol composition over 120 000 years. Tellus 46B:390–418

Hansson ME, Saltzman ES (1993) The firs Greenland ice core record of methanesulfonate and sulfate over a full glacial cycle. Geophys res lett 20:1163–1166

Hao WM, Liu M-H (1994) Spatial and temporal distribution of tropical biomass burning. Global biogeochem cy 8:495

Hao W-M, Liu M-H, Crutzen PJ (1990) Estimates of annual and regional releases of CO_2 and other trace gases to the atmosphere from fires in the Tropics, based on the FAO statistics for the period 1975–1980. In: Goldammer JG (ed) Fire in the tropical biota: Ecosystem processes and global challenges. Springer-Verlag, Berlin, pp 440–462

Happell JD, Wallace DWR (1996) Methyl iodide in the Greenland/Norwegian Seas and the tropical Atlantic Ocean: Evidence for photochemical production. Geophys res lett 23:2105–2108

Hard TM, George LA, O'Brien RJ (1995) FAGE determination of tropospheric HO and HO_2. J atmos sci 52:3354–3372

Hargreaves KJ, Wienhold FG, Klemedtsson L, Arah JRM, Beverland IJ, Fowler D, Galle B, Griffith DWT, Skriba U, Smith KA, Welling M, Harris GW (1996) Measurement of nitrous oxide emission from agricultural land using micrometeorological methods. Atmos environ 30:1563–1571

Harley P, Monson R, Lerdau M (1999) Ecological and evolutionary aspects of isoprene emission from plants. Oecologia 118:109–123

Harned HS, Owen BB (1958) The physical chemistry of electrolyte solutions. Reinhold, New York

Harris GW, Wienhold FG, Zenker T (1996) Airborne observations of strong biogenic NO_x emissions from the Namibian savanna at the end of the dry season. J geophys res 101(D19):23707–23712

Harrison KG (2000) Role of increased marine silica input on paleo CO_2 levels. Paleoceanography 15:292–298

Hart TJ (1934) On the phytoplankton of the Southwest Atlantic and the Bellinghausen Sea 1929–1931. Discovery Reports 8:1–268

Hartley DE, Prinn RG (1993) On the feasibility of determining surface emissions of trace gases using an inverse method in a three-dimensional chemical transport model. J geophys res 98:5183–5198

Harvey HW (1933) On the rate of diatom growth. J mar biol assoc uk 19:253–276

Hasselblad V, Eddy DM, Kotchmar DJ (1992) Synthesis of environmental evidence: Nitrogen dioxide epidemiology studies. Japca J air waste ma 42:662–671

Hauglustaine DA, Brasseur GP (2001) Evolution of tropospheric ozone under anthropogenic activities and associated radiative forcing of climate. J geophys res 106:32337–32360

Hauglustaine DA, Ridley BA, Solomon S, Hess PG, Madronich S (1996) HNO_x/NO_x ratio in the remote troposphere during MLOPEX 2: Evidence for nitric acid reduction on carbonaceous aerosols. Geophys res lett 23:2609–2612

Hauglustaine DA, Brasseur GP, Walters S, Rasch PJ, Muller J-F, Emmons LK, Carroll MA (1998) MOZART, a global chemical transport model for ozone and related chemical tracers 2. Model results and evaluation. J geophys res 103:28291–28335

Hauglustaine DA, Madronich S, Ridley BA, Flocke SJ, Cantrell CA, Eisele FL, Shetter RE, Tanner DJ, Ginoux P, Atlas EL (1999) Photochemistry and budget of ozone during the Mauna Loa Observatory Photochemistry Experiment (MLOPEX 2). J geophys res 104(D23):30275–30307

Hausmann M, Platt U (1994) Spectroscopic measurement of bromine oxide and ozone in the high Arctic during Polar Sunrise Experiment 1992. J geophys res 99:25399–25414

Haußecke H (1996) Messung und Simulation von kleinskaligen Austauschvorgängen an der Ozeanoberfläche mittels Thermographie

Haynes PH, Marks CJ, McIntyre ME, Shepherd TG, Shine KP (1991) On the "downward control" of extratropical diabatic circulations by eddy-induced mean zonal forces. J atmos sci 48:651–687

Haywood JM, Ramaswamy V (1998) Global sensitivity studies of the direct radiative forcing due to anthropogenic sulphate and black carbon aerosols. J geophys res 103:6043–6058

Haywood JM, Stouffer RJ, Wetherald RT, Manabe S, Ramaswamy V (1997) Transient response of a coupled model to estimated changes in greenhouse gas and sulfate concentrations. Geophys res lett 24:1335–1338

Haywood JM, Ramaswamy V, Soden BJ (1999) Tropospheric aerosol climate forcing in clear-sky satellite observations over the oceans. Science 283:1299–1303

He F, Hopke PK (1995) SO_2 oxidation and H_2O-H_2SO_4 binary nucleation by radon decay. Aerosol sci tech 23:411–421

He S, Carmichael GR (1999) Sensitivity of photolysis rates and ozone production in the troposphere to aerosol properties. J geophys res 104:26307–26324

Heal MR, Heard DE, Pilling MJ, Whitaker BJ (1995) On the development and validation of FAGE for local measurement of tropospheric OH and HO_2. J atmos sci 52:3428–3448

Heath DF, Krueger AJ, Roeder HR, Henderson BD (1975) The solar backscatter ultraviolet and total ozone mapping spectrometer (SBUV/TOMS) for Nimbus G. Opt eng 14:323–331

Hebestreit K, Stutz J, Rosen D, Matveev V, Peleg M, Luria M, Platt U (1999) DOAS measurements of tropospheric bromine oxide in mid-latitudes. Science 283:55–57

Hedberg K, Hedberg CW, Iber C, White KE, Osterholm MT, Jones DBW, Flink JR, MacDonald KL (1989) An outbreak of nitrogen dioxide-induced respiratory illness among ice hockey players. Jama-j am med assoc 262:3014-3017

Hegels E, Crutzen PJ, Klüpfel T, Perner D, Burrows PJ (1998) Global distribution of atmospheric bromine monoxide from GOME on Earth-observing satellite ERS 2. Geophys res lett 25:3127–3130

Hegg DA (1994) Cloud condensation nucleus-sulphate mass relationship and cloud albedo. J geophys res 99:25903–25907

Hegg DA, Larson TV (1990) The effects of microphysical parameterization on model predictions of sulphate production in clouds. Tellus 42B:272–284

Hegg DA, Radke LF, Hobbs PV (1990) Particle production associated with marine clouds. J geophys res 95:13917–13926

Hegg DA, Livingston J, Hobbs PV, Novakov T, Russell P (1997) Chemical apportionment of aerosol column depth off the mid-Atlantic coast of the United States. J geophys res 102:25293–25303

Heikes BG (1992) Formaldehyde and hydroperoxides at Mauna Loa Observatory. J geophys res 97:18001–18013

Heil A, Goldammer JG (2001) Smoke-haze pollution: A review of the 1997 episode in Southeast Asia. Reg environ change j 2(1):24–37

Hein R, Crutzen PJ, Heimann M (1997) An inverse modeling approach to investigate the global atmospheric methane cycle. Global biogeochem cy 11:43–76

Heintz F, Platt U, Flentje H, Dubois R (1996) Long-term observation of nitrate radicals at the TOR Station, Kap Arkona (Rügen). J geophys res 101:22891–22910

Heintzenberg J, Charlson J (1996) Design and applications of the integrating nephelometer: A review. J atmos ocean tech 13:987–1000

Heintzenberg J, Erfurt G (2000) Modification of a commercial integrating nephelometer for outdoor measurements. J atmos ocean tech 17:1645–1650

Heintzenberg J, Leck C (1994) Seasonal variation of the atmospheric aerosol near the top of the marine boundary layer over Spitsbergen related to the Arctic sulphur cycle. Tellus 46B:52–67

Heintzenberg J, Ström J, Ogren JA, Fimpel H-P (1991) Vertical profiles of aerosol properties in the summer troposphere of the European Arctic. Atmos environ 25A:621–628

Heintzenberg J, Müller K, Birmili W, Spindler G, Wiedensohler A (1998) Mass-related aerosol properties over the Leipzig Basin. J geophys res 103:13125–13136

Heintzenberg J, Covert DS, Van Dingenen R (2000) Size distribution and chemical composition of marine aerosols: A compilation and review. Tellus 52B:1104–1122

Heitlinger M, Volz-Thomas A, Mihelcic D, Müsgen P, Burrows JP, Andres MDHernandez, Stöbener D, Perner D, Arnold T, Seuwen R, Clemitshaw KC, Penkett SA, Laverdet G, El K-Boudali, Teton S, Hjorth J, Poulida O, Hastie D, Arias MC, Borrell P, Borrell PM (1996) Peroxy Radical InterComparison Exercise (PRICE II). In: Larsen B, Versino B, Angeletti G (eds) Proceedings of the 7th European Symposium on Physico-Chemical Behaviour of Atmospheric Pollutants, the Oxidising Capacity of the Troposphere. European Commission, Luxembourg, pp 177–181

Helas G, Slanina J, Steinbrecher R (1997) Biogenic volatile organic compounds in the atmosphere. SPB Academic Publishing

Herman JR, Larko D (1994) Low ozone amounts during 1992–1993 from Nimbus 7 and Meteor 3 total ozone mapping spectrometer. J geophys res 99:3483–3496

Herman JR, Barthia PK, Torres O, Hsu C, Seftor C, Celarier E (1997) Global distribution of UV absorbing aerosol from Nimbus 7/TOMS data. J geophys res 102:16911–16922

Hermann M (1999) Development and application of an aerosol measurement system for use on commercial aircraft. Ph.D. thesis, University of Leipzig, Leipzig

Herrmann H, Zellner R (1998) Reactions of NO_3-radicals in aqueous solution. In: Alfassi ZB (ed) N-centered radicals. J. Wiley, New York, pp 291–343

Herrmann H, Ervens B, Nowacki P, Wolke R, Zellner R (1999) A chemical aqueous phase radical mechanism for tropospheric chemistry. Chemosphere 38:1223–1232

Herrmann H, Ervens B, Jacobi HW, Wolke R, Nowacki P, Zellner R (2000) CAPRAM23: A chemical aqueous phase radical mechanism for tropospheric chemistry. J atmos chem 36:231–284

Hess PG, Madronich S (1997) On tropospheric oscillations. J geophys res 93:15949–15965

Hess PG, Flocke S, Lamarque J-F, Barth MC, Madronich S (2000) Episodic modeling of the chemical structure of the troposphere as revealed during the spring MLOPEX 2 intensive. J geophys res 105:26809

Hicks BB, Meyers TP, Vong RJ, Wesely ML (1990) Air surface exchange. In: Irving PM (ed) Atmospheric process research and process model development. National Acid Precipitation Assessment Program, Washington, DC

Higurashi A, Nakajima T (1999) Development of a two-channel aerosol retrieval algorithm on a global scale using NOAA AVHRR. J atmos sci 56:924–941

Hill RW, White BA, Cottrell MT, Dacey JWH (1998) Virus-mediated total release of dimethylsulfoniopropionate from marine phytoplankton: A potential climate process. Aquat microb ecol 14:1–6

Hirokawa J, Onaka K, Kajii Y, Akimoto H (1998) Heterogeneous processes involving sodium halide particles and ozone: Molecular bromine release in the marine boundary layer in the absence of nitrogen oxides. Geophys res lett 25:2449–2452
Hobbs PV, Yates JC (1985) Atmospheric aerosol measurements over North America and the North Atlantic Ocean. Atmos environ 19:163–179
Hobbs PV, Reid JS, Kotchenruther RA, Ferek RJ, Weiss R (1997) Direct radiative forcing by smoke from biomass burning. Science 275:1776–1778
Hoell JM, Jr., Davis DD, Gregory GL, McNeal RJ, Bendura RJ, Drewry JW, Barrick JD, Kirchhoff VWJH, Motta AG, Navarro RL, Dorko WD, Owen DW (1993) Operational overview of the NASA GTE/CITE 3 airborne instrument intercomparisons for sulfur dioxide, hydrogen sulfide, carbonyl sulfide, dimethyl sulfide, and carbon disulfide. J geophys res 98:23291–23304
Hoell JM, Davis DD, Jacob DJ, Rogers MO, Newell RE, Fuelberg HE, McNeal RJ, Raper JL, Bendura RJ (1999) Pacific Exploratory Mission in the tropical Pacific: PEM Tropics A, August-September 1996. J geophys res 104:5567–5583
Hoffmann T, Odum JR, Bowman F, Collins D, Klockow D, Flagan RC, Seinfeld JH (1997) Formation of organic aerosols from the oxidation of biogenic hydrocarbons. J atmos chem 26:189–222
Hoffmann T, Bandur R, Marggraf U, Linscheid M (1998) Molecular composition of organic aerosols formed in the alpha-pinene/ozone reaction: Implications for new particle formation processes. J geophys res 103:25569–25578
Hofmann D (1993) Twenty years of balloon-borne tropospheric aerosol measurements at Laramie, Wyoming. J geophys res 98:12753–12766
Hofzumahaus A, Dorn H-P, Callies J, Platt U, Ehhalt DH (1991) Tropospheric OH concentration measurements by laser long-path absorption spectroscopy. Atmos environ 25A:2017–2022
Hofzumahaus A, Aschmutat U, Hebling M, Holland F, Ehhalt DH (1996) The measurement of tropospheric OH radicals by laser-induced fluorescence spectroscopy during the POPCORN field campaign. Geophys res lett 23:2541–2545
Hofzumahaus A, Aschmutat U, Brandenburger U, Brauers T, Dorn H-P, Hausmann M, Hebling M, Holland F, Plass-Dülmer C, Ehhalt DH (1998) Intercomparison of tropospheric OH measurements by different laser techniques during the POPCORN campaign 1994. J atmos chem 31:227–246
Högfeldt E (1982) Stability constants of metal-Ion complexes, part A: Inorganic ligands. Pergamon Press, Oxford
Holben BN, Eck TF, Slutsker I, Tanré D, Buis JP, Setzer A, Vermote E, Reagan JA, Kaufman Y, Nakajima T, Lavenu F, Jankowiak I, Smirnov A (1998) AERONET – A federated instrument network and data archive for aerosol characterization. Remote sens environ 66:1–16
Holland EA, Dentener FJ, Brasswell BH, Sulzman JM (1999) Contemporary and pre-industrial global reactive nitrogen budget. Biogeochemistry 46:7–43
Höller H, Schumann U (2000) Final report to the Commission of the European Communities. In: European Lightning Nitrogen Oxides Project
Holligan PM, Fernandez E, Aiken J, Balch W, Boyd P, Burkill P, Finch M, Groom S, Malin G, Muller K, Purdie D, Robinson C, Trees C, Turner S, van der Wal P (1993) A biogeochemical study of the Coccolithophore, *Emiliania huxleyi*, in the North Atlantic. Global biogeochem cy 7:879–900
Holton JR, Haynes PH, McIntyre ME, Douglass AR, Rood RB, Pfister L (1995) Stratosphere-troposphere exchange. Rev geophys 33:403–439
Holtslag AAM, Boville BA (1993) Local versus nonlocal boundary layer diffusion in a global climate model. J climate 6:1825–1842
Honrath RE, Peterson MC, Guo S, Dibb JE, Shepson PB, Campbell B (1999) Evidence of NO_x production within or upon ice particles in the Greenland snowpack. Geophys res lett 26:695–698
Hoppel WA (1988) The role of non-precipitating cloud cycles and gas-to-particle conversion in the maintenance of the submicron aerosol size distribution over the tropical oceans. In: Hobbs PV, McCormick MP (eds) Aerosols and climate. A. Deepak Publishing, Hampton, Virginia, pp 9–19
Hoppel WA, Frick GM, Larson RE (1986) Effect of nonprecipitating clouds on the aerosol size distribution in the marine boundary layer. Geophys res lett 13:125–128
Hoppel WA, Fitzgerald JW, Frick GM, Larson RE, Mack EJ (1990) Aerosol size distributions and optical properties found in the marine boundary layer over the Atlantic Ocean. J geophys res 95:3659–3686
Hoppel WA, Frick GM, Fitzgerald JW, Larson RE (1994) Marine boundary layer measurements of new particle formation and the effects nonprecipitating clouds have on aerosol size distribution. J geophys res 99:14443–14459
Horowitz LW, Liang J, Gardner GM, Jacob DJ (1998) Export of reactive nitrogen from North America during summertime: Sensitivity to hydrocarbon chemistry. J geophys res 103:13451–13476
Hõrrak U, Salm J, Tammet H (1998) Bursts of intermediate ions in atmospheric air. J geophys res-atmos 103:13909
Horvath L, Pinto J, Weidinger T (2000) Estimate of the dry deposition of atmospheric nitrogen and sulfur species to spruce forest. In: Sixth International Conference on Air-Surface Exchange of Gases and Particles. Edinburgh, Scotland
Hosker RP, Lindberg SE (1982) Review: Atmospheric deposition and plant assimilation of gases and particles. Atmos environ 16:889
Hough AM (1988) The calculation of photolusis rates for use in global tropospheric modelling studies. Oxford
Houweling S, Dentener F, Lelieveld J (1998) The impact of nonmethane hydrocarbon compounds on tropospheric photochemistry. J geophys res 103:10673–10696
Houweling S, Kaminski FDT, Lelieveld J, Heimann M (1999) Inverse modeling of methane sources and sinks using the adjoint of a global transport model. J geophys res 104:26137–26160
Hov Ø, Zlatev Z, Berkowicz R, Eliassen A, Prahm LP (1989) Comparison of numerical techniques for use in air pollution models with non-linear chemical reactions. Atmos environ 23:967
Hov Ø, Flatøy F, Foot JS, Dewey K, Monks PS, Edwards GD, Green TJ, Brough N, Mills GP, Bauguitte S, Penkett SA (1998) Testing atmospheric chemistry in anticyclones (TACIA). CEC, Brussels
Hov Ø, Flatoy F, Solberg S, Schmidbauer N, Dye C, Foot JS, Dewey K, Kent J, Richter H, Kaye A, Kley D, Schmitgen S, Penkett SA, Reeves C, Bandy B, Brough N, Green T, Mills G, Law K, Plantevin P-H, Savage N, Arnold S, Lobb M, Evans M, O'Connor F, Pyle J, Beekmann M, Kowol J-Santen, Dugault E, Monks PS, Edwards GD, McQuaid J (2000) Maximum Oxidation Rates in the Free Troposphere (MAXOX). Commission of the European Communities
Howarth RW, Billen G, Swaney D, Townsend A, Jaworski N, Lajtha K, Downing JA, Elmgren R, Caraco N, Jordan T, Berendse F, Freney J, Kudeyarov V, Murdoch P, Zhao-Liang Z (1996) Regional nitrogen budgets and riverine N & P fluxes for the drainages to the North Atlantic Ocean: Natural and human influences. Biogeochemistry 35:75–139
Huang Y, Sass RL, Fisher FM (1997) Methane emission from Texas rice paddy soils. 1. Quantitative multi-year dependence of CH_4 emission on soil, cultivar and grain yield. Glob change biol 3:479–489
Huber C (1997) Untersuchungen zur Ammoniakimmission und zum Stoffhaushalt auf ungekalkten und neugekalkten Flächen in einem stickstoffübersättigten Fichtenökosystem (Höglwald). In: (ed) Reiche Ökologie. Hieronymus-Verlag, München, Germany, pp 1–183
Huebert BJ, Lee G (1990) Airborne aerosol inlet passing efficiency measurement. J geophys res 95:16369–16381
Huebert BJ, Lenschow DH (1999) What have Lagrangian experiments accomplished? IGACtivities newsletter 4–8
Huebert BJ, Pszenny AAP, Blomquist B (1996) The ASTEX/MAGE experiment. J geophys res 101(D2):4319–4329
Huebert BJ, Howell SG, Zhuang L, Heath JA, Litchy MR, Wylie DJ, Kreidler-Moss JL, Cöppicus S, Pfeiffer JE (1998) Filter and impactor measurements of anions and cations during the First Aerosol Characterization Experiment (ACE-1). J geophys res 103:16493–16509
Huey LG, Hanson DR, Howard CJ (1995) Reactions of SF6– and I– with atmospheric trace gases. J chem phys 99:5001–5008

Huie RE (1995) Free radical chemistry of the atmospheric aqueous phase. In: Barker JR (ed) Progress and problems in atmospheric chemistry. World Scientific, Singapore
Huntrieser H, Feigl CH, Schlager H, Schröder F, Gerbig CH, van Velthoven P, Flatøy F, Théry C, Höller H, Schumann U (2000) Contribution of lightning-produced NO_x to the European and global NO_x budget: Results and estimates from airborne EULINOX measurements. In: Hoeller H, Schumann U (eds) EULINOX - The European LIghtning Oxides Project. 43–73, DLR-Forschungsbericht, pp 2000–2028
Husar RB, Prospero JM, Stowe LL (1997) Characterization of tropospheric aerosols over the oceans with the NOAA advanced very high resolution radiometer optical thickness operational product. J geophys res 102:16889–16909
Hynes AJ, Stoker RB, Pounds AJ, McKay T, Bradshaw JD, Nicovich JM, Wine PH (1995) A mechanistic study of the reaction of OH with dimethyl-d_6 sulfide. Direct observation of adduct formation and the kinetics of the adduct reaction with O_2. J chem phys 99:16967–16975
Imre D, Xu J, Tang IN, McGraw R (1997) Ammonium bisulfate/water equilibrium and metastability phase diagrams. Phys chem chem phys 101:4919–4195
Institute TNM (1998) EMEP MSC-W status report: Transboundary acidifying air pollution in Europe, Part 1: Calculation of acidifying and eutrophying compounds and comparison with observations. The Norwegian Meteorological Institute
Intergovernmental Panel on Climate Change (IPCC) (1990) Climate change 1990: The Intergovernmental Panel on Climate Change scientific assessment. In: Houghton JT, Jenkins GJ, Ephraums JJ (eds) IPCC first assessment report 1990. Cambridge University Press, Cambridge, p 365
Intergovernmental Panel on Climate Change (IPCC) (1992) Climate change 1992: The supplementary report to the Intergovernmental Panel on Climate Change scientific assessment. Edited by Houghton JT, Callander BA, Varney SK. Cambridge University Press, Cambridge
Intergovernmental Panel on Climate Change (IPCC) (1995) Climate change 1994: Radiative forcing of climate change and an evaluation of the IPCC IS92 emission scenarios. Edited by Houghton JT, Filho LGM, Bruce J, Lee H, Callander BA, Haites E, Harris N, Maskell K. Cambridge University Press, Cambridge
Intergovernmental Panel on Climate Change (IPCC) (1996) Climate change 1995: The science of climate change. In: Houghton JT, Meira Filho LG, Callander BA, Harriss N, Kattenberg A, Maskell K (eds) IPCC second assessment report. Contribution of Working Group I, Cambridge University Press, Cambridge, p 572
Intergovernmental Panel on Climate Change (IPCC) (1999a) Report of IPCC working group on direct nitrous oxide emissions from agricultural soils. OECD, Wageningen, The Netherlands
Intergovernmental Panel on Climate Change (IPCC) (1999b) Aviation and the global atmosphere. A special report of IPCC Working Groups I and III in collaboration with the Scientific Assessment Panel to the Montreal Protocol on Substances that Deplete the Ozone Layer. In: Penner JE, Lister DH, Griggs DJ, Dokken DJ, McFarland M (eds) IPCC special reports. Cambridge University Press, Cambridge, p 373
Intergovernmental Panel on Climate Change (IPCC) (2000) Special report on emissions scenarios. In: Nakicenovic N, Swart R (eds) IPCC special reports. Cambridge University Press, Cambridge, p 570
Intergovernmental Panel on Climate Change (IPCC) (2001) Climate change 2001: The scientific Basis. In: Houghton JT, Ding Y, Griggs DJ, Noguer M, van der Linden PJ, Dai X, Maskell K, Johnson CA (eds) IPCC third assessment report: Climate change 2001. Cambridge University Press, Cambridge
Jacob DJ (1986) Chemistry of OH in remote clouds and its role in the production of formic acid and peroxymonosulphate. J geophys res 91:9807–9826
Jacob DJ (2000) Heterogeneous chemistry and tropospheric ozone. Atmos environ 34:2131–2159
Jacob DJ, Heikes BG, Fan S-M, Logan JA, Mauzerall DL, Bradshaw JD, Singh HB, Gregory GL, Talbot RW, Blake DR, Sachse GW (1996) Origin of ozone and NO_x in the tropical troposphere: A photochemical analysis of aircraft observations over the South Atlantic basin. J geophys res 101:24235–24250
Jacob DJ, Prather MJ, Rasch PJ, Shia R-L, Balkanski YJ, Beagley SR, Bergmann DJ, Blackshear WT, Brown M, Chiba M, Chipperfield MP, de Grandpre J, Dignon JE, Feichter J, Genthon C, Grose WL, Kasibhatla PS, Kohler I, Kritz MA, Law K, Penner JE, Ramonet M, Reeves CE, Rotman DA, Stockwell DZ, Van Velthoven PFJ, Verver G, Wild O, Yang H, Zimmermann P (1997) Evaluation and intercomparison of global atmospheric transport models using ^{222}Rn and other short-lived tracers. J geophys res 102:5953–5970
Jacobi H-W, Wicktor F, Herrmann H, Zellner R (1999) A laser flash photolysis kinetic study of reaction of the Cl_2-radical anion with oxygenated hydrocarbons in aqueous solution. Int J chem kinet 31:169–181
Jacobson MZ (1994) Developing, coupling, and applying a gas, aerosol, transport, and radiation model to study urban and regional air pollution Ph.D. thesis, UCLA
Jacobson MZ (1995) Computation of global photochemistry with SMVGEAR II. Atmos environ 29A:2541–2546
Jacobson MZ (1998) Vector and scalar improvements through absolute error tolerance control. Atmos environ 32:791–796
Jacobson MZ (1999) Fundamentals of atmospheric modeling Cambridge University Press, Cambridge
Jacobson MZ (2001) Global direct radiative forcing due to multicomponent anthropogenic and natural aerosols. J geophys res 106:1551–1568
Jacobson MZ, Turco RP (1994) SMVGEAR: A sparse matrix, vectorized Gear code for atmospheric models. Atmos environ 28:273–284
Jaeglé L, Jacob DJ, Wennberg PO, Spivakovsky CM, Hanisco TF, Lanzendor EJ, Hintsa EJ, Fahey DW, Keim ER, Proffitt MH, Atlas EL, Flocke F, Schauffler S, McElroy CT, Midwinter C, Pfister L, Wilson JC (1997) Observed OH and HO_2 in the upper troposphere suggest a major source from convective injection of peroxides. Geophys res lett 24:3181–3184
Jaeglé L, Jacob DJ, Brune WH, Faloona IC, Tan D, Kondo Y, Sachse GW, Anderson B, Gregory GL, Vay S, Singh HB, Blake DR, Shetter R (1999) Ozone production in the upper troposphere and the influence of aircraft during SONEX: Approach of NO_x-saturated conditions. Geophys res lett 26:3081–3084
Jaffe DA, Honrath RE, Zhang L, Akimoto H, Shimizu A, Mukai H, Murano K, Hatakeyama S, Merrill J (1996) Measurements of NO, NO_y, CO and O_3 and estimation of the ozone production rate at Oki Island, Japan, during PEM-West. J geophys res 101:2037–2048
Jäger H, Carnuth W (1994) The decay of the El Chichon stratospheric perturbation, observed by lidar at northern midlatitudes. Geophys res lett 14:696–699
Jähne B (1982) Dry deposition of gases over water surfaces (gas exchange). In: Flothmann D (ed) Exchange of air pollutants at the air/earth interface (dry deposition). Battelle Institute, Frankfurt
Jähne B, Haußecke H (1998) Air-water gas exchange. Annu rev fluid mech 30:443–468
Jain AK, Briegleb BP, Minschwaner K, Wuebbles DJ (2000) Radiative forcings and global warming potentials of 39 greenhouse gases. J geophys res 105:20773–20790
Jang M, Kamens RM (1999) Newly characterized products and composition of secondary aerosols from the reaction of α-pinene with ozone. Atmos environ 33:459–474
Jaworski NA, Howarth RW, Hetling LJ (1997) Atmospheric deposition of nitrogen oxides onto the landscape contributes to coastal eutrophication in the northeast United States. Environ sci technol 31:1995–2004
Jayne JT, Duan SX, Davidovits P, Worsnop DR, Zahniser MS, Kolb CE (1992) Uptake of gas-phase aldehydes by water surfaces. J chem phys 96:5452–5460

Jayne JT, Leard DC, Zhang X, Davidovits P, Smith KA, Kolb CE, Worsnop DR (2000) Development of an aerosol mass spectrometer for size and composition analysis of submicron particles. Aerosol sci tech 33:49–70

Jeffers PM, Wolfe NL (1996) On the degradation of methyl bromide in sea water. Geophys res lett 23:1773–1776

Jeker DP, Pfister L, Thompson AM, Brunner D, Boccippio DJ, Pickering KE, Wernli H, Kondo Y, Staehelin J (2000) Measurements of nitrogen oxides at the tropopause: Attribution to convection and correlation with lightning. J geophys res 105(D3):3679–3700

Jenkin ME, Sauders SM, Pilling MJ (1997) The tropospheric degradation of volatile organic compounds: A protocol for mechanism development. Atmos environ 31:81–104

Jensen EJ, Toon OB (1997) The potential impact of soot particles from aircraft exhaust on cirrus clouds. Geophys res lett 24:249–252

Jeuken ABM, Siegmund P, Heijboer L, Feichter J, Bengtsson L (1996) On the potential of assimilating meteorological analyses in a global climate model for the purpose of model validation. J geophys res 101:16939–16950

Jickells TD, Spokes LJ (2001) Atmospheric iron inputs to the oceans. In: Turner DR, Hunter KA (eds) The biogeochemistry of iron in sea water. J. Wiley, New York

Jobson BT, Niki H, Yokouchi Y, Bottenheim J, Hopper F, Leaitch R (1994) Measurements of C_2-C_6 hydrocarbons during the Polar Sunrise 1992 Experiment: Evidence for Cl atom and Br atom chemistry. J geophys res 99(D12):25355–25368

Jöckel P (2000) Cosmogenic ^{14}CO as tracer for atmospheric chemistry and transport. Dissertation thesis, Rupertus Carola University, Heidelberg, Germany

Jöckel P, Brenninkmeijer CAM (2001) The seasonal cycle of cosmogenic ^{14}CO at the surface level: A solar cycle adjusted, zonal average climatology based on observations. J geophys res

Jöckel P, Brenninkmeijer CAM, Lawrence MG (2000) Atmospheric response time of cosmogenic ^{14}CO to changes in solar activity. J geophys res 105:6737–6744

Jöckel P, Brenninkmeijer CAM, Lawrence MG, Jeuken ABM, van Velthoven PF (2002) Evaluation of stratosphere-troposphere exchange and the hydroxyl radical distribution in 3-dimensional global atmospheric models using observations of cosmogenic ^{14}CO. J geophys res 107:4446

Jodwalis CM, Benner RL, Eslinger DL (2000) Modeling of dimethyl sulfide, ocean mixing, biological production sea-to-air flux for high latitudes. J geophys res 105D:14387–14399

Johansson SAE, Campbell JL, Malmqvist KG (1995) Particle-Induced X-ray Emission Spectrometry (PIXE). J. Wiley, New York

Johnson CE, Collins WJ, Stevenson DS, Derwent RG (1999) Relative roles of climate and emission changes on future tropospheric oxidant concentrations. J geophys res 104:18631–18645

Johnson KS, Coale KH, Elrod VA, Tindale NW (1994) Iron photochemistry in seawater from the equatorial Pacific. Mar chem 46:319–334

Johnson TJ, Weinhold FG, Burrows JP, Harris GW, Burkhard H (1991) Measurements of line strengths in the HO_2 ν_1 overtone band at 1.5 μm using an InGaAsP laser. J chem phys 95:6499–6502

Jones A, Slingo A (1996) Predicting cloud-droplet effective radius and indirect sulphate aerosol forcing using a general circulation model. Q j roy meteor soc 122:1573–1595

Jones A, Roberts DL, Slingo A (1994) A climate model study of indirect radiative forcing by anthropogenic sulphate aerosols. Nature 370:450–453

Jones A, Roberts DL, Woodage MJ (1999) The indirect effects of anthropogenic sulphate aerosol forcing using a general circulation model. Q j roy meteor soc 122:1573–1595

Jones PD, New N, Parker DE, Martin S, Rigor IG (1999) Surface air temperature and its change over the past 150 years. Rev geophys 37:173–199

Jonquieres I, Marenco A, Maalej A, Rohrer F (1998) Study of ozone formation and transatlantic transport from biomass burning emissions over West Africa during the airborne Tropospheric Ozone Campaigns TROPOZ I and TROPOZ II. J geophys res 103(D15):19059–19073

Jonsson A, Persson KA, Grigoriadis V (1985) Measurement of some low molecular weight oxygenated, aromatic and chlorinated hydrocarbons in ambient air and in vehicle emissions. Environ int 11:383–392

Junge CE (1962) Global ozone budget and exchange between stratosphere and troposphere. Tellus 14:363–377

Kahn R, West R, McDonald D, Rheingans B, Mishchenko MI (1997) Sensitivity of multiangle remote sensing observations to aerosol sphericity. J geophys res 102:16861–16870

Kaiser E-A, Hohrs K, Kücke M, Schnug E, Heinemeyer O, Munch JC (1998) Nitrous oxide release form arable soil: Importance of N-fertilisation, crops and temporal variation. Soil biol biochem 30:1553–1563

Kalberer M, Ammann M, Arens F, Gäggeler HW, Baltensperger U (1999) Heterogeneous formation of nitrous acid (HONO) on soot aerosol particles. J geophys res 104:13825–13832

Kalnay E, Kanamitsu M, Kistler R, Collins W, Deaven D, Gandin L, Iredell M, Saha S, White G, Woollen J, Zhu Y, Chelliah M, Ebisuzaki W, Higgins W, Janowiak J, Mo KC, Ropelewski C, Wang J, Leetmaa A, Reynolds R, Jenne R, Joseph D (1996) The NMC/NCAR 40-year reanalysis project. B am meteorol soc 77:437–471

Kamens RM, Jaoui M (2001) Modeling aerosol formation from α-pinene + NO_x in the presence of natural sunlight using gas-phase kinetics and gas-particle partitioning theory. Environ sci technol 35:1394–1405

Kamm S, Möhler O, Naumann K-H, Saathoff H, Schurath U (1999) The heterogeneous reaction of ozone with soot aerosol. Atmos environ 33:4651–4661

Kanakidou M, Tsigaridis K, Dentener FJ, Crutzen PJ (1999) Human activity enhances the formation of organic aerosols by biogenic hydrocarbon oxidation. J geophys res 105: 9243–9254

Kanaya Y, Sadanaga Y, Matsumoto J, Sharma U, Hirokawa J, Kajii Y, Akimoto H (1999) Nighttime observation of the HO_2 radical by an LIF instrument at Oki Island, Japan, and its possible origins. Geophys res lett 26(14):2179–2182

Kaplan WA, Wofsy SC, Keller M, Da Costa JM (1988) Emission of NO and deposition of O_3 in a tropical forest system. J geophys res 93:1389–1395

Kärcher B (1999) Aviation-produced aerosols and contrails. Surv geophys 20:113–167

Karlsdottir S, Isaksen ISA (2000) Changing methane lifetime. Possible cause for reduced growth. Geophys res lett 27:93–97

Kasibhatla P, Levy H II, Moxim WJ, Pandis SN, Corbett JJ, Peterson MC, Honrath RE, Frost GJ, Knapp K, Parrish DD, Ryerson TB (2000) Do emissions from ships have a significant impact on concentrations of nitrogen oxides in the marine boundary layer? Geophys res lett 27:2229–2232

Kato N, Akimoto H (1992) Anthropogenic emissions of SO_2 and NO_x in Asia: Emissions inventories (plus errata). Atmos environ 26A:2997–3017

Katoshevski D, Nenes A, Seinfeld JH (1999) A study of processes governing the maintenance of aerosols in the marine boundary layer. J aerosol sci 30:503–532

Kaufman YJ, Tanré D, Gordon HR, Nakajima T, Lenoble J, Frouin R, Grassl H, Herman BM, King MD, Teillet PM (1997a) Passive remote sensing of tropospheric aerosol and atmospheric correction for the aerosol effect. J geophys res 102:16815–16830

Kaufman YJ, Tanré D, Remer LA, Vermote EF, Chu A, Holben BN (1997b) Operational remote sensing of tropospheric aerosol over land from EOS-Moderate Resolution Imaging Spectroradiometer. J geophys res 102:17051–17067

Kaufman YJ, Hobbs PV, Kirchhoff VWJH, Artaxo P, Remer LA, Holben BN, King MD, Ward DE, Prins EM, Longo KM, Mattos LF, Nobre CA, Spinhirne JD, Ji Q, Thompson AM, Gleason LF, Christopher SA, Tsay SC (1998a) Smoke, Clouds, Radiation-Brazil (SCAR-B) experiment. J geophys res 103:31783–31808

Kaufman YJ, Justice CO, Flynn L, Kendall J, Prins E, Ward DE, Menzel P, Setzer A (1998b) Potential global fire monitoring from EOS-MODIS. J geophys res 103:32215–32238

Keeling RF (1993) On the role of large bubbles in air-sea gas exchange and super saturation in the ocean. J mar res 51:237–271

Keene WC, Pszenney AAP, Jacob DJ, Duce RA, Galloway JN, Schultz-Tokos J, Sievering H, Boatman JF (1990) The geochemical cycling of reactive chlorine through the marine troposphere. Global biogeochem cy 4:407–430
Keene WC, Sander R, Pszenny AAP, Vogt R, Crutzen PJ, Galloway JN (1998) Aerosol pH in the marine boundary layer: A review and model evaluation. J aerosol sci 29:339–356
Keller M, Reiners WA (1994) Soil-atmosphere exchange of nitrous oxide, nitric oxide, and methane under secondary succession of pasture to forest in the Atlantic lowlands of Costa Rica. Global biogeochem cy 8:399–409
Keller M, Kaplan WA, Wofsy SC (1986) Emissions of N_2O and CH_4 and CO_2 from tropical forest soils. J geophys res 91:11791–11802
Keller M, Veldkamp E, Weitz AM, Reiners WA (1993) Effect of pasture age on soil trace-gas emissions from a deforested area of Costa Rica. Nature 365:244–246
Keller MD, Bellows WK, Guillard RRL (1989) Dimethyl sulfide production in marine phytoplankton. In: Saltzman ES, Cooper WJ (eds) Biogenic sulphur in the environment. American Chemical Society, Washington, DC, pp 167–182
Keller MD, Kiene R, Matrai PA (1999a) Production of glycine betaine (GBT) and dimethylsulfoniopropionate (DMSP) in batch cultures of marine phytoplankton. Mar biol 135:237–248
Keller MD, Kiene R, Matrai PA (1999b) Production of glycine betaine (GBT) and dimethylsulfonio propionate (DMSP) in nitrogen-limited chemostats cultures of marine phytoplankton. Mar biol 135:249–257
Kennedy V (1983) The estuary as a filter. Academic Press, New York
Kennett JP, Cannariato KG, Hendy IL, Behl RJ (2000) Carbon isotopic evidence for methane hydrate instability during Quaternary interstadials. Science 288:128–133
Kent GS, Trepte CR, Skeens KM, Winker DM (1998) LITE and SAGE II measurements of aerosols in the southern hemisphere upper troposphere. J geophys res 103:19111–19127
Kesselmeier J, Bode K, Schaefer L, Schebeske G, Wolf A, Brancaleoni E, Cecinato A, Ciccioli P, Frattoni M, Dutaur L, Fugit JL, Simon V, Torres L (1998) Simultaneous field measurements of terpene and isoprene emissions from two dominant Mediterranean oak species in relation to a North American species. Atmos environ 32:1947–1953
Kettle AJ (1994) A model of the temporal and spatial distribution of carbon monoxide in the mixed layer
Kettle AJ, Andreae MO (1998) Predicting the production rate of carbonyl sulfide and other photochemical tracers in the global oceans. In: Gordon Research Conference. Il Ciocco, Italy
Kettle AJ, Andreae MO, Amouroux D, Andreae TW, Bates TS, Berresheim H, Bingemer H, Boniforti R, Curran MAJ, DiTullio GR, Helas G, Jones GB, Keller MD, Kiene RP, Leck C, Levasseur M, Maspero M, Matrai P, McTaggart AR, Mihalopoulos N, Nguyen BC, Novo A, Putaud JP, Rapsomanikis S, Roberts G, Schebeske G, Sharma S, Simo R, Staubes R, Turner S, Uher G (1999) A global database of sea surface dimethylsulphide (DMS) measurements and a simple model to predict sea surface DMS as a function of latitude, longitude and month. Global biogeochem cy 13:399–444
Khalil MAK, Rasmussen R, Gunawardena R (1993) Atmospheric methyl bromide: Trends and global mass balance. J geophys res 98(D2):2887–2896
Khattatov BV, Gille JC, Lyjak LV, Brasseur GP, Dvortsov VL, Roche AE, Waters J (1999) Assimilation of photochemically active species in a case analysis of UARS data. J geophys res 104:18715–18737
Kieber DJ, Jiao J, Kiene RP, Bates TS (1996) Impact of dimethylsulfide photochemistry on methyl sulphur cycling in the Equatorial Pacific Ocean. J geophys res-oceans 101(C2):3715–3722
Kiehl JT, Schneider TL, Rasch PJ, Barth MC, Wong J (2000) Radiative forcing due to sulfate aerosols from simulations with the NCAR Community Climate Model (CCM3). J geophys res 105:1441–1457
Kiene RP (1996) Production of methanethiol from dimethylsulfoniopropionate in marine surface waters. Mar chem 54:69–83
Kiene RP, Bates TS (1990) Biological removal of dimethyl sulphide from seawater. Nature 345:702–705
Kiene RP, Linn LJ, Gonzalez J, Moran MA, Bruton JA (1999) Dimethylsulfoniopropionate and methanethiol are important precursors of methionine and protein-sulphur in marine bacterioplankton. Appl environ microb 65:4549–4558
Kim TO, Ishida T, Adachi M, Okuyama K, Seinfeld JH (1998) Nanometer-sized particle formation from $NH_3/SO_2/H_2O$/air mixtures by ionizing irradiation. Aerosol sci tech 29:111–125
Kim Y, Cess RD (1993) Effect of anthropogenic sulfate aerosols on low level cloud albedo over oceans. J geophys res 98:14883–14885
King DB, Saltzman ES (1997) Removal of methyl bromide in coastal seawater: Chemical and biological rates. J geophys res-oceans 102:18715–18721
King DB, Butler JH, Montzka SA, Yvon-Lewis SA, Elkins JW (2000) Methyl bromide saturations in the North Atlantic and coastal, Eastern Pacific Oceans. J geophys res 105:19763–19769
King MD, Radke LF, Hobbs PV (1993) Optical properties of marine stratocumulus clouds modified by ships. J geophys res 98:2729–2739
King MD, Kaufman YJ, Tanré D, Nakajima T (1999) Remote sensing of tropospheric aerosols from space: Past, present, and future. B am meteorol soc 80:2229–2259
Kirchner U, Scheer V, Vogt R (2000) FTIR spectroscopic investigation of the mechanism and kinetics of the heterogeneous reactions of NO_2 and HNO_3 with soot. J phys chem a 104:8908–8915
Kirchoff VWJH, Rasmussen RA (1990) Time variation of CO and ozone concentration in a region subject to biomass burning. J geophys res 95:7521–7532
Kirstensen L, Jorgensen CE, Kirkegaard P, Pilegaard K (1997) First-order chemistry in the surface flux layer. J atmos chem 27:249
Kirstine W, Galbally I, Ye Y, Hooper M (1998) Emissions of volatile organic compounds (including oxygenated species) from pasture. J geophys res 103:10605–10619
Kita K, Fujiwara M, Kawakami S (2000) Total ozone increase associated with forest fires over the Indonesian region and its relation to the El Niño-Southern Oscillation. Atmos environ 34(17):2681–2690
Kita K, Kawakami S, Miyazaki Y, Higashi Y, Kondo Y, Nishi N, Koike M, Blake DR, Machida T, Sano T, Hu W, Ko M, Ogawa T (2001) Photochemical production of ozone in the upper troposphere in association with cumulus convection over Indonesia. J geophys res (in press)
Kley D, Beck J, Grennfelt PI, Hov Ø, Penkett SA (1997) Tropospheric ozone research (TOR). A sub-project of EUROTRAC. J atmos chem 28:1–9
Kley PJC, Smit HGJ, Vomel H, Oltmans SJ, Grassl H, Ramanathan V (1996) Observations of near-zero ozone concentrations over the convective Pacific: Effects on air chemistry. Science 274:230–233
Kleypas JA, Buddemeier RW, Archer D, Gattuso J-P, Langdon C, Opdyke BN (1999) Geochemical consequences of increased atmospheric carbon dioxide on coral reefs. Science 284:118–120
Klinger L, Greenberg J, Guenther A, Tyndall G, Zimmerman P, Bangui M, Mutsambote JM, Kenfack D (1998) Patterns in volatile organic compound emissions along a savanna-rainforest gradient in Central Africa. J geophys res 102:1443–1454
Klonecki A, Levy H (1997) Tropospheric chemical ozone tendencies in CO-CH_4-NO_y-H_2O system: Their sensitivity to variations in environmental parameters and their application to a global chemistry transport model study. J geophys res 102(D17):21221–21237
Klotz B, Sørensen S, Barnes I, Becker KH, Etzkorn T, Volkamer R, Platt U, Wirtz K, Martín-Reviejo M (1998) Atmospheric oxidation of toluene in a large volume outdoor photoreactor: *In situ* determination of ring-retaining product yields. Phys chem chem phys 102:10289–10299
Koch D, Jacob D, Tegen I, Rind D, Chin M (1999) Tropospheric sulfur simulation and sulfate direct radiative forcing in the Goddard Institute for Space Studies general circulation model. J geophys res 104:23799–23822
Koike M, Kondo Y, Akutagawa D, Kita K, Nishi N, Liu SC, Blake DR, Kawakami S, Takegawa N, Ko M, Zhao Y, Ogawa T (2001) Reactive nitrogen over the tropical Western Pacific: Influence from lightning and biomass burning. J geophys res (in press)

Kondo Y, Koike M, Kita K, Ikeda H, Takegawa N, Kawakami S, Blake D, Liu SC, Ko M, Miyazaki Y, Irie H, Higashi Y, Liley B, Nishi N, Zhao Y, Ogawa T (2001) Effects of biomass burning, lightning, and convection on O_3, CO, and NO_y over the tropical Pacific and Australia in August-October 1998 and 1999. J geophys res (in press)

Koschorreck M, Conrad R (1993) Oxidation of atmospheric methane in soil: Measurements in the field, in soil and in soil samples. Global biogeochem cy 3(1):109–121

Kotchenruther RA, Hobbs PV, Hegg DA (1999) Humidification factors for atmospheric aerosols off the mid-Atlantic coast of the United States. J geophys res 104:2239–2251

Koutsenogii PK, Jaenicke R (1994) Number concentration and size distribution of atmospheric aerosol in Siberia. J aerosol sci 25:377–383

Kramm G, Mueller H, Fowler D, Hoefken KD, Meixner F, Schaller E (1991) A modified profile method for determining vertical fluxes of NO, NO_2, ozone and HNO_3 in the atmospheric surface layer. J atmos chem 13:265

Kreher K, Johnston SWWPV, Platt U (1997) Ground-based measurements of tropospheric and stratospheric BrO at Arrival Heights (78° S), Antarctica. Geophys res lett 24:3021–3024

Kreidenweis SM, Seinfeld JH (1988) Nucleation of sulfuric acid-water and methanesulfonic acid-Water solution particles: Implications for the atmospheric chemistry of organosulfur species. Atmos environ 22:283–296

Kreidenweis SM, Flagan RC, Seinfeld JH, Okuyama K (1989) Binary nucleation of methansulfonic acid and water. J aerosol sci 20:585–607

Kreidenweis SM, Yin F, Wang S-C, Grosjean D, Flagan RC, Seinfeld JH (1991) Aerosol formation during photooxidation of organosulfur species. Atmos environ 25A:2491–2500

Kreidenweis SM, Zhang Y, Taylor GR (1997) The effects of clouds on aerosol and chemical species production and distribution: 2. Chemistry model description and sensitivity analysis. J geophys res 102:23867–23882

Kreidenweis SM, McInnes LM, Brechtel FJ (1998) Observations of aerosol volatility and elemental composition at Maquarie Island during the First Aerosol Characterization Experiment (ACE-1). J geophys res 103:16511–16524

Kroeze C, Seitzinger SP (1998) Nitrogen inputs to rivers, estuaries and continental shelves and related nitrous oxide emissions in 1990 and 2050: A global model. Nutr cycl agroecosys 52:195–212

Kroeze C, Mosier AR, Bouwman L (1999) Closing the global N_2O budget: A retrospective analysis 1500–1994. Global biogeochem cy 13:1–8

Krol M, van Leeuwen P-J, Lelieveld J (1998) Global OH trend inferred from methylchloroform measurements. J geophys res 103:10697–10711

Krol MC, Weele MV (1997) Implications of variations in photodissociation rates for global tropospheric chemistry. Atmos environ 31:1257–1273

Krüger M, Frenzel P, Conrad R (2001) Microbial processes influencing methane emission from rice fields. Glob change biol 7:49–63

Kuang Z, Yung YL (2000) Reflectivity variations off the Peru coast: Evidence for indirect effect of anthropogenic sulfate aerosols on clouds. Geophys res lett 27:2501

Kuhlbusch TA, Lobert JM, Crutzen PJ, Warnek P (1991) Molecular nitrogen emissions from denitrification during biomass burning. Nature 351:135–137

Kuhn M, Builtjes PJH, Poppe D, Simpson D, Stockwell WR, Anderson-Skold Y, Baart A, Das M, Fielder F, Hov O, Kirchner F, Makar PA, Milford JB, Roemer MGM, Ruhnke R, Strand A, Vogel B, Vogel H (1998) Intercomparison of the gas-phase chemistry in several chemistry and transport models. Atmos environ 32:693–709

Kulmala M, Laaksonen A, Korhonen P, Vesala T, Ahonen T, Barrett JC (1993) The effect of atmospheric nitric acid vapor on cloud condensation nucleus activation. J geophys res 98:22949–22958

Kulmala M, Vehkamäki H, Vesala T, Barrett JC, Clement CF (1995) Aerosol formation in diffusive boundary layer: Binary homogeneous nucleation of ammonia and water vapours. J aerosol sci 26(4):547–558

Kulmala M, Laaksonen A, Pirjola L (1998) Parameterizations for sulfuric acid/water nucleation rates. J geophys res 103:8301–8307

Kulmala M, Pirjola L, Mäkelä JM (2000) Stable sulphonate clusters as a source of new atmospheric particles. Nature 404:66–69

Kurylo MJ, Rodriguez JM, Andreae MO, Atlas EL, Blake DR, Butler JH, Lal S, Lary DJ, Midgley PM, Montzka SA, Novelli PC, Reeves CE, Simmonds PG, Steele LP, Sturges WT, Weiss RF, Yokouichi Y (1999) Short-lived ozone-related compounds. In: Ennis CA (ed) Scientific assessment of the ozone layer: 1998. World Meteorological Organization, Geneva, Switzerland

Kvenvolden KA, Lilley MD, Lorenson TD, Barnes PW, McLaughlin E (1993) The Beaufort Sea continental shelf as a seasonal source of atmospheric methane. Geophys res lett 20:2459–2462

Kwint RLJ, Kramer KJM (1996) Annual cycle of the production and fate of DMS and DMSP in a marine coastal system. Mar ecol-prog ser 134:217–224

Kwint RLJ, Quist P, Hansen TA, Dijkhuizen L, Kramer KJM (1996) Turnover of dimethylsulfoniopropionate and dimethylsulfide in the marine environment: A mesocosm experiment. Mar ecol-prog ser 145:223–232

Kwok ESC, Atkinson R, Arey J (1995) Observation of hydroxycarbonyls from the OH radical-initiated reaction of isoprene. Environ sci technol 29:2467–2469

Laaksonen A, Hienola J, Kulmala M, Arnold F (1997) Supercooled cirrus cloud formation modified by nitric acid pollution of the upper troposphere. Geophys res lett 24:3009–3012

Laaksonen A, Korhonen P, Kulmala M, Charlson RJ (1998) Modification of the Köhler equation to include soluble trace gases and slightly soluble substances. J atmos sci 55:853–862

Labatut CA (1997) Ozone and heat fluxes over a Mediterranean pseudosteppe. Atmos environ 31:177–184

Lacaux J-P (1999) DEBITS activity in Africa: Atmospheric deposition in northern hemisphere of tropical Africa. In: Proceeding IGAC 99 Conference. Bologna, Itally

Lacaux J-P, Delmas R, Kouadio G, Cros B, Andreae MO (1991) Precipitation chemistry in the Mayombé forest of equatorial Africa. J geophys res 97:6195–6206

Lacaux J-P, Cachier H, Delmas R (1993) Biomass burning in Africa: An overview of its impact on atmospheric chemistry. In: Crutzen PJ, Goldammer JG (eds) Fire in the environment: The ecological, atmospheric, and climatic importance of vegetation fires. J. Wiley, Chichester, pp 159–188

Lacis AA, Wuebbles DJ, Logan JA (1990) Radiative forcing of climate by changes in the vertical-distribution of ozone. J geophys res 95(D7):9971–9981

Lagrange J, Pallares C, Wenger G, Lagrange P (1993) Electrolyte effects on aqueous atmospheric oxidation of sulphur dioxide by hydrogen peroxide. Atmos environ 27A:129–137

Lagrange J, Lagrange P, Pallares C, Wenger G (1995) Electrolyte effects on aqueous oxidation of sulfur dioxide by strong oxidants. In: Warneck P (ed) Transport and chemical transformation of pollutants in the troposphere. Springer-Verlag, Berlin, pp 89–96

Lagrange J, Lagrange P, Lahoutifard N (1999a) Kinetics of oxidation and acidification reactions in the atmospheric aqueous phase: From pure water to deliquescent aerosol. In: Schurath U (ed) CMD Annual Report, GSF, München, Germany, pp 140–143

Lagrange J, Wenger G, Lagrange P (1999b) Kinetic study of HMSA formation and decomposition: Tropospheric relevance. J chem phys 96:610–633

Laj P, Fuzzi S, Facchini MC, Orsi G, Berner A, Kruisz C, Wobrock W, Beswick KM, Gallagher MW, Colvile RN, Choularton TW, Nason P, Jones B (1997) Experimental evidence for in-cloud production of aerosol sulphate. Atmos environ 31:2503–2514

Lamarque J-F, Khattatov BV, Gille JC (1999) Assimilation of Measurements of Air Pollution from Space (MAPS) CO in a global three-dimensional model. J geophys res 104:26209–26218

Lambert G, LeCloarec M-F, Pennisi MM (1988) Volcanic output of SO_2 and trace metals: A new approach. Geochim cosmochim ac 52:39–42

Lammers S, Suess E, Mansurov MN, Anikev VV (1995) Variations of atmospheric methane supply from the Sea of Okhotsk induced by the seasonal ice cover. Global biogeochem cy 9:351–358

Landgraf J, Crutzen PJ (1998) An efficient method for on-line calculations of photolysis and heating rates. J atmos sci 55:863–878
Langford AO (1999) Stratosphere-troposphere exchange at the subtropical jet: Contribution to the tropospheric ozone budget at midlatitudes. Geophys res lett 26(16):2449–2452
Langner J (1991) Sulfur in the troposphere: A global three-dimensional model study. Ph.D. thesis, Stockholm University, Stockholm
Langner J, Rodhe H (1991) A global three-dimensional model of the tropospheric sulfur cycle. J atmos chem 13:225–264
Langway CC Jr., Osada K, Clausen HB, Hammer CU, Shoji H, Mitani A (1994) New chemical stratigraphy over the last millennium for Byrd Station, Antarctica. Tellus 46B:40–51
Laroche D, Vezina AF, Levasseur M, Gosselin M, Stefels J, Keller MD, Matrai PA, Kwint RLJ (1999) DMSP synthesis and exudation in phytoplankton: A modeling approach. Mar ecol-prog ser 180:37–49
Lary DJ, Lee AM, Toumi R, Newchurch MJ, Pirre M, Renard JB (1997) Carbon aerosols and atmospheric photochemistry. J geophys res 102:3671–3682
Lassey KR, Ulyatt MJ (2000) Methane emissions by grazing livestock: A synopsis of 1000 direct measurements. In: van Ham J, Baede APM, Meyer LA, Ybema R (eds) Non-CO_2 greenhouse gases: Scientific understanding, control and implementation. Kluwer Academic Publishers, Dordrecht, pp 101–106
Lassey KR, Lowe DC, Manning MR (2000) The trend in atmospheric methane $\Delta^{13}C$ and implications for isotopic constraints on the global methane budget. Global biogeochem cy 14(1):41–49
Lavanchy VMH, Gäggeler HW, Schotterer U, Schwikowski M, Baltensperger U (1999) Historical record of carbonaceous particle concentrations from a European high-alpine glacier (Colle Gnifetti, Switzerland). J geophys res 104:21227–21236
Lavoué D, Liousse C, Cachier H, Stocks BJ, Goldhammer JG (2000) Modeling of carbocaneous particles emitted by boreal and temperate fires at northern latitudes. J geophys res 105:26871–26890
Law KS, Plantevin P-H, Shallcross DE, Rogers H, Grouhel C, Thouret V, Marenco A, Pyle JA (1998) Evaluation of modeled O_3 using MOZAIC data. J geophys res 103:25721–25740
Law KS, Plantevin P-H, Thouret V, Marenco A, Asman WAH, Lawrence M, Crutzen PJ, Muller J-F, Hauglustaine DA, Kanakidou M (2000) Comparison between global chemistry transport model results and measurement of ozone and water vapour by airbus in-service aircraft (MOZAIC) data. J geophys res 105:1503–1525
Lawrence MG, Crutzen PJ (1998) The impact of cloud particle gravitational settling on soluble trace gas distributions. Tellus 50B:263–289
Lawrence MG, Chameides WL, Kasibhatla PS, Levy H, Moxim W (1995) Lightning and atmospheric chemistry: The rate of atmospheric NO production. In: Volland H (ed) Handbook of atmospheric electrodynamics. CRC Press, Boca Raton, pp 197–211
Lawrence MG, Crutzen PJ, Rasch PJ, Eaton BE, Mahowald NM (1999a) A model for studies of tropospheric chemistry: Description, global distributions, and evaluation. J geophys res 104:26245–26277
Lawrence MG, Landgraf J, Jockel P, Eaton B (1999b) Artifacts in global atmospheric modeling: Two recent examples. EOS 80:123
Lawrence MG, Crutzen PJ, Rasch PJ (1999c) Analysis of the CEPEX ozone data using a 3D chemistry-meteorology model. Q j roy meteor soc 125:2987–3009
Lawrence MG, Williams J, Rasch PJ, Crutzen PJ (2000) Photochemistry and plume tracer forecasts during the INDOEX 1999 field phase. J geophys res (in preparation for INDOEX special issue)
Le Canut P, Andreae MO, Harris GW, Wienhold FG, Zenker T (1996) Airborne studies of emissions from savanna fires in southern Africa, 1. Aerosol emissions measured with a laser optical particle counter. J geophys res 101:23615–23630
Leaitch WR, Isaac GA, Strapp JW, Banic CM, Wiebe HA (1992) The relationship between cloud droplet number concentrations and anthropogenic pollution: Observations and climatic implications. J geophys res 97:2463–2474
Leck C (1990) Dimethyl sulfide in the Baltic Sea: Annual variability in relation to biological activity. J geophys res 95C:3353–3363
Leck C, Bigg EK (1999) Aerosol production over remote marine areas: A new route. Geophys res lett 23:3577–3581
Leck C, Persson C (1996) Seasonal and short-term variability in dimethylsulfide, sulphur dioxide and biogenic sulphur and sea salt aerosol particles in the arctic marine boundary layer during summer and autumn. Tellus 48B:272–299
Leck C, Rodhe H (1991) Emissions of marine biogenic sulphur to the atmosphere of Northern Europe. J atmos chem 12:63–86
Leck C, Bagander LE, Larsson U, Johansson S, Hadju S (1990) DMS in the Baltic Sea: Annual variability in relation to biological activity. J geophys res 95:3353–3364
Leck C, Bigg EK, Covert DS, Heintzenberg J, Maenhaut W, Nilsson ED, Wiedensohler A (1996) Overview of the atmospheric research programme during the International Arctic Ocean Expedition of 1991 (IAOE-91) and its scientific results. Tellus 48B:136–155
Leck C, Nilsson ED, Backlin L, Bigg EK (2001) Atmospheric program on the Arctic Ocean Expedition 1996 (AOE-96): An overview of scientific goals, experimental approach, and instruments. J geophys res-atmos 106
Ledyard KM, Dacey JWH (1994) Dimethylsulfide production from dimethylsulfoniopropionate by a marine bacterium. Mar ecol-prog ser 110:95–103
Ledyard KM, Dacey JWH (1996) Microbial cycling of DMSP and DMS in coastal and oligotrophic seawater. Limnol oceanogr 41:33–40
Lee AM, Carver GD, Chipperfield MP, Pyle JA (1997) Three-dimensional chemical forecasting: A methodology. J geophys res 102:3905–3919
Lee M, Heikes BG, Jacob DJ (1998) Enhancements of hydroperoxides and formaldehyde in biomass burning impacted air and their effect on atmospheric oxidant cycles. J geophys res 103:13201–13212
Lee S-H, Akimoto H, Nakane H, Kurnosenko S, Kinjo Y (1998) Lower tropospheric ozone trend observed in 1989–1997 at Okinawa, Japan. Geophys res lett 25:1637–1640
Legrand M, Feniet-Saigne C (1991) Strong El Niño revealed by methanesulphonic acid in South Polar snow layers. Geophys res lett 21:187–190
Legrand M, Delmas RJ, Charlson RD (1988) Climatic forcing implications from Vostok ice core sulphate data. Nature 334:418–419
Legrand M, Feniet-Saigne C, Saltzman ES, Germain C, Barkov NI, Petrov VN (1991) Ice-core record of oceanic emissions of dimethylsulphide during the last climate cycle. Nature 350:144–146
Legrand M, Feniet-Saigne C, Saltzman ES, Germain C (1992) Spatial and temporal variations of methanesulfonic acid and non-sea salt sulphate in Antarctic ice. J atmos chem 14:245–260
Legrand M, Hammer CU, Angelis MD, Savarino J, Clausen DRJH, Johnsen SJ (1997) Sulphur derived species (methanesulfonate and SO_4) over the last climatic cycle in the Greenland Ice Core Project (Central Greenland) ice core. J geophys res 102:26663–26679
Lelieveld J, Crutzen PJ (1991) The role of clouds in tropospheric photochemistry. J atmos chem 12:229–267
Lelieveld J, Crutzen PJ (1994) Role of deep cloud convection in the ozone budget of the troposphere. Science 264(5166):1759–1761
Lelieveld J, Dentener F (2000) What controls tropsopheric ozone? J geophys res 105:3531–3551
Lelieveld J, van Dorland R (1995) Ozone chemistry changes in the troposphere and consequent radiative forcing of climate. In: Wang WC, Isaksen ISA (eds) Atmospheric ozone as a climate gas. Springer-Verlag, New York, pp 227–258
Lelieveld J, Roelofs GJ, Ganzeveld L, Feichter J, Rodhe H (1997) Terrestrial sources and distribution of atmospheric sulfur. Philos t roy soc b 352:149–158
Lelieveld J, Crutzen PJ, Dentener FJ (1998) Changing concentration, lifetime and climate forcing of atmospheric methane. Tellus 50:128
Lelieveld J, Crutzen PJ, Ramanathan V, Andreae MO, Brenninkmeijer CAM, Campos T, Cass GR, Dickerson RR, Fischer H, de JAGouw, Hansel A, Jefferson A, Kley D, de Laat A, Lal S, Lawrence MG, Lobert JM, Mayol-Bracero OL, Mitra AP, Novakov T, Oltmans SJ, Prather KA, Reiner T, Rodhe H, Scheeren HA, Sikka D, Williams J (2001) The Indian Ocean Experiment: Widespread air pollution from South and Southeast Asia. Science 291:1031–1036

Lerdau M, Keller M (1997) Controls on isoprene emission from trees in a subtropical dry forest. Plant cell environ 20:569–578
Lerdau M, Guenther A, Monson R (1997) Plant production and emission of volatile organic compounds. Science 47(6):373–383
Leuenberger M, Siegenthaler U (1992) Ice-age atmospheric concentration of nitrous oxide from an Antarctic ice core. Nature 360:449–451
Levasseur M, Gosselin M, Michaud S (1994) A new source of dimethyl sulfide for the Arctic atmosphere. Mar biol 121:381–387
Levine JS, Winstead EL, Parsons DAB, Scholes MC, Scholes RJ, Cofer WR, Cahoon DR, Sebacher DI (1996) Biogenic soil emissions of nitric oxide (NO) and nitrous oxide (N_2O) from savannas in South Africa: The impact of wetting and burning. J geophys res 101(D19):23689–23697
Levy H II (1971) Normal atmosphere: Large radical and formaldehyde concentrations predicted. Science 173:141–143
Levy H II (1972) Photochemistry of the lower troposphere. Planet space sci 20:919–935
Levy H II, Mahlman JD, Moxim WJ, Liu SC (1985) Tropospheric ozone: The role of transport. J geophys res 90:3753–3772
Levy H II, Moxim WJ, Kasibhatla PS (1996) A global three-dimensional time-dependent lightning source of tropospheric NO_x. J geophys res 101:22911–22922
Levy H II, Kasibhatla PS, Moxim WJ, Klonecki AA, Hirsch AI, Oltmans SJ, Chameides WL (1997) The global impact of human activity on tropospheric ozone. Geophys res lett 24:791–794
Levy H II, Moxim WJ, Klonecki AA, Kasibhatla PS (1999) Simulated tropospheric NO_x: Its evaluation, global distribution, and individual source contributions. J geophys res 104:26279–26306
Levy JI, Carrothers TJ, Tuomisto J, Hammitt JK, Evans JS (2001) Assessing the public health benefits of reduced ozone concentrations. Environ health persp 109:1215–1226
Li C, Frolking S, Frolking T (1992) A model of nitrous oxide evolution from soil driven by rainfall events: 1. Model structure and sensitivity. J geophys res 97:9759–9776
Li X, Maring H, Savoie D, Voss K, Prospero JM (1996) Dominance of mineral dust in aerosol light-scattering in the North Atlantic trade winds. Nature 380:416–419
Likens G, Borman F, Johnson M (1974) Acid rain. Environment 14:33–40
Lin E (1993) Agricultural techniques: Factors controlling methane emissions. In: Gao L, Wu L, Zhen D, Han X (eds) Proceedings of International Symposium on Climate Change, Natural Disasters and Agricultural Strategies. China Meteorological Press, Beijing, pp 120–126
Lin SJ, Rood RB (1996) Multidimensional flux-form semi-lagrangian transport schemes. Mon weather rev 124:2046–2070
Lindesay JA, Andreae MO, Goldhammer JG, Harris G, Annegarn HJ, Garstan M, Scholes RJ, van Wilgen BW (1996) International Geosphere-Biosphere Programme/International Global Atmospheric Chemistry SAFARI-92 field experiment: Background and overview (Paper 96JD01512). J geophys res 101(D19):23521–23530
Lindinger W, Hansel A, Jordan A (1998a) On-line monitoring of volatile organic compounds at pptv levels by means of proton-transfer-reaction mass spectrometry (PTR-MS). Medical applications, food control and environmental research. Int J mass spectrom ion proc 173:191–241
Lindinger W, Hansel A, Jordan A (1998b) Proton-transfer-reaction mass spectrometry (PTR-MS): On-line monitoring of volatile organic compounds at pptv levels. Chem soc rev 27:347–354
Lindqvist O, Johansson K, Aastrup M, Anderson A, Bringmark L, Hovsenius G, Håkanson L, Iverfeldt Å, Meili M, Timm B (1991) Mercury in the Swedish environment: Recent research on causes, consequences and corrective measures. Water air soil poll 55:1-261
Liousse C, Penner JE, Chuang C, Walton JJ, Eddleman H (1996) A global three-dimensional model study of carbonaceous aerosols. J geophys res 101:19411–19432
Lippmann M, Maynard RL (1999) Air quality guidelines and standards. In: Holgate ST, Samet JM, Maynard RL, Koren HS (eds) Air pollution and health. Academic Press, London, pp 749–796
Liss PS, Duce RA (1997) The sea surface and global change. Cambridge University Press, Cambridge
Liss PS, Merlivat L (1986) Air-sea gas exchange rates: Introduction and synthesis. In: Buat-Menard P (ed) The role of air-sea exchange in geochemical cycling. D. Reidel, Boston, pp 113–129
Liss PS, Malin G, Turner SM, Holligan PM (1994) Dimethyl sulphide and *Phaeocystis*: A review. J marine syst 5:41–53
Liss PS, Hatton AD, Malin G, Nightingale PD, Turner SM (1997) Marine sulphur emissions. Philos t roy soc b 352:159–168
Litherland AE (1980) Ultrasensitive mass spectrometry with accelerators. Annu rev nucl part s 30:437–473
Liu SC, Trainer M, Fehsenfeld FC, Parrish DD, Williams EJ, Fahey DW, Hübler G, Murphy PC (1987) Ozone production in the rural atmosphere and the implications for regional and global ozone distribution. J geophys res 92:4191–4207
Liu SG, Reiners WA, Keller M, Schimel DS (1999) Model simulation of changes in N_2O and NO emissions with conversion of tropical rain forests to pastures in the Costa Rica Atlantic Zone. Global biogeochem cy 13:663–677
Liu X, Adams F, Cafmeyer J, Maenhaut W (2000) Biomass burning in Southern Africa: Individual particle characterization of atmospheric aerosols and savanna fire samples. J atmos chem 36:135–155
Livingston JM, Kapustin V, Schmid B, Russell PB, Quinn PK, Bates TS, Durkee PA, Smith PJ, Freudenthaler V, Wiegner M, Covert DS, Gasso S, Hegg D, Collins DR, Flagan RC, Seinfeld JH, Vitale V, Tomasi C (2000) Shipboard sunphotometer measurements of aerosol optical depth spectra and columnar water vapor during ACE-2 and comparison with selected land, ship, aircraft, and satellite measurements. Tellus 52B:594–619
Lobert JM, Scharffe DH, Hao W-M, Kuhlbusch TA, Seuwen R, Warneck P, Crutzen PJ (1991) Experimental evaluation of biomass burning emissions: Nitrogen and carbon containing compounds. In: Levine JS (ed) Global biomass burning: Atmospheric, climatic and biospheric implications. MIT Press, Cambridge, Mass., pp 289–304
Lobert JM, Butler JH, Montzka SA, Geller LS, Myers RC, Elkins JW (1995) A net sink for atmospheric CH_3Br in the East Pacific Ocean. Science 267:1002–1005
Lobert JM, Butler JH, Geller LS, Yvon SA, Montzka SA, Myers RC, Clarke AD, Elkins JW (1996) BLAST94: Bromine latitudinal air/sea transect 1994: Report on oceanic measurements of methyl bromide and other compounds. Climate Monitoring and Diagnostics Laboratory, National Oceanic and Atmospheric Administration, Boulder, Colorado
Lobert JM, Yvon-Lewis SA, Butler JH, Montzka SA, Myers RC (1997) Undersaturation of CH_3Br in the Southern Ocean. Geophys res lett 24:171–172
Lobert JM, Keene WC, Logan JA, Yevich R (1999) Global chlorine emissions from biomass burning: Reactive chlorine emissions inventory. J geophys res 104:8373–8389
Logan JA (1985) Tropospheric ozone: Seasonal behavior, trends, anthropogenic influence. J geophys res 90:10463–10482
Logan JA (1994) Trends in the vertical distribution of ozone: An analysis of ozonesonde data. J geophys res 99:25553–25585
Logan JA (1999) An analysis of ozone sonde data for the troposphere: Recommendations for testing 3-D models and development of a gridded climatology for tropospheric ozone. J geophys res 104:16115–16149
Logan JA, Kirchoff VWJH (1986) Seasonal variations of tropospheric ozone at Natal, Brazil. J geophys res 91:7875–7881
Logan JA, Megretskaia IA, Miller AJ, Tiao GC, Choi D, Zhang L, Bishop L, Stolarski R, Labow GJ, Hollandsworth SM, Bodeker GE, Claude H, DeMuer D, Kerr JB, Tarasick DW, Oltmans SJ, Johnson B, Schmidlin F, Staehelin J, Viatte P, Uchino O (1999) Trends in the vertical distribution of ozone: A comparison of two analyses of ozonesonde data. J geophys res 104:26373–26399
Lohmann U, Feichter J (1997) Impact of sulphate aerosols on albedo and lifetime of clouds: A sensitivity study with the ECHAM4 GCM. J geophys res 102:13685–13700
Lohmann U, Feichter J, Chuang C, Penner J (1999a) Prediction of the number of cloud droplets in the ECHAM GCM. J geophys res 104:9169–9198
Lohmann U, von Salzen K, McFarlane N, Leighton HG, Feichter J (1999b) Tropospheric sulfur cycle in the Canadian general circulation model. J geophys res 104:26833–26858

Lohmann U, Feichter J, Penner JE, Leaitch R (2000) Indirect effect of sulphate and carbonaceous aerosols: A mechanistic treatment. J geophys res 105:12193–12206

Loreto F, Cicciolo P, Brancaleoni E, Cecinato A, Frattoni M, Sharkey TD (1996) Different sources of reduced carbon contribute to form three classes of terpenoid emitted by *Quercus ilex* L. leaves. P natl acad sci usa 93:9966–9969

Louis JF (1979) A parametric model of vertical eddy fluxes in the atmosphere. Bound-lay meteorol 17:187–202

Lovejoy ER, Hanson DR (1995) Measurement of the kinetics of reactive uptake by submicron sulfuric acid particles. J chem phys 99:2080–2087

Lovelock JE (1975) Natural halocarbons in the air and in the sea. Nature 256:193–194

Lovelock JE (1979) Gaia: A new look at life on Earth. Oxford University Press, Oxford

Lovelock JE (1988) The ages of gaia. W.W. Norton and Company, New York

Lovelock JE, Maggs RJ, Rasmussen RA (1972) Atmospheric dimethyl sulphide and the natural sulphur cycle. Nature 237:452–453

Lowe DC, Brenninkmeijer CAM, Manning MR, Sparks RJ, Wallace GW (1988) Radiocarbon determination of atmospheric methane at Baring Head, New Zealand. Nature 372:522–525

Lübkert-Alcano B, Krzyzanowski M (1995) Estimate of health impacts of acidifying air pollutants and tropospheric ozone in Europe. Water air soil poll 85:167–176

Ludwig J, Marufu LT, Huber B, Andreae MO, Helas G (2002) Combustion of biomass fuels in developing countries – A major source of atmospheric pollutants. Environ pollut (in press)

Luizao F, Matson P, Livingston G, Luizao R, Vitousek P (1989) Nitrous oxide flux following tropical land clearing. Global biogeochem cy 3:281–285

Luke W, Dickerson R (1987) Flux of reactive nitrogen compounds from eastern North America to the western Atlantic Ocean. Global biogeochem cy 1:329–343

Lurmann FW, Loyd AC, Atkinson R (1986) A chemical mechanism for use in long range transport/acid deposition computer modeling. J geophys res 91:10905–10936

Lyster PM, Cohn SE, Menard R, Chang L-P, Lin S-J, Olsen R (1997) An implementation of a two-dimensional filter for atmospheric chemical constituent assimilation of massively parallel computers. Mon weather rev 125:1674

Machida T, Nakazawa T, Fujii Y, Aoki S, Watanabe O (1995) Increase in the atmospheric nitrous oxide concentration during the last 250 years. Geophys res lett 22:2921–2924

Mackay GI, Karecki DR, Schiff HL (1996) Tunable diode laser absorption measurements of H_2O_2 and HCHO during the Mauna Loa Observatory Photochemistry Experiment. J geophys res 101:14721–14728

Madronich S, Calvert JG (1989) The NCAR master mechanism of the gas phase chemistry-Version 2.0. NCAR Technical note NCAR/TN-333+STR, National Center for Atmospheric Research, Boulder, CO

Madronich S, Granier C (1992) Impact of recent total ozone changes on tropospheric ozone photodissociation, hydroxyl radicals, and methane trends. Geophys res lett 19:465–467

Maenhaut W (1990) Recent advances in nuclear and atomic spectrometric techniques for trace element analysis: A new look at the position of PIXE. Nucl instrum methods B49:518–532

Maenhaut W (1992) Trace element analysis of environmental samples by nuclear analytical techniques. I j pixe 2:609–635

Maher BA, Hounslow MW (1999) Paleomonsoons II: Magnetic records of aeolian dust in Quaternary sediments of the Indian Ocean. In: Maber B, Thompson R (eds) Quaternary climates, environments and magnetism. Cambridge University Press, Cambridge, pp 126–162

Mahowald NM, Rasch PJ, Prinn RG (1995) Cumulus parameterizations in chemical transport models. J geophys res 100:26173–26189

Mahowald NM, Rasch PJ, Eaton BE, Whittlestone S, Prinn RG (1997a) Transport of ^{222}Rn to the remote troposphere using MATCH and assimilated winds from ECMWF and NCEP/NCAR. J geophys res 102:28139–28151

Mahowald NM, Prinn RG, Rasch PJ (1997b) Deducing CCl_3F emissions using an inverse method and chemical transport models with assimilated winds. J geophys res 102:28153–28168

Mak JE, Brenninkmeijer CAM, Manning MR (1992) Evidence for a missing carbon-monoxide sink based on tropospheric measurements of (CO)-C-14. Geophys res lett 19:1467–1470

Mak JE, Brenninkmeijer CAM, Tamaresis J (1994) Atmospheric (CO)-C-14 observations and their use for estimating carbon monoxide removal rates. J geophys res 99:22915–22922

Mäkelä J (1992) Studies on irradiation induced aerosol particle formation processes in air containing sulphur dioxide. Ph.D. thesis, University of Helsinki

Malin G, Turner S, Liss PS, Holligan P, Harbour D (1993) Dimethyl sulfide and dimethylsulphoniopropionate in the Northeast Atlantic during the summer coccolithophore bloom. Deep-sea res 40:1487–1508

Malin G, Wilson WH, Bratbak G, Liss PS, Mann NH (1998) Elevated production of dimethylsulfide resulting from viral infection of cultures of *Phaeocystis pouchetii*. Limnol oceanogr 43:1389–1393

Mallina RV, Wexler AS, Rhoads KP, Johnston MV (2000) High speed particle beam generation: A dynamic focusing mechanism for selecting ultrafine particles. Aerosol sci tech 33:87–104

Malm WC, Molenar JV, Eldred RA, Sisler JF (1996) Examining the relationship among atmospheric aerosols and light scattering and extinction in the Grand Canyon area. J geophys res 101:19251–19265

Mann ME, Bradley RS, Hughes MK (1998) Global-scale temperature patterns and climate forcing over the past six centuries. Nature 392:779–787

Manning MR, Lowe DC, Melhuish WH, Sparks RJ, Wallace G, Brenninkmeijer CAM, McGill RC (1990) The use of radiocarbon measurements in atmospheric studies. Radiocarbon 32:37–58(1)

Mannion AM (1992) Acidification and eutrophication. In: Mannion AM, Bowlby BR (eds) Environmental issues in the 1990s. J. Wiley, Chichester, pp 177–195

Mansfield T, Goulding K, Sheppard L (1998) Major biological issues resulting from anthropogenic disturbance of the nitrogen cycle. In: Third New Phytologist Symposium. New Phytol., Lancaster University, UK, pp 1–229

Marenco A, Gouget H, Nedelec P, Pages JP, Karcher F (1994) Evidence of a long-term increase in tropospheric ozone from Pic du Midi data series: Consequences: Positive radiative forcing. J geophys res 99:16617–16632

Marenco A, Thouret V, Nedelec P, Smit H, Helten M, Kley D, Karcher F, Simon P, Law K, Pyle J, Poschmann G, Von RWrede, Hume C, Cook T (1998) Measurement of ozone and water vapor by Airbus in-service aircraft: The MOZAIC airborne program, an overview. J geophys res 103:25631–25642

Mari C, Jacob DJ, Bechtold P (2000) Transport and scavenging of soluble gases in a deep convective cloud. J geophys res 105(D17):22255–22267

Marshall SF, Covert DS, Charlson RJ (1995) Relationship between asymmetry factor and backscatter ratio: Implications for aerosol climate forcing. Appl optics 34:6306–6311

Marti J (1990) Diurnal variation in the undisturbed continental aerosol: Results from a measurement program in Arizona. Atmos res 25:351–362

Marticorena B, Bergametti G (1995) Modeling the atmospheric dust cycle: 1. Design of a soil-derived dust emission scheme. J geophys res D100:16415–16430

Martin GM, Jonas PR (1997) A simple-model study of the interaction between cumulus and stratocumulus clouds in the marine boundary layer. Q j roy meteor soc 123:1199–1225

Martin JH, Fitzwater SE (1988) Iron deficiency limits phytoplankton growth in the northeast Pacific subarctic. Nature 331:341–343

Martin JH, Coale KH, Johnson KS, Fitzwater SE, Gordon RM, Tanner SJ, Hunter CN, Elrod VA, Nowicki JL, Coley TL, Barber RT, Lindley S, Watson AJ, Van Scoy K, Law CS, Liddicoat MI, Ling R, Stanton T, Stockel J, Collins C, Anderson A, Bidigare R, Ondrusek M, Latasa M, Millero FJ, Lee K, Yao W, Zhang JZ, Friederich G, Sakamoto C, Chavez F, Buck K, Kolber Z, Greene R, Falkowski P, Chisholm SW, Hoge F, Swift R, Yungel J, Turner S, Nightingale P, Hatton A, Liss P, Tindale NW (1994) Testing the iron hypothesis in ecosystems of the equatorial Pacific Ocean. Nature 371:123–129

Martinerie P, Brasseur GP, Granier C (1995) The chemical composition of ancient atmospheres: A model study constrained by ice core data. J geophys res 100:14291–14304

Martonchik JV (1997) Determination of aerosol optical depth and land surface directional reflectances using multiangle imagery. J geophys res 102:17015–17022

Marufu L, Dentener F, Lelieveld J, Andreae MO, Helas G (2000) Photochemistry of the African troposphere: The influence of biomass-burning emissions. J geophys res 105:14513

Massman WJ (1993) Partitioning ozone fluxes to sparse grass and soil and the inferred resistances to dry deposition. Atmos environ 27:167

Mather JH, Stevens PS, Brune WH (1997) OH and HO_2 measurements using laser-induced fluorescence. J geophys res 102:6427–6436

Mathias-Maser S (1998) Primary biological particles: Their significance, sources, sampling methods and size distribution in the atmosphere. In: Harrison RM, van Grieken R (eds) Atmospheric particles. J. Wiley, Chichester, pp 349–368

Matrai PA, Keller MD (1993) Dimethylsulfide in a large-scale cocolithophore bloom in the Gulf of Maine. Cont shelf res 13:831–843

Matrai PA, Keller MD (1994) Total organic sulphur and dimethylsulfoniopropionate in marine phytoplankton: Intracellular variations. Mar biol 119:61–68

Matrai PA, Vernet M (1997) Dynamics of the vernal bloom in the marginal ice zone of the Barents Sea: Dimethylsulfide and dimethylsulfoniopropionate budgets. J geophys res 102:22965–22979

Matrai PA, Balch WM, Cooper DJ, Saltzman ES (1993) Ocean color and atmospheric DMS: On their mesoscale variability. J geophys res 98:23469–23476

Matrai PA, Vernet M, Hood R, Jennings A, Saemundsdottir S, Brody E (1995) Light-dependent production of DMS and carbon incorporation by polar strains of *Phaeocystis* spp.. Mar biol 124:157–167

Matson PA, Harriss RC (1995) Biogenic trace gases: Measuring emissions from soil and water. Blackwell Science, London

Matson PA, Ojima DS (1990) Terrestrial biosphere exchange with global atmospheric chemistry. Stockholm

Matson PA, Vitousek PM, Schimel DS (1989) Regional extrapolation of trace gas flux based on soils and ecosystems in exchange of trace gases between terrestrial ecosystems and the atmosphere. In: Andreae MO, Schimel DS (eds) Dahlem Konferenzen. J. Wiley, Chichester, pp 97–108

Matson PA, Billow C, Hall S, Zachariassen J (1996) Fertlization practices and soil variations control nitrogen oxide emissions from tropical sugar cane. J geophys res 101(13):18533–18545

Matson PA, Naylor R, Ortiz-Monasterio I (1998) Integration of environmental agronomic and economic aspects of fertilizer management. Science 280:112–115

Matsueda H, Innoue HY (1996) Measurement of atmospheric CO_2 and CH_4 using a commercial airliner from 1993 to 1994. Atmos environ 30:1647–1655

Matsumi Y, Comes FJ, Hancock G, Hofzumahaus A, Hynes AJ, Kawasaki M, Ravishankara AR (2002) Quantum yields for production of $O(^1D)$ in the ultraviolet photolysis of ozone: Recommandation based on evaluation of laboratory data. J geophys res (in press)

Matthews E (1993) Wetlands. In: Khalil MAK (ed) Atmospheric methane: Sources, sinks, and role in global change. Springer-Verlag, Berlin, pp 314–359

Matthews E, Fung I (1987) Methane emissions from natural wetlands: Global distribution, area, and environmental characteristics of sources. Global biogeochem cy 1:61–86

Matthews RB, Wassmann R, Arah J (2000) Using a crop-soil simulation model and GIS techniques to assess methane emissions from rice fields in Asia. I. Model development. Nutr cycl agroecosys 58:141–159

Mauldin R, Tanner D, Eisele F (1999) Measurements of OH during PEM-Tropics A. J geophys res 104:5817–5827

Mauldin RL, Madronich S, Flocke SJ, Eisele FL, Frost GJ, Prevot ASH (1997) New insights on OH: Measurements around and in clouds. Geophys res lett 24:3033–3036

Mauldin RL III, Frost GJ, Chen G, Tanner DJ, Prevot ASH, Davis DD, Eisele FL (1998a) OH measurements during the First Aerosol Characterization Experiment (ACE-1): Observations and model comparisons. J geophys res 103:16713–16729

Mauldin RL III, Tanner DJ, Eisele FL (1998b) A new chemical ionization mass spectrometer technique for the fast measurement of gas phase nitric acid in the atmosphere. J geophys res 103:3361–3367

Mayewski PA, Lyons WB, Spencer MJ, Twickler MS, Buck CF, Whitlow S (1990) An ice-core record of atmospheric response to anthropogenic sulphate and nitrate. Nature 346:554–556

Maynard AD, Brown LM (1992) Electron energy loss spectroscopy of ultrafine particles in the scanning transmission electron microscope. J aerosol sci 23:S433–S436

Maynard RL, Waller R (1999) Carbon monoxide. In: Holgate ST, Samet JM, Koren HS, Maynard RL (eds) Air pollution and health. Academic Press, London, pp 749–796

McCormick MP, Zawodny JM, Veiga RE, Larson JC, Wang PH (1989) An overview of SAGE I and SAGE II ozone measurements. Planet space sci 37:1567–1586

McGillis WR, Edson JB, Wanninkhof R (1999) Direct air-sea flux measurement of CO_2 over the North Atlantic Ocean and the comparison to indirect methods. In: 2nd Int. Symp. CO_2 in the Oceans. Tsukuba, Japan

McGillis WR, Dacey JWH, Frew NM, Bock EJ, Nelson BK (2000) Water-air flux of dimethylsulfide. J geophys res 105:1187–1193

McGillis WR, Edson JB, Hare JE, Fairall CW (2001) Direct covariance air-sea CO_2 fluxes. J geophys res 106:16729–16746

McGraw R (1997) Description of aerosol dynamics by the quadrature method of moments. Aerosol sci tech 27:255–265

McGraw R, Huang PI, Schwartz SE (1995) Description of atmospheric aerosols from moments of the particle size distribution. Geophys res lett 22:2929–2932

McInnes LM, Bergin MH, Ogren JA, Schwartz SE (1998) Apportionment of light scattering and hygroscopic growth to aerosol composition. Geophys res lett 25:513–516

McMurry PH (2000) A review of atmospheric aerosol measurements. Atmos environ 34:1959–1999

McNeely JA, Gadgil M, Leveque C, Padoch C, Redford K (1995) Human influences on biodiversity. In: Heywood VH, Watson RT (eds) Global biodiversity assessment. Cambridge University Press, Cambridge

Meehl GA, Washington W, Erickson DJ III, Briegleb BP, Jaumann P (1996) Climate change from increased CO_2 and the direct and indirect effects of sulfate aerosols. Geophys res lett 23:3755–3758

Mellouki A, Jourdain JL, Le Bras G (1988) Discharge flow study of the $CH_3S + NO_2$ reaction mechanism using $Cl + CH_3SH$ as the CH_3S source. Chem phys lett 148:231–236

Menaut J-C, Abbadie L, Lavenu F, Loudjani P, Podaire A (1991) Biomass burning in West African savannas. In: Levine JS (ed) Global biomass burning: Atmospheric, climatic, and biospheric implications. MIT Press, Cambridge, Mass., pp 133–142

Meng Z, Dabdub D, Seinfeld JH (1998) Size-resolved and chemically resolved model of atmospheric aerosol dynamics. J geophys res 103:3419–3436

Mentel T, Sohn M, Wahner A (1999) Nitrate effect in the heterogeneous hydrolysis of dinitrogen pentoxide on aqueous aerosols. Phys chem chem phys 1:5451–5457

Merrill JT, Newell RE, Bachmeier AS (1997) A meteorological overview for the Pacific Exploratory Mission-West Phase B. J geophys res 102:28241–28254

Metzger S, Dentener F, Lelieveld J (1999) Aerosol multiphase equilibrium composition: A parameterization for global modeling, Institute for Marine and Atmospheric Research Utrecht (IMAU), Princetonplein 5

Mihelcic D, Volz-Thomas A, Pätz HW, Kley D, Mihelcic M (1990) Numerical analysis of ESR spectra from atmospheric samples. J atmos chem 11:271–297

Mihelic D, Klemp D, üsgen P, Pätz HW, Volz-Thomas A (1993) Simultaneous measurements of peroxy and nitrate radicals at Schauinsland. J atmos chem 16:313–335

Milich L (1999) The role of methane in global warming: Where might mitigation strategies be focused? Global environ chang 9:179–201

Miller RL, Tegen I (1998) Climate response to soil dust aerosols. J climate 11:3247–3267

Minami K, Neue HU (1994) Rice paddies as a methane source. In: White DH, Howden SM (eds) Climate change: Significance for agriculture and forestry-system approaches arising from an IPPCC meeting. Kluwer Academic, Dordrecht, pp 13–26

Mishchenko MI, Geogdzhayev IV, Cairns B, Rossow WB, Lacis AA (1999) Aerosol retrievals over the ocean by use of channels 1 and 2 AVHRR data: Sensitivity analysis and preliminary results. Appl optics 38:7325–7341

Miyoshi A, Hatakeyama S, Washida N (1994) OH radical-initiated photooxidation of isoprene: An estimate of global CO production. J geophys res 99:18779–18787

Mo KC, Higgins RW (1996) Large scale atmospheric moisture transport as evaluated in the NCEP/NCAR and NASA/DAO reanalyses. Mon weather rev 1531–1545

Monahan EC (1986) The ocean as a source of atmospheric particles. In: Buat-Menard P (ed) The role of air-sea exchange in geochemical cycling. Reid, Dordrecht, pp 129–136

Monks PS (2000) A review of the observations and origins of the spring ozone maximum. Atmos environ 34:3545–3561

Monks PS, Carpenter LJ, Penkett SA, Ayers GP (1996) Night-time peroxy radical chemistry in the remote marine boundary layer over the Southern Ocean. Geophys res lett 23:535–538

Montzka SA, Spivakovsky CM, Butler JH, Elkins JW, Lock LT, Mondeel DJ (2000) New observational constraints for atmospheric hydroxyl on global and hemispheric scales. Science 288(5465):500–503

Mooney HA, Canadell J, Chapin FS III, Ehleringer JR, Körner CH, McMurtrie RE, Parton WJ, Pitelka LF, Schulze E-D (1999) Ecosystem physiology responses to global change. In: Walker B, Steffen W, Canadell J, Ingram J (eds) The terrestrial biosphere and global change. Implications for natural and managed ecosystems. Cambridge University Press, Cambridge, pp 141–189

Moore RM, Groszko W (1999) Methyl iodide distribution in the ocean and fluxes to the atmosphere. J geophys res 104: 11163–11171

Moore RM, Webb M (1996) The relationships between methyl bromide and chlorophyll a in high latitude ocean waters. Geophys res lett 23:2951–2954

Moore RM, Milley JE, Chatt A (1984) The potential for biological mobilisation of trace elements from aeolian dust in the ocean and its importance in the case of iron. Oceanol acta 7:221–228

Moore RM, Tokarczyk R, Tait VK, Poulin M, Geen C (1995) Marine phytoplankton as a natural source of volatile organohalogens. In: Grimvall A, deLeer EWB (eds) Naturally-produced organohalogens. Kluwer Academic, Dordrecht, pp 283–294

Moore RM, Groszko W, Niven SJ (1996) Ocean-atmosphere exchange of methyl chloride: Results from NW Atlantic and Pacific Ocean studies. J geophys res 101:28529–28538

Mopper K, Zika RG (1987) Free amino acids in marine rains: Evidence for oxidation and potential role in nitrogen cycling. Nature 325:246–249

Morawska L, Thomas S, Bofinger ND, Wainwright D, Neale D (1998) Comprehensive characterisation of aerosols in a subtropical urban atmosphere: Particle size distribution and correlation with gaseous pollutants. Atmos environ 32:2461–2478

Mosier AR, Kroeze C (1999) Contribution of agroecosystems to the global atmospheric N_2O budget. In: Proceedings of International Workshop on Reducing N_2O Emission from Agroecosystems.Banff, Canada

Mosier AR, Duxbury JM, Freney JR, Heinmeyer O, Minami K, Johnson DE (1998) Mitigating agricultural emissions of methane. Climatic change 40:39–80

Moxim WJ, Levy H II, Kasibhatla PS (1996) Simulated global tropospheric PAN: Its transport and impact on NO_x. J geophys res 101:12621–12638

Mozurkewich M, Calvert JG (1988) Reaction probability of N_2O_5 on aqueous aerosols. J geophys res 93:15889–15896

Mudway IS, Kelly FJ (2000) Ozone and the lung: A sensitivity issue. Mol aspects med 21:1–48

Müller J-F (1992) Geographical distribution and seasonal variation of surface emissions and deposition velocities of atmospheric gases. J geophys res 97:3787–3804

Müller J-F, Brasseur G (1995) IMAGES: A three-dimensional chemical transport model of the global troposphere. J geophys res 100:16445–16490

Murphy DM, Fahey DW (1987) Mathematical treatment of the wall loss of a trace species in denuder and catalytic-converter tubes. Anal chem 59(23):2753–2759

Murphy DM, Thomson DS (1995) Laser ionization mass spectroscopy of single aerosol particles. Aerosol sci tech 22:237–249

Murphy DM, Anderson JR, Quinn PK, McInnes LM, Brechtel FJ, Kreidenweis SM, Middlebrook AM, Posfai M, Thomson DS, Buseck PR (1998a) Sea-salt particles and aerosol radiative properties in the Southern Ocean marine boundary layer. Nature 392:62–65

Murphy DM, Thomson DS, Mahoney MJ (1998b) *In situ* measurements of organics, meteoritic material, mercury and other elements in aerosols at 5 to 19 kilometers. Science 282:1664–1669

Nagao I, Matsumoto K, Tanaka H (1999) Sunrise ozone destruction found in the subtropical marine. Geophys res lett 26:3377–3380

Najjar RG, Erickson DJ III, Madronich S, Sonntag C (1995) Modeling the air-sea fluxes of gases formed from the decomposition of dissolved organic matter: Carbonyl sulfide and carbon monoxide. In: Zepp RG (ed) Role of nonliving organic matter in the earth's carbon cycle. J. Wiley, New York, pp 107–132

Nakajima T, Higurashi A, Kawamoto K, Penner JE (2001) A possible correlation between satellite-derived cloud and aerosol microphysical parameters. Geophys res lett 28:1171–1174

Naqvi SW, Yoshinari T, Jayakumar A, Devol AH, Codispoti LA (2000) Seasonal emissions of isotopically heavy N_2O from the suboxic Arabian Sea. EOS 80(49):259

Nastrom GD (1979) Ozone in the upper troposphere from the GASP measurements. J geophys res 84:3638–3688

National Acad. of Sciences (1985) Report on acid rain formation. National Academy Press, Washington, D.C.

National Research Council (1986) Acid deposition: Long-term trends. National Academy Press, Washington, D.C.

National Research Council (1992) Rethinking the ozone problem in urban and regional air pollution. National Academy Press, Washington, D.C.

National Research Council (1999a) Global environmental change: Research pathways for the next decade. Edited by Committee on Global Change Research, National Academy Press, Washington, D.C.

National Research Council (1999b) Our common journey: A transition toward sustainability. Edited by Board on Sustainable Development, National Academy Press, Washington, D.C.

National Research Council (2001) Global air quality. An imperative for long-term observational strategies. National Academy Press

Neil DO, Gordley LL, Marshall BT, Sachse GW (2000) Tropospheric carbon monoxide measurements from geostationary orbit. In: (ed) Satellite remote sensing of clouds and the atmosphere VI. SPIE, Barcelona

Neilson A, Lewin R (1974) The uptake and utilization of organic carbon by algae: An essay in in comparative biochemistry. Phycologia 13:227–264

Nemeruyk GE (1970) Migration of salts into the atmosphere during transpiration. Sov plant physiol+ 17:560–566

Nemesure S, Wagener R, Schwartz SE (1995) Direct shortwave forcing of climate by anthropogenic sulfate aerosol: Sensitivity to particle size, composition, and relative humidity. J geophys res 100:26105–26116

Nemitz E, Fowler D, Dorsey JR, Theobold MR, McDonald A, Gallagher MW (2000) Direct measurements of size-dependent particle fluxes above a city: Interpretation and source apportion. In: Sixth International Conference on Air-Surface Exchange of Gases and Particles. Edinburgh, Scotland

Nenes A, Pilinis C, Pandis SN (1998) Isoroppia: A new thermodynamic equilibrium model for multicomponent inorganic aerosols. Aquat geochem 4:123–125

Nenes A, Flagan RC, Chuang PY, Seinfeld JH (2001) A theoretical analysis of cloud condensation nucleus (CCN) instruments. J geophys res 106(D4):3449–3474

Nepstad DC, erissimo A, Alencar A, Nobre C, Lima E, Lefebvre P, Schlesinger P, Potter C, Moutinho P, Mendoza E, Cochrane M, Brooks V (1999) Large-scale impoverishment of Amazonian forests by logging and fire. Nature 398:505–508

Neue HU (1997) Fluxes of methane from rice fields and potential for mitigation. Soil use manage 13:258–267

Neue HU, Boonjawat J, Melillo J (1998) Methane emission from ricefields. In: Galloway J (ed) Asian change in the context of global change. Cambridge University Press, Cambridge, pp 187–209

Neusüß C, Weise D, Birmili W, Wex H, Wiedensohler A, Covert D (2000) Size-segregated chemical, gravimetric and number distribution-derived mass closure of the aerosol in Sagres, Portugal during ACE-2. Tellus 52B:169–184

Nevison C, Weiss R, Erickson DJ III (1995) Global oceanic emissions of nitrous oxide. J geophys res 100:15809–15820

Newman M, Sardeshmukh PD, Bergman J (2000) An assessment of the NCEP, NASA and ECMWF reanalyses over the tropical west pacific warm pool. B am meteorol soc 81:41–48

Newman SM, Lane IC, Orr-Ewing AJ, Newnham DA, Ballard J (1999) Integrated absorption intensity and Einstein coefficients for the O_2 $a^1\Delta_g$–$X^3\Sigma_g$–(0,0) transition: A comparison of cavity ringdown and high resolution Fourier transform spectroscopy with a long-path absorption cell. J chem phys 110:10749–10757

Nganga DA, Minga B, Cros B, Biona CB, Fishman J, Grant WB (1996) The vertical distribution of ozone measured at Brazzaville, Congo during TRACE-A. J geophys res 101:24095–24103

Nguyen BC, Gaudry A, Bonsang B, Lambert G (1978) Reevaluation of the role of dimethyl sulphide in the sulphur budget. Nature 275:637–639

Nguyen BC, Belviso S, Mihalopoulos N (1988) Dimethyl sulfide production during natural phytoplankton blooms. Mar chem 24:133–141

Nicholson KW (2001) Wind tunnel measurements of particle deposition to building materials. Atmos environ

Nickus U, Kuhn M, Baltensperger U, Delmas R, Gäggeler H, Kasper A, Kromp-Kolb H, Maupetit F, Novo A, Pichlmayer F, Preunkert S, Puxbaum H, Rossi G, Schöner W, Schwikowski M, Seibert P, Staudinger M, Trockner V, Wagenbach D, Winiwarter W (1997) SNOSP: Ion deposition and concentration in high alpine snow packs. Tellus 49B:56–71

Niewiadomski M (1989) Sulphur dioxide and sulphate in a three-dimensional field of convective clouds: Numerical simulations. Atmos environ 23:477–487

Nightingale PD, Malin G, Liss PS (1995) Production of chloroform and other low-molecular-weight halocarbons by some species of macroalgae. Limnol oceanogr 40:680–689

Nightingale PD, Malin G, Law CS, Watson AJ, Liss PS, Liddicoat MI, Boutin J, Upstill-Goddard RC (2000a) *In situ* evaluation of air-sea gas exchange parameterizations using novel conservative and volatile tracers. Global biogeochem cy 14(1):373–387

Nightingale PD, Liss PS, Schlosser P (2000b) Measurements of air-sea gas transfer during an open ocean algal bloom. Geophys res lett

Nisbet EG (1992) Sources of atmospheric CH_4 in early postglacial time. J geophys res 97:12859–12867

Noble CA, Prather KA (1996) Real-time measurement of correlated size and composition profiles of individual atmospheric aerosol particles. Environ sci technol 30:2667–2680

Noone KJ, Hansson H-C, Mallant KAM (1992a) Droplet sampling from crosswinds: An inlet efficiency calibration. J aerosol sci 23:153–164

Noone KJ, Ogren JA, Hallberg A, Heintzenberg J, Ström J, Hansson H-C, Svenningsson IB, Wiedensohler A, Fuzzi S, Facchini MC, Arends BG, Berner A (1992b) Changes in aerosol size- and phase distributions due to physical and chemical processes in fog. Tellus 44B:489–504

Noone KJ, Öström E, Ferek RJ, Garrett T, Hobbs PV, Johnson DW, Taylor JP, Russell LM, Flagan RC, Seinfeld JH, O'Dowd CD, Smith MH, Durkee PA, Nielson K, Hudson JG, Pockanly RA, DeBock L, Grieken RV, Gasparovic RF, Brooks I (2000) A case study of ships forming and not forming tracks in moderately polluted clouds. J atmos sci

Noordkamp DJB, Schotten M, Gieskes WWC, Forney LJ, Gottschal JC, van Rijssel M (1998) High acrylate concentrations in the mucous of *Phaeocystis globosa* colonies. Aquat microb ecol 16:45–52

Nouchi I, Mariko S, Aoki K (1990) Mechanism of methane transport from the rhizosphere to the atmosphere through rice plants. Plant physiol 94:59–66

Novakov T, Hegg DA, Hobbs PV (1997) Airborne measurements of carbonaceous aerosol on the East Coast of the United States. J geophys res 102:30023–30030

Novelli PC, Masarie KA, Lang PM (1998) Distributions and recent changes of carbon monoxide in the lower troposphere. J geophys res 103:19015–19033

O'Dowd CD, Davison B, Lowe JA, Smith MH, Harrison RM, Hewitt CN (1997a) Biogenic sulphur emissions and inferred sulphate CCN concentrations in and around Antarctica. J geophys res 102:12839–12854

O'Dowd CD, Smith MH, Consterdine IE, Lowe JA (1997b) Marine aerosol, sea-salt and the marine sulphur cycle: A short review. Atmos environ 31:73–80

O'Dowd CD, Lowe J, Smith MH, Kaye AD (1999) The relative importance of sea-salt and nss-sulphate aerosol to the marine CCN population: An improved multi-component aerosol-droplet parameterisation. Q j roy meteor soc 125:1295–1313

O'Dowd CD, Lowe JA, Clegg SL, Clegg N, Smith MH (2000) Modelling heterogeneous sulphate production in maritime stratiform clouds: Model development and application. J geophys res 150:7143–7160

O'Halloran I (1993) Ammonia volatilization from liquid hog manure: Influence of aeration and trapping systems. Soil sci soc am J 57:1300–1303

Odum JR, Hoffmann T, Bowman F, Collins D, Flagan RC, Seinfeld JH (1996) Gas/particle partitioning and secondary organic aerosol yields. Environ sci technol 30:2580–2585

Odum JR, Jungkamp PW, Griffin RJ, Flagan RC, Seinfeld JH (1997) The atmospheric aerosol-forming potential of whole gasoline vapor. Science 276:96–99

Ogawa T, Miyata A (1985) Seasonal variation of the tropospheric ozone: A summer minimum in Japan. J meteorol soc jpn 63(5):937–946

Ogawa T, Shimoda H, Hayashi M, Imasu R, Ono A, Nishinomiya S, Kobayashi H (1994) IMG, interferometric measurements of greenhouse gases from space. Adv space res 14:25–28

Ogren J (1995) *In situ* observations of aerosol properties. In: Charlson RJ, Heintzenberg J (eds) Aerosol forcing of climate. J. Wiley, New York, pp 215–226

Ogren JA, Charlson RJ (1992) Implications for models and measurements of chemical inhomogeneities among cloud droplets. Tellus 44B:208–225

Ojima D, Mosier A, Del Grosso S, Parton WJ (2000) TRAGNET analysis and synthesis of trace gas fluxes. Global biogeochem cy 14(4):995–997

Okazaki K, Wiener RW, Willeke K (1987) The combined effect of aspiration and transmission on aerosol sampling accuracy for horizontal isoaxial sampling. Atmos environ 21:1181–1185

Olivier JGJ, Bouwmann AF, Van der Maas CWM, Berdowski JJM, Veldt C, Bloos JPJ, Visschedijk AJH, Zandveld PYJ, Haverlag JL (1996) Description of EDGAR version 2.0: A set of global emission inventories of greenhouse gases and ozone-depleting substances for all anthropogenic and most natural sources on a per country basis and on a $1 \times 1°$ grid. National Institute of Public Health and the Environment, Bilthoven, The Netherlands

Olson J, Prather M, Berntsen T, Carmichael G, Chatfield R, Connell P, Derwent R, Horowitz L, Jin SX, Kanakidou M, Kasibhatla P, Kotamarthi R, Kuhn M, Law K, Penner J, Perilski L, Sillman S, Stordal F, Thompson A, Wild O (1997) Results from the Intergovernmental Panel on Climatic Change photochemical model intercomparison (Photocomp). J geophys res 102:5979–5991

Olson JR, Fishman J, Kirchhoff VWJH, Nganga D, Cros B (1996) Analysis of the distribution of ozone over the southern Atlantic region. J geophys res 101:24083–24094

Olson R, Simonson TR (1949) Rates of ionic reactions in aqueous solution. J chem phys 17:1167–1173

Oltmans SJ, Levy H II (1992) Seasonal cycle of surface ozone over the western north Atlantic. Nature 358:392–394

Oltmans SJ, Levy H II (1994) Surface O_3 measurements from a global network. Atmos environ 28:9–24
Oltmans SJ, Schnell RC, Sheridan PJ, Peterson RE, Li SM, Winchester JW, Tans PP, Sturges WT, Kahl JD, Barrie LA (1989) Seasonal surface ozone and filterable bromine relationship in the high Arctic. Atmos environ 23:2431–2441
Oltmans SJ, Lefohn AS, Scheel HE, Harris JM, Levy H II, Galbally IE, Brunke E-G, Meyer CP, Lathrop JA, Johnson BJ, Shadwick DS, Cuevas E, Schmidlin FJ, Tarasick DW, Claude H, Kerr JB, Uchino O, Mohnen V (1998) Trends of ozone in the troposphere. Geophys res lett 25(2):139–142
Orlando JJ, Tyndall GS, Bilde M, Ferronato C, Wallington TJ, Vereecken L, Peeters J (1998) Laboratory and theoretical study of the oxy radicals in the OH- and Cl-initiated oxidation of ethene. Phys chem chem phys 102:8116–8123
Ortiz-Monasterio JL, Matson PA, Panek J, Naylor RL (1996) Nitrogen fertiliser management for N_2O and NO emissions in Mexican irrigated wheat. In: Transactions 9th Nitrogen Workshop. Braunschweig Tech. Univ., Braunschweig, pp 531–534
Otter LB (1992) Soil carbon fractionation of sand and clay soils under different burning regimes. Honours Dissertation, University of the Witwatersrand, Johannesburg, South Africa
Otter LB, Scholes MC (2000) Methane sources and sinks in a periodically flooded South African savanna. Global biogeochem cy 14:97
Otter LB, WenXing XY, Scholes MC, Meixner FX (1999) Nitric oxide emissions from a southern African savanna. J geophys res 104(D15):18471–18485
Ouimette JR, Flagan RC (1982) The extinction coefficient of multicomponent aerosols. Atmos environ 16:2405–2419
Oum KW, Lakin MJ, Finlayson-Pitts BJ (1998) Bromine activation in the troposphere by the dark reaction of O_3 with seawater ice. Geophys res lett 25:3923–3926
Owens NJP, Law CS, Mantoura RFC, Burkill PH, Llewellyn CA (1991) Methane flux to the atmosphere from the Arabian Sea. Nature 354:293–296
Padro J (1996) Summary of ozone dry deposition measurements and model estimates. Atmos environ 30:2363
Paerl H (1995) Coastal eutrophication in relation to atmospheric nitrogen deposition: Current perspectives. Ophelia 41:237–259
Paerl H (1997) Coastal eutrophication and harmful algal blooms: Importance of atmospheric deposition and groundwater as "new" nitrogen and other nutrient sources. Limnol oceanogr 42:1154–1165
Paerl H, Fogel M (1994) Isotopic characterization of atmospheric nitrogen inputs as sources of enhanced primary production in coastal Atlantic Ocean waters. Mar biol 119:635–645
Paerl HW (1985) Enhancement of marine primary production by nitrogen-enriched acid rain. Nature 315:747–749
Paerl HW, Whitall DR (1999) Anthropogenically-derived atmospheric nitrogen deposition, marine eutrophication and harmful algal bloom expansion: Is there a link? Ambio 28:307–311
Paige RC, Plopper CG (1999) Acute and chronic effects of ozone in animal models. In: Holgate ST, Samet JM, Koren HS, Maynard RL (eds) Air pollution and health. Academic Press, London, pp 531–557
Pan HL, Wu WS (1995) Implementing a mass flux convection parameterization package for the NMC medium-range forecast model. National Meteorological Center
Pandis SN, Seinfeld JH, Pilinis C (1990) Chemical composition differences in fog and cloud droplets of different sizes. Atmos environ 24A:1957–1969
Pandis SN, Paulson SE, Seinfeld JH, Flagan C (1991) Aerosol formation in the photooxidation of isoprene and α-pinene. Atmos environ 25A:997–1008
Panek JA, Matson PA, Ortiz-Monasterio I, Brooks P (2000) Distinguishing nitrification and denitrification sources of N_2O in a Mexican wheat system using ^{15}N. Ecol appl 10(2):506–514
Papen H, Butterbach-Bahl K (1999) Three years continuous record of N-Trace gas fluxes from untreated and limed soils of a N-saturated spruce and beech forest ecosystem in Germany: I. N_2O emissions. J geophys res 104(D15):18487–18504
Parrish DD (1993) Carbon monoxide and light alkanes as tropospheric tracers of anthropogenic ozone. In: Niki H, Becker KH (eds) The tropospheric chemistry of ozone in the polar regions. Springer-Verlag, Berlin, pp 155–169
Parrish DD, Norton RB, Bollinger MJ, Liu SC, Murphy PC, Albritton DL, Fehsenfeld FC, Huebert BJ (1986) Measurements of HNO_3 and NO_3-particulates at a rural site in the Colorado Mountains. J geophys res 91:5379–5393
Parrish DD, Buhr M, Trainer M, Norton R, Shimshock J, Fehsenfeld F, Anlauf K, Bottenheim J, Tang Y, Wiebe H, Roberts J, Tanner R, Newman L, Bowersox V, Olszyna K, Bailey E, Rodgers M, Wang T, Berresheim H, Roychowdhury U, Demerjian K (1993a) The total reactive oxidized nitrogen levels and the partitioning between the individual species at six rural sites in Eastern North America. J geophys res 98(D2):2927–2939
Parrish DD, Holloway JS, Trainer M, Murphy PC, Forbes GL, Fehsenfeld FC (1993b) Export of North American ozone pollution to the North Atlantic Ocean. Science 259:1436–1439
Parrish DD, Trainer M, Holloway JS, Yee JE, Warshawsky MS, Fehsenfeld FC, Forbes GL, Moody JL (1998) Relationships between ozone and carbon monoxide at surface sites in the North Atlantic region. J geophys res 103:13357–13376
Parsons DAB, Scholes MC, Scholes RJ, Levine JS (1996) Biogenic NO emissions from savanna soils as a function of fire regime, soil type, soil nitrogen and water status. J geophys res 101(D19):23683–23688
Parton WJ, Mosier AR, Ojima DS, Valentine DW, Schimel DS, Weier K, Kulmala AE (1996) Generalized model for N_2 and N_2O production from nitrification and denitrification. Global biogeochem cy 10:401–412
Parton WJ, Holland EA, del Grosso SJ, Hartman MD, Martin RE, Mosier AR, Ojima DS, Schmiel DS (2001) Generalized model for NO_x and N_2O emissions from soils. J geophys res
Pasteur EC, Mulvaney R, Peel DA, Saltzman S, Whung P-Y (1995) A 340-year record of biogenic sulphur from the Weddell Sea area, Antarctica. Ann glaciol 21:169–174
Patroescu IV, Barnes I, Becker KH, Mihalopoulos N (1999) FT-IR product study of the OH-initiated oxidation of DMS in the presence of NO_x. Atmos environ 33:25–35
Paull CK, Ussler W, Dillon WP (1991) Is the extent of glaciation limited by marine gas hydrates? Geophys res lett 18:432–434
Paulson SE, Flagan RC, Seinfeld JH (1992) Atmospheric photooxidation of isoprene, Part I: The hydroxyl radical and ground state atomic oxygen reactions. Int J chem kinet 24:79–101
Peierls BL, Paerl HW (1997) The bioavailability of atmospheric organic nitrogen deposition to coastal phytoplankton. Limnol oceanogr 42:1819–1880
Peng TH, Takahashi T, Broecker WS (1974) Surface radon measurements in the North Pacific Ocean station PAPA. J geophys res 79:1772–1780
Penkett SA, Brice KA (1986) The spring maximum in photooxi-dants in the Northern Hemisphere troposphere. Nature 319:655–657
Penkett SA, Jones BMR, Brice KA, Eggleton AEJ (1979) The importance of atmospheric ozone and hydrogen peroxide in oxidising sulphur dioxide in cloud and rainwater. Atmos environ 13:123–127
Penkett SA, Blake NJ, Lightman P, Marsh ARW, Anwyl P, Butcher G (1993) The seasonal variation of non-methane hydrocarbons in the free troposphere over the North Atlantic Ocean: Possible evidence for extensive reaction of hydrocarbons with the nitrate radical. J geophys res 98:2865–2885
Penkett SA, Bandy BJ, Reeves CE, McKenna D, Hignett P (1995) Measurements of peroxides in the atmosphere and their relevance to the understanding of global tropospheric chemistry. Faraday discuss 100:155–174
Penkett SA, Monks PS, Carpenter LJ, Clemitshaw KC, Ayers GP, Gillet RW, Galbally IE, Meyer CP (1997) Relationships between ozone photolysis rates and peroxy radical concentrations in clean marine air over the Southern Ocean. J geophys res 102:12805–12817
Penkett SA, Reeves CE, Bandy BJ, Kent JM, Richer HR (1998) Comparison of calculated and measured peroxide data collected in marine air to investigate prominent features of the annual cycle of ozone in the troposphere. J geophys res 104:13377–13388

Penner JE, Atherton CS, Dignon J, Ghan SJ, Walton JJ, Hameed S (1991) Tropospheric nitrogen: A three-dimensional study of sources, distributions, and deposition. J geophys res 96:959–990

Penner JE, Eddleman H, Novakov T (1993) Towards the development of a global inventory for black carbon emissions. Atmos environ 27A:1277–1295

Penner JE, Bergmann DJ, Walton JJ, Kinnison D, Prather MJ, Rotman D, Price C, Pickering KE, Baughcum SL (1998a) An evaluation of upper troposphere NO_x with two models. J geophys res 103:22097–2213

Penner JE, Chuang CC, Grant K (1998b) Climate forcing by carbonaceous and sulphate aerosols. Clim dynam 14:839–851

Penner JE, Andreae M, Annegarn H, Barrie L, Feichter J, Hegg D, Leaitch RJA, Murphy D, Nganga J, Pitari G (2001) Aerosols, their direct and indirect effects. In: Houghton JT, Ding Y, Griggs DJ, Noguer M, van der Linden PJ, Dai X, Maskell K, Johnson CA (eds) Climate change 2001: The scientific basis. Contribution of Working Group I to the third assessment report of the Intergovernmental Panel on Climate Change (IPCC). Cambridge University Press, Cambridge

Peppler RA, Bahrmann CP, Barnard JC, Campbell JR, Cheng M-D, Ferrare RA, Halthore RN, Heilman LA, Hlavka DL, Laulainen NS, Lin C-J, Ogren JA, Poellot MR, Remer LA, Sassen K, Spinhirne JD, Splitt ME, Turner DD (2000) ARM Southern Great Plains site observations of the Smoke Pall associated with the 1998 Central American fires. B am meteorol soc 81:2563–2591

Perner D, Ehhalt DH, Paetz HW, Platt U, Roth EP, Voltz A (1976) OH radicals in the lower troposphere. Geophys res lett 3:466–468

Perner D, Arnold T, Crowley J, Klüpfel T, Martinez M, Seuwen R (1999) The measurements of active chlorine in the atmosphere by chemical amplification. J atmos chem 34:9–20

Perrin DD (1979) Stability constants of metal-ion complexes, part B: Organic ligands. Pergamon Press, Oxford

Perry KD, Hobbs PV (1994) Further evidence for particle nucleation in clear air adjacent to marine cumulus clouds. J geophys res 99:22803–22818

Petit J-R, Jouzel J, Raynaud D, Barkov NI, Barnola JM, Basile I, Benders M, Chappellaz J, Davis M, Delaygue G, Delmotte M, Kotlyakov VM, Legrand M, Lipenkov VY, Lorius C, Pepin L, Ritz C, Saltzman E, Stievenard M (1999) Climate and atmospheric history of the past 420 000 years from the Vostok ice core, Antarctica. Nature 399:429–436

Phadnis MJ, Carmichael GR (2000) Numerical investigation of the influence of mineral dust on the tropospheric chemistry of East Asia. J atmos chem 36:285–323

Pham M, Müller J-F, Brasseur GP, Granier C, G Mégie (1995) A three-dimensional study of the tropospheric sulfur cycle. J geophys res 100:26061–26092

Phillips NA, Shukla J (1973) On the strategy of combining coarse and fine grid meshes in numerical weather prediction. J appl meteorol 12:763–770

Pickering KE, Dickerson RR, Huffman GJ, Boatman JF, Schanot A (1988) Trace gas transport in the vicinity of frontal convective clouds. J geophys res 93:759–773

Pickering KE, Thompson AM, Dickerson RR, Luke WT, McNamara DP, Greenberg JP, Zimmerman PR (1990) Model calculations of tropospheric ozone production potential following observed convective events. J geophys res 95:14049–14062

Pickering KE, Thompson AM, Scala JR, Tao W-K, Simpson J, Garstang M (1991) Photochemical ozone production in tropical squall line convection during NASA Global Tropospheric Experiment/Amazon Boundary Layer Experiment 2A. J geophys res 96:3099–3114

Pickering KE, Thompson AM, Scala JR, Tao W-K, Dickerson RR, Simpson J (1992) Free tropospheric ozone production following entrainment of urban plumes into deep convection. J geophys res 97:17985–18000

Pierotti D, Wofsy SC, Jacob D, Rasmussen R (1990) Isoprene and its oxidation products: Methacrolein and methyl vinyl ketone. J geophys res 95:1871–1881

Piketh SJ, Tyson PD, Steffen W (2000) Aeolian transport from southern Africa and iron fertilisation of marine biota in the South Indian Ocean. S afr J sci 96:244–246

Pilinis C, Pandis SN, Seinfeld JH (1995) Sensitivity of direct climate forcing by atmospheric aerosols to aerosol size and composition. J geophys res 100:18739–18754

Pincus R, Baker MB (1994) Effect of precipitation on the albedo susceptibility of clouds in the marine boundary layer. Nature 372:250–252

Piotrowicz SR, Bezek HF, Harvey GR, Springer-Young M, Hanson KJ (1991) On the ozone minimum over the equatorial Pacific Ocean. J geophys res 96:18679–18687

Pitzer KS (1991) Activity coefficients in electrolyte solutions. CRC Press, Boca Raton

Plantevin P-H, Law KS, Shallcross DE, Chipperfield MP, Pyle JA, Maren A (2000) The contribution of stratospheric O_3 to the O_3 burden of the troposphere: A model study using MOZAIC data. In: Proceedings of the Quadrennial Ozone Symposium, Sapporo, Japan

Plass-Dulmer C, Koppmann R, Ratte M, Rudolph J (1995) Light nonmethane hydrocarbons in seawater. Global biogeochem cy 9:79–100

Plass-Dulmer C, Brauers T, Rudolph J (1998) POPCORN: A field study of photochemistry in North-Eastern Germany. J atmos chem 31:5–31

Platt U, Heintz F (1994) Nitrate radicals in tropospheric chemistry. Israel J chem 34:289–300

Pochanart P, Kreasuwun J, Sukasem P, Geeratithadaniyom W, Tabucanon MS, Hirokawa J, Kajii Y, Akimoto H (2001) Tropical tropospheric ozone observed in Thailand. Atmos environ 35(15):2657–2668

Poisson N, Kanakidou M, Crutzen PJ (2000) Impact of non-methane hydrocarbons on tropospheric chemistry and the oxidizing power of the global troposphere: 3-dimensional modelling results. J atmos chem 36:157–230

Pope CA III, Dockery DW (1999) Epidemiology of particle effects. In: Holgate ST, Samet JM, Koren HS, Maynard RL (eds) Air pollution and health. Academic Press, London, pp 673–706

Pos WH, Riemer DD, Zika RG (1998) Carbonyl sulfide OCS and carbon monoxide CO in natural waters: Evidence of a coupled production pathway. Mar chem 62:89–101

Pöschl U, Letzel T, Schauer C, Niessner R (2001) Interaction of ozone and water vapor with spark discharge soot aerosol particles coated with benzo[a]pyrene: O_3 and H_2O adsorption, benzo[a]pyrene degradation, and atmospheric implications. J phys chem a 105:4029–4041

Potter CS, Davidson EA, Verchot LV (1996a) Estimation of global biogeochemical controls and seasonality in soil methane consumption. Chemosphere 32:2219–2246

Potter CS, Matson PA, Vitousek PM, Davidson EA (1996b) Process modelling of controls on nitrogen gas emissions from soils worldwide. J geophys res 101(D1):1361–1377

Prather MJ (1986) Numerical advection by conservation of second order moments. J geophys res 91:6671–6681

Prather MJ, Jacob DJ (1997) A persistent imbalance in HO_x and NO_x photochemistry of the upper troposphere driven by deep tropical convection. Geophys res lett 24:3189–3192

Prather MJ, Remsberg EE (1993) The atmospheric effects of stratospheric aircraft: Report of the 1992 models and measurements workshop. In: National Aeronautics and Space (ed) NASA reference publication 1292. National Aeronautics and Space Administration, Washington, D.C.

Prather MJ, McElroy MB, Wofsy SC, Logan JA (1979) Stratospheric chemistry: Multiple solutions. Geophys res lett 6:163–164

Preiswerk D, Najjar R (1998) An Analysis of a global model of air-sea fluxes of carbonyl sulfide. EOS 79(17):S40

Price C, Rind D (1992) A simple lightning parameterization for calculating global lightning distributions. J geophys res 97:9919–9933

Price C, Penner J, Prather M (1997) NO_x from lightning: 1. Global distribution based on lightning physics. J geophys res 102:5929–5941

Price JD, Vaughan G (1993) The potential for stratosphere troposphere exchange in cut-off low systems. Q j roy meteor soc 119(510):Part B, 343–365

Priemé A, Christensen S, Dobbie KE, Smith KA (1997) Slow increase in rate of methane oxidation in soils with time following land use change from arable agriculture to woodland. Soil biol biochem 29:1269–1273

Prinn RG (1994) Global atmospheric-biospheric chemistry: An overview. In: Prinn RG (ed) Global atmospheric-biospheric chemistry. Plenum Press, New York
Prinn RG, Weiss RF, Miller BR, Huang J, Alyea FN, Cunnold DM, Fraser PJ, Hartley DE, Simmons PG (1995) Atmospheric trend and lifetime of CH_3CCl_3 and global OH concentrations. Science 269:187–192
Prinn RG, Weiss RF, Fraser PJ, Simmonds PG, Cunnold DM, Alyea FN, O'Doherty S, Salameh P, Miller BR, Huang J, Wang RHJ, Hartley DE, Harth C, Steele LP, Sturrock G, Midgeley P, McCulloch A (2000) A history of chemically and radiatively important gases in air deduced from ALE/GAGE/AGAGE. J geophys res 105:17751–17793
Prinn RG, Huang J, Weiss RF, Cunnold DM, Fraser PJ, Simmonds PG, Harth C, Salameh P, O'Doherty S, Wang RHJ, Porter L, Miller BR, McCulloch A (2001) Evidence for significant variations of atmospheric hydroxyl radicals in the last two decades. Science 292:1882–1888
Prins EM, Feltz JM, Menzel WP, Ward DE (1998) An overview of GOES 8 diurnal fire and smoke results for SCAR-B and 1995 fire season in South America. J geophys res 103:31821–31835
Prospero JM (1996) The atmospheric transport of particles to the ocean. In: Ittekkot V, Schäfer P, Honjo S, Depetris PJ (eds) Particle flux in the ocean. J. Wiley, pp 19–52
Prospero JM, Charlson RJ, Mohnen V, Jaenicke R, Delany AC, Moyers J, Zoller W, Rahn K (1983) The atmospheric aerosol system: An overview. Rev geophys space phys 21:1607–1629
Prospero JM, Uematsu M, Savoie DL (1989) Mineral aerosol transport to the Pacific Ocean. In: Riley JP, Chester R, Duce RA (eds) Chemical oceanography 10. Academic Press, London, pp 188–218
Prospero JM, Barrett K, Church T, Dentener F, Duce RA, Galloway JN, Levy H, Moody J, Quinn P (1996) Nitrogen dynamics of the North Atlantic Ocean: Atmospheric deposition of nutrients to the North Atlantic Basin. Biogeochemistry 35:27–73
Pruppacher HR, Klett JD (1980) Microphysics of clouds and precipitation. Reidel, Hingham, MA
Pszenny AAP, Keene WC, Jacob DJ, Fan S, Maben JR, Zetwo MP, Springer M-Young, Galloway JN (1993) Evidence of inorganic chlorine gases other than hydrogen chloride in marine surface air. Geophys res lett 20:699–702
Pszenny AAP, Prinn RG, Kleiman GL, Shi X, Bates TS (1999) Nonmethane hydrocarbons in surface waters, their sea-air fluxes and impact on OH in the marine boundary layer during ACE-1. J geophys res 104:21785–21901
Putaud JP, van Dingenen RMM, Virkkula A, Raes F, Maring H, Prospero JM, Swietlicki E, Berg OH, Hillamo R, Mäkelä T (2000) Chemical mass closure and assessment of the origin of the submicron aerosol in the marine boundary layer and the free troposphere at Tenerife during ACE-2. Tellus 52B:141–168
Quay P, Stutsman J, Wilbur D, Snover A, Dlugokencky E, Brown T (1999) The isotopic composition of atmospheric methane. Global biogeochem cy 13(2):445–461
Quinn PK, Coffman DJ (1998) Local closure during the First Aerosol Characterization Experiment (ACE-1): Aerosol mass concentration and scattering and backscattering coefficients. J geophys res 103:16575–16596
Quinn PK, Coffman DJ (1999) Comment on "Contribution of different aerosol species to the global aerosol extinction optical thickness: Estimates from model results" by Tegen et al. J geophys res 104:4241–4248
Quinn PK, Charlson RJ, Bates TS (1988) Simultaneous observations of ammonia in the atmosphere and ocean. Nature 335:336–338
Quinn PK, Bates TS, Johnson JE, Covert DS, Charlson RJ (1990) Interactions between the sulphur and reduced nitrogen cycles over the central Pacific Ocean. J geophys res 95:16405–16416
Quinn PK, Anderson TL, Bates TS, Dlugi R, Heintzenberg J, von Hoyningen-Huene W, Kulmala M, Russell PB, Swietlicki E (1996a) Closure in tropospheric aerosol-climate research: A review and future needs for addressing aerosol direct shortwave radiative forcing. Contributions to atmospheric physics 69:547–577
Quinn PK, Barrett KJ, Dentener FJ, Lipschultz F, Six KD (1996b) Estimation of the air/sea exchange of ammonia for the North Atlantic basic. Biogeochemistry 35:275–304
Quinn PK, Kapustin VN, Bates TS, Covert DS (1996c) Chemical and optical properties of marine boundary layer aerosol particles of the mid-Pacific in relation to sources and meteorological transport. J geophys res 101:6931–6951
Quinn PK, Coffmann DJ, Kapustin VN, Bates TS, Covert DS (1998) Aerosol optical properties during the First Aerosol Characterisation Experiment (ACE-1) and the underlying chemical and physical aerosol properties. J geophys res 103:16547–16564
Quinn PK, Bates TS, Coffman DJ, Miller TL, Johnson JE, Covert DS, Putaud J-P, Neusüß C, Novakov T (2000) A comparison of aerosol chemical and optical properties from the first and second Aerosol Characterization Experiments. Tellus B 52:239–257
Rader DJ, McMurry PH (1986) Application of the tandem differential mobility analyzer to studies of droplet growth or evaporation. J aerosol sci 17:771–787
Radford-Knoery J, Cutter G (1994) Biogeochemistry of dissolved hydrogen sulfide species and carbonyl sulfide in the western North Atlantic. Geochim cosmochim ac 58:5421–5431
Radke LF, Hobbs PV (1991) Humidity and particle fields around some small cumulus clouds. J atmos sci 48:1190–1193
Radke LF, Stith JL, Hegg DA, Hobbs PV (1978) Airborne studies of particles and gases from forest fires. Japca J air waste ma 28:30–34
Radke LF, Coakley JA Jr., King MD (1989) Direct and remote sensing observations of the effects of ships on clouds. Science 246:1146–1149
Raes F (1995) Entrainment of free tropospheric aerosols as a regulating mechanism for cloud condensation nuclei in the remote marine boundary layer. J geophys res 100:2893–2903
Raes F, Saltelli A, van Dingenen R (1992) Modelling formation and growth of H_2SO_4-H_2O aerosols: Uncertainty analysis and experimental evaluation. J aerosol sci 23:759–771
Raes F, Wilson J, van Dingenen R (1995) Aerosol dynamics and its implication for the global aerosol climatology. In: Charlson RJ, Heintzenberg J (eds) Aerosol forcing of climate. J. Wiley, New York, pp 153–169
Raes F, van Dingenen R, Cuevas E, van Velthoven PFJ, Prospero JM (1997) Observations of aerosols in the free troposphere and marine boundary layer of the subtropical Northeast Atlantic: Discussion of processes determining their size distribution. J geophys res 102:21315–21328
Raes F, Bates T, Vogelenzang D, van Liedekerke M, Verver G (2000a) The second aerosol characterization experiment (ACE-2): General overview and main results. Tellus 52B:111–125
Raes F, van Dingenen R, Vignati E, Wilson J, Putaud J-P, Seinfeld JH, Adams PJ (2000b) Formation and cycling of aerosols in the global troposphere. Atmos environ 34:4215–4240
Ramacher B, Rudolph J, Koppmann R (1999) Hydrocarbon measurements during tropospheric ozone depletion events. J geophys res 104:3633–3653
Ramanathan V, Andreae MO, Anderson J, Cass G, Clarke T, Collins WD, Coakley J, Gandrud B, Heymsfield A, Jayaraman A, Kiehl J, Ogren J, Prospero J, Novakov T, Prather K (2000) The Indian Ocean Experiment: An Overview of the aerosol climate forcing with implications to climate change. Science (in prep)
Ramaswamy V, Boucher O, Haigh J, Hauglustaine D, Haywood J, Myhre G, Nakajima T, Shi GY, Solomon S, Betts R, Charlson R, Chuang C, Daniel JS, Del Genio A, van Dorland R, Feichter J, Fuglestvedt J, de F. Forster PM, Ghan SJ, Jones A, Kiehl JT, Koch D, Land C, Lean J, Lohmann U, Minschwaner K, Penner JE, Roberts DL, Rodhe H, Roelofs GJ, Rotstayn LD, Schneider TL, Schumann U, Schwartz SE, Schwarzkopf MD, Shine KP, Smith S, Stevenson DS, Stordal F, Tegen I, Zhang Y (2001) Radiative forcing of climate change. In: Houghton JT, Ding Y, Griggs DJ, Noguer M, van der Linden PJ, Dai X, Maskell K, Johnson CA (eds) Climate change 2001: The scientific basis. Contribution of Working Group I to the Third Assessment Report of the Intergovernmental Panel on Climate Change. Cambridge University Press, Cambridge, pp 350–416
Ramonet M, Le Roulley J-C, Bousquet P, Monfray P (1996) Radon222 measurements during the TROPOZ II campaign and comparison with a global atmospheric transport model. J atmos chem 23:107–136

Randall DA, Coakley JA Jr., Fairall CW, Kropfli RA, Lenschow DH (1984) Outlook for research on subtropical marine stratiform clouds. B am meteorol soc 65:1290–1301

Rasch PJ, Lawrence M (1998) Recent development in transport methods at NCAR. In: Machenhaue B (ed) MPI workshop on conservative transport schemes. Max-Planck-Institute for Meteorology, Hamburg, pp 65–75

Rasch PJ, Mahowald NM, Eaton BE (1997) Representations of transport, convection and the hydrologic cycle in chemical transport models: Implications for the modeling of short lived and soluble species. J geophys res 102:28127–28138

Rasch PJ, Williamson DL (1990) Computational aspects of moisture transport in global models of the atmosphere. Q j roy meteor soc 116:1071–1090

Rasch PJ, Feichter J, Law K, Mahowald N, Penner J, Benkovitz C, Genthon C, Giannakopoulos C, Kasibhatla P, Koch D, Levy H, Maki T, Prather M, Roberts DL, Roelofs G-J, Stevenson D, Stockwell Z, Taguchi S, Kritz M, Chipperfield M, Baldocchi D, McMurry P, Barrie L, Balkanski Y, Chatfield R, Kjellstrom E, Lawrence M, Lee HN, Lelieveld J, Noone KJ, Seinfeld J, Stenchikov G, Schwartz S, Walcek C, Williamson D (2000) A comparison of scavenging and deposition processes in global models: Results from the WCRP Cambridge Workshop of 1995. Tellus 52B:1025–1956

Ratte M, Plass-Dulmer C, Koppmann R, Rudolph J (1995) Horizontal and vertical profiles of light hydrocarbons in sea water related to biological, chemical, and physical parameters. Tellus 47B:607–623

Ravishankara AR (1997) Heterogeneous and multiphase chemistry in the troposphere. Science 276:1058–1065

Ravishankara AR, Longfellow CA (1999) Reactions on tropospheric condensed matter. Phys chem chem phys 1:5433–5441

Ravishankara AR, Hancock G, Kawasaki M, Matsumi Y (1998) Photochemistry of ozone: Surprises and recent lessons. Science 280:60–61

Ray EA, Moore FL, Elkins JW, Dutton GS, Fahey DW, Vömel H, Oltmans SJ, Rosenlof KH (1999) Transport into the Northern Hemisphere lowermost stratosphere revealed by *in situ* trace measurements. J geophys res 104:26565–26580

Raynaud D, Chappellaz J, Barnola J-M, Korotkevich YS, Lorius C (1988) Climatic and CH_4-cycle implications of glacial-interglacial CH_4 change in the Vostok ice core. Nature 333:655–657

Raynaud D, Jouzel J, Barnola JM, Chappellaz JA, Delmas RJ, Lorius C (1993) The ice record of greenhouse gases. Science 259:926–934

Rea DK (1994) The paleoclimatic record provided by eolian deposition in the deep sea: The geologic history of wind. Rev geophys 32(2):159–195

Regener VH (1975) The vertical flux of atmospheric ozone. J geophys res 62:221

Reichle HG, Conners VS, Holland JA, Hypes WD, Wallio HA, Casas JC, Gormsen BB, Saylor MS, Hasketh WD (1986) Middle and upper tropospheric carbon monoxide mixing ratios as measured by a satellite borne remote sensor during November 1981. J geophys res 91:10865–10888

Reid JS, Hobbs PV, Liousse C, Martins JV, Weiss RE, Eck TF (1998) Comparisons of techniques for measuring shortwave absorption and black carbon content of aerosols from biomass burning in Brazil. J geophys res 103:32031–32040

Reilly PTA, Gieray RA, Whitten WB, Ramsey JM (1998) Real-time characterization of the organic composition and size of individual diesel engine smoke particles. Environ sci technol 32:2672–2679

Reiner T, Hanke M, Arnold F (1998) Aircraft-borne measurements of peroxy radicals in the middle troposphere. Geophys res lett 25:47–50

Reiter R, Jäger H (1986) Results of 8-year continuous measurements of aerosol profiles in the stratosphere with discussion of the importance of stratospheric aerosols to an estimate of effects on the global climate. Meteorol atmos phys 35:19–48

Remer LA, Kaufman YJ (1998) Dynamical aerosol model: Urban/industrial aerosol. J geophys res 103:13859–13871

Remer LA, Kaufman YJ, Holben BN, Thompson AM, McNamara D (1998) Biomass burning aerosol size distribution and modeled optical properties. J geophys res 103:31879–31891

Restad K, Isaksen ISA, Berntsen TK (1998) Global distribution of sulphate in the troposphere. A 3-dimensional model study. Atmos environ 32:3593–3609

Rhew RC, Miller BR, Weiss RF (2000) Natural methyl bromide and methyl chloride emissions from coastal salt marshes. Nature 403:292–295

Rich S, Waggoner PE, Tomlinson H (1970) Ozone uptake by bean leaves. Science 169:79

Richter A, Wittrock F, Eisinger M, Burrows JP (1998) GOME observations of tropospheric BrO in northern hemispheric spring and summer 1997. Geophys res lett 25:2683–2686

Richter A, Wittrock F, Ladstätter-Weißenmayer A, Burrows JP (2001) GOME measurements of stratospheric and tropospheric BrO. Adv space res (accepted)

Ridgwell A, Marshall SJ, Gregson K (1999) Consumption of atmospheric methane by soils: A process-based model. Global biogeochem cy 13:59–70

Ridley BA, Carroll MA, Gregory GL (1987) Measurements of nitric oxide in the boundary layer and free troposphere over the Pacific Ocean. J geophys res 92:2025–2047

Ridley BA, Walega J, Hübler G, Montzka D, Atlas E, Hauglustaine DA, Grahek F, Lind J, Campos T, Norton R, Greenberg J, Schauffler S, Oltmans S, Whittlestone S (1998) Measurements of NO_x and PAN and estimates of O_3 production over the seasons during Mauna Loa Observatory Photochemistry Experiment 2. J geophys res 103:8323–8339

Ridley BA, Walega J, Montzka D, Grahek F, Atlas E, Flocke F, Stroud V, Deary J, Gallant A, Bottenheim J, Anlauf K, Worthy D, Sumner AL, Splawn B, Shepson PB (2000) Is the Arctic surface layer a source and sink of NO_x in winter/spring? J atmos chem 36: 1–22

Riley WJ, Matson PA (2000) NLOSS: A mechanistic model of denitrified N_2O and N_2 evolution from soil. Soil sci 165(3):237–249

Rind D, Lerner J (1996) Use of on-line tracers as a diagnostic tool in general circulation model development: 1. Horizontal and vertical transport in the troposphere. J geophys res 101: 12667–12683

Rind D, Lerner J, Shah K, Suozzo R (1999) Use of on-line tracers as a diagnostic tool in general circulation model development: 2. Transport between the troposphere and stratosphere. J geophys res 104:9151–9167

Rinne HJI, Delany AC, Greenberg JP, Guenther AB (2000) A true eddy accumulation system for trace gas fluxes using disjunct eddy sampling method. J geophys res 105:24791–24798

Rinsland CP, Levine JS, Miles T (1985) Concentration of methane in the troposphere deduced from 1951 infrared solar spectra. Nature 330:245–249

Ro C-U, Osan J, Grieken RV (1999) Determination of low-Z elements in individual environmental particles using windowless EPMA. Anal chem 71:1521–1528

Roberts JM, Fajer RW, Springston SR (1989) The capillary gas chromatographic separations of alkyl nitrates and peroxycarboxylic nitric anhydrides. Anal chem 61:771–772

Roberts JM, Parrish DD, Norton RB, Bertman SB, Holloway JS, Trainer M, Fehsenfeld FC, Carroll MA, Albercock GM, Wang T, Forbes G (1996) Episodic removal of NO_y species from the marine boundary layer over the North Atlantic. J geophys res 101:28947–28960

Roberts JM, Flocke F, Weinheimer A, Tanimoto H, Jobson BT, Riemer D, Apel E, Atlas E, Donnelly S, Stroud V, Johnson K, Weaver R, Fehsenfeld FC (2001) Observations of peroxyacrylic nitric anhydride, APAN, during the TexAQS 2000 Intensive Experiment. Geophys res lett 28:4195–4198

Robinson RA, Stokes RH (1959) Electrolyte solutions. Butterworth. London

Robock A, Mao J (1992) Winter warming from large volcanic eruptions. Geophys res lett 12:2405–2408

Rodhe H (1999) Human impact on the atmospheric sulfur budget. Tellus 51A-B:110–122

Rodhe H, Grennfelt P, Wisniewski J, Ågren C, Bengtsson G, Johansson K, Kauppi P, Kucera V, Rasmussen L, Rosseland B, Schytte L, Selldén G (1995) Acid rain '95? Summary statement from 5th International Conference on Acid Deposition – Science and Policy, Göteborg, Sweden, 1995. Water air soil poll 85:1–14

Roelofs G-J, Lelieveld J (1995) Distribution and budget of O_3 in the troposphere calculated with a chemistry general circulation model. J geophys res 100:20983–20998
Roelofs G-J, Lelieveld J (1997) Model study of the influence of cross-tropopause O_3 transports on tropospheric O_3 levels. Tellus 49B:38–55
Roelofs G-J, Lelieveld J, van Dorland R (1997) A three-dimensional chemistry/general circulation transport model simulation of anthropogenically derived ozone in the troposphere and its radiative climate forcing. J geophys res 102:23389–23401
Roelofs G-J, Lelieveld J, Ganzeveld L (1998) Simulation of global sulfate distribution and the influence on effective cloud drop radii with a coupled photochemistry-sulfur cycle model. Tellus 50B:224–242
Rogaski CA, Golden DM, Williams LR (1997) Reactive uptake and hydration experiments on amorphous carbon treated with NO_2, SO_2, O_3, $HNOO_3$, and H_2SO_4. Geophys res lett 24:381–384
Romieu I (1999) Epidemiological studies of health effects arising from motor vehicle air pollution. In: Schwela D, Zali O (eds) Urban traffic pollution. E & FN Spon, London, pp 9–69
Rood RB (1987) Numerical advection algorithms and their role in atmospheric transport and chemistry models. Rev geophys 25:71
Rosenfeld D (2000) Suppression of rain and snow by urban and industrial air pollution. Science 287:1793–1796
Rosenfeld D, Lensky IM (1998) Satellite-based insights into precipitation formation processes in continental and maritime convective clouds. B am meteorol soc 79:2457–2476
Roths J, Zenker T, Parchatka U, Wienhold F, Harris GW (1996) Four-laser airborne infrared spectrometer for atmospheric trace gas measurements. Appl optics 35:7075–7084
Rotman DA, Atherton CS, Bergmann D, Chuang C, Connell P, Dignon J, Franz A, Kinnison D, Proctor D, Tannahill J (2000) IMPACT: A global three-dimensional model on the troposphere and stratosphere. J geophys res (to be submitted)
Rotstayn LD (1999a) Climate sensitivity of the CSIRO GCM: Effect of cloud modeling assumptions. J climate 12:334–356
Rotstayn LD (1999b) Indirect forcing by anthropogenic aerosols: A global climate model calculation of the effective-radius and cloud-lifetime effects. J geophys res 104:9369–9380
Roulet NT, Ash R, Quinton W, Moore TR (1993) Methane flux from drained northern peatlands: The effect of a persistent lowering of the water table on flux. Global biogeochem cy 7(4):749–769
Roy DP, Giglio L, Kendall J, Justice CO (1999) Multitemporal active-fire based burn scar detection algorithm. Int J remote sens 20:1031–1038
Rudek J, Paerl HW, Mallin MA, Bates PW (1991) Seasonal and hydrological control of phytoplankton nutrient limitation in the lower Neuse River Estuary, North Carolina. Mar ecol-prog ser 75:133–142
Rudolph J, Ehhalt DH (1981) Measurements of C_2-C_5 hydrocarbons over the North Atlantic. J geophys res 86:11959–11964
Rudolph J, Johnen FJ (1990) Measurements of light atmospheric hydrocarbons over the Atlantic in regions of low biological activity. J geophys res 95:20583–20591
Rudolph J, Koppmann R, Plass-Dulmer C (1996) The budgets of ethane and tetrachloroethene: Is there evidence for an impact of reactions with chlorine atoms in the troposphere? Atmos environ 30(10–11):1887–1894
Russel GL, Lerner JA (1981) A new finite-differencing scheme for the tracer transport equation. J appl meteorol 20:1483–1498
Russell LM, Zhang S-H, Flagan RC, Seinfeld JH, Stolzenburg MR, Caldow R (1996) Radially classified aerosol detector for aircraft-based submicron aerosol measurements. J atmos ocean tech 13:598–609
Russell PB, Heintzenberg J (2000) An overview of the ACE-2 Clear Sky Column Closure Experiment (CLEARCOLUMN). Tellus 52B:463–483
Russell PB, Hobbs PV, Stowe LL (1999) Aerosol properties and radiative effects in the United States east coast haze plume: An overview of the tropospheric aerosol radiative forcing observational experiment (TARFOX). J geophys res 104:2213–2222
Saathoff H, Naumann K-H, Riemer N, Kamm S, Möhler O, Schurath U, Vogel VHB (2001) The loss of NO_2, HNO_3, NO_3/N_2O_5, and $HO_2/HOONO_2$ on soot aerosol: A chamber and modeling study. Geophys res lett 28:1957–1960
Saeger M, Langstaff J, Walters R, Modica L, Zimmerman D, Fratt D, Dulleba D, Ryan R, Demmy J, Tax W, Sprague D, Mudgett D, Werner AS (1989) The 1985 NAPAP emissions inventory (Version 2): Development of the annual data and modelers' tape. US Environmental Protection Agency
Saemundsdottir S, Matrai PA (1998) Biological production of methyl bromide by cultures of marine phytoplankton. Limnol oceanogr 43:81–87
Sah RN (1995) Plasma source mass spectrometric analysis of biological and environmental samples: Dealing with potential interferences. Appl spectrosc rev 30:35–80
Sakka A, Gosselin M, Levasseur M, Michaud S, Montfort P, Demers S (1997) Effects of reduced ultraviolet radiation on aqueous concentrations of dimethylsulfoniopropionate and dimethylsulfide during a microcosm study in the lower St. Lawrence Estuary. Mar ecol-prog ser 149:227–238
Sala OE, Chapin FS III, Gardner RH, Lauenroth WK, Mooney HA, Ramakrishnan PS (1999) Global change, biodiversity and ecological complexity. In: Walker B, Steffen W, Canadell J, Ingram J (eds) The terrestrial biosphere and global change. Implications for natural and managed ecosystems. Cambridge University Press, Cambridge, pp 304–328
Salisbury G, Rickard AR, Monks PS, Allan BJ, Bauguitte S, Penkett SA, Carslaw N, Lewis AC, Creasey DJ, Heard DE, Jacobs PJ, Lee JD (2001) The production of peroxy radicals at night via reactions of ozone and the nitrate radical in the marine boundary layer. J geophys res 106(D12):12669–12688
Saltzman ES, Savoie DL, Prospero JM, Zika RG (1986) Methanesulfonic acid and non-sea-salt sulphate in Pacific air: Regional and seasonal. J atmos chem 4:227–240
Saltzman ES, Whung PY, Mayewski PA (1997) Methanesulfonate in the Greenland Ice Sheet Project 2 ice core. J geophys res 102:26649–26657
Sander R (1999) Modeling atmospheric chemistry: Interactions between gas-phase species and liquid cloud/aerosol particles. Surv geophys 20:1–31
Sander R, Crutzen PJ (1996) Model study indicating halogen activation and ozone destruction in polluted air masses transported to the sea. J geophys res 101D:9121–9138
Sander SP, Friedl RR, DeMore WB, Golden DM, Kurylo MJ, Hampson RF, Huie RE, Moortgat GK, Ravishankara AR, Kolb CE, Molina MJ (1997) Chemical kinetics and photochemical data for use in stratospheric modeling, supplement to evaluation 12: Update of key reactions. NASA Jet Propulsion Laboratory 00-3, Pasadena, CA
Sandholm S, Bradshaw JD, Dorris KS, Rodgers MO, Davis DD (1990) An airborne-compatible photofragmentation two-photon laser-induced fluorescence instrument for measuring background tropospheric NO, NO_x, and NO_2. J geophys res 95:10155–10161
Sass RL, Fisher FM (1995) Methane emissions from Texas rice fields: A five year study. In: Peng S, Ingram KT, Neue HU, Ziska LH (eds) Climate change and rice. Springer-Verlag, Berlin, pp 46–59
Satheesh SK, Ramanathan V (2000) Observational determination of the direct aerosol solar radiative forcing in the Tropics. Nature 405:60
Saxena P, Hildemann LM (1996) Water-soluble organics in atmospheric particles: A critical review of the literature and application of thermodynamics to identify candidate compounds. J atmos chem 24:57–109
Scala JR, Garstang M, ao W-K, Pickering KE, Thompson AM, Simpson J, Kirchhoff VWJH, Browell EV, Sachse GW, Torres AL, Gregory GL, Rasmussen RA, Khalil MAK (1990) Cloud draft structure and trace gas transport. J geophys res 95:17015–17030
Scarratt MG, Moore RM (1996) Production of methyl chloride and methyl bromide in laboratory cultures of marine phytoplankton. Mar chem 54:263–272
Schall C, Heumann KG (1993) GC determination of volatile organoiodine and organobromine compounds in Arctic seawater and air samples. Fresen J anal chem 346:717–722
Scheel HE, Areskoug H, Geiss H, Gomiscel B, Granby K, Haszpra L, Klasinc L, Kley D, Laurila T, Lindkog T, Roemer M, Schmitt M, Simmonds P, Solberg S, Toupance G (1997) On the spatial distribution and seasonal variation of lower-troposphere ozone over Europe. J atmos chem 28:11–28

Schiffman A, Nelson DD, Robinson MS, Nesbitt DJ (1991) High-resolution infrared flash kinetic spectroscopy of OH radicals. J chem phys 95:2629–2636

Schimel D, Alves D, Enting I, Heimann M, Joos F, Raynaud D, Wigley T, Prather M, Derwent R, Ehhalt D, Fraser P, Sauhueza E, Zhou X, Jonas P, Charlson R, Rodhe H, Sadasivan S, Shine KP, Fouquart Y, Ramaswamy V, Solomon S, Srinivasan J, Albritton D, Derwent R, Isaksen I, Lal M, Wuebbles D (1996) Radiative forcing of climate change. In: Houghton JT, Filho LGM, Callander BA, Harris N, Kattenberg A, Maskell K (eds) Climate change 1995: The science of climate change. Cambridge University Press, Cambridge, pp 65–131

Schimel DS (1995) Terrestrial ecosystems and the carbon cycle. Glob change biol 1:77–91

Schlager H, Schulte P, Flatøy F, Slemr F, van Velthoven P, Ziereis H, Schumann U (1999) Regional nitric oxide enhancements in the North Atlantic flight corridor observed and modeled during POLINAT-2. Geophys res lett 26(20):3061–3064

Schlesinger WH (1997) Biogeochemistry: An analysis of global change. Academic Press, San Diego

Schlesinger WH, Hartley AE (1992) A global budget for atmospheric NH_3. Biogeochemistry 15:191–211

Schmid B, Livingston JM, Russell PB, Durkee PA, Jonsson HH, Collins DR, Flagan RC, Seinfeld JH, Gassó S, Hegg DA, Öström E, Noone KJ, Welton EJ, Voss KJ, Gordon HR, Formenti P, Andreae MO (2000) Clear sky closure studies of lower tropospheric aerosol and water vapor during ACE-2 using airborne sunphotometer, airborne *in situ*, space-borne, and ground-based measurements. Tellus 52B:568–593

Schmidt M, Graul R, Sartorius H, Levin I (1996) Carbon dioxide and methane in continental Europe: A climatology, and ^{222}radon-based emission estimates. Tellus 48(4):457–473

Schnitzler J, Lehning A, Steinbrecher R (1997) Seasonal pattern of isoprene synthase activity in *Quercus robur* leaves and its significance for modeling isoprene emission rates. Bot acta 110:240–243

Scholes RJ, Kendall J, Justice CO (1996) The quantity of biomass burned in Southern Africa. J geophys res 101:23667–23676

Scholes RJ, Schulze E-D, Pitelka LF, Hall DO (1999) Biogeochemistry of terrestrial ecosystems. In: Walker B, Steffen W, Canadell J, Ingram J (eds) The terrestrial biosphere and global change. Implications for natural and managed ecosystems. Cambridge University Press, Cambridge, pp 271–303

Schröder F, Ström J (1997) Aircraft measurements of sub micrometer aerosol particles (>7 nm) in the midlatitude free troposphere and tropopause region. Atmos res 44:333–356

Schroeder WH, Anlauf KG, Barrie LA, Lu JY, Steffen A, Schneeberger DR, Berg T (1998) Mercury vapour depletion in Arctic air during springtime. Nature 394:331

Schulz M, Balkanski YJ, Guelle W, Dulac F (1998) Role of aerosol size distribution and source location in a three-dimensional simulation of a Saharan dust episode tested against satellite-derived optical thickness. J geophys res 103:10579–10592

Schultz M, Jacob DJ, Logan JA, Wang Y, Blake DR, Bradshaw JD, Gregory G, Sachse G, Heikes B, Shetter R, Singh HB, Talbot R (1999) On the origin of tropospheric ozone and NO_x over the tropical South Pacific. J geophys res 104:5829–5844

Schultz M, Jacob DJ, Bradshaw JD, Sandholm ST, Dibb JE, Talbot RW, Singh HB (2000) Chemical NO_x budget in the upper troposphere over the tropical South Pacific. J geophys res 105:6669–6679

Schumann U, Schlager H, Arnold F, Ovarlez J, Kelder H, Hov Ø, Hayman G, Isaksen ISA, Staehelin J, Whitefield PD (2000) Pollution from aircraft emissions in the North Atlantic flight corridor: Overview on the POLINAT projects. J geophys res 105(D3):3605–3631

Schwartz J, Dockery DW (1992) Increased mortality in Philadelphia associated with daily air pollutant concentrations. Am rev respir dis 145:600–604

Schwartz SE (1988) Are global cloud albedo and climate controled by marine phytoplankton? Nature 336:441–445

Schwartz SE (1996) The Whitehouse effect: Shortwave radiative forcing of climate by anthropogenic aerosols: An overview. J aerosol sci 27:359–382

Schwartz SE, Slingo A (1996) Enhanced shortwave cloud radiative forcing due to anthropogenic aerosols. In: Crutzen P, Ramanathan V (eds) Clouds, chemistry and climate: Proceedings of NATO Advanced Research Workshop. Springer-Verlag, Heidelberg, pp 191–236

Schwarzenböck A, Heintzenberg J, Mertes S (2000) Incorporation of aerosol particles between 25 and 850 nanometers into cloud elements: Measurement with a new complementary sampling system. Atmos res 52(4):241–260

Schwela D (1995) Public health implications of urban air pollution in developing countries. In: Proc. 10th World Clean Air Congress. Espoo, Finland

Schwela D (2000) Air pollution and health in urban areas. Rev environ health 15:13–42

Scott WD, Hobbs PV (1967) The formation of sulphate in water droplets. J atmos sci 24:54–57

Scranton MI, McShane K (1991) Methane fluxes in the southern North Sea: The role of European rivers. Cont shelf res 11:37–52

Scudlark JR, Church TM (1993) Atmospheric input of inorganic nitrogen to Delaware Bay. Estuaries 16(4):747–759

Seeley JV, Jayne JT, Molina MJ (1996) Kinetic studies of chlorine atom reactions using the turbulent flow tube technique. J chem phys 100:4019–4025

Sehmel GA (1980) Particle and gas dry deposition: A review. Atmos environ 14:983

Seinfeld JH, Pandis SN (1998) Atmospheric chemistry and physics. J. Wiley, New York

Seitzinger SP, Sanders RW (1999) Atmospheric inputs of dissolved organic nitrogen stimulate estuarine bacteria and phytoplankton. Limnol oceanogr 44:721–730

Seitzinger SP, Styles RV, Kroeze C (2000) Global distribution of N_2O emissions from aquatic systems: Natural emissions and anthropogenic effects. Chemosphere – global change science 2:267–279

Serça D, Guenther A, Klinger L, Helmig D, Zimmerman P, Hereid D (1998) Methyl bromide deposition to soils. Atmos environ 32:1581–1586

Severinghaus JP, Brook EJ (1999) Abrupt climate change at the end of the last glacial period inferred from trapped air in polar ice. Science 286:930–934

Sharkey T (1996) Emission of low molecular mass hydrocarbons from plants. Trends plant sci 1:78–82

Shaw GE (1989) Production of condensation nuclei in clean air by nucleation of H_2SO_4. Atmos environ 23:2841–2846

Shaw GE, Benner RL, Cantrell W, Clarke AD (1998a) On the regulation of climate: A sulphate particle feedback loop involving deep convection. Climatic change 39(1):23–33

Shaw WJ, Spicer CW, Kenny DV (1998b) Eddy correlation fluxes of trace gases using a tandem mass spectrometer. Atmos environ 32:2887–2898

Sheridan PJ, Norton RB (1998) Determination of the passing efficiency for aerosol chemical species through a typical aircraft-mounted, diffuser-type aerosol inlet system. J geophys res 103:8215–8225

Shetter RE, Cantrell CA, Lantz KO, Flocke SJ, Orlando JJ, Tyndall GS, Gilpin TM, Fischer CA, Madronich S, Calvert JG (1996) Actinometric and radiometric measurement and modeling of the photolysis rate coefficient of ozone to $O(^1D)$ during Mauna Loa Observatory Photochemistry Experiment 2. J geophys res 101:14631–14641

Shine KP, PM de F Forster PM (1999) The effects of human activity on radiative forcing of climate change: A review of recent developments. Global Planet Change 20:205–225

Shorter JH, Kolb CE, Crill PM, Kerwin RA, Talbot RW, Hines ME, Harriss RC (1995) Rapid degradation of atmospheric methyl bromide in soils. Nature 377:717–719

Shulman ML, Jacobson MC, Charlson RJ, Synovec RE, Young TE (1996) Dissolution behavior and surface tension effects of organic compounds in nucleating cloud droplets. Geophys res lett 23:277–280

Sickles JE II, Hodson LL, McClenny WA, Paur RJ, Ellestad TG, Mulik JD, Anlauf KG, Wiebe HA, Mackay GI, Schiff HI, Bubacz DK (1990) Field comparison of methods for the measurement of gaseous and particulate contributors to acidic dry deposition. Atmos environ 24A:155–165

Sievering H (1984) Small particle dry deposition on natural waters: Modeling uncertainty. J geophys res 89:9679–9681

Sievering H, Boatman J, Galloway J, Keene W, Kim Y, Luria M, Ray J (1991) Heterogeneous sulfur conversion in sea-salt aerosol particles: The role of aerosol water content and size distribution. Atmos environ 25A:1479–1487

Sievering H, Boatman J, Gorman E, Kim Y, Anderson L, Ennis G, Lurai M, Pandis S (1992) Removal of sulphur from the marine boundary layer by ozone oxidation in sea-salt aerosols. Nature 360:571–573

Sievering H, Lerner B, Slavich J, Anderson J, Posfai M, Cainey J (1999) O_3 oxidation of SO_2 in sea-salt aerosol water: Size distribution of non-sea-salt sulphate during ACE-1. J geophys res 104:21707–21717

Silver G, Fall R (1995) Characterization of aspen isoprene synthase, an enzyme responsible for leaf isoprene emission to the atmosphere. J biol chem 270:13010–13016

Silver JA, Bomse DS, Stanton AC (1991) Diode laser measurements of trace concentrations of ammonia in an entrained-flow coal reactor. Appl optics 30:1505–1511

Simmonds PG, Seuring S, Nickless G, Derwent RG (1997) Segregation and interpretation of ozone and carbon monoxide measurements by air mass origin at the TOR station Mace Head, Ireland from 1987 to 1995. J atmos chem 28(1–3):45–59

Simo R, Pedros-Alio C (1999) Role of vertical mixing in controlling the oceanic production of dimethylsulphide. Nature 402:396–399

Simpson D, Anderson-Skold Y, Jenkin ME (1993) Updating the chemical scheme for the EMEP MSC-W oxidant model: Current status. EMEP MSC-W Note 2/93

Simpson D, Winiwarter W, Cinderby S, Ferreiro A, Guenther A, Hewitt CN, Janson R, Khalil A, Owen S, Pierce T, Puxbaum H, Shearer M, Steinbrecher R, Svennson B, Tarrason L, Oquist M (1999) Inventorying emissions from nature in Europe. J geophys res 104:8113–8152

Singh HB (1977) Atmospheric halocarbons: Evidence in favor of reduced averaged hydroxyl radical concentration in the troposphere. Geophys res lett 4:101–104

Singh HB (1987) Reactive nitrogen in the troposphere. Environ sci technol 21:320–327

Singh HB, Kasting JF (1988) Chlorine-hydrocarbon photochemistry in the marine troposphere and stratosphere. J atmos chem 7:261–285

Singh HB, Salas LJ (1982) Measurement of selected light hydrocarbons over the Pacific Ocean: Latitudinal and seasonal variations. Geophys res lett 9:842–845

Singh HB, Salas LJ, Stiles RE (1983) Selected man-made halogenated chemicals in the air and oceanic environment. J geophys res 88:3675–3683

Singh HB, Herlth D, Kolyer R, Salas L, Bradshaw JD, Sandholm ST, Davis DD, Kondo Y, Koike M, Talbot R, Gregory GL, Sachse GW, Browell E, Blake D, Rowland F, Newell R, Merrill J, Heikes B, Liu S, Crutzen P, Kanakidou M (1996a) Reactive nitrogen and ozone over the western Pacific: Distribution, partitioning and sources. J geophys res 101:1793–1808

Singh HB, Gregory GL, Anderson B, Browell E, Sachse GW, Davis DD, Crawford J, Bradshaw JD, Talbot R, Blake DR, Thornton D, Newell R, Merrill J (1996b) Low ozone in the marine boundary layer of the tropical Pacific Ocean: Photochemical loss, chlorine atoms, and entrainment. J geophys res 101:1907–1917

Singh HB, Thakur AN, Chen YE, Kanakidou M (1996c) Tetrachloroethylene as an indicator of extremely low Cl atom concentrations in the troposphere. Geophys res lett 23:1529–1532

Singh HB, Herlth D, Kolyer R, Viezee W, Salas LJ, Bradshaw JD, Sandholm ST, Talbot RW, Gregory GL, Anderson B, Sachse GW, Browell E, Bachmeier AS, Blake DR, Heikes B, Jacob D, Fuelberg HE (1996d) Impact of biomass burning emissions on the composition of the south Atlantic troposphere: Reactive nitrogen and ozone. J geophys res 101(D19):24203–24219

Singh HB, Chen Y, Thakur AN, Kondo Y, Talbot RW, Gregory GL, Sachse GW, Blake DR, Bradshaw JD, Wang Y, Jacob DJ (1998) Latitudinal distribution of reactive nitrogen in the free troposphere over the Pacific Ocean in late winter/early spring. J geophys res 103:28237–28246

Singh HB, Thompson A, Schlager H (1999) SONEX airborne mission and coordinated POLINAT-2 activity: Overview and accomplishments. Geophys res lett 26:3053–3056

Singh HB, iezee W, Chen Y, Bradshaw J, Sandholm S, Blake D, Blake N, Heikes B, Snow J, Talbot R, Gregory G, Sachse G, Vay S (2000) Biomass burning influences on the composition of the remote south Pacific troposphere: Analysis based on observations from PEM-Tropics-A. Atmos environ 34:635–644

Singh HB, Chen Y, Staudt A, Jacob D, Blake D, Heikes B, Snow J (2001) Evidence from the Pacific troposphere for large global sources of oxygenated organic compounds. Nature 410:1078–1081

Sirju A-P, Shepson PB (1995) Laboratory and field investigation of the DNPH cartridge technique for the measurement of atmospheric carbonyl compounds. Environ sci technol 29:384–392

Sirois A, Barrie LA (1999) Arctic lower tropospheric aerosol trends and composition at Alert, Canada: 1980–1995. J geophys res 104:11599–11618

Skamarock W, Oliger J, Street RL (1989) Adaptive grid refinement for numerical weather prediction. J comput phys 80:27–60

Skamarock WC, Powers JG, Barth M, Dye JE, Matejka T, Bartels D, Baumann K, Stith J, Parrish DD, Hubler G (2000) Numerical simulations of the 10 July STERAO/Deep Convection Experiment convective system: Kinematics and transport. J geophys res 105:19973–19990

Skiba U, Smith KA (2000) The control of nitrous oxide emissions from agricultural and natural soils. Global change science 2: 379–386

Skiba U, Fowler D, Smith KA (1997) Nitric oxide emisions from agricultural soils in temperate and tropical climates: Sources, controls and mitigation options. Nutr cycl agroecosys 48:139–153

Skudlark J, Church T (1993) Atmospheric input of inorganic nitrogen to Delaware Bay. Estuaries 16:747–759

Slingo A (1990) Sensitivity of the earth's radiation budget to changes in low clouds. Nature 343:49–51

Slinn SA, Slinn WGN (1980) Predictions for particle deposition on natural waters. Atmos environ 14:1013–1016

Sloane CS (1983) Optical properties of aerosols: Comparison of measurements with model calculations. Atmos environ 17:409–416

Smethie WM Jr., Takahashi T, Chipman DW, Ledwell JR (1985) Gas exchange and CO_2 flux in the tropical Atlantic Ocean determined from ^{222}Rn and pCO_2 measurements. J geophys res 90:7005–7022

Smith KA, Dobbie KE, Ball BC, Bakken LR, Sitaula BK, Hansen S, Brumme R, Borken W, Christensen S, Priemé A, Fowler D, MacDonald JA, Skiba U, Klemedtsson L, Kasimir-Klemedtsson A, Degórska A, Orlanski P (2000) Oxidation of atmospheric methane in N. European soils, comparison with other ecosystems uncertainties in the global terrestrial sink. Glob change biol 6:791–803

Smith N, Plane JMC, Nien C-F, Solomon PA (1995) Nighttime radical chemistry in the San Joaquin Valley. Atmos environ 29:2887–2897

Smith W, Harrison W, Hinton D, Parsons V, Larar A, Revercomb H, Huang A, Velden C, Menzel P, Peterson R, Bingham G, Huppi R (2001) GIFTS – A system for wind profiling from geostationary satellites. In: 5th International Winds Workshop. EUMETSAT, Lorne, Australia, pp 253–258

Sokolik IN, Toon OB (1996) Direct radiative forcing by anthropogenic airborne mineral aerosols. Nature 381:681–683

Sokolik IN, Toon OB, Bergstrom RW (1998) Modeling the radiative characteristics of airborne mineral aerosols at infrared wavelenghts. J geophys res 103:8813–8826

Solberg S, Schmidtbauer N, Semb A, Stordal F (1996) Boundary-layer ozone depletion as seen in the Norwegian Arctic in spring. J atmos chem 23:301–332

Solomon R, Baker M (1998) Lightning flash rate and type in convective storms. J geophys res 103:14041–14057

Sørensen S, Falbe-Hansen H, Mangoni M, Hjorth J, Jensen NR (1996) Observation of DMSO and $CH_3S(O)OH$ from the gas phase reaction between DMS and OH. J atmos chem 24:299–315

SPARC/IOC/GAW (1998) Assessment of trends in the vertical distribution of ozone. In: Harris RHN, Phillips C (eds) SPARC report no. 1. WMO ozone research and monitoring: Project report no. 43. SPARC Office, Verrières le Buisson Cedex, France

Spicer CW, Chapman EG, Finlayson-Pitts BJ, Plastridge RA, Hubbe JM, Fast JD, Berkowitz CM (1998) Observations of molecular chlorine in coastal air. Nature 394:352–356

Spiro PA, Jacob DJ, Logan JA (1992) Global inventory of sulfur emissions with $1 \times 1°$ resolution. J geophys res 97:6023–6036

Spivakovsky C, Logan JA, Montzka SA, Balkanski YJ, Foreman-Fowler M, Jones DBA, Horowitz LW, Fusco AC, Brenninkmeijer CAM, Prather MJ, Wofsy SC, McElroy MB (2000) Three-dimensional climatological distribution of tropospheric OH: Update and evaluation. J geophys res 105:8931–8980

Spix C, Wichmann HE (1996) Daily mortality in Köln, West Germany 1975–85: Process of epidemiological model building and results. J epidemiol commun h 50(Suppl 1):52–57

Spix C, Heinrich J, Dockery DW, Schwartz J, Volksch G, Schwinkowski K, Collen C, Wichmann HE (1993) Air pollution and daily mortality in Erfurt, East Germany, 1980–1989. Environ health persp 101:518–526

Staehelin J, Thudium J, Buehler R, Voltz-Thomas A, Graber W (1994) Trends in surface ozone concentrations at Arosa (Switzerland). Atmos environ 28:75–87

Staffelbach T, Neftel A, Stauffer B, Jacob D (1991) A record of the atmospheric methane sink from the formaldehyde in polar ice cores. Nature 349:603–605

Staudt AC, Jacob DJ, Logan JA, Bachiochi D, Krishnamurti TN, Sachse GW (2001) Continental sources, transoceanic transport, and interhemispheric exchange of carbon monoxide over the Pacific. J geophys res 106:32571–32589

Stauffer B, Lochbronner E, Oeschger H, Schwander J (1988) Methane concentration in the glacial atmosphere was only half that of the preindustrial Holocene. Nature 332:812–814

Stecher HA III, Luther GW III, MacTaggart DL, Farwell SO, Crossley DR, Dorko WD, Goldan PD, Beltz N, Krischke U, Luke WT, Thornton DC, Talbot RW, Lefer BL, Scheuer EM, Benner RL, Wu J, Saltzman ES, Gallagher MS, Ferek RJ (1997) Results of the Gas-Phase Sulfur Intercomparison Experiment (GASIE): Overview of experimental setup, results and general conclusions. J geophys res 102:16219–16236

Stefels J, Dijkhuizen L (1996) Characteristics of DMSP-lyase in *Phaeocystis* sp. (Prymnesiophyceae). Mar ecol-prog ser 131:307–313

Steinbrecher R, Ziegler H (1997) Monoterpene production by trees. In: Rennenberg H (ed) Trees: Contributions to modern tree physiology. Backhuys, pp 119–138

Steinke M, Daniel C, Kirst GO (1996) DMSP lyase in marine micro- and macroalgae: Intraspecific differences in cleavage activity. In: Kiene RP, Visscher PT, Keller MD (eds) Biological and environmental chemistry of DMSP and related sulfonium compounds. Plenum Press, pp 317–324

Stenchikov G, Dickerson R, Pickering K, Ellis W Jr., Doddridge B, Kondragunta S, Poulida O (1996) Stratosphere-troposphere exchange in a midlatitude mesoscale convective complex, 2. Numerical simulations. J geophys res 101:6837–6851

Stern DI, Kaufmann RK (1996) Estimates of global anthropogenic methane emissions 1860–1993. Chemosphere 33:159–176

Stern JE, Flagan RC, Grosjean D, Seinfeld JH (1987) Aerosol formation and growth in atmospheric aromatic hydrocarbon photooxidation. Environ sci technol 21:1224–1230

Stevenson DS, Johnson CE, Collins WJ, Derwent RG, Shine KP, Edwards JM (1998) Evolution of tropospheric ozone radiative forcing. Geophys res lett 25:3819–3822

Stevenson DS, Johnson CE, Derwent RG, Edwards JM (2000) Future tropospheric ozone radiative forcing and methane turnover: The impact of climate change. Geophys res lett 27:2073–2076

Stewart RW, Thompson AM (1995) Kinetic data imprecisions in photochemical rate calculations: Means, medians and temperature dependence. J geophys res 101:11695–11698

Stocks BJ, Wotton BM, Flannigan MD, Fosberg MA, Cahoon DR, Goldammer JG (1999) Boreal forest fire regimes and climate change. In: Wengen-98 Workshop on Global Change Research: Biomass burning and its inter-relationships with the climate system. Wengen, Switzerland

Stocks BJ, Fosberg MA, Wotton BM, Lynham TJ, Ryan KC (2000) Climate change and forest fire activity in North American boreal forests. In: Kasischke ES, Stocks BJ (eds) Fire, climate change, and carbon cycling in the boreal forest. Springer-Verlag, Berlin, pp 368–376

Stocks BJ, Wotton BM, Flannigan MD, Fosberg MA, Cahoon DR, Goldammer JG (2001) Boreal forest fire regimes and climate change. In: Beniston M, Verstraete MM (eds) Remote sensing and climate modeling: Synergies and limitations. Kluwer Academic, Dordrecht, Boston, pp 233–246

Stockwell DZ, Chipperfield MP (1999) A tropospheric chemical-transport model: Development and validation of the model transport schemes. Q j roy meteor soc 125:1747–1783

Stockwell WR (1986) A homogeneous gas phase mechanism for use in a regional acid deposition model. Atmos environ 20:1615–1632

Stohl A, Trickl T (1999) A textbook example of long-range transport: Simultananeous observation of ozone maxima of stratospheric and North American origin in the free troposphere over Europe. J geophys res 104:30445–30462

Stolzenburg MR, McMurry PH (1991) An ultrafine aerosol condensation nucleus counter. Aerosol sci tech 14:48–65

Stott PA, Tett SFB, Jones GS, Allen MR, Mitchell JFB, Jenkins GJ (2000) External control of 20th century temperature by natural and anthropogenic forcings. Science 290:2133–2137

Strey R, Viisanen Y (1995) Measurement of the molecular content of binary nuclei. 3. Use of the nucleation rate surfaces for the water-alcohol series. J chem phys 103:4333–4315

Ström J, Ohlsson S (1998) Real-time measurement of absorbing material in contrail ice using a counterflow virtual impactor. J geophys res 103:8737–8741

Stroppiana D, Pinnock S, Grégoire J-M (2000) The global fire product: Daily fire occurrence, from April 1992 to December 1993, derived from NOAA-AVHRR data. Int J remote sens 21:1279–1288

Stull RB (1988) An introduction to boundary layer meteorology. Kluwer Academic, Dordrecht

Sturges WT, Cota GF, Buckley PT (1992) Bromoform emission from Arctic ice algae. Nature 358:660–662

Sturges WT, Sullivan CW, Schnell RC, Heidt LE, Pollock WH (1993) Bromoalkane production by Antarctic ice algae. Tellus 45B:120–126

Sturges WT, McIntyre HP, Penkett SA, Chapellaz J, Barnola JM, Mulvaney R, Atlas E, Stroud V (2001) Methyl bromide, other brominated methanes, and methyl iodide in polar firn air. J geophys res 106D:1595–1606

Stutz J, Hebestreit K, Alicke B, Platt U (1999) Chemistry of halogen oxides in the troposphere: Comparison of model calculations with recent field data. J atmos chem 34(1):65–85

Sugawara S, Nakazawa T, Inoue G, Machida T, Mukai H, Vinnichenko NK, Khattatov VU (1996) Aircraft measurements of the stable carbon isotopic ratio of atmospheric methane over Siberia. Global biogeochem cy 10(2):223–231

Suhre K, Cammas J-P, Nedelec P, Rosset R, Marenco A, Smit HGJ (1997) Ozone-rich transients in the upper equatorial Atlantic troposphere. Nature 388:661–663

Sumner AL, Shepson PB (1999) Snow pack production of formaldehyde and its effect on the arctic troposphere. Nature 398:230–233

Sun J, Massman W (1999) Ozone transport during the California ozone deposition experiment. J geophys res 104:11939

Sunyer J, Castellsague J, Saez M, Tobias A, Anto JM (1996) Air pollution and mortality in Barcelona. J epidemiol commun h 50(Suppl 1):76–80

Tabazadeh A, Santee ML, Danilin MY, Pumphrey HC, Newman PA, Hamil PJ, Mergenthaler JL (2000) Quantifying denitrification and its effect on ozone recover. Science 288:1407–1411

Takahashi K, Matsumi Y, Kawasaki M (1996) Photodissociation processes of ozone in the Huggins band at 308–326 nm: Direct observation of O(1D2) and O(3Pj) products. J chem phys 100:4084–4089

Talbot RW, Dibb JE, Lefer BL, Bradshaw JD, Sandholm ST, Blake DR, Blake NJ, Sachse GW, Collins JE Jr., Heikes BG, Merrill JT, Gregory GL, Anderson BE, Singh HB, Thornton DC, Bandy AR, Pueschel RF (1997) Chemical characteristics of continental outflow from Asia to the troposphere over the western Pacific Ocean during February–March 1994: Results from PEM-West B. J geophys res 102:28255–28274

Talbot RW, Dibb JE, Scheuer EM, Kondo Y, Koike M, Singh HB, Salas L, Fukui Y, Ballenthin JO, Meads RF, Miller TM, Hunton DE, Viggiano AA, Blake DR, Blake NJ, Atlas E, Flocke F, Jacob DJ, Jaeglé L (1999) Reactive nitrogen budget during the NASA SONEX mission. Geophys res lett 26:3057–3060

Tan D, Faloona I, Simpas JB, Shepson PB, Couch TK, Sumner AL, Carroll MA, Thornberry T, Apel E, Riemer D, Stockwell W (2001) HO_x budgets in a deciduous forest: Results from the PROPHET summer 1998 campaign. J geophys res 106:24407–24427
Tang IN, Munkelwitz HR (1994) Water activities, densities and refractive indices of aqueous sulfate and nitrate droplets of atmospheric importance. J geophys res 99:18801–18808
Tang IN, Tridico AC, Fung KH (1997) Thermodynamic and optical properties of sea salt aerosols. J geophys res 102:23269–23275
Tang KW (2000) Dynamics of dimethylsulfoniopropionate (DMSP) in a migratory grazer: A laboratory simulation study. J exp mar biol ecol 243:283–293
Tanner DJ, Jefferson A, Eisele FL (1997) Selected ion chemical ionization mass spectrometric measurement of OH. J geophys res 102:6415–6425
Tanré DL, Remer A, Kaufman YJ, Herman M, Mattoo S (1997) Remote sensing of aerosol properties over oceans using the MODIS/EOS spectral radiances. J geophys res 102:16971–16988
Tanré DL, Remer A, Kaufman YJ, attoo S, Hobbs PV, Livingston JM, Russell PB, Smirnov A (1999) Retrieval of aerosol optical thickness and size distribution over ocean from the MODIS Airborne Simulator during TARFOX. J geophys res 104:2261–2278
Taylor KE, Penner JE (1994) Response of the climate system to atmospheric aerosols and greenhouse gases. Nature 369:734–737
Taylor SR, McLennan SM (1985) The continental crust: Its composition. Blackwell Science, Oxford
Tegen I (1999) Reply. J geophys res 104:4249–4250
Tegen I, Fung I (1994) Modeling of mineral dust in the atmosphere: Sources, transport and optical thickness. J geophys res 99: 22897–22914
Tegen I, Fung I (1995) Contribution to the atmospheric mineral aerosol load from land surface modification. J geophys res 100:18707–18726
Tegen I, Miller R (1998) A general circulation model study on the interannual variability of soil dust aerosol. J geophys res 103:25975–25995
Tegen I, Hollrig P, Chin M, Fung I, Jacob D, Penner J (1997) Contribution of different aerosol species to the global aerosol extinction optical thickness: Estimates from model results. J geophys res 102:23895–23915
Thakur AN, Singh HB, Mariani P, Chen Y, Wang Y, Jacob DJ, Brasseur GP, Muller JF, Lawrence M (1999) Distribution of reactive nitrogen species in the remote free troposphere: Data and model comparisons. Atmos environ 33(9):1403–1422
Thompson AM (1992) The oxidizing capacity of the earth's atmosphere: Probable past and future changes. Science 256:1157–1168
Thompson AM, Cicerone RJ (1986) Possible perturbations to atmospheric CO, CH_4 and OH. J geophys res 91:10853–10864
Thompson AM, Hudson RD (1999) Tropical tropospheric ozone (TTO) maps from Nimus 7 and Earth Probe TOMS by the modified-residual method: Evaluation with soundes, ENSO signals, and trends from Atlantic regional time series. J geophys res 104:26961–26975
Thompson AM, Chappellaz JA, Fung IY, Kucsera TL (1993) Atmospheric methane increase since the last glacial maximum. 2: Interactions with oxidants. Tellus 45B:242–257
Thompson AM, Pickering KE, McNamara DP, Schoeberl MR, Hudson RD, Kim J-H, Browell EV, Kirchhoff VWJH, Nganga D (1996a) Where did tropospheric ozone over southern Africa and the tropical Atlantic come from in October 1992? Insights from TOMS, GTE TRACE-A, and SAFARI 1992. J geophys res 101:24251–24278
Thompson AM, Diab R, Bodeker G, M Z, Coetzee G, Archer C, McNamara D, Pickering K, Combrink J, Fishman J, Nganga D (1996b) Ozone over southern Africa during SAFARI-92 TRACE A. J geophys res 101(D19):23793–23807
Thompson AM, Tao W-K, Pickering KE, Scala JR, Simpson J (1997) Tropical deep convection and ozone formation. B am meteorol soc 78:1043–1054
Thompson AM, Sparling LC, Kondo Y, Anderson BE, Gregory GL, Sachse GW (1999) Perspectives on NO, NO_y and fine aerosol sources and variability during SONEX. Geophys res lett 26(20):3073–3076
Thorpe RB, Law KS, Bekki S, Pyle JA (1996) Nisbet – Is methane-driven deglaciation consistent with the ice core record? J geophys res 101:28627–28635
Thouret V, Marenco A, Nedelec P, Grouhel C (1998a) Ozone climatologies at 9–12 km altitude as seen by the MOZAIC airborne program between September 1994 and August 1996. J geophys res 103:25653–25679
Thouret V, Marenco A, Logan J, Nedelec P, Grouhel C (1998b) Comparisons of ozone measurements from the MOZAIC airborne program and the ozone sounding network at eight locations. J geophys res 103:25695–25720
Thunis P, Cuvelier C (2000) Impact of biogenic emissions on ozone formation in the Mediterranean area: A BEMA modelling study. Atmos environ 34:467–481
Thurston GD, Ito K (1999) Epidemiological studies of ozone exposure effects. In: Holgate ST, Samet JM, Koren HS, Maynard RL (eds) Air pollutants and effects on health. Academic Press, London, pp 485–510
Tiao GC, Reisel GC, Pedrick JH, Allenby GM, Mateer CL, Miller AJ, Deluisi JJ (1986) A statistical analysis of ozonesonde data. J geophys res 91:13121–13136
Tiedtke M (1989) A comprehensive mass flux scheme of cumulus parameterization in large-scale models. Mon weather rev 23:2505
Timmreck C (2001) Three dimensional simulation of stratospheric background aerosol: First results of a multi-annual GCM simulation. J geophys res 106:28313–28332
Timmreck C, Graf H-F, Feichter J (1999) Simulation of Mt. Pinatubo volcanic aerosol with the Hamburg climate model ECHAM4. Theor appl climatol 62:85–108
Tingey DT, Taylor GE Jr. (1982) Variation in plant response to ozone: A conceptual model of physiological events. In: Unsworth MH, Ormrod DF (eds) Effects of gaseous air pollution in agriculture and horticulture. Butterworth Scientific, London, pp 113–138
Tobias HJ, Kooiman PM, Docherty KS, Ziemann PJ (2000) Real-time chemical analysis of organic aerosols using a thermal desorption particle beam mass spectrometer. Aerosol sci tech 33:170–190
Toon OB, Miake-Lye RC (1998) Subsonic aircraft: Contrail and cloud effects special study (SUCCESS). Geophys res lett 25:1109–1112
Torres AL, Thompson AM (1993) Nitric oxide in the equatorial Pacific boundary layer: SAGA 3 measurements. J geophys res 98:16949–16954
Torres O, Bhartia PK, Herman JR, Ahmad Z, Gleason J (1998) Derivation of aerosol properties from satellite measurements of backscattered ultraviolet radiation: Theoretical basis. J geophys res 103:17099–17110
Touloumi G, Pocock J, Katsouyanni K, Trichopoulos D (1994) Short-term effects of air pollution on daily mortality in Athens: A time-series analysis. Int J chem kinet 23:957–967
Touloumi G, Samoli E, Katsouyanni K (1996) Daily mortality and "winter type" air pollution in Athens, Greece: A time-series analysis within the APHEA project. J epidemiol commun h 50(Suppl 1):47–51
Trainer M, Williams EJ, Parrish DD, Buhr MP, Allwine EJ, Westberg HH, Fehsenfeld FC, Liu SC (1987) Models and observations of the impact of natural hydrocarbons on rural ozone. Nature 329:705–707
Treguer P, Pondaven P (2000) Silica control of carbon dioxide. Nature 406:358–359
Tremblay A, Leighton H (1986) A three-dimensional cloud chemistry model. J clim appl meteorol 25:652–671
Trenberth K, Guillemot C (1998) Evaluation of the atmospheric moisture and hydrological cycle in the NCEP/NCAR reanalyses. Clim dynam
Tuazon EC, Atkinson R (1990) A product study of the gas-phase reaction of isoprene with the OH radical in the presence of NO_x. Int J chem kinet 22:1221–1236
Tuckermann M, Gölz ARC, Lorenzen-Schmidt H, Senne T, Stutz J, Trost B, Unold W, Platt U (1997) DOAS-observation of halogen radical-catalysed arctic boundary layer ozone destruction during the ARCTOC-campaigns 1995 and 1996 in Ny-Alesund, Spitsbergen. Tellus 49B:533–555

Tuovinen J-P, Barrett K, Styve H (1994) Transboundary acidifying pollution in Europe: Calculated fields and budgets 1985–93. Report EMEP/MSC-W, The UN Economic Commission for Europe (ECE) Co-operative Programme for Monitoring and Evaluation of the Long-range Transmission of Air Pollutants in Europe (EMEP)

Turner CR, Wells RB, Olbrich KA (1996c) Deposition chemistry in South Africa: Air pollution and its impacts on the South African Highveld Clevelant. 80–85

Turner DR, Hunter KA (2001) The biogeochemistry of iron in seawater. Lupac Series on Analytical and Physical Chemistry of the Environmental Systems, Vol 7, J. Wiley, London

Turner NC, Waggoner PE, Rich S (1974) Removal of ozone from the atmosphere by soil and vegetation. Nature 250:486

Turner SM, Nightingale PD, Broadgate W, Liss PS (1995) The distribution of dimethyl sulphide in Antarctic waters and sea ice. Deep-sea res II 42:1059–1080

Turner SM, Malin G, Nightingale PD, Liss PS (1996a) Seasonal variation of dimethyl sulphide in the North Sea and an assessment of fluxes to the atmosphere. Mar chem 54:245–262

Turner SM, Nightingale PD, Spokes LJ, Liddicoat MI, Liss PS (1996b) Increased dimethyl sulphide concentrations in sea water from *in situ* iron enrichment. Nature 383:513–516

Turnipseed AA, Ravishankara AR (1993) The atmospheric oxidation of dimethylsulfide: Elementary steps in a complex mechanism. In: Restelli G, Angeletti G (eds) Dimethylsulfide: Oceans, atmosphere and climate. Kluwer Academic, Norwell, Mass., pp 185–195

Turnipseed AA, Barone SB, Ravishankara AR (1992) Observation of CH_3S addition to O_2 in the gas phase. J chem phys 96:7502–7505

Twohy CH, Austin PH, Charlson RJ (1989) Chemical consequences of the initial diffusional growth of cloud droplets: A clean marine case. Tellus 41B:51–60

Twomey S (1959) The nuclei of natural cloud formation, Part II: The supersaturation on natural clouds and variation of cloud droplet concentration. Geofis pura appl 43:243–249

Twomey S (1974) Pollution and the planetary albedo. Atmos environ 8:1251–1256

Tyler SC, Brailsford GW, Yagi K, Minami K, Cicerone RJ (1994) Seasonal variations in methane flux and $^{13}CH_4$ values for rice paddies in Japan and their implications. Global biogeochem cy 8:1–12

Tyndall GS, Ravishankara AR (1989) Kinetics and mechanism of the reactions of CH_3S with O_2 and NO_2 at 298 K. J chem phys 93:2426–2435

Tyndall GS, Cox RA, Granier C, Lesclaux R, Moortgat GK, Pilling MJ, Ravishankara AR, Wallington TJ (2001) Atmospheric chemistry of small organic peroxy radicals. J geophys res 106: 12157–12182

Uher G, Andreae MO (1997) Photochemical production of carbonyl sulfide in North Sea water: A process study. Limnol oceanogr 42:432–442

Ulshöfer VS (1995) Photochemische Produktion von Carbonylsulfid im Oberflächenwasser der Ozeane und Gasaustausch mit der Atmosphaere

Ulshöfer VS, Floeck O, Uher G, Andreae MO (1996) Photochemical production and air-sea exchange of carbonyl sulfide in the eastern Mediterranean Sea. Mar chem 53:25–39

United Nations Environment Programme Staff (1998) Environmental effects of ozone depletion: 1998 assessment. United Nations Environment Programme (UNEP), Nairobi

US Environmental Protection Agency (1991) Air quality criteria for carbon monoxide. USEPA, Office of Research and Development, Washington D.C.

US Environmental Protection Agency (1993) Air quality criteria for oxides of nitrogen. USEPA, Environmental Criteria and Assessment Office, Research Triangle Park, NC

US Environmental Protection Agency (1997) National air pollutant emissions trends 1990–1996. Office of Air Quality and Standards, US Environmental Protection Agency, Washington, D.C.

US Environmental Protection Agency (1998) National air pollutants emission trends report Procedures, 1900–1996. Office of Air Quality and Standards, US Environmental Protection Agency, Research Triangle Park, NC

US Environmental Protection Agency (2000) National air pollutants emission trends report, 1900–1998. Office of Air Quality and Standards, US Environmental Protection Agency, Research Triangle Park, NC

US Environmental Protection Agency (2001a) Procedures document for national emission inventory, criteria air pollutants, 1985–1999. Office of Air Quality and Standards, US Environmental Protection Agency, Research Triangle Park, NC

US Environmental Protection Agency (2001b) Latest findings on national air quality: 2000 status and trends. Office of Air Quality and Standards, US Environmental Protection Agency, Research Triangle Park, NC

Vaghjiani GL, Ravishankara AR (1991) Rate coefficient for the reaction of OH with CH_4: Implications to the atmospheric lifetime and budget of methane. Nature 350:406–409

Valentini R, Greco S, Seufert G, Bertin N, Ciccioli P, Cecinato A, Brancaleoni E, Frattoni M (1997) Fluxes of biogenic VOC from Mediterranean vegetation by trap enrichment relaxed eddy accumulation. Atmos environ 31(SI):229–238

Valero FP, Herman J, Minnis P, Collins WD, Sadourny R, Wiscombe W, Lubin D, Ogilvie K (1999) Triana: A deep space earth and solar observatory. National Academy of Sciences

Valigura RA, Luke WT, Artz RS, Hicks BB (1996) Atmospheric nutrient inputs to coastal areas: Reducing the uncertainties. NOAA, Silver Spring, MD

van der Gon HU Neue (1995) Influence of organic matter incorporation on the methane emission from a wetland rice field. Global biogeochem cy 9:11–22

van Dingenen R, Raes F (1993) Ternary nucleation of methane sulphonic acid, sulphuric acid and water vapour. J aerosol sci 24:1–17

van Dorland R, Dentener FJ, Lelieveld J (1997) Radiative forcing due to tropospheric ozone and sulfate aerosols. J geophys res 102:28079–28100

van Duyl FC, Gieskes WWC, Kop AJ, Lewis WE (1998) Biological control of short-term variation inthe concentration of DMSP and DMS during a *Phaeocystis* spring bloom. J sea res 40:221–231

Varner RK, Crill PM, Talbot RW, Shorter JH (1999) An estimate of the uptake of atmospheric methyl bromide by agricultural soils. Geophys res lett 26:727–730

Verchot LV, Davidson EA, Cattanio JH, Ackerman IL, Erickson HE, Keller M (1999) Land use change and biogeochemical controls of nitrogen oxide emissions from soils in eastern Amazon. Global biogeochem cy 13(1):31–46

Verhoeff AP, Hoek G, Schwartz J, van Wijnen JH (1996) Air pollution and daily mortality in Amsterdam, The Netherlands. Epidemiology 7:225–230

Vetter Y, Sharp JH (1993) The influence of light intensity on dimethyl sulfide production by a marine diatom. Limnol oceanogr 38:419–425

Vigotti MA, Rossi G, Bisanti L, Zanobeti A, Schwartz J (1996) Short-term effects of urban air pollution on respiratory health in Milan: 1980–1989. J epidemiol commun h 50(Suppl 1):571–575

Viisanen Y, Kulmala M, Laaksonen A (1997) Experiments on gas-liquid nucleation of sulfuric acid and water. J chem phys 107:920–926

Vila-Guerau J, Duynkerke PG, Zeller K (1995) Atmospheric surface-layer similarity theory applied to chemically reactive species. J geophys res 100:1397

Villalta PW, Howard CJ (1996) Direct kinetics study of the $CH_3C(O)O_2 + NO$ reaction using chemical ion mass spectrometry. J chem phys 100:13624–13628

Vincent JH (1989) Aerosol sampling: Science and practice. J. Wiley, Chichester

Vitousek PM, Sanford RL (1986) Nutrient cycling in moist tropical forest. Annu rev ecol syst 17:137–167

Vitousek PM, Mooney HA, Lubchenco J, Melillo JM (1997) Human domination of Earth's ecosystems. Science 277:494–499

Vogt R, Finlayson-Pitts FP (1994) A diffuse reflectance infrared Fourier-transform spectroscopic (drifts) study of the surface reaction of NaCl with gaseous NO_2 and HNO_3. J phys chem-us 98:3747–3755

Vogt R, Crutzen PJ, Sanders R (1996) A mechanism for halogen release from sea-salt aerosol in the remote marine boundary. Nature 383:327–330

Vogt R, Sander R, Von Glasow R, Crutzen PJ (1999) Iodine chemistry and its role in halogen activation and ozone loss in the marine boundary layer: A model study. J atmos chem 32:375–395

Voisin D, Legrand M, Chaumerliac N (2000) Scavenging of acidic gases (HCOOH, CH_3COOH, HNO_3, HCl, and SO_2) and ammonia in mixed liquid-solid water clouds at the Puy de Dôme mountain (France). J geophys res 105:6817–6835

Voldner EC, Barrie LA, Sirois A (1986) A literature review of dry deposition. Atmos environ 20:2101

Voldner EC, Li YF, Scholtz MT, Davidson KA (2001) $1 \times 1°$ global SO_2 and NO_x 2-level inventory resolved seasonally into emissions sectors and point and area emission sources. IGAC/GEIA Web page

Volz A, Kley D (1988) Evaluation of the Montsouris series of ozone measurements made in the nineteenth century. Nature 332:240–242

Volz-Thomas A, Arnold T, Behmann T, Borrell P, Borrell PM, Burrows JP, Cantrell CA, Carpenter LJ, Clemitshaw KC, Gilge S, Heitlinger M, Klüpfel T, Kramp F, Mihelcic D, Müsgen P, Pätz H-W, Penkett SA, Perner D, Schultz M, Shetter R, Slemr J, Wissenmayer M (1998) Peroxy radical inter comparison exercise: A formal comparison of methods for ambient measurements of peroxy radicals (PRICE-2). In: Final report to the Commission of the European Communities (EV5V-CT91-0042)

von Hobe M (1999) The behavior of carbonyl sulfide in the ocean: Field and modelling studies

Vourlitis GL, Oechel WC, Hastings SJ, Jenkins MA (1993) The effect of soil moisture and thaw depth on CH_4 flux from wet coastal tundra ecosystems on the North Slope of Alaska, Atmospheric methane: Sources, sinks and role in global change. In: Khalil MAK, Shrearer MJ (eds) Proceedings of the NATO Advanced Research Workshop. Chemosphere, Mount Hood, Oregan, pp 329–337

Wadge A (1999) Lead. In: Holgate ST, Samet JM, Koren HS, Maynard RL (eds) Air pollution and health. Academic Press, London, pp 797–812

Wagener R, Nemesure S, Schwartz SE (1997) Aerosol optical depth over oceans: High space and time resolution retrieval and error budget from satellite radiometry. J atmos ocean tech 14:577–590

Wagner T, Platt U (1998) Observation of tropospheric BrO from the GOME satellite. Nature 395:486–490

Wagner T, Leue C, Wenig M, Pfeilsticker K, Platt U (2001) Spatial and temporal distribution of enhanced boundary layer BrO concentrations measured by the GOME instrument aboard ERS-2. J geophys res-atmos 106:24225–24236

Wagnon P, Delmas RJ, Legrand M (1999) Loss of volatile acid species from upper firn layers at Vostok, Antarctica. J geophys res 104:3423–3431

Wahlen M, Tanaka N, Henry R, Deck B, Zeglen J, Vogel JS, Southon J, Shemesh A, Fairbanks A, Broecker W (1989) Carbon-14 in methane sources and in atmospheric methane: The contribution from fossil carbon. Science 245:286–290

Wahlen SC, Reeburgh WS (1992) Interannual variations in tundra methane flux: A 4-year time series at fixed sites. Global biogeochem cy 6:139–160

Wahlen SC, Reeburgh WS, Barber V (1992) Oxidation of methane in boreal forest soils: A comparison of seven measures. Biogeochemistry 16:181–211

Waibel AE, Peter T, Carslaw KS, Oelhaf H, Wetzel G, Crutzen PJ, Poschl U, Tsias A, Reimer E, Fischer H (1999) Arctic ozone loss due to denitrification. Science 283(5410):2064–2069

Wallace JM, Hobbs PV (1977) Atmospheric science: An introductory survey. Academic Press, New York

Walter BP, Heimann M (2000) A process-based, climate sensitive model to derive methane emissions from natural wetlands: Application to five wetland sites, sensitivity to model parameters, and climate. Global biogeochem cy 14:745–765

Walter BP, Matthews E (2000) Modeling global methane emissions from wetlands: Climate-induced variability 1980–present. EOS 81:F75

Walter H (1973) Coagulation and size distribution of condensation aerosols. J aerosol sci 4:1–15

Wang C, Chang JS (1993a) A three-dimensional numerical model of cloud dynamics, microphysics: 1. Concepts and formulation. J geophys res 98:14827–14844

Wang C, Chang JS (1993b) A three-dimensional numerical model of cloud dynamics, microphysics: 3. Redistribution of pollutants. J geophys res 98:16787–16798

Wang C, Chang JS (1993c) A three-dimensional numerical model of cloud dynamics, microphysics: 4. Cloud chemistry and precipitation chemistry. J geophys res 98:16799–16808

Wang C, Crutzen PJ (1995) Impact of a simulated severe local storm on the redistribution of sulfur dioxide. J geophys res 100:11357–11367

Wang C, Prinn RG (2000) On the roles of deep convective clouds in tropospheric chemistry. J geophys res 105:22269–22297

Wang C, Crutzen PJ, Ramanathan V, Williams SF (1995) The role of a deep convective storm over the tropical Pacific Ocean in the redistribution of atmospheric chemical species. J geophys res 100:11509–11516

Wang J, Deeter MN, Gille JC, Bailey PL (1999) Retrieval of tropospheric carbon monoxide profiles from MOPITT: Algorithm description and retrieval simulation. In: Optical spectroscopic techniques and instrumentation for atmospheric and space research III. SPIE Proceedings

Wang S-C, Paulson SE, Grosjean D, Flagan RC, Seinfeld JH (1992) Aerosol formation and growth in atmospheric organic NO_x systems – I. Outdoor smog chamber studies of C7- and C8-hydrocarbons. Atmos environ 26A:403–420

Wang T, Carroll MA, Albercook GM, Owens KR, Duderstadt KA, Markevitch AN, Parrish DD, Holloway JS, Fehsenfeld FC, Forbes G, Ogren J (1996a) Ground-based measurements of NO_x and total reactive oxidized nitrogen (NO_y) at Sable Island, Nova Scotia, during the NARE 1993 summer intensive. J geophys res 101:28991–29004

Wang W-C, Mao H, Isaksen ISA, Fuglestvedt JS, Karlsdottir S (1996b) Effects of climate-chemistry interactions on the radiative forcing of increasing atmospheric methane. In: Bojkov RD, Visconti G (eds) Proceedings from the XVIII Quadrennial Ozone Symposium.pp 821–826

Wang Y, Jacob DJ (1998) Anthropogenic forcing on tropospheric ozone and OH since pre-industrial times. J geophys res 103:31123–31135

Wang Y, Jacob DJ, Logan JA (1998a) Global simulation of tropospheric O_3-NO_x-hydrocarbon chemistry 1. Model formulation. J geophys res 103:10713–10725

Wang Y, Jacob DJ, Logan JA (1998b) Global simulation of tropospheric O_3-NO_x-hydrocarbon chemistry 2. Model evaluation and global ozone budget. J geophys res 103:10727–10755

Wang Y, Jacob DJ, Logan JA (1998c) Global simulation of tropospheric O_3-NO_x-hydrocarbon chemistry 3. Origin of tropospheric ozone and effects of nonmethane hydrocarbons. J geophys res 103:10757–10767

Wang Y-P, Meyer CP, Galbally IE, Smith CJ (1997) Comparisons of field measurements of carbon dioxide and nitrous oxide fluxes with model simulations for a legume pasture in southeast Australia. J geophys res 102:28013–28024

Wanninkhof R (1992) Relationship between gas exchange and wind speed over the ocean. J geophys res 97:7373–7381

Wanninkhof R, McGillis WR (1999) A cubic relationship between air-sea CO_2 exchange and windspeed. Geophys res lett 26:1889–1892

Wanninkhof R, Ledwell JR, Broecker WS (1985) Gas exchange-wind speed relationship measured with sulphur hexafluoride on a lake. Science 227:1224–1226

Wanninkhof R, Ledwell JR, Broecker WS, Hamilton M (1987) Gas exchange on Mono Lake and Crowley Lake, California. J geophys res 92:14567–80

Wanninkhof RW, Asher WE, Weppernig R, Chen H, Schlosser R, Langdon C, Sambrotto R (1993) Gas transfer experiment on Georges Bank using two volatile deliberate tracers. J geophys res 98:20237–48

Ward DE, Hao W-M, Susott RA, Babbitt RA, Shea RW, Kauffman JB, Justice CO (1996) Effect of fuel composition on combustion efficiency and emission factors for African savanna ecosystems. J geophys res 101:23569–23576

Warneck P (1991) Chemical reactions in clouds. Fresen J anal chem 340:585–590

Warneke C, Holzinger R, Hansel A, Jordan A, Lindinger W, Pöschl U, Williams J, Hoor P, Fischer H, Crutzen PJ, Scheeren HA, Lelieveld J (2001) Isoprene and its oxidation products methyl vinyl ketone, methacrolein, and isoprene related peroxides measured online over the tropical rain forest of Surinam in March 1998. J atmos chem 38:167–185

Washington DC (1991) Air quality criteria for carbon monoxide. EPA-600/B-90/045F, Office of Research and Development, Washington DC

Washington DC (1992) Natural sources of atmospheric aerosol particles. In: Schwartz SE, Slinn WGN (eds) Precipitation scavenging and atmosphere-surface exchange. Hemisphere Publishing Corporation, Washington, DC, pp 1617–1639

Washington DC (1996) Scheduled civil aircraft emissions inventories for 1992: Database development and analysis. US National Aeronautics and Space Administration, Washington, D.C.

Watson AJ (1997) Surface Ocean-Lower Atmosphere Study (SOLAS). Global Change Newsletter (IGBP) 31

Watson AJ, Liss PS (1998) Marine biological controls on climate via the carbon and sulphur geochemical cycles. Philos t roy soc b 353:41–51

Watson AJ, Liss PS, Duce R (1991a) Design of a small-scale *in situ* iron fertilization experiment. Limnol oceanogr 36:1960–1965

Watson AJ, Upstill-Goddard RC, Liss PS (1991b) Air-sea exchange in rough and stormy seas, measured by a dual tracer technique. Nature 349:145–147

Watson RT, Rodhe H, Oeschger H, Siegenthaler U (1990) Greenhouse gases and aerosols. In: Houghton JT (ed) Climate change: The IPCC assessment. Cambridge University Press, Cambridge, pp 25–46

Weber RJ, McMurry PH, Eisele FL, Tanner DJ (1995) Measurement of expected nucleation precursor species and 3–500-nm diameter particles at Mauna Loa Observatory, Hawaii. J atmos sci 52:2242–2257

Weber RJ, Marti JJ, McMurry PH, Eisele FL, Tanner DJ, Jefferson A (1997) Measurements of new particle formation and ultrafine particle growth rates at a clean continental site. J geophys res 102:4375–4385

Weber RJ, Stolzenburg MR, Pandis SN, McMurry PH (1998) Inversion of ultrafine condensation nucleus counter pulse height distributions to obtain nanoparticle (~3–10 nm) size distributions. J aerosol sci 29:601–615

Weber RJ, McMurry PH, Mauldin RL III, Tanner DJ, Eisele FL, Clarke AD, Kapustin VN (1999) New particle formation in the remote troposphere: A comparison of observations at various sites. Geophys res lett 26:307–310

Weingartner E, Nyeki S, Baltensperger U (1999) Seasonal and diurnal variation of aerosol size distributions ($10 < D < 750$ nm) at a high-alpine site (Jungfraujoch 3580 m asl). J geophys res 104:26809–26820

Weinstock B, Niki H (1972) Carbon monoxide balance in nature. Science 176:290–292

Weiss PS, Andrews SS, Johnson JE, Zafiriou OC (1995) Photoproduction of carbonyl sulfide in south Pacific Ocean waters as a function of irradiation wavelength. Geophys res lett 22:215–218

Welch KA, Mayewski PA, Whitlow SI (1993) Methane sulfonic acid in coastal Antarctic snow related to sea ice extent. Geophys res lett 20:443–446

Wennberg PO, Hanisco TF, Jaegle L, Jacob DJ, Hintsa EJ, Lanzendorf EJ, Anderson JG, Gao R-S, Keim ER, Donnelly SG, Del LANegro, Fahey DW, McKeen SA, Salawitch RJ, Webster CR, May RD, Herman RL, Proffitt MH, Margitan JJ, Atlas EL, Schauffler SM, Flocke F, McElroy CT, Bui TP (1998) Hydrogen radicals, nitrogen radicals, and the production of O_3 in the upper troposphere. Science 279:49–53

Wennberg PO, Salawitch RJ, Donaldson DJ, Hanisco TF, Lazendorf EJ, Perkins KK, Lloyd SA, Vaida V, Gao RS, Hintsa EJ, Cohen RC, Swartz WH, Kusterer TL, Anderson DE, Done IM (1999) Twilight observations of OH and HO_2 suggest unknown HO_x sources. Geophys res lett 26:1373–1376

Wesely MA, Hicks BB (2000) A review of the current status of knowledge on dry deposition. Atmos environ 34:2261–2282

Wesely ML (1989) Parameterization of surface resistances to gaseous dry deposition in regional-scale numerical models. Atmos environ 23:1293

Wesely ML, Eastman JA, Cook DR, Hicks BB (1978) Daytime variations of ozone eddy fluxes to maize. Bound-lay meteorol 15:361

Wesely ML, Eastman JA, Stedman DH, Yalvac ED (1982) An eddy correlation measurement of NO_2, flux to vegetation and comparison to O_3 flux. Atmos environ 16:815

Wesley ML, Sisterson DL, Hart RL, Drapcho DL, Lee IY (1989) Observations of nitric oxide over grass. J atmos chem 9:447

Wetzel MA, Stowe LL (1999) Satellite-observed patterns in stratus cloud microphysics, aerosol optical thickness and shortwave radiative forcing. J geophys res 104:31286–31299

Wexler A, Prather K (2000) Introduction: Online single particle analysis. Aerosol sci tech 33:1–2

Whitby KT (1978) The physical characteristics of sulfur aerosols. Atmos environ 12:135–159

White WH (1986) On the theoretical and empirical basis for apportioning extinction by aerosols: A critical review. Atmos environ 20:1659–1672

White WH, Dietz D (1984) Does the photochemistry of the troposphere admit more than one steady-state? Nature 309:242–244

Whung P-Y, Saltzman ES, Spencer MJ, Mayewski PA, Gundestrup N (1994) Two-hundred-year record of biogenic sulphur in a south Greenland ice core. J geophys res 99:1147–1156

Wiedensohler A, Aalto P, Covert D, Heintzenberg J, McMurry PH (1994) Intercomparison of four methods to determine size distributions of low concentration (~100 cm-3), ultrafine aerosols ($3 < Dp < 10$ nm) with illustrative data from the Arctic. Aerosol sci tech 21:95–109

Wiedensohler A, Hansson H-C, Orsini D, Wendisch M, Wagner F, Bower KN, Choularton TW, Wells M, Parkin M, Acker A, Wieprecht W, Fachini MC, Lind JA, Fuzzi S, Arends BG, Kulmala M (1997) Night-time formation and occurrence of new particles associated with orographic clouds. Atmos environ 31:2545–2559

Wietlisbach V, Pope CA III, Ackermann-Liebrich U (1996) Air pollution and daily mortality in three Swiss urban areas. Social and preventive medicine 41:107–115

Wild O, Law KS, McKenna DS, Bandy BJ, Penkett SA, Pyle JA (1996) Photochemical trajectory modelling studies of the North Atlantic region during August 1993. J geophys res 101:29269–29288

Wild O, Zhu X, Prather MJ (2000) Fast-J: Accurate simulation of in- and below-cloud photolysis in tropospheric chemical models. J atmos chem 37:245–282

Wildt J, Kley D, Rockel A, Rockel P, Segschneider HJ (1997) Emission of NO from several higher plant species. J geophys res 102:5919

Willeke K, Baron PA (1993) *Aerosol measurement: Principles, techniques, and applications.* Van Nostrand Reinhold, New York

Willey JD, Peaerl HW (1993) Enhancement of chlorophyll a production in Gulf Stream seawater by synthetic vs. natural rain. Mar biol 116:329–334

Williams EJ, Hutchinson GL, Fehesenfeld FC (1989) NO_x and N_2O emissions from soils. Global biogeochem cy 6:447

Williams EJ, Sandholm ST, Bradshaw JD, Schendel JS, Langford AO, Quinn PK, LeBel PJ, Vay SA, Roberts PD, Norton RB, Watkins BA, Buhr MP, Parrish DD, Calvert JG, Fehsenfeld FC (1992) An intercomparison of five ammonia measurement techniques. J geophys res 97:11591–11611

Williams J, Roberts JM, Bertman SB, Stroud CA, Fehsenfeld FC, Baumann K, Buhr MP, Knapp K, Murphy PC, Nowick M, Williams EJ (2000) A method for the airborne measurement of PAN, PPN, and MPAN. J geophys res 105:28943–28960

Williams MR, Fischer TM, Melack JM (1997) Chemical composition and deposition of rain in the Central Amazon, Brazil. Atmos environ 31:207–217

Wilson JJN, Raes F, Saltelli A (1999) M3, a multi modal model for aerosol dynamics. J aerosol sci (submitted)

Wilson R, Spengler JD (1996) Particles in our air: Concentrations and health effects. Harvard University Press, Cambridge, Mass

Wilson SR, Dick AL, Fraser PJ, Whittlestone S (1997) Nitrous oxide flux estimates for South-Eastern Australia. J atmos chem 26(2):169–188

Wingenter OW, Kubo MK, Blake NJ, Smith TW Jr., Blake DR, Rowland FS (1996) Hydrocarbon and halocarbon measurements as photochemical and dynamical indicators of atmospheric hydroxyl, atomic chlorine, and vertical mixing obtained during Lagrangian flights. J geophys res 101:4331–4340

Winker DM, Wielicki BA (1999) The PICASSO-CENA mission. In: Sensors, systems, and next generation satellites. Proc. SPIE, pp 26–36

Winker DM, Couch RH, McCormick MP (1996) An overview of LITE: NASA's Lidar In-space Technology Experiment. Proc IEEE 84:164–180

Winklmayr W, Reischl GP, Lindner AO, Berner A (1991) A new electromobility spectrometer for the measurement of aerosol size distributions in the size range from 1 to 1000 nm. J aerosol sci 22:289–296

Wobrock W, Schell D, Maser R, Jaeschke W, Georgii HW, Wieprecht W, Arends BG, Mols JJ, Kos GPA, Fuzzi S, Facchini MC, Orsi G, Berner A, Solly I, Kruisz C, Svenningsson IB, Wiedensohler A, Hansson HC, Ogren JA, Noone KJ, Hallberg A, Pahle S, Schneider T, Winkler P, Winiwarter W, Colcile R, Choularton TW, Flossmann AI, Borrman S (1994) The Kleiner Feldberg cloud experiment 1990: An overview. J atmos chem 19:3–36

Wolfe GV, Kiene RP (1993) Radio isotope and chemical inhibitor measurements of dimethyl sulfide consumption rates and kinetics in estuarine waters. Mar ecol-prog ser 99:261–269

Wolfe GV, Steinke M (1996) Grazing-activated production of dimethyl sulfide (DMS) by two clones of *Emiliania huxleyi*. Limnol oceanogr 41:1151–1160

Wolfe GV, Steinke M, Kirst GO (1997) Grazing-activated chemical defence in a unicellular marine alga. Nature 387:894–897

Woodcock AH (1953) Salt nuclei in marine air as a function of altitude and wind force. J meteorol 10(5):362–371

Woolf DK (1993) Bubbles and the air-sea transfer velocity of gases. Atmos ocean 31:517–540

Woolf DK, Bowyer PA, Monahan EC (1987) Discriminating between the film drops and jet drops produced by a simulated whitecap. J geophys res 92:5142–5150

World Health Organization (1995) Inorganic lead. In: Environmental health criteria. World Health Organization, Geneva

World Health Organization (1997) Nitrogen oxides. In: Environmental health criteria 188. Geneva

World Health Organization (1999a) Air quality guidelines. World Health Organization, Geneva

World Health Organization (1999b) Carbon monoxide. World Health Organization, Geneva

World Health Organization (2000) Air quality guidelines for Europe. WHO Regional Publications, Copenhagen

World Meteorological Organization (1993) Report of the WMO meeting of experts to consider the aerosol component of GAW: Boulder, Colorado, US, 16–19 December 1991, AREP/ENV, (1992). (GAW-79). (WMO/TD 485), Geneva

World Meteorological Organization (1995) Scientific assessment of ozone depletion: 1994. Global Ozone Research and Monitoring Project, Geneva, Switzerland

World Meteorological Organization (1999) Scientific assessment of ozone Depletion: 1998. Global Ozone Research and Monitoring Project, Geneva, Switzerland

Wright DL (2000) Retrieval of optical properties of atmospheric aerosols from moments of the particle size distribution. J aerosol sci 1–18

Wright DL, McGraw R, Benkovitz CM, Schwartz SE (2000) Six-moment representation of multiple aerosol populations in a sub-hemispheric chemical transformation model. Geophys res lett 27:967–970

Wyser K, Ström J (1998) A possible change in cloud radiative forcing due to aircraft exhaust. Geophys res lett 25:1673–1676

Xiao H, Carmichael GR, Durchenwald JN, Yhornton D, Brady A (1997) Long range transport of SO_x and dust in East Asia during the PEM West-B experiment. J geophys res 102: 28589–28612

Xu J, Imre D, McGraw R, Tang I (1998) Ammonium sulfate: Equilibrium and metastability phase diagrams from 40 to −50 °C. J phys chem b 102:7462–7469

Xu X, Gao J, Dockery DW, Chen Y (1994) Air pollution and daily mortality in residential areas of Beijing, China. Archives of Environmental Health 49:216–222

Yao H, Conrad R (1999) Thermodynamics of methane production in different rice paddy soils from China, the Philippines and Italy. Soil biol biochem 31:463–473

Yi SM, Holsen TM, Zhu X, Noll KE (1997) Sulfate dry deposition measured with a water surface sampler: A comparison to modeled results. J geophys res 102:19695–19705

Yienger JJ, Levy H (1995) Empirical model of global soil-biogenic NO_x emissions. J geophys res 100:11447–11464

Yin F, Grosjean D, Flagan RC, Seinfeld JH (1990a) Photooxidation of dimethyl sulfide and dimethyl disulfide. II Mechanism evaluation. J atmos chem 11:365–399

Yin F, Grosjean D, Seinfeld JH (1990b) Photooxidation of dimethyl sulfide and dimethyl disulfide. I Mechanism development. J atmos chem 11:309–364

Yoch DC, Ansede JH, Rabinowitz KS (1997) Evidence for intracellular and extracellular dimethylsulfoniopropionate (DMSP) lyases and DMSP uptake sites in two species of marine bacteria. Appl environ microb 63:3182–3188

Yokelson RJ, Susott R, Ward DE, Reardon J, Griffith DWT (1997) Emissions from smoldering combustion of biomass measured by open-path Fourier transform infrared spectroscopy. J geophys res 102:18865–18877

Young RW, Carder KL, Betzer PR, Costello DK, Duce RA, DiTullio GR, Tindale NW, Laws EA, ematsu M, Merrill JT, Feely RA (1991) Atmospheric iron inputs and primary productivity: Phytoplankton responses in the North Pacific. Global biogeochem cy 5:119–134

Yu F, Turco RP (1997) The role of ions in the formation and evolution of aircraft plumes. Geophys res lett 24:1927–1930

Yu F, Turco RP (2000) Ultrafine aerosol formation via ion-mediated nucleation. Geophys res lett 27(6):883–886

Yu J, Flagan RC, Seinfeld JH (1998) Identification of products containing -COOH, -OH, and -C=O in atmospheric oxidation of hydrocarbons. Environ sci technol 32:2357–2370

Yu J, Griffin RJ, Cocker DR III, Flagan RC, Seinfeld JH (1999) Observation of gaseous and particulate products of monoterpene oxidation in forest atmospheres. Geophys res lett 26:1145–1148

Yu J, Jeffries HE, Le Lacheur RM (1995) Identifying airborne carbonyl compounds in isoprene atmospheric photooxidation products by their PFBHA oximes using gas chromatography/ ion trap mass spectrometry. Environ sci technol 29:1923–1932

Yuskiewicz BA, Stratmann F, Birmili W, Wiedensohler A, Swietlicki E, Berg O, Zhou J (1999) The effects of in-cloud mass production on atmospheric light scatter. Atmos res 50:265–288

Yvon SA, Plane JMC, Nien C-F, Cooper DJ, Saltzman ES (1996a) Interactions between nitrogen and sulfur cycles in the polluted marine boundary layer. J geophys res 101:1379–1386

Yvon SA, Saltzman ES, Cooper DJ, Bates TS, Thompson AM (1996b) Atmospheric dimethylsulfide cycling at a tropical South Pacific station (12 S, 135 W): A comparison of field and model results. J geophys res 101:6899–6910

Yvon-Lewis SA, Butler JH (1997) The potential effect of oceanic biological degradation on the lifetime of atmospheric CH_3Br. Geophys res lett 24:1227–1230

Zahn A, Brenninkmeijer CAM, Maiss M, Scharffe DH, Crutzen PJ, Hermann M, Wiedensohler A, Heintzenberg J, Güsten H, Heinrich G, Fischer H, Cuijpers JWM, Velthoven PFJV (2000) Identification of extratropical two-way troposphere-stratosphere mixing based on CARIBIC measurements of O_3, CO, and ultrafine particles. J geophys res 105:1527–1535

Zander R, Demoulin P, Ehhalt DH, Schmidt U (1989) Secular increases of the vertical abundance of methane derived from IR solar spectra recorded at the Jungfraujoch Station. J geophys res 94:1129–1139

Zanis P, Monks PS, Schuepbach E, Penkett SA (1999) On the relationship of $HO_2 + RO_2$ during the Free Tropospheric Experiment (FREETEX '96) at the Jungfraujoch Observatory (3580 m above sea level) in the Swiss Alps. J geophys res 104:26913–26926

Zappoli S, Andracchio A, Fuzzi S, Facchini MC, Gelencser A, Kiss G, Krivacsy Z, Molnár A, Mészaros E, Hansson HC, Rosman K, Zebühr Y (1999) Inorganic, organic and macromolecular components of fine aerosol in different areas of Europe in relation to their water solubility. Atmos environ 33:2733–2743

Zellner R, Herrmann H (1995) Free radical chemistry of the aqueous atmospheric phase. In: Clark RJH, Hester RE (eds) Advances in spectroscopy, spectroscopy in environmental science. J. Wiley, London, pp 381–451

Zemp E, Elsasser S, Schindler C, Künzli N, Perruchoud AP, Domenighetti G, Medici T, Ackermann-Liebrich U, Leuenberger P, Monn C, Bolognini G, Bongard J-P, Brändli O, Karrer W, Keller R, Schöni MH, Tschopp J-M, Villiger B, Zellweger J-P, SALPALDIA Team (1999) Long-term ambient air pollution and respiratory symptoms in adults (SAPALDIA Study). Am J resp crit care 159:1257–1266

Zepp RG, Andreae MO (1994) Factors affecting the photochemical production of carbonyl sulfide in seawater. Geophys res lett 21:2813–2816

Zepp RG, Miller WL, Burke RA, Parsons DAB, Scholes MC (1996) Effects of moisture and burning on soil-atmosphere exchange of trace carbon gases in a southern African savanna. J geophys res 101(D19):23699–23706

Zetzsch C, Behnke W (1992) Heterogeneous sources of atomic Cl in the troposphere. Ber bunsen phys chem 96:488–493

Zhang Y, Sunwoo Y, Kotamarthi V, Carmichael GR (1994a) Photochemical oxidant processes in the presence of dust: An evaluation of the impact of dust on particulate nitrate and ozone formation. J appl meteorol 33:813–824

Zhang X, Turpin BJ, McMurry PH, Hering SV, Stolzenburg MR (1994b) Mie theory evaluation of species contributions to 1990 wintertime visibility reduction in the Grand Canyon. Japca J air waste ma 44:153–162

Zhang XQ, McMurry PH (1987) Theoretical analysis of evaporative losses from impactor and filter deposits. Atmos environ 21:1779–1789

Zhao Y (2000) Line pair selections for remote sensing of atmospheric ammonia by use of a coherent CO_2 differential absorption lidar system. Appl optics 39:997–1007

Zhu L, Johnston G (1995) Kinetics and products of the reaction of the vinoxy radical with O_2. J chem phys 99:15114–15119

Zhu X, Prospero JM, Millero FJ, Savoie DL, Brass GW (1992) The solubility of ferric iron in marine aerosol solutions at ambient relative humidities. Mar chem 38:91–107

Zhuang L, Huebert BJ (1996) Lagrangian analysis of the total ammonia budget during Atlantic Stratocumulus Transition Experiment/Marine Aerosol and Gas Exchange. J geophys res 101:4341–4350

Zielinski GA, Mayewski PA, Meeker LD, Whitlow S, Twickler MS (1996) A 110 000-yr record of explosive volcanism from the GISP2 (Greenland) ice core. Quaternary res 45(2):109–118

Ziemke JR, Chandra S, Bhartia PK (1998) Two new methods for deriving tropospheric column ozone from TOMS measurements: Assimilated UARS MLS/HALOE and convective-cloud differential techniques. J geophys res 103(D17):22115–22127

Zimmerman PR, Greenberg JP, Westberg CE (1988) Measurements of atmospheric hydrocarbons and biogenic emission fluxes in the Amazon boundary layer. J geophys res 93:1407–1416

Zmirou D, Barumandzadeh T, Balducci F, Ritter P, Laham G, Ghilardi JP (1996) Short term effects of air pollution on mortality in the city of Lyon, France, 1985–90. J epidemiol commun h 50(Suppl 1):30–35

Zmirou D, Schwartz J, Saez M, Zanobetti A, Wojtyniak B, Touloumi G, Spix C, Ponce de Leon A, Le Moullec Y, Bacharova L, Schouten J, Pönkä KKA (1998) Time-series analysis of air pollution and cause-specific mortality. Epidemiology 9:495–503

Zondlo MA, Barone SB, Tolbert MA (1997) Uptake of HNO_3 on ice under upper tropospheric conditions. Geophys res lett 24:1391–1394

Zufall MJ, Davidson CI (1998) Dry deposition of particles. In: Harrison RM, Grieken REV (eds) Atmospheric particles. J. Wiley, Chichester, pp 425–473

Appendix

A.1 Authors and Reviewers

A.1.1 Authors

Ackermann, Ingmar J.
Ford Forschungszentrum, Germany
Akimoto, Hajime
Frontier Research System for Global Change (IGCR), Japan
Anderson, Tad L.
University of Washington, US
Andreae, Meinrat O.
Max-Planck-Institut für Chemie, Germany
Apel, Eric C.
National Center for Atmospheric Research, US
Artaxo, Paulo
Universidade de São Paulo, Brazil
Atherton, Cynthia S.
Lawrence Livermore National Laboratory, US
Atlas, Elliot L.
National Center for Atmospheric Research (NCAR), US
Barrie, Leonard A.
Pacific Northwest National Laboratory, US
Bates, Timothy S.
NOAA/PMEL-R/E/PM/OCRD, US
Benkovitz, Carmen
Brookhaven National Laboratory, US
Bigg, Keith
Australian Antarctic Society, Australia
Bond, Tami C.
NOAA-PMEL, US
Brasseur, Guy P.
Max-Planck-Institut für Meteorologie, Germany; and National Center for Atmospheric Research (NCAR), US
Brenguier, Jean-Louis
CNRM/MEI/MNP
Brenninkmeijer, Carl A. M.
National Institute of Water and Atmospheric Research (NIWA), New Zealand
Burrows, John P.
Universität Bremen, Germany
Butler, James H.
NOAA Climate Monitoring and Diagnostics Laboratory, US
Calvert, Jack G.
National Center for Atmospheric Research, US
Carslaw, Nicola
University of Leeds, United Kingdom
Ciccioli, Paolo
Istituto ISAO (CNR), Italy
Cieslik, Stanislaw A.
European Commission, Environment Institute, Italy
Clerbaux, Cathy
Universite Paris VI, France
Covert, David S.
University of Washington, US
Cox, R. Anthony
University of Cambridge, United Kingdom
Crosley, David R.
SRI International, US
Crutzen, Paul J.
Max-Planck-Institut für Chemie, Germany
Davidson, Cliff I.
Carnegie Mellon University, US
Delmas, Robert J.
Lab. de Glaciologie et Geophysique de l'Environnement, France
Dentener, Frank J.
Utrecht University, The Netherlands
Derwent, Richard G.
UK Meteorological Office, United Kingdom
Drummond, James R.
University of Toronto, Canada
Duce, Robert A.
Texas A&M University, US
Eisele, Fred L.
National Center for Atmospheric Research (NCAR), US
Erickson III, David J.
National Center for Atmospheric Research (NCAR), US
Fehsenfeld, Frederick C.
NOAA Aeronomy Laboratory, US
Feichter, Johann
Max-Planck-Institut für Meteorologie, Germany

Fishman, Jack
NASA Langley Research Center, US
Flossman, Andrea I.
Université Blaise Pascal, C.N.R.S., France
Fuzzi, Sandro
Istituto ISAO (CNR), Italy
Galbally, Ian E.
CSIRO Division of Atmospheric Research, Australia
Gallardo-Klenner, Laura
Chilean Environmental Agency, Chile
Graf, Hans-F.
Max-Planck-Institut für Meteorologie, Germany
Granier, Claire
Université Pierre et Marie Curie, France; and NOAA Aeronomy Laboratory, US
Grant, William B.
NASA Langley Research Center, US
Guenther, Alex B.
National Center for Atmospheric Research (NCAR), US
Hales, Jeremy M.
ENVAIR, US
Hao, Wei Min
USDA Forest Service, US
Harriss, Robert C.
National Center for Atmospheric Research (NCAR), US
Heard, Dwayne
University of Leeds, United Kingdom
Heintzenberg, Jost
Institute for Tropospheric Research, Germany
Herrmann, Hartmut
Inst. für Troposphärenforschung e. V., Germany
Hoffman, Thorsten
Gesellscaft zur forderung der Spektrochemie, Germany
Houweling, Sandor
Econometric Institute, The Netherlands
Hov, Øystein
Norwegian Institute for Air Research (NILU), Norway
Huebert, Barry J.
University of Hawaii at Manoa, US
Husar, Rudolf B.
Washington University, US
Isaksen, Ivar S. A.
Universitetet I Oslo, Norway
Jacob, Daniel J.
Harvard University, US
Jaenicke, Ruprecht
Johannes Gutenberg-Universitat Mainz, Germany
Jähne, Bernd
Universität Heidelberg, Germany
Jöckel, Patrick
Max-Planck-Institut für Meteorologie, Germany
Kanakidou, Maria
University of Crete, Greece
Kärcher, Bernd
Institut für Physik der Atmosphäre, DLR, Germany
Kaufman, Yoram
NASA Goddard Space Flight Center, US
Kent, Geoffrey S.
Science and Technology Corp, US
Kettle, Anthony
University of East Anglia, Norwich, United Kingdom
Khattatov, Vyacheslav
Central Aerological Observatory, Russia
Kiene, Ronald P.
University of South Alabama, US
Koike, Makoto
University of Tokyo, Japan
Kolb, Charles E.
Aerodyne Research Inc., US
Kondo, Yutaka
University of Tokyo, Japan
Kulmala, Markku
University of Helsinki, Finland
Lacaux, Jean-Pierre
Medias-France, France
Lamarque, Jean-François
National Center for Atmospheric Research (NCAR), US
Law, Katharine S.
University of Cambridge, United Kingdom
Leck, Caroline
Stockholm University, Sweden
Lelieveld, Jos
Max-Planck-Institut für Chemie, Germany
Liousse, Catherine
Observatoire Midi-Pyrénées, France
Liss, Peter S.
University of East Anglia, United Kingdom
Lohmann, Ulrike
Dalhousie University, Canada
Lowe, David C.
National Institute of Water and Atmospheric Research (NIWA), New Zealand
Madronich, Sasha
National Center for Atmospheric Research (NCAR), US
Mahowald, Natalie
University of California at Santa Barbara and National Center for Atmospheric Research (NCAR), US
Malin, Gill
University of East Anglia, United Kingdom
Manning, Martin R.
National Institute of Water and Atmospheric Research (NIWA), New Zealand
Marenco, Alain
Université Paul Sabatier, France
Marticorena, Beatrice
Université de Paris XII, France
Matrai, Patricia A.
Bigelow Laboratory for Ocean Sciences, US

Matson, Pamela A.
Stanford University, US
McGillis, W. R.
Woods Hole Oceanographic Institution, US
McMurry, Peter H.
University of Minnesota, US
Miller, John M.
Air Resources Laboratory, NOAA, US
Monks, Paul S.
University of Leicester, United Kingdom
Montzka, Steve A.
NOAA Climate Monitoring and Diagnostics Laboratory, US
Moody, Jennifer L.
University of Virginia, US
Mosier, Arvin R.
US Department of Agriculture, US
Neue, Heinz-Ulrich
UFZ, Umweltforschungszentrum, Germany
Noone, Kevin
Stockholm University, Sweden
O'Conner, Fiona
Cambridge University, United Kingdom
O'Dowd, Colin
University of Helsinki, Finland
Orlando, John J.
National Center for Atmospheric Research, US
Paerl, Hans W.
The University of North Carolina at Chapel Hill, US
Palmer, Paul I.
Harvard University, US
Parrish, David D.
NOAA Aeronomy Laboratory, US
Penkett, Stuart A.
University of East Anglia, United Kingdom
Penner, Joyce E.
University of Michigan, US
Pickering, Ken
NASA Goddard Space Flight Center, US
Plane, John
University of East Anglia, United Kingdom
Platt, Ulrich F.
Universität Heidelberg, Germany
Prather, Michael J.
University of California at Irvine, US
Pszenny, Alex
University of New Hampshire, US
Putaud, J.-P.
Grenoble, France
Quinn, Patricia K.
NOAA Aeronomy Laboratory, US
Raes, Frank
European Commission, Environment Institute, Italy
Rodhe, Henning
Stockholm University, Sweden
Roelofs, Geert-Jan
Utrecht University, The Netherlands
Sahagian, Dork
University of New Hampshire, US
Schlager, Hans
DLR Institut für Physik der Atmosphäre, Germany
Scholes, Mary C.
University of the Witwatersrand, South Africa
Schurath, Ulrich
Institut für Meteorologie und Klimaforschung, Germany
Schwartz, Stephen E.
Brookhaven National Laboratory, US
Schwela, Dieter
World Health Organization, Geneva
Seakins, Paul
University of Leeds, United Kingdom
Seiler, Wolfgang
Fraunhofer Inst. für Atmosphärische Umweltforschung, Germany
Seinfeld, John H.
California Institute of Technology, US
Sievering, Herman
University of Colorado, Denver, US
Singh, Hanwant B.
NASA Goddard Space Flight Center, US
Smith, Keith A.
University of Edinburgh, United Kingdom
Snider, Jeff
University of Wyoming, US
Sokolik, Irina
University of Colorado, Boulder, US
Steffen, Will
IGBP Secretariat, Sweden
Stohl, Andreas
Technische Universitaet Muenchen, Germany
Stratmann, Frank
Institute for Tropospheric Research, Germany
Swietlicki, Erik
Lund University, Sweden
Thompson, Anne M.
NASA Goddard Space Flight Center, US
Trivett, Neil B. A.
Environmental Systems Research, Canada
Tyndall, Geoffrey S.
National Center for Atmospheric Research, US
Van Dingenen, Rita
European Commission, Environment Institute, Italy
Walters, Stacy
National Center for Atmospheric Research (NCAR), US
Wang, Chien
Massachusetts Institute of Technology, US
Weiss, Ray F.
Scripps Institution of Oceanography, US

Westphal, Douglas L.
Naval Research Laboratory, US
Wexler, Anthony S.
University of California at Davis, US
Wiedensohler, Alfred
Institute for Tropospheric Research, Germany
Wilson, Julian
European Commission, Environment Institute, Italy
Winker, David M.
NASA Langley Research Center, US
Wuebbles, Donald J.
University of Illinois, Urbana-Champaign, US

A.1.2 Reviewers

Ayers, Greg
Commonwealth Scientific and Industrial Research Organisation, Australia
Brune, William
The Pennsylvania State University, US
Buat-Ménard, Patrick
Université de Bordeaux, France
Carroll, Mary Anne
University of Michigan, US
Charlson, Robert
University of Washington, US
Clarke, Anthony
University of Hawaii, US
Crutzen, Paul
Max-Planck-Institut für Chemie, Mainz, Germany
Curtius, Joaquim
NOAA Aeronomy Laboratory, US
Dentener, Frank
European Commission Joint Research Centre, Italy
Duce, Robert
Texas A&M University, US
Ehhalt, Dieter
Forschungszentrum Jülich, Germany
Fahey, David
NOAA Aeronomy Laboratory, US
Gong, Sun-Ling
Meteorological Service of Canada, Canada
Hales, Jake
Envair, Inc., US
Heimann, Martin
Max-Planck-Institut für Biogeochemie, Jena, Germany
Hobbs, Peter
University of Washington, US
Kaye, Jack
NASA Headquarters, US
Keller, Michael
US Department of Agriculture, US
Kiehl, Jeffrey
National Center for Atmospheric Research, US
Kim, Y.
Kwang Ju Institute for Science and Technology, Korea
Kreidenweis, Sonia
Colorado State University, US
Levine, Joel S.
NASA Langley Research Center, US
Liu, Shaw
Academica Sinica, Taiwan
Lowenthal, Douglas
Desert Research Institute, US
Matson, Pamela
Stanford University, US
McConnell, Jack
York University, Canada
Neue, Heinz-Ulrich
Umweltforschungszentrum, Halle, Germany
Ogawa, Toshihiro
National Space Development Agency, Japan
Reeburgh, William
University of California, Irvine, US
Rodriguez, Jose
University of Miami, US
Sander, Rolf
Max-Planck-Institut für Chemie, Mainz, Germany
Simmonds, Peter
University of Bristol, United Kingdom
Størdal, Frode
Norwegian Institute for Air Research, Norway
Thompson, Anne
NASA Goddard Space Flight Center, US

A.2 IGAC Foci and Activities and their Convenors, 2001

Focus 1: Biosphere-Atmosphere Interaction

1.1 Biosphere-Atmosphere Trace Gas Exchange (BATREX). Conveners: A. Mosier, US and H.-U. Neue, Germany.
1.2 Deposition of Biogeochemically Important Trace Species (DEBITS). Convener: J.-P. Lacaux, France.
1.3 Biomass Burning Experiment: Impact of Fire on the Atmosphere and Biosphere. Conveners: J. Goldammer, Germany and J. Penner, US.
1.4 Marine Aerosol and Gas Exchange (MAGE). Convener: B. Huebert, US.

Focus 2: Oxidants and Photochemistry

2.1 Intercontinental Transport and Chemical Transformations (ITCT). Conveners: H. Akimoto, Japan, F. Fehsenfeld, US and S. Penkett, UK.
2.2 North Atlantic Regional Experiment (NARE). Conveners: D. Parrish, US and S. Penkett, UK.
2.3 East Asian/North Pacific Regional Experiment (APARE). Conveners: M. Uematsu, Japan.

2.4 Global Tropospheric Ozone Project (GTOP). Conveners: J. Fishman, US and S. Lal (India).

Focus 3: Atmospheric Aerosols

3.1 Aerosol Characterisation and Process Studies (ACAPS). Conveners: T. Bates, US and J. Gras, Australia.
3.2 Aerosol-Cloud Interactions (ACI). Conveners: T. Choularton, UK and D. Hegg, US.
3.3 Properties of Aerosols in the Stratosphere and Troposphere: Origin and Radiative Effects (PASTORE). Conveners: H.-F. Graf, Germany and M. P. McCormick, US.

Focus 4: Capacity Building

4.1 Atmospheric Chemistry and Environmental Education in Global Change (ACE^{ED}). Conveners: D. Jaffe, US and J. Boonjawat, Thailand.

Focus 5: Fundamental and Cross-Cutting Activities

5.1 Polar Atmospheric and Snow Chemistry (PASC). Conveners: L. Barrie, Canada and R. J. Delmas, France.
5.2 Global Emission Inventory Activity (GEIA). Conveners: D. Cunnold, US and J. Olivier, The Netherlands.
5.3 Global Integration and Modeling (GIM). Conveners: M. Kanakidou, Greece and P. Kasibhatla, US.
5.4 Intercalibrations/Intercomparisons:
  5.4.1 Nitrous Oxide and Halocarbons Intercalibrations Experiment (NOHALICE). Conveners: P. Fraser, Australia.
  5.4.2 Non-Methane Hydrocarbon Intercomparison Experiment (NOMHICE). Conveners: E. Apel, US.
5.5 Laboratory Studies (LABS). Conveners: R. Cox, UK and A. Ravishankara, US.

A.3 IGAC Field Campaigns

A.3.1 Short Term

AASE-1
AASE-2
ABLE-2A
ABLE-2B
ABLE-3A
ABLE-3B
ACE-1[1]
ACE-2[1]
ACE-Asia[1]
ASTEX/MAGE[1]
BIBLE[1]
CITE-1-Ames
CITE-1-Hawaii
CITE-2
CITE-3
ELCHEM
EXPRESSO[1]
INDOEX
INSTAC-1
MLOPEX-2[1]
NARE 1993 Summer Intensive[1]
OCTA-
OCTA-2
OCTA-3
OCTA-4
PEM-Tropics-A[1]
PEM-West A AND B[1]
Polarstern Expedition 1988
PSE-1992[1]
SAFARI-92[1]
STARE[1]
STRATOZ-3
STRATOZ-3S
SUCCESS
TARFOX[1]
TOTE
TRACE-P
TROPOZ-1
TROPOZ-2
VOTE

[1] Conducted as part of or endorsed formally by IGAC or one or more of its Activities.

A.3.2 Long Term (Multi-Year)

MOZAIC[1]
AEROCE[1]
TRAGNET[1]

A.4 Publications

A.4.1 Books Presenting Material Related to IGAC Activities

Clark JS, Cachier H, Goldammer JG, Stocks BJ (eds) Sediment records of biomass burning and global change. Springer-Verlag, Berlin-Heidelberg-New York
Crutzen PJ, Goldammer JG (eds) (1993) Fire in the environment: The ecological, atmospheric, and climatic importance of vegetation fires. Dahlem Workshop Reports, Environmental Sciences Research Report 13, J. Wiley & Sons, Chichester, England
Delmas RJ (ed) (1995) Ice core studies of global biogeochemical cycles. NATO ASI Series 1, Vol. 30, Springer-Verlag, New York
Goldammer JG (ed) (1993) Feuer in Waldökosystemen der Tropen und Subtropen. Birkhäuser-Verlag, Basel-Boston
Goldammer JG, Furyaev VV (eds) (1996) Fire in ecosystems of boreal eurasia. Kluwer Academic Publishers, Dordrecht

Levine JS (ed) (1996) Biomass burning and global change. MIT Press, Cambridge, Mass.
Minami K, Mosier A, Sass RL (eds) (1994) CH_4 and N_2O global emissions and controls from rice fields and other agricultural and industrial sources. National Institute of Agro-Environmental Sciences, Japan
Prinn RG (ed) (1994) Global atmospheric-biospheric chemistry. Plenum Press, New York
van Wilgen BW, Andreae MO, Goldammer JG, Lindesay JA (eds) (1997) Savanna: Ecological and environmental perspectives. Witwatersrand University Press, Johannesburg, South Africa
Wolff E, Bales R (eds) (1996) Chemical exchange between the atmosphere and polar snow. NATO ASI Series 1, Vol. 43, Springer-Verlag, New York

A.4.2 Journal Special Sections/Issues and Overview Papers

Polarstern Expedition 1988

Objectives: To investigate the distribution and concentration levels of atmospheric oxidants such as O_3 and H_2O_2; to assess the influence of long-range transport on the trace gas composition of the marine boundary layer; to study the contribution of air-sea exchange to the budgets of hydrocarbons and sulfur species; and to determine the radiation budget and photolysis frequencies of O_3 and NO_2 in the Tropics.
Publication: J atmos chem 15(3–4), 1992

Polar Sunrise Experiment 1992 (PSE-1992)

Objective: To better understand the role of marine halogens and chemical destruction mechanisms in lower tropospheric ozone depletion at polar sunrise.
Publication: J geophys res 99(D12), 1994

Nonmethane Hydrocarbon Intercomparison Experiment (NOMHICE)

Objective: To assess how accurately a large range of hydrocarbons (from C_2 to C_{10}) can be measured by individual scientists and laboratories.
Publications: J geophys res 99(D8):16651–16664, 1994; J geophys res 104(D21):26069–26086, 1999

Fire of Savannas/Dynamique et Chimie Atmosphéique en Forêt Equatoriale (FOS/DECAFE)

Objective: To study the gas and particle emissions from savanna fires.
Publication: J atmos chem 22(1–2), 1995

International Conference on Regional Environment and Climate Changes in East Asia

Objective: To study problems related to the long-range transport of trace species, regional air quality and air pollution, acid deposition, climatic changes and variations, and the sources and sinks of trace species, mainly in the Asian region.
Publication: Terr atmos ocean sci 6(3), 1995

Second Mauna Loa Photochemistry Experiment (MLOPEX-2)

Objective: To understand the complex factors that control the chemistry and seasonal evolution of reactive gases and radical species in the remote troposphere over the North Pacific Ocean.
Publication: J geophys res 101(D9), 1996

Southern Tropical Atlantic Regional Experiment (STARE)

Objective: To provide a comprehensive investigation of the chemical composition, transport and chemistry of the atmosphere over the tropical South Atlantic Ocean and the adjacent South American and African continents.
Publication: J geophys res 101(D19), 1996

Atlantic Stratocumulus Transition Experiment/Marine Aerosol and Gas Exchange (ASTEX/MAGE)

Objective: To improve the capability for studying cloud-aerosol-chemistry interactions and the air-sea fluxes that affect them, particularly through developing and testing a Lagrangian strategy for deriving air-sea exchange rates from chemical budgets in moving air masses.
Publication: J geophys res 101(D2), 1996

Pacific Exploratory Mission – West A and B (PEM-West A and B)

Objective: To study chemical processes and long range transport over the northwestern Pacific Ocean and to estimate the magnitude of human impact on the oceanic atmosphere over this region, particularly for tropospheric ozone and its precursors as well as for sulfur species.
Publications: J geophys res 101(D1), 1996; J geophys res 102(D23), 1997

North Atlantic Regional Experiment (NARE) 1993 Summer Intensive

Objective: To understand the processes responsible for either *in situ* production or transport of ozone over the Atlantic, including the more remote areas.
Publications: J geophys res 101(D22), 1996; J geophys res 103(D11), 1998

Workshop on NO_x Emission from Soils and its Influence on Atmospheric Chemistry

Objective: To integrate existing field and laboratory studies on NO and N_2O emissions from soils in the world, and to make clear what we know about NO production and consumption processes and their controls.

Publication: Nutr cycl agroecosys 48(1–2), 1997

Measurements of Ozone and Water Vapor by Airbus In-Service Aircraft (MOZAIC)

Objective: To help understand the atmosphere and how it is changing under the influence of human activity, with particular interest in the effects of aircraft.

Publication: J geophys res 103(D19), 1998

First Aerosol Characterization Experiment (ACE-1)

Objective: To quantify the chemical and physical processes controlling the evolution and properties of the atmospheric aerosol over the remote oceans of the Southern Hemisphere and to understand this multiphase atmospheric chemical system sufficiently to be able to provide a prognostic analysis of future radiative forcing and climate.

Publications: J geophys res 103(D13), 1998; J geophys res 104(D17), 1999

Tropospheric Aerosol Radiative Forcing Observational Experiment (TARFOX)

Objective: To reduce uncertainties in the effects of anthropogenic aerosols on climate by determining the direct radiative impacts, as well as the chemical, physical, and optical properties of the aerosol carried over the western Atlantic Ocean from the United States.

Publications: J geophys res 104(D2), 1999; J geophys res 105(D8), 2000

GEIA Reactive Chlorine Emissions Inventory (RCEI)

Objective: To develop and verify individual gridded global emission inventories as a function of source type for major reactive chlorine species in the troposphere and, thence, (*1*) to develop species-specific composite inventories by integrating the individual emission fields over source type, (*2*) to evaluate budget closure for each species by comparing composite emissions from major known sources with total fluxes inferred from inversion modeling and related approaches, (*3*) to differentiate the relative contributions of natural and anthropogenic sources, and (*4*) to assess associated uncertainties in the tropospheric chlorine cycle.

Publication: J geophys res 104(D27), 1999

Experiment for Regional Sources and Sinks of Oxidants (EXPRESSO)

Objective: To quantify and better understand the processes controlling surface fluxes of photochemical precursors along a tropical forest-to-savanna gradient in central Africa.

Publication: J geophys res 104(D23), 1999

3-D Global Simulations of Tropospheric Chemistry with Focus on Ozone Distributions – Results of the GIM/IGAC Intercomparison Exercise

Objective: To evaluate the capabilities of various models to simulate tropospheric ozone and to identify key areas of uncertainty in understanding of the tropospheric ozone budget.

Publication: European Commission Report EUR 18842, Office for Official Publications of the European Communities, Luxembourg, 79 pp, 1999.

3-D Global Simulations of Tropospheric CO Distributions – Results of the GIM/IGAC Intercomparison 1997 Exercise

Objective: To evaluate the capabilities of various models to simulate tropospheric ozone and to identify key areas of uncertainty in understanding of the tropospheric ozone budget.

Publication: Chemosphere: Global change sci 1, 1999

The Second Aerosol Characterization Experiment (ACE-2)

Objective: To determine and understand the properties and controlling factors of the aerosol in the anthropogenically modified atmosphere of the North Atlantic and assess their relevance for radiative forcing.

Publication: Tellus 52B, 2000

A.4.3 IGAC Scientific Conference Proceedings

The Chemistry of the Global Atmosphere, 5–11 September 1990, CACGP Conference, Chamrousse, France

Buat-Ménard P, Delmas RJ (eds) (1991) Special issue related to the 7th International Symposium of the Commission for Atmospheric Chemistry and Global Pollution on the Chemistry of the Global Atmosphere. J atmos chem 14(1–4)

1st IGAC Scientific Conference, 18–22 April 1993, Eilat, Israel

Prinn RG (ed) (1994) Global atmospheric-biospheric chemistry. Plenum Press, New York

Prinn RG, Pszenny AAP, Crutzen PJ (eds) (1994) Contributed papers from the First IGAC Scientific Conference. J geophys res 99(D8)

2nd IGAC Scientific Conference and 8th CACGP Symposium on Global Atmospheric Chemistry, 5–9 September 1994, Fuji-Yoshida, Japan

Penkett SA, Akimoto H, Ayers GP (eds) (1996) Contributed papers from the 2nd IGAC Scientific Conference and 8th CACGP Symposium on Global Atmospheric Chemistry. Atmos environ 30(10/11)

WMO-IGAC Conference on the Measurement and Assessment of Atmospheric Composition, 9–14 October 1995, Beijing, China

WMO (1995) Extended abstracts of the WMO-IGAC Conference on the Measurement and Assessment of Atmospheric Composition. WMO/GAW Report No. 107, World Meteorological Organization, Geneva, Switzerland

International Symposium on Atmospheric Chemistry and Future Global Environment, 11–13 November 1997, Nagoya, Japan

Science Council of Japan and National Space Development Agency (1997) Program and extended abstracts of the International Symposium on Atmospheric Chemistry and Future Global Environment, 11–13 November 1997, Nagoya, Japan. Science Council of Japan and National Space Development Agency

IGAC/SPARC/GAW Conference on Global Measurement Systems for Atmospheric Composition, 20–22 May 1997, Toronto, Canada

Drummond J (ed) (1999) Proceedings of the IGAC/SPARC/GAW Conference on Global Measurement Systems for Atmospheric Composition. J atmos sci 56(2)

5th IGAC Scientific Conference and 9th CACGP Symposium on Global Atmospheric Chemistry, 19–25 August 1998, Seattle, Washington, US

Bates TS, Quinn PK (eds) (1999) Contributed papers from the 5th IGAC Scientific Conference and 9th CACGP Symposium on Global Atmospheric Chemistry. J geophys res 104(D21)

6th IGAC Scientific Conference on Global Atmospheric Chemistry, 13–17 September 1999, Bologna, Italy

Fuzzi S (ed) (2000) Special issue: Sixth Scientific Conference of the International Global Atmospheric Chemistry Project (IGAC). Atmos environ 34(29–30):4961–5338

A.4.4 IGAC Programmatic and Other Publications

Galbally IE (ed) (1989) The International Global Atmospheric Chemistry (IGAC) Programme. Commission on Atmospheric Chemistry and Global Pollution of the International Association of Meteorology and Atmospheric Physics, ISBN 0-643-05062-0

Matson PA, Ojima DS (1990) Terrestrial biosphere exchange with global atmospheric chemistry. Global Change Report No. 32, International Geosphere-Biosphere Programme, Stockholm, Sweden, 105 pp

Molina LT (ed) (1992) The International Global Atmospheric Chemistry (IGAC) Programme: An overview. IGAC Core Project Office, Cambridge, Mass., 26 pp

Pszenny AAP, Prinn RG (eds) (1994) International Global Atmospheric Chemistry (IGAC) Project: The operational plan. Global Change Report No. 32, International Geosphere-Biosphere Programme, Stockholm, Sweden, 126 pp

Sass RL, Neue H-U (eds) (1994) Global measurement standardization of methane emissions from irrigated rice cultivation. IGAC Core Project Office, Cambridge, Mass., 10 pp

Hobbs PV, Huebert BJ (eds) (1996) Atmospheric aerosols: A new focus of the International Global Atmospheric Chemistry Project. IGAC Core Project Office, Cambridge, Mass. 46 pp

Pszenny A (ed) IGACtivities NewsLetter. Issued quarterly since June 1995 by the IGAC Core Project Office, Cambridge, Mass.

Pszenny A (1997) The International Global Atmospheric Chemistry (IGAC) Project. EUROTRAC Newsletter, No. 18, EUROTRAC International Scientific Secretariat, Garmisch-Partenkirchen, Germany

Pszenny A, Brasseur G (1997) Tropospheric ozone: An emphasis of IGAC research. Global Change NewsLetter No. 30, International Geosphere-Biosphere Programme, Stockholm, Sweden

Smith KA, Bogner J (1997) Joint North American-European Workshop on Measurement and Modeling of Methane Fluxes from Landfills. IGAC Core Project Office, Cambridge, Mass. 46 pp

A.5 Acronyms

AASE	Airborne Arctic Stratospheric Expedition
ABL	Atmospheric boundary layer
ABLE	Amazon Boundary Layer Experiment
ACE	Aerosol Characterisation Experiment (1, 2, and Asia)
ACSOE	Atmospheric Chemistry Studies in the Oceanic Environment
ADEOS	Advanced Earth Observing Satellite
AEROBIC	Aerosol Formation from Biogenic Organic Carbon
AERONET	Aerosol Robotic Network
AFARI	African Fire-Atmosphere Research Initiative
AGAGE	Advanced Global Atmospheric Gases Experiment
ALE	Atmospheric Lifetime Experiment
AMS	Accelerator mass spectrometry
ANCAT	Abatement of Nuisances Caused by Air Transport
AOM	Agricultural Optimisation Model
APARE	East Asia-North Pacific Regional Study
API	Atmospheric pressure ionisation
APS	Aerodynamic particle sizer
APSS	Albemarle-Pamlico Sound System
ASGA-EX	Air-Sea Gas Exchange Experiment
ASHOE	Airborne Southern Hemisphere Experiment
ASTEX	Atlantic Stratocumulus Experiment
AVHRR	Advanced Very High Resolution Radiometry
BATGE	Biosphere-Atmosphere Trace Gas Exchange in Tropics
BATREX	Biosphere-Atmosphere Trace Gas Exchange

BEMA	Biogenic Emissions from the Mediterranean Area
BC	Black carbon
BIBEX	Biomass Burning Experiment
BIBLE	Biomass Burning and Lightning Experiment
BIPHOREP	Biogenic VOC Emissions and Photochemistry in the Boreal Regions of Europe
BMBF	Bundesministerium für Bildung und Forschung
BOREAS	Boreal Ecosystem Atmosphere Study
BP	British Petroleum
CAAP	Composition and Acidity of the Asian Precipitation programme
CALIPSO	Cloud-Aerosol Lidar and Infrared Pathfinder Satellite Observations (formerly PICASSO-CENA)
CAPMon	Canadian Air and Precipitation Monitoring Network
CARIBIC	Civil Aircraft for Remote Sensing and *In situ* Measurements in the Troposphere and Lower Stratosphere Based on the Instrumentation Container Concept
CASA	Carnegie-Ames-Stanford Approach model
CaSTNET	Clean Air Status & Trends Network
CCB	Cold conveyor belt
CCD	Convective-cloud differential
CCG	Carbon Cycle Group
CCM	Community Climate Model
CCN	Cloud condensation nucleus
CDOM	Coloured dissolved organic matter
CEPEX	Central Equatorial Pacific Experiment
CIMS	Chemical ion mass spectrometry
CITE	Chemical Instrumentation Test and Evaluation Experiment
CLAES	Cryogenic Limb Array Etalon Spectrometer
CMDL	Climate Monitoring and Diagnostics Laboratory
CNRS	Centre National de la Recherche Scientifique
COSAM	Comparison of Large Scale Sulphate Aerosol Models study
CPC	Condensation particle counter
CPU	Central processing unit
CTM	Chemistry transport model
DAO	Data Assimilation Office
DEBITS	Deposition of Biogeochemically Important Trace Species Activity
DIAL	Differential absorption lidar
DMPS	Differential mobility particle sizer
DMS	Dimethylsulphide
DMSO	Dimethylsulphoxide
$DMSO_2$	Dimethylsulphone
DMSP	Dimethylsulphoniopropionate
DNDC	Denitrification and Decomposition Model
DNPH	Dintrophenylhydrazine
DOAS	Differential Optical Absorption Spectroscopy
DOE	Department of Energy
DON	Derived dissolved organic nitrogen
DU	Dobson units
EARLINET	European Aerosol Research Lidar Network to Establish Aerosol Climatology
ECMWF	European Centre for Medium-Range Weather Forecasting
EDGAR	Emission Database for Global Atmospheric Research
EELS	Electron energy loss spectrometry
EIPO	European IGAC Project Office
EMEP	European Model and Evaluation Plan
EMICs	Earth system models of intermediate complexity
ENSO	El Niño-Southern Oscillation
ENVISAT	Environmental Satellite
EOS	Earth Observing Satellite
EPA	Environmental Protection Agency
EPIC	Equatorial Processes Including Coupling
EPMA	Electron probe microanalysis
ERS	Earth Resources Satellite, and European Remote Sensing Satellite
ESA	European Space Agency
EULINOX	European Lightning Oxides Project
EUMETSAT	European Meteorological Satellite
EUPHORE	European Photoreactor Programme
EUROTRAC	European Experiment for Transporting and Transforming Environmentally Relevant Trace Construction in the Troposphere
EUSTACHE	European Studies on Trace Gases and Atmospheric Chemistry
EXPRESSO	Experiment for Regional Sources and Sinks of Oxidants
FAGE	Fluorescence assay by gas expansion
FAO	Food and Agriculture Organisation
FCCC	Framework Convention on Climate Change
FEV1	Forced expiratory volume over one second
FIRESCAN	Fire Research Campaign Asia-North
FLOP	Floating-point operation per second
FT	Free troposphere
FTIR	Fourier transform infrared
FTS	Fourier transform spectrometer
FVC	Forced vital capacity

GAGE	Global Atmospheric Gases Experiment
GAIM	Global Analysis, Integration and Modelling
GASIE	Gaseous Sulfur Intercomparison Experiment
GASP	Gas in Peat
GAW	Global Atmosphere Watch
GC	Gas chromatography
GCM	General circulation model
GC-MS	Gas chromograpy-mass spectroscopy
GCOS	Global Climate Observing System
GCTE	Global Change and Terrestrial Ecosystems
GEIA	Global Emissions Inventory Activity
GEM	Global Environmental Multi-scale
GFDL	Geophysical Fluid Dynamics Laboratory
GIFTS	Geosynchronous Imaging Fourier Transform Spectrometer
GIM	Global Integration and Modelling
GISP 2	Greenland Ice Sheet Program
GISS	Goddard Institute for Space Studies
GLOBE	Global Learning and Observations to Benefit the Environment
GOME	Global Ozone Monitoring Experiment
GPS	Global Positioning System
GRIP	Greenland Ice Project
GTE	Global Tropospheric Experiment
HALOE	Halogen Occultation Experiment
HIP	Hole-in-the-pipe
IAEA	International Atomic Energy Agency
ICFME	International Crown Fire Modeling Experiment
ICP-MS	Inductively coupled plasma mass spectrometry
IGAC	International Global Atmospheric Chemistry
IGBP	International Geosphere-Biosphere Programme
IHDP	International Human Dimensions Programme
IMG	Interferometric monitor for greenhouse gases
IMPROVE	Improvement of Microphysical Parameterization through Observational Verifcation Experiment
INDOEX	Indian Ocean Experiment
IPCC	Intergovernmental Panel on Climate Change
IRMS	Isotope ratio mass spectrometry
IRONEX	Iron Enrichment Experiment
ITCT	Intercontinental Transport and Chemical Transformation
ITCZ	Inter-Tropical Convergence Zone
IUPAC	International Union of Pure and Applied Chemistry
JGOFS	Joint Global Ocean Flux Study
JPL	Jet Propulsion Laboratory
LAC	Land Aerial Cover
LAMMS	Laser microprobe mass spectrometry
LASE	Laser Atmospheric Sounding Experiment
LASE	Lidar Atmospheric Sensing Experiment
LBA	Large-Scale Biosphere-Atmosphere Experiment in Amazonia
LES	Large eddy simulation
LIDAR	Light detection and ranging
LIF	Laser-induced fluorescence
LIS	Lightning images sensor
LITE	Lidar In-space Technology Experiment
LOICZ	Land Ocean Interactions in the Coastal Zone
LPAS	Long path absorption spectroscopy
LUCC	Land Use and Cover Change
MAGE	Marine Aerosol and Gas Exchange
MAPS	Measurement of Air Pollution from Satellites
MATCH	Model for Atmospheric Chemistry and Transport
MAXOX	Maximum Oxidation Rates in the Free Troposphere
MBL	Marine boundary layer
MCM	Master chemical mechanisms
METOP	Meteorological Operational (satellites)
MIE	Multi-step implicit-explicit method
MIESR	Matrix isolation-electron spin resonance
MISR	Multi-angle imaging spectroradiometer
MLOPEX	Mauna Loa Photochemistry Experiment
MODIS	Moderate resolution imaging spectroradiometer
MOPITT	Measurement of Pollution in the Troposphere
MOZAIC	Measurement of Ozone Aboard In-service Aircraft
MS	Mass spectrometry
MSA	Methanesulphonate
NADAP	National Atmospheric Deposition Assessment Programme
NAFC	North Atlantic Flight Corridor
NAPAP	National Acid Precipitation Assessment Program
NARE	North Atlantic Regional Experiment
NAS	National Academy of Sciences
NASA	National Aeronautics and Space Administration
NASA-DAO	NASA Data Assimilation Office
NCAR	National Center for Atmospheric Research

NCEP	National Centers for Environmental Prediction
NEP	Net ecosystem productivity
NMHC	Non-methane hydrocarbon
NOAA	National Oceanic and Atmospheric Administration
NOHALICE	Nitrous Oxide and Halocarbons Intercomparison Experiment
NOMHICE	Nonmethane Hydrocarbon Intercomparison Experiment
NOXAR	Nitrogen Oxide and Ozone along Air Routes
NPOESS	National Polar Orbiting Environmental Satellite System
NRC	National Research Council
NSS	Non-sea-salt
OC	Organic carbon
OECD	Organization for Economic Cooperation and Development
OMI	Ozone Monitoring Instrument
OPC	Optical particle counters
OTD	Optical transient detector
PAGES	Past Global Changes
PAH	Polycyclic aromatic hydrocarbons
PAN	Peroxyacetyl nitrate
PASC	Polar Air and Snow Chemistry
PBL	Planetary boundary layer
PCB	Polychlorinated biphenyls
PEELS	Parallel-EELS
PEM	Pacific Exploratory Missions
PERCA	Peroxy radical chemical amplification
PFBHA	Pentafluorobenzyl hydroxylamine hydrochloride
PICASSO-CENA	Pathfinder Instruments for Cloud and Aerosol Spaceborne Observations – Climatologie Etendue des Nuages et des Aerosols
PIXE	Particle induced x-ray emission
PM	Particulate matter
POLDER	Polarization and Directionality of the Earth's Reflectances
POLINAT	Pollution from Aircraft Emissions in the North Atlantic Flight Corridor
POPCORN	Photo-Oxidant Formation by Plant Emitted Compounds and OH Radicals
PPN	Peroxypropionyl nitrate
PTRMS	Proton transfer reaction mass spectroscopy
QA/SAC	Quality Assurance/Science Activity Centres
QMOM	Quadrature method of moments
RADM	Regional Acid Deposition Model
REMPI	Resonance enhanced multi-photon ionisation
RH	Relative humidity
ROSE	Rural Oxidants in the Southern Environment
SAFARI	Southern Africa Fire-Atmosphere Research Initiative
SAGE	Stratospheric Aerosol and Gas Experiment
SAPHIR	Simulation of Atmospheric Photochemistry in a Large Reaction Chamber
SASS	Subsonic Assessment
SBUV	Solar Backscattered Ultraviolet
SCIAMACHY	Scanning Imaging Absorption Spectrometer for Atmospheric Chartography
SEI	Stockholm Environment Institute
SEM	Scanning electron micrometry
SESAME	Second European Stratospheric Arctic and Mid-latitude Experiment
SICIMS	Selected ion chemical ionisation mass spectrometry
SIMS	Secondary ion mass spectrometry
SMPS	Scanning mobility particle sizer
SOA	Secondary organic aerosols
SOAPEX	Southern Ocean Atmospheric Photochemistry Experiment
SOIREE	Southern Ocean Iron Enrichment Experiment
SOLAS	Surface Ocean-Lower Atmosphere Study
SONEX	SASS Ozone and Nitrogen Oxide Experiment
SOS	Southern Oxidants Study
SPA	Single particle analysis
SPARC	Stratospheric Processes and their Role in Climate
SPCZ	South Pacific Convergence Zone
SPM	Suspended particulate matter
SRES	Special Report on Emission Scenarios
STARE	Southern Tropical Atlantic Regional Experiment
START	System for Analysis, Research and Training
STE	Stratosphere-troposphere exchange
STEM	Scanning transmission electron micrometry
STEREO-A	Solar Terrestrial Relations Observatory
STRATOZ	Stratospheric-Tropospheric Experiments: Readiation, Aerosols, and Ozone
STREAM	Stratosphere-Troposphere Experiment by Aircraft Measurements
SUCCESS	Subsonic Aircraft: Contrail and Cloud Effects Special Study

TACIA Testing Atmospheric Chemistry in Anticyclones
TARFOX Tropospheric Aerosol Radiative Forcing Observational Experiment
TAS Total aerosol sampler
TCO Tropospheric column ozone
TDLAS Tuneable diode laser absorption spectrometry
TES Tropospheric emission spectrometer
TOF-SIMS Time-of-flight secondary ion mass spectrometry
TOHPE Tropospheric OH Photochemistry Experiment
TOMS Total ozone mapping spectrometer
TOR Tropospheric ozone residual
TRACE-A Transport and Atmospheric Chemistry Near the Equator-Atlantic
TRAGNET Trace Gas network
TRMM Tropical Rainfall Measuring Mission
TROPOZ Tropical Ozone Experiment

UARS Upper Atmosphere Research Satellite
UNEP United Nations Environment Programme
UNFCCC United Nations Framework Convention for Climate Change
USDA US Department of Agriculture
UT Upper troposphere
UV Ultraviolet

VOC Volatile organic carbon
VWM Volume-weighted mean

WCB Warm conveyor belt
WCRP World Climate Research Programme
WFPS Water-filled pore space
WMO World Meteorological Organisation
WOCE World Ocean Circulation Experiment

ZIBBEE Zambian International Biomass Burning Emissions Experiment

Index

A

B

C

D

E

H

I

N

O

P

Q

R

S

T

U

V

W

X, Y, Z